SCHAUM'S OUTLINE OF

THEORY AND PROBLEMS

OF

DYNAMIC STRUCTURAL ANALYSIS

•

by

JAN J. TUMA, Ph.D.

Arizona State University
Tempe, Arizona

and

FRANKLIN Y. CHENG, Ph.D.

University of Missouri
Rolla, Missouri

•

SCHAUM'S OUTLINE SERIES

McGRAW-HILL BOOK COMPANY

New York St. Louis San Francisco Auckland Bogotá Guatemala Hamburg Johannesburg Lisbon London Madrid Mexico Montreal New Delhi Panama Paris San Juan São Paulo Singapore Sydney Tokyo Toronto

JAN J. TUMA is currently Professor of Civil Engineering at Arizona State University. He was formerly Professor and Head of the School of Civil Engineering at Oklahoma State University and engineering consultant specializing in writing reference works and textbooks in engineering. Professor Tuma received his Ph.D. from the University of Colorado. Among his publications are the widely used *Engineering Mathematics Handbook*, *Technology Mathematics Handbook*, *Handbook of Physical Calculations*, and *Engineering Soil Mechanics* (with A. M. Hady). His other books in the Schaum's Outline Series are *Structural Analysis*, *Advanced Structural Analysis* (with R. K. Munshi) and *Space Structural Analysis* (with M. N. Reddy).

FRANKLIN Y. CHENG received his Ph.D from the University of Wisconsin and is currently Professor of Civil Engineering at the University of Missouri at Rolla. He has taught a variety of courses in structural engineering and has carried out several NSF research projects focusing on earthquake engineering and structural optimization. He has over ninety publications and is an active consultant for government and industry. Dr. Cheng is a committee member of Task Group 4 of the Structural Stability Research Council and a member of several ASCE committees. He served as chairman of the International Symposium on Earthquake Structural Engineering.

Schaum's Outline of Theory and Problems of
DYNAMIC STRUCTURAL ANALYSIS

1 2 3 4 5 6 7 8 9 10 11 12 13 14 15 16 17 18 19 20 SH SH 8 7 6 5 4 3 2

ISBN 0-07-065437-9

Sponsoring Editor, John Aliano
Editing Supervisor, Marthe Grice
Production Manager, Nick Monti

Library of Congress Cataloging in Publication Data

Tuma, Jan J.
Schaum's outline of theory and problems of dynamic structural analysis.

(Schaum's outline series)
Includes index.
1. Structural dynamics. 2. Structures, Theory of.
I. Cheng, Franklin Y. II. Title. III. Series.
TA654.T85 624.1'71 82-216
ISBN 0-07-065437-9 AACR2

Preface

This book presents in one volume the major theorems, definitions, and methods of the dynamic analysis of engineering structures. It was prepared to serve as a college outline in the first standard course in structural dynamics. The statements of theory are sufficiently complete and supported by derivations and proofs so that the book may be used as a textbook in a formal course or as a supplement to all current texts on this subject.

The numerous solved problems with conclusions underlining the major points of solution constitute the best means of understanding the basic principles. As in the other books prepared by the senior author, the supplementary problems with answers offer a complete review of the material of each chapter and also serve as a reference handbook, providing the practicing engineer with a wealth of ready-made solutions to many important problems.

The subject matter is divided into ten chapters arranged in a logical sequence. Since the dynamic analysis of structures is only a special application of the dynamics of deformable bodies, a close relationship is established in this book between the theorems of dynamic structural analysis and those of mechanics of engineering materials.

The first chapter introduces the basic concepts of engineering dynamics applied to structural systems, including the restatement of Newton's laws, virtual work, Lagrange's and Hamilton's functions. The second and third chapters show the applications of the principles of free and forced vibration in the analysis of single and multidegree freedom systems. Particular attention is given in these chapters to the evaluation of convolution integrals and their inclusion in the dynamic matrices.

Like static analysis, the dynamic analysis of structures developed along two distinct paths: vectorial analysis and virtual work analysis. The fourth chapter follows the latter path, introducing the energy methods of dynamic analysis, with particular applications to the lumped mass systems moving in normal coordinates. Finally, the fifth chapter shows the application of the transport matrix method, the flexibility matrix method, and the stiffness matrix method in the analysis of lumped mass systems.

The second part of the book covers the dynamic analysis of distributed mass systems. The longitudinal and transverse vibration of free bars, bars encased in elastic mediums, and beam-columns is investigated in the sixth and seventh chapters. The particular matrices of these basic and special cases are then assembled in the respective transport matrices in the eighth chapter. The closed forms and the series forms of these matrices are displayed and used as the base for the construction of the flexibility and stiffness matrices in the ninth chapter. The last chapter investigates the longitudinal and transverse vibration of straight bars of variable cross section by lumped mass methods, stepped mass methods, and substitute function methods; thus summarizing all methods presented in the preceding chapters.

PREFACE

The organization and preparation of the material presented in this book spanned a period of several years, during which the authors received assistance from several institutions. In particular, the assistance of Oklahoma State University, Stillwater, Oklahoma; Federal Institute of Technology, Zurich, Switzerland; University of Colorado, Boulder, Colorado; University of Missouri, Rolla, Missouri; and Arizona State University, Tempe, Arizona are gratefully acknowledged. Although certain portions of this book are believed to be new, the majority of the information is based on an extensive wealth of reference material. Space limitations prevent the inclusion of a complete list of references, but an effort was made to credit those references which were directly used, and for additional information the reader is directed to the specific pages indicated in the bibliography.

The authors solicit comments and recommendations by users of this book for improvements in future editions. Finally, gratitude is expressed to Thomas Dembofsky, Nicola Monti, and Marthe Grice for their excellent editorial cooperation, encouragement, and continued interest.

J. J. TUMA
F. Y. CHENG

Contents

Chapter 1

Concepts of Analysis

1.1 INTRODUCTION

Static System

The *planar structures* investigated in the preceding volumes* (I and II) have been defined as systems of bars forming with their loads and reactions a coplanar system. Since their gradually applied loads have been assumed to be stationary or slow moving, but always of constant magnitude (independent of time), their analysis involved the *principles of static equilibrium* and *conditions of time-independent deformation compatibility.* Such systems are designed as static systems and their analysis is called static structural analysis.

Dynamic System

Although many structural systems fall in the former category, there are a large number of planar structures subjected to loads which are applied suddenly and/or the magnitude of which varies rapidly with time (dynamic loads, dynamic excitations).

In the analysis of the latter systems, the *principles of dynamic equilibrium* and the *conditions of time-dependent deformation compatibility* must be employed. Such systems are classified as dynamic systems and their analysis is called dynamic structural analysis.

Essentially, all structural systems are *dynamic systems*, since the assumed gradually applied loads are not necessarily applied gradually and the assumed slow moving loads are not always moving slowly. Consequently, the static structural analysis is always an approximation, yielding good results where the assumption of static loads is justified.

Mathematical Model

The study presented in this volume is restricted to the dynamic analysis of planar beams, arches, rings, frames, and trusses. Only *numerical analysis* based on the principles of linear small displacement elasticity is introduced.

For the analysis, the real structure is replaced by an *ideal structure* (mathematical model) represented by centroidal axes of its members, supported by idealized constraints and acted upon by assumed, symbolic *load vectors, damping vectors, restoring vectors,* and *inertia vectors.*

Assumptions

The following properties assigned to this model are introduced as the *general assumptions* of the analysis:

(*a*) The model is in a state of dynamic equilibrium.

(*b*) All loads are applied in the plane of symmetry of the cross section.

(*c*) The structural material is homogeneous, isotropic, continuous, and remains continuous during the loading and unloading.

(*d*) The cause-effect relationships follow definite laws given analytically by the constitutive equations.

---

* All references to the preceding volumes, I, II, and III, refer to the following books:
I: Jan J. Tuma, *Schaum's Outline of Structural Analysis*, McGraw-Hill, New York, 1969.
II: Jan J. Tuma and R. K. Munshi, *Schaum's Outline of Advanced Structural Analysis*, McGraw-Hill, New York, 1971.
III: Jan J. Tuma and M. N. Reddy, *Schaum's Outline of Space Structural Analysis*, McGraw-Hill, New York, 1982.

Unless specifically stated otherwise, it may be assumed that:

(*e*) All deformations are small and do not alter (significantly) the initial geometry of the structure.

(*f*) The structural material is elasto-plastic and, in the elastic range, the linear superposition of causes and effects is admissible.

The last assumption implies that all governing differential equations are linear equations.

Functions and Constants

Five types of quantities are utilized in the analysis: (*a*) *geometric quantities* (coordinates, segments, slopes, angles, properties of cross section), (*b*) *kinematic quantities* (displacements, velocities, accelerations), (*c*) *dynamic quantities* (mass, loads, stresses, reactions, inertia forces and moments), (*d*) *material constants* (moduli of structural material, spring and viscosity constants, volume change coefficients), (*e*) *scalar quantity of change* (temperature, time).

Their definitions and analytical representations are given in the following sections.

1.2 GEOMETRIC VECTORS

Coordinates and Angles

As in the static analysis, the *geometry of the ideal planar structure* is defined by the centroidal axes of its members, by the position of joints, and by the location of constraints and releases, in terms of the *initial position coordinates* $x, y, z = 0$ and the *initial position angles* $\phi_x = 0$, $\phi_y = 0$, ϕ_z.

Since the *datum of the dynamic analysis* is the geometry of the deformed structure under the action of static loads (dead load, stationary live load, etc.), the *datum position coordinates* are $x + \delta_x, y + \delta_y, z = 0$, and the *datum position angles* are $\phi_x = 0, \phi_y = 0, \phi_z + \theta_z$, where δ_x and δ_y are the linear displacements in the X, Y directions, respectively, and θ_z is the angular displacement about the Z axis.

As the static displacements have been assumed to be too small to alter significantly the initial geometry, the datum geometry of the dynamic analysis in this book is approximated by the initial geometry.

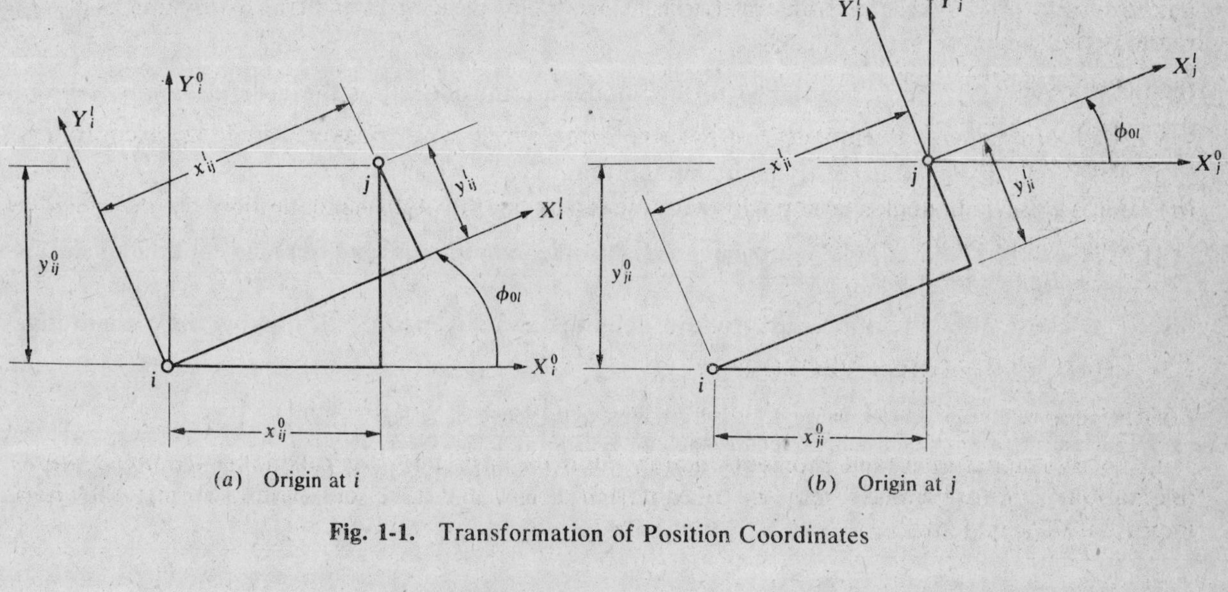

Fig. 1-1. Transformation of Position Coordinates

Coordinate Systems

Two systems of coordinate axes are introduced in the analysis:

(*a*) *Reference system* (datum, global system, 0-system), arbitrarily selected set of orthogonal axes X^0, Y^0, Z^0, the direction of which is fixed and common for all parts of the structure.

(*b*) *Member system* (local system, *l*-system) given by the principal axes X^l, Y^l, Z^l of the cross section at the station of investigation. Consequently, this system must translate and rotate with the cross section, taking a specific position at each station of investigation.

The position coordinates of a joint, support or cross section related to these systems are *directed segments*, so that

$$x_{ij}^0 = -x_{ji}^0, \qquad y_{ij}^0 = -y_{ji}^0, \qquad z_{ij}^0 = -z_{ji}^0 = 0 \tag{1-1}$$

$$x_{ij}^l = -x_{ji}^l, \qquad y_{ij}^l = -y_{ji}^l, \qquad z_{ij}^l = -z_{ji}^l = 0 \tag{1-2}$$

where the superscript, first subscript, and second subscript identify the system, origin, and position, respectively (Fig. 1-1).

Transformations

The transformation relationships between these two systems are given by the *angular transformation matrices*, π^{0l} and π^{l0}, which are

$$\pi^{0l} = \begin{bmatrix} \cos\phi_{0l} & -\sin\phi_{0l} & 0 \\ \sin\phi_{0l} & \cos\phi_{0l} & 0 \\ 0 & 0 & 1 \end{bmatrix} \qquad \pi^{l0} = \begin{bmatrix} \cos\phi_{0l} & \sin\phi_{0l} & 0 \\ -\sin\phi_{0l} & \cos\phi_{0l} & 0 \\ 0 & 0 & 1 \end{bmatrix} \tag{1-3, 4}$$

These matrices are *orthogonal* since

$$\pi^{0l} = \pi^{l0)T} = \pi^{l0)-1} \qquad \text{and} \qquad \pi^{l0} = \pi^{0l)T} = \pi^{0l)-1} \tag{1-5}$$

where $\pi^{)T}$ is the *transpose* and $\pi^{)-1}$ is the *inverse* of the respective matrix.

A vector V_i (any vector) can be resolved into three directed segments in each system, given by the column matrices

$$V_i^0 = \{V_{ix}^0, V_{iy}^0, V_{iz}^0\} \qquad \text{and} \qquad V_i^l = \{V_{ix}^l, V_{iy}^l, V_{iz}^l\}$$

Their relationships in terms of (*1-4*) are then

$$V_i^0 = \pi^{0l} V_i^l \qquad V_i^l = \pi^{l0} V_i^0 \tag{1-6}$$

and are valid for any vector transformation at any point as long as the transformation angle ϕ_{0l} remains the same.

Sign Conventions

(*a*) All coordinates and segments are positive if measured in the positive direction of coordinate axes.

(*b*) All slopes and angles are positive if measured in the right-hand motion on the positive direction of coordinate axes.

1.3 LOAD AND STRESS VECTORS

Load Vector

The dynamic forces and moments acting upon the structure are called the *dynamic causes* (excitations). These causes may be of different origin and take on various forms, the most important of which are:

(*a*) *Harmonic excitations* produced by oscillating machinery operating within the structure.
(*b*) *Impact* and *vibrations* produced by vehicles or machinery moving over or along the structure.
(*c*) *Impulsive loads* produced by wind gusts, water waves, and blasts.
(*d*) *Excitations* produced by earthquake shocks and sonic booms.

In general, a *dynamic load vector* may be represented by a matrix function of time (t) as

$$W_m^m(t) = \{P_{mx}^m(t), P_{my}^m(t), Q_{mz}^m(t)\} \tag{1-7}$$

where $W_m^m(t)$ is the load vector at m in the m-system, consisting of two force components (P) and one moment component (Q), and the subscripts x, y, z give the component's direction (Fig. 1-2a).

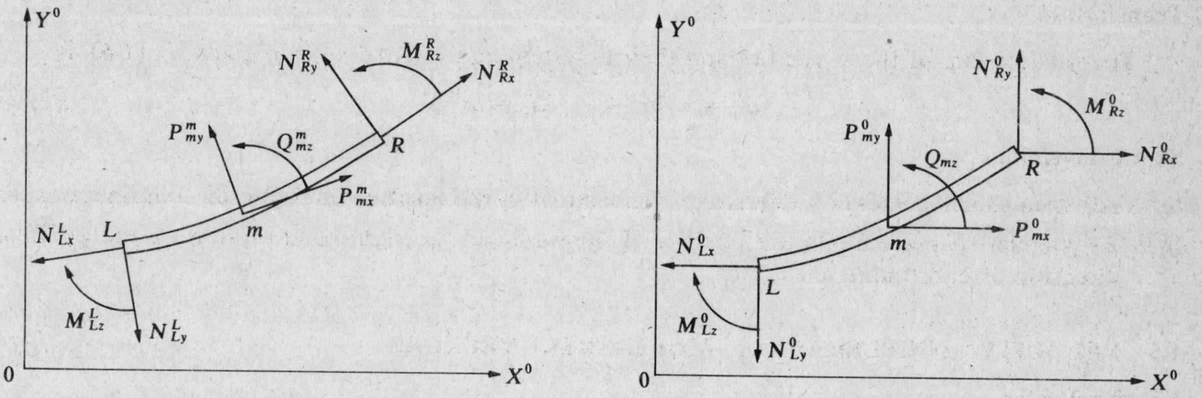

(*a*) Causes and Effects in *l*-System (*b*) Causes and Effects in 0-System

Fig. 1-2. Free Body *LR*

Stress Vectors

The dynamic effect of these causes at the centroid of an arbitrarily selected section is known as the *dynamic stress*, which again may be represented as a matrix function of time (t).

At the left end (L) and right end (R) of the bar in Fig. 1-2a the stress vectors are:

$$\sigma_L^L(t) = \{N_{Lx}^L(t), N_{Ly}^L(t), M_{Lz}^L(t)\} \qquad \sigma_R^R(t) = \{N_{Rx}^R(t), N_{Ry}^R(t), M_{Rz}^R(t)\} \tag{1-8, 1-9}$$

where σ_L^L, σ_R^R are the stresses at the respective end and in the respective system consisting of the normal forces (N_{Lx}^L or N_{Rx}^R), the shearing forces (N_{Ly}^L or N_{Ry}^R), and the bending moments (M_{Lz}^L or M_{Rz}^R), where again the subscripts x, y, z indicate the component's direction and the superscripts L and R identify the angular position of the member system.

Transformations

The relationship of these vectors and their 0-system equivalents (Fig. 1-2b) is given by (*1-6*) as

$$\sigma_L^0 = \pi^{0L}\sigma_L^L \qquad W_m^0 = \pi^{0m}W_m^m \qquad \sigma_R^0 = \pi^{0R}\sigma_R^R \tag{1-10}$$

where π^{0L}, π^{0m}, π^{0R} are special cases of π^{0l} given by (*1-4*).

Sign Conventions

(*a*) All loads and right end (far end) stresses are governed by the sign convention of geometry (Sec. 1.2).
(*b*) All left end (near end) stresses are governed by the opposite sign convention.

1.4 DISPLACEMENT VECTORS

Deformations and Displacements

The *dynamic deformation* of a bar is defined as the change in its shape produced by dynamic causes. As the bar deforms, its centroidal axis takes on a new shape, known as the *dynamic elastic curve.* The ordinates and slopes of this curve, measured with respect to the datum, are designated as the *dynamic displacements.* In general, they are functions of time and may be assembled into a matrix function as

$$\Delta_j^l(t) = \{u_j^l(t), v_j^l(t), \theta_j^l(t)\} \tag{1-11}$$

where $\Delta_j^l(t)$ is the displacement vector at j in the l-system consisting of the linear displacements u_j^l, v_j^l and the angular displacement θ_j^l, acting along the X_j^l, Y_j^l, and Z_j^l axes respectively.

Transformations

The relationship of these vectors and their 0-system equivalents is again given by (*1-6*) as

$$\Delta_j^0 = \pi^{0l}\Delta_j^l \qquad \Delta_j^l = \pi^{l0}\Delta_j^0 \tag{1-12}$$

Sign Conventions

(*a*) All linear displacements are positive if measured in the positive direction of coordinate axes.

(*b*) All angular displacements are positive if measured in the right-hand motion on the positive direction of coordinate axes.

1.5 VELOCITY, ACCELERATION, AND INERTIA VECTORS

Velocity Vector

The *time rate of change of the displacement vector* at a given point on the centroidal axial of a structural member is called the *velocity vector* of that point. With the notation

$$d\Delta_j^l(t)/dt = \dot{\Delta}_j^l \qquad du_j^l/dt = \dot{u}_j^l \qquad dv_j^l/dt = \dot{v}_j^l \qquad d\theta_j^l/dt = \dot{\theta}_j^l \tag{1-13}$$

the velocity vector is

$$\dot{\Delta}_j^l = \{\dot{u}_j^l, \dot{v}_j^l, \dot{\theta}_j^l\} \tag{1-14}$$

where j designates the position and l identifies the system.

Acceleration Vector

The *time rate of change of the velocity vector* at the same point is called the *acceleration vector* of that point. With the notation

$$d^2u_j^l/dt^2 = d\dot{u}_j^l/dt = \ddot{u}_j^l \qquad d^2v_j^l/dt^2 = d\dot{v}_j^l/dt = \ddot{v}_j^l \qquad d^2\theta_j^l/dt^2 = d\dot{\theta}_j^l/dt = \ddot{\theta}_j^l \tag{1-15}$$

the acceleration vector is

$$\ddot{\Delta}_j^l = \{\ddot{u}_j^l, \ddot{v}_j^l, \ddot{\theta}_j^l\} \tag{1-16}$$

where j and l have the same meaning as before.

Inertia Vector

The motion of the mass of a structural member defined at j by $\Delta_j^l, \dot{\Delta}_j^l, \ddot{\Delta}_j^l$ produces at j a new vector, called the *inertia vector,*

$$S_j^l = \{\mu_j\ddot{u}_j^l, \mu_j\ddot{v}_j^l, \rho_j\ddot{\theta}_j^l\} \tag{1-17}$$

where μ_j is the unit length mass, ρ_j is the unit length mass moment of inertia, and $\ddot{u}_j^l, \ddot{v}_j^l, \ddot{\theta}_j^l$ are the components of the acceleration vector $\ddot{\Delta}_j^l$ given by (*1-16*). The first two components of S_j^l are called the *inertia forces* and the third one is denoted as the *inertia moment.*

Transformations

The transformation of the velocity, acceleration, and inertia vectors follows the pattern of (*1-6*), i.e.,

$$\left.\begin{aligned}\dot{\Delta}_j^0 &= \pi^{01}\dot{\Delta}_j^l\\ \ddot{\Delta}_j^0 &= \pi^{01}\ddot{\Delta}_j^l\\ S_j^0 &= \pi^{01}S_j^l\end{aligned}\right\} \quad (1\text{-}18) \qquad \left.\begin{aligned}\dot{\Delta}_j^l &= \pi^{10}\dot{\Delta}_j^0\\ \ddot{\Delta}_j^l &= \pi^{10}\ddot{\Delta}_j^0\\ S_j^l &= \pi^{10}S_j^0\end{aligned}\right\} \quad (1\text{-}19)$$

Sign Conventions

(*a*) The signs of the velocities and accelerations are governed by the sign conventions of geometry.

(*b*) For the signs of inertia forces and moments the opposite is true.

1.6 CLASSIFICATION

Criteria

Four criteria are commonly used in the classification of planar dynamic structural systems: (1) *geometric form*, (2) *degree of stress indeterminacy*, (3) *degree of displacement indeterminacy*, and (4) *degree of order.*

Geometric Form

According to their forms (configuration of members), the planar structures are classified as: *cables, beams, columns, arches, rings, frames, trusses*, and/or their combinations. The dynamic analysis of cables is not discussed in this book.

Degree of Stress Indeterminacy

A planar structure is *dynamically stress determinate* if its stresses (and reactions) can be computed from the conditions of dynamic equilibrium alone.

A planar structure is *dynamically stress indeterminate* if its stresses (and reactions) cannot be computed from the conditions of dynamic equilibrium alone and the deformation conditions must be considered.

The superfluous forces and moments (which are not necessary for dynamic equilibrium) are called *redundants* and their number defines the *degree of stress indeterminacy* of the structure. For a planar system of bars connected together by joints, the total number of independent conditions of dynamic equilibrium is

$$e = 3m + 3j + f \tag{1-20}$$

where m = number of members, j = number of joints, and f = number of all special conditions (releases such as free ends, hinges, or guides).

Since there are six unknown end stresses in each member and r unknown reactions at supports, the number of redundants in terms of (*1-20*) is

$$n = 6m + r - e = 3m + r - 3j - f > 0 \tag{1-21}$$

If $n = 0$, the system is stress determinate, and if $n < 0$, the system is geometrically unstable and unsuitable to carry loads. In (*1-20, 1-21*) the number of joints j includes all internal joints and all points of support.

Degree of Displacement Indeterminacy

Any system of bars can be always visualized as a *system of joints* connected together by elastic springs and the accounting of its *degrees of freedom* related solely to these joints. Then instead of working with redundant stresses, the joint displacements are introduced as the redundants.

For a system of m members connected together by j joints with s internal releases, g internal constraints, and r reactive constraints, the total number of admissible and independent joint displacements (linear and angular) is

$$d = 3j + s - g - r \tag{1-22}$$

where the internal releases are new and independent displacements are introduced by hinges, guides, etc. The number of joints includes all internal joints and points of support. Internal constraints are given by the definition of the system and the number of reactive constraints equals the number of reactions.

Degree of Order

The *first-order theory* (classical theory) of structures neglects the displacements of structure in the formulation of the equilibrium conditions and the dynamic effects are linear combinations of the dynamic causes.

The *second-order theory* of structures includes the displacements of structure in the formulation of the equilibrium conditions and the dynamic effects are no longer linear combination of the dynamic causes.

The first- and second-order theory of structures must not be confused with the designation of structures of order one, two, or three used in space structural analysis (III).

1.7 PRINCIPLES OF DYNAMIC ANALYSIS

Concepts

The dynamic analysis of structures, like the static analysis, developed historically along two distinct paths:

(*a*) The *vectorial dynamics*, based on Newton's laws.
(*b*) The *virtual work dynamics*, based on the principle of conservation of energy.

Although both concepts lead ultimately to the same equations of motion, concept (*b*), appearing in the form of Lagrange's equations and Hamilton's principles, is more easily applicable; it minimizes the thought and effort involved in the preparation of the equations of motion.

Newton's Laws

The following three laws of motion, known as *Newton's Laws*, are considered the axioms of vectorial mechanics:

(I) *First Law*
A particle (body) remains in its state of rest, or of uniform rectilinear motion, unless compelled by force to change that state.

(II) *Second Law*
The acceleration of a given element is proportional to the force applied to it and acts in the direction of that force.

$$P^l_{jx} = \mu_j \frac{d\dot{u}_j}{dt} = \mu_j \ddot{u}^l_j \qquad P^l_{jy} = \mu_j \frac{d\dot{v}^l_j}{dt} = \mu_j \ddot{v}^l_j \qquad Q^l_{jz} = \rho_j \frac{d\dot{\theta}^l_j}{dt} = \rho_j \ddot{\theta}^l_j \tag{1-23}$$

(III) *Third Law*
To every action there is an equal and opposite reaction.

The postulation of these laws is based on the assumption that all quantities are related to a fixed frame of reference, which is absolutely at rest, or to any other frame of reference which is in a state of uniform motion. Such frames, called *Newtonian frames of reference*, are the coordinate axes systems of Fig. 1-1.

Virtual Work Principles

The following two *principles of virtual work* are considered the axioms of virtual work mechanics:

(I) *Bernoulli's Principle (static equilibrium)*

A mechanical plane system is in a state of static equilibrium if the virtual work done by all real forces and/or moments is zero for every virtual displacement consistent with the constraints.

$$\delta W = \Sigma(\mathbf{W}_i \cdot \delta\mathbf{\Delta}_i) + \Sigma(\mathbf{R}_j \cdot \delta\mathbf{\Delta}_j) = 0 \tag{1-24}$$

where $\mathbf{W}_i$ and $\mathbf{R}_j$ are the real load and reaction vectors, $\delta\mathbf{\Delta}_i$ and $\delta\mathbf{\Delta}_j$ are the virtual displacement vectors at i and j respectively, $(\mathbf{W}_i \cdot \delta\mathbf{\Delta}_i)$ and $(\mathbf{R}_j \cdot \delta\mathbf{\Delta}_j)$ are scalar products of two vectors, and Σ indicates the sum.

(II) *d'Alambert's Principle (dynamic equilibrium)*

A mechanical plane system is in a state of dynamic equilibrium at any instant if the virtual work done at that instant by all real and inertia forces and moments is zero for every virtual displacement consistent with the constraints.

$$\delta W = \Sigma(\mathbf{W}_i \cdot \delta\mathbf{\Delta}_i) + \Sigma(\mathbf{R}_j \cdot \delta\mathbf{\Delta}_j) - \Sigma(\mathbf{S}_k \cdot \delta\mathbf{\Delta}_k) = 0 \tag{1-25}$$

which is *(1-24)* complemented by the negative virtual work of the inertia vector $\mathbf{S}_k$ *(1-17)*.

The postulation of these principles is based on the assumption of Newtonian frame of reference, imaginary (virtual) displacements, thermal isolation, and absence of dissipation. Furthermore, the system involved is assumed to be rigid (nondeformable). The extension of *(1-24)* and *(1-25)* to deformable systems and the inclusion of dissipative functions are presented in Chapters 4 and 7.

Energy Theorems

The energy defined as the capacity of the mechanical system to do work occurs in a structural system in two distinct forms, as the *kinetic energy T* (energy of motion) and the *potential energy V* (energy of position).

(I) *Law of Kinetic Energy*

The work W produced by moving a mechanical system acted upon by forces and moments from a given position 1 to a new position 2 equals the change in kinetic energy $(T_2 - T_1)$.

$$W = \Sigma \int_1^2 \mathbf{F}_i \cdot d\mathbf{\Delta}_i = \tfrac{1}{2}\Sigma[\mu_i(\dot{u}_{i,2}^2 - \dot{u}_{i,1}^2) + \mu_i(\dot{v}_{i,2}^2 - \dot{v}_{i,1}^2) + \rho_i(\dot{\theta}_{i,2}^2 - \dot{\theta}_{i,1}^2)] \tag{1-26}$$

where $\mathbf{F}_i$ represents load, reaction, and stress vectors, $\mathbf{\Delta}_i$ is their displacement vector, and

$$T_1 = \tfrac{1}{2}\Sigma(\mu_i\dot{u}_{i,1}^2 + \mu_i\dot{v}_{i,1}^2 + \rho_i\dot{\theta}_{i,1}^2)$$
$$T_2 = \tfrac{1}{2}\Sigma(\mu_i\dot{u}_{i,2}^2 + \mu_i\dot{v}_{i,2}^2 + \rho_i\dot{\theta}_{i,2}^2)$$

are the kinetic energies at 1 and 2 respectively.

(II) *Conservative Force Field*

A vector field $\mathbf{F}$ is conservative if and only if there exists a continuously differentiable scalar field V such that

$$\mathbf{F} = -\nabla V \tag{1-27}$$

and if and only if

$$\nabla \times \mathbf{F} = 0 \tag{1-28}$$

(III) *Law of Potential Energy*

For a conservative vector field the work defined by (*1-27*) and (*1-28*) equals also the negative change in potential energy, $-(V_2 - V_1)$.

$$W = \Sigma \int \mathbf{F}_i \cdot d\mathbf{\Delta}_i = -\Sigma(V_{i,2} - V_{i,1}) \qquad (1\text{-}29)$$

where $V_1 = \Sigma V_{i,1}$ and $V_2 = \Sigma V_{i,2}$ are the total potential energies at positions 1 and 2 respectively.

(IV) *Law of Conservation of Energy*

In a conservative vector field the total energy is constant.

$$T + V = \text{constant} \qquad (1\text{-}30)$$

where T and V are respectively the total kinetic and potential energies of the system.

The postulation of these laws is based on the same assumptions as those of virtual work. Their extension to deformable systems and the inclusion of dissipative functions is possible as shown in Chapters 4 and 7.

Variational Methods

Finally, the last class of methods used in this volume is based on the variational treatment which dispenses with vectors and deals exclusively in terms of fundamental scalars, the *Lagrangian function*,

$$L = T - V \qquad (1\text{-}31)$$

and the *Hamiltonian function*,

$$H = T + V \qquad (1\text{-}32)$$

The development and the application of the variational treatment of these functions are covered in Chapter 4.

Chapter 2

Systems with One Degree of Freedom

2.1 MECHANICAL MODELS

Concept

For the dynamic analysis, it is often advantageous to represent the *real structural system* by an *ideal mechanical model* which is better adaptable to the mathematical treatment and the dynamic behavior of which closely approximates the behavior of the real structure.

Basic Elements

Three types of basic elements are used in the construction of this model:

(*a*) The *lumped mass element*, assumed to be perfectly rigid and represented by a solid block.

(*b*) The *elastic element*, assumed to be massless, elastic, and represented by a spring.

(*c*) The *viscoelastic element*, assumed to be massless, viscoelastic, and represented by a dashpot (damper).

The combination of these elements leads to a large variety of models which are classified as *single* or *multi-degree freedom systems* according to the number of springs they contain.

Basic Models

Two types of single degree freedom models, called the *basic models*, are constructed and their dynamic behavior is analyzed in this chapter: (*a*) the *Hookean model* (Figs. 2-1, 2-2, and 2-3) and (*b*) the *Kelvin model* (Figs. 2-6, 2-7 and 2-8).

In their construction the mass, spring, and dashpot constants must be assumed, computed, or determined by experiments.

2.2 HOOKEAN MODELS

Physical Characteristics

The *simplest ideal mechanical model* is the *Hookean model* consisting of a lumped mass block attached by an elastic spring to a rigid foundation. By definition, this model has *one degree of freedom* allowing a small displacement of the lumped mass along or about one axis only. Consequently, three types of this model are possible:

(*1*) The *longitudinal motion model* (Fig. 2-1) allows a small linear displacement u of the mass along the X axis. The model is defined by the mass μ, the linear spring constant k_u, and the load function $P_x(t)$.

(*2*) The *transverse motion model* (Fig. 2-2) allows a small linear displacement v of the mass along the Y axis. The model is defined by the mass μ, the linear spring constant k_v, and the load function $P_y(t)$.

(*3*) The *angular motion model* (Fig. 2-3) allows a small angular displacement θ of the mass about the Z axis. The model is defined by the mass moment of inertia ρ, the angular spring constant k_θ, and the load function $Q_z(t)$.

The absence of the dashpot element in these models indicates that *no damping forces* are present.

Equations of Motion

The dynamic load acting on the mass of the respective model produces a motion governed by the corresponding differential equation derived from the *conditions of dynamic equilibrium* (Problem 2.1).

Three such differential equations, each designated by the number of the model, are:

$$\mu \ddot{u} + k_u u = P_x(t) \tag{2-1a}$$

$$\mu \ddot{v} + k_v v = P_y(t) \tag{2-2a}$$

$$\rho \ddot{\theta} + k_\theta \theta = Q_z(t) \tag{2-3a}$$

where the displacement coordinates u, v, θ and the load functions are related to the initial system X, Y, Z.

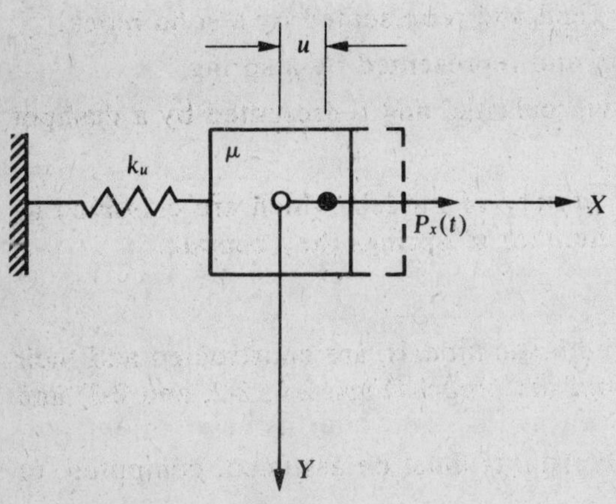

Fig. 2-1. Longitudinal Motion

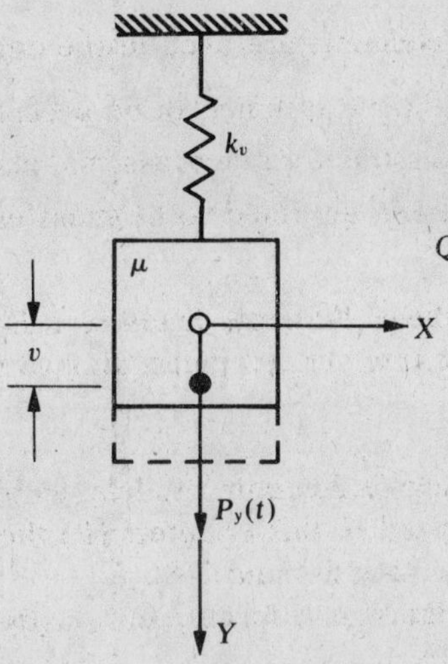

Fig. 2-2. Transverse Motion

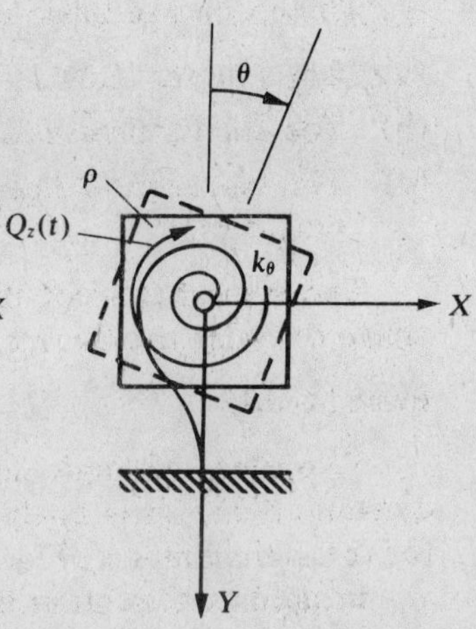

Fig. 2-3. Angular Motion

The *integrals of these equations* obtained by the method of Laplace transforms are respectively:

$$u = u_0 \cos \omega_u t + \frac{\dot{u}_0}{\omega_u} \sin \omega_u t + \frac{1}{\mu \omega_u} \int_0^t \sin [\omega_u(t - \tau)] P_x(\tau)\, d\tau \tag{2-1b}$$

$$v = v_0 \cos \omega_v t + \frac{\dot{v}_0}{\omega_v} \sin \omega_v t + \frac{1}{\mu \omega_v} \int_0^t \sin [\omega_v(t - \tau)] P_y(\tau)\, d\tau \tag{2-2b}$$

$$\theta = \theta_0 \cos \omega_\theta t + \frac{\dot{\theta}_0}{\omega_\theta} \sin \omega_\theta t + \frac{1}{\rho \omega_\theta} \int_0^t \sin [\omega_\theta(t - \tau)] Q_z(\tau)\, d\tau \tag{2-3b}$$

where
$$\omega_u = \sqrt{\frac{k_u}{\mu}}, \qquad \omega_v = \sqrt{\frac{k_v}{\mu}}, \qquad \omega_\theta = \sqrt{\frac{k_\theta}{\rho}}$$

are the *natural circular frequencies* and u_0, v_0, θ_0, and $\dot{u}_0, \dot{v}_0, \dot{\theta}_0$, called the *initial conditions*, are the respective *displacements* and *velocities* at $t = 0$ (Problem 2.2).

These equations (*2-1b, 2b, 3b*) define the motion of the respective models and are known as the *equations of motion*. Their trigonometric functions accompanying the initial values are called the *shape functions* and the convolution integrals involving the load functions are designated as the *particular functions*.

2.3 FREE VIBRATION WITHOUT DAMPING

Equations of Motion

When the load function equals zero and the motion is produced by the respective initial values only, the resulting motion is called *free vibration* (harmonic motion, harmonic oscillation). Since the equations of free vibration of the Hookean models (Figs. 2-1, 2-2, 2-3) are the general integrals (*2-1b, 2b, 3b*) with the particular functions deleted, they are symbolically expressed below by one equation as

$$\Delta = \underbrace{\Delta_0 \cos \omega_\Delta t}_{\text{curve }(a)} + \underbrace{\frac{\dot{\Delta}_0}{\omega_\Delta} \sin \omega_\Delta t}_{\text{curve }(b)} \tag{2-4}$$

where Δ is u, v, or θ and ω_Δ is ω_u, ω_v, or ω_θ respectively.

Graphical Representation

The graph of the time-displacement curve of free vibration may be constructed by superposing the curve (a) and the curve (b) described by (*2-4*) in Figs. 2-4*a*, *b* or by either of the *modified equations* (*2-4*) given below.

$$\Delta = R_\Delta \cos (\omega_\Delta t - \alpha_\Delta) \qquad \text{or} \qquad \Delta = R_\Delta \sin (\omega_\Delta t + \beta_\Delta) \tag{2-5a, b}$$

where the *natural amplitude* is

$$R_\Delta = \sqrt{(\Delta_0)^2 + (\dot{\Delta}_0/\omega_\Delta)^2} \tag{2-6}$$

and the *phase angles* representing the *shift of the harmonic curve* on the $\omega_\Delta t$ axis are

$$\alpha_\Delta = \tan^{-1} \frac{\dot{\Delta}_0}{\omega_\Delta \Delta_0} \qquad \text{and} \qquad \beta_\Delta = \tan^{-1} \frac{\omega_\Delta \Delta_0}{\dot{\Delta}_0} \tag{2-7a, b}$$

The graphical representation of (*2-5*) to (*2-7*) is shown in Fig. 2-4*c* and their derivations and applications are given in Problems 2.3 to 2.6.

Special Cases

If one of the *initial values equals zero*, the graph of motion reduces to the curve (a) or the curve (b) and

$$R_\Delta = \Delta_0, \quad \alpha_\Delta = 0 \qquad \text{or} \qquad R_\Delta = \dot{\Delta}_0/\omega_\Delta, \quad \beta_\Delta = 0 \tag{2-8a, b}$$

as shown in Fig. 2-4*a* and *b* respectively.

Parameters of Motion

The free vibration discussed above is said to be *harmonic*, i.e., the displacement *varies sinusoidally* with time and *repeats itself* after a definite time interval. The parameters of this motion are defined below.

(a) The *natural amplitude* R_Δ is the greatest displacement of the lumped mass (from its equilibrium position) given by (2-6).

(b) The *natural period* T_Δ is the time required for one complete cycle, after which a repetition of the same motion takes place.

$$T_\Delta = 2\pi/\omega_\Delta \tag{2-9}$$

where ω_Δ is the *natural circular frequency* given by (*2-1b, 2b, 3b*).

(c) The *natural frequency* f_Δ is the number of complete cycles per unit time, i.e.,

$$f_\Delta = 1/T_\Delta = \omega_\Delta/2\pi \tag{2-10}$$

The study of the preceding results reveals that

(1) The *natural amplitude* R_Δ is affected by the mass and spring constants and depends on the initial values.

(2) The *natural period* T_Δ and the *natural frequency* f_Δ are not affected by the initial values and depend on the mass and spring constant only.

Finally, it should be noted that the *sum of the phase angles* $\alpha_\Delta + \beta_\Delta = \pi/2$ which is $T_\Delta\omega_\Delta/4$.

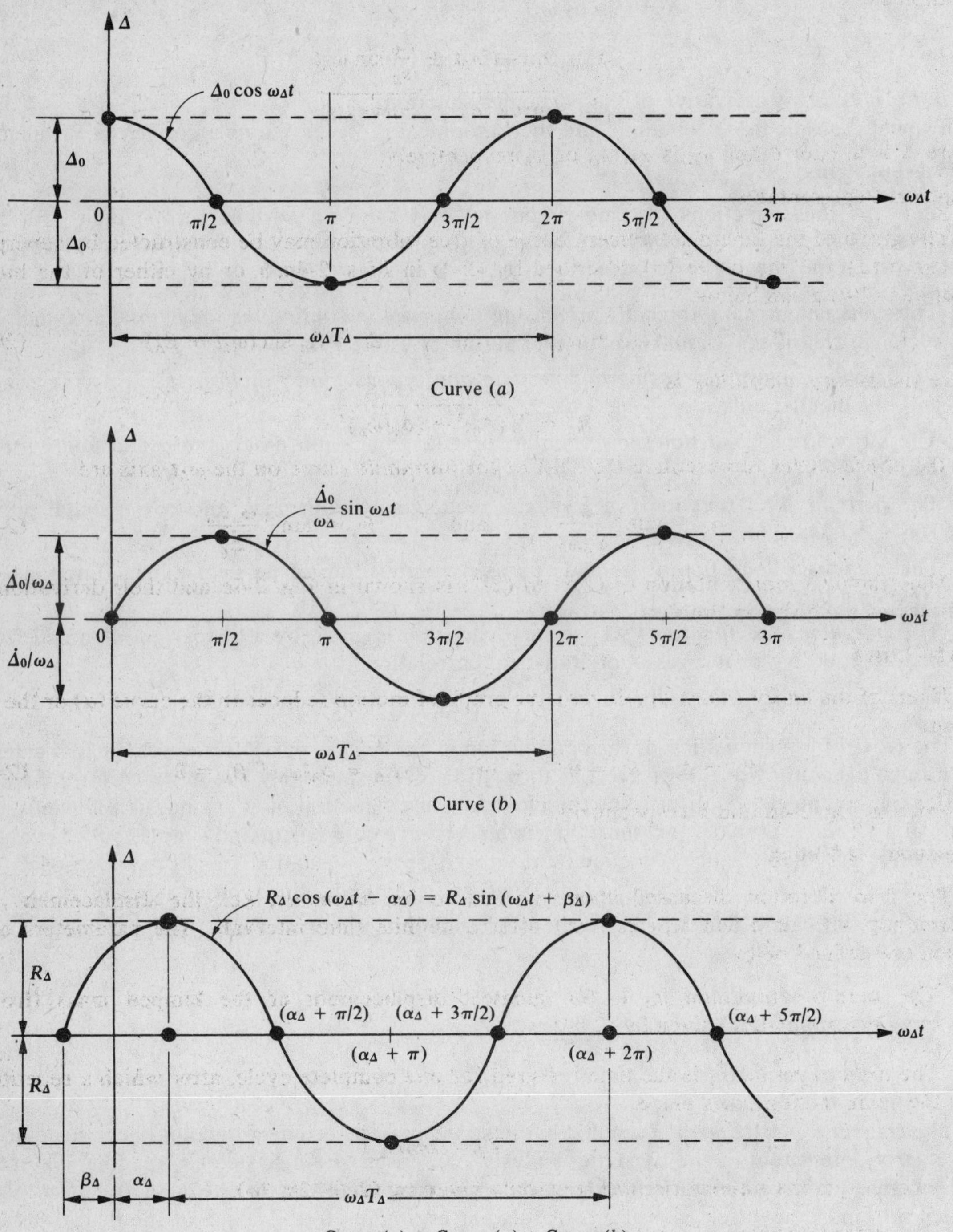

Curve (c) = Curve (a) + Curve (b)

Fig. 2-4. Displacement Curves—Free Vibration Without Damping

2.4 FORCED VIBRATION WITHOUT DAMPING

If the load function in (*2-1b, 2b, 3b*) is not zero and the initial values Δ_0, $\dot{\Delta}_0$ are or are not zero, the resulting motion is called a *forced vibration*, the graph of which is in general the superposition of curve (*c*) in Fig. 2-4 and a new curve defined by the particular function Φ_Δ.

Symbolically, the equation of motion (*2-1b, 2b, 3b*) and its first derivative in this case are respectively

$$\begin{aligned} \Delta &= \Delta_0 \cos \omega_\Delta t + \frac{\dot{\Delta}_0}{\omega_\Delta} \sin \omega_\Delta t + \Phi_\Delta \\ \dot{\Delta} &= -\Delta_0 \omega_\Delta \sin \omega_\Delta t + \dot{\Delta}_0 \cos \omega_\Delta t + \dot{\Phi}_\Delta \end{aligned} \tag{2-11}$$

where $\dot{\Phi}_\Delta$ is the first derivative of the particular function.

Typical cases of these functions and their graphical representations are given in Problem 2.9.

Forcing Functions

Since the load functions (forcing functions) may take on various forms, their orderly and complete classification is very difficult; they can be however described by broad categories defined below.

(*1*) The *constant* or *monotonically increasing nonperiodic continuous load functions* such as the rectangular, triangular, and exponential pulses of Table P-2.31.

(*2*) The *monotonically decreasing nonperiodic continuous load functions* such as the triangular and exponential pulses of Table P-2.32.

(*3*) The *symmetrical load functions* such as the triangular, sinusoidal, and cosinusoidal pulses of Table P-2.33.

(*4*) The *periodic load functions* such as the rectangular, sinusoidal, and cosinusoidal pulses of Table P-2.34.

(*5*) The *composite load functions* made up as a composition of particular load functions of the preceding types.

(*6*) The *irregular load functions* whose time variation is given by a histograph obtained from an in-situ recording or is a result of statistical correlation.

Methods of Evaluation

The *closed form evaluation* of the convolution integral Φ_Δ is possible and practical in categories (*1*) through (*4*) as shown in Tables P-2.31 through P-2.34 respectively and in Problems 2.9 and 2.10.

For the composite load functions the closed form evaluation of Φ_Δ is possible but not always practical. In such cases the *transport matrix method* is a convenient and simple tool of evaluation.

Finally, for the irregular functions the closed form evaluation of Φ_Δ is not possible and approximate methods must be employed. Two such methods, the *Fourier series method* and the *numerical step-by-step method* applied in the evaluation of Φ_Δ, are shown in Problems 2.12 and 2.13.

2.5 DYNAMIC TRANSPORT MATRIX

Concept

The *transport matrix method* applied in the static analysis of linear and nonlinear structures is a particularly potent tool of the dynamic analysis. In cases of single degree freedom systems in a state of free or forced vibration without damping, the time axis of the load function $W_\Delta(t)$ is divided into judiciously selected time intervals corresponding to the shape of the load function.

For the interval *ij*, $t_{ji} = t_j - t_i$, the *dynamic transport matrix equation* constructed by (*2-11*) is

$$\underbrace{\begin{bmatrix} +1 \\ \hline \Delta_j \\ \dfrac{\dot{\Delta}_j}{\omega} \end{bmatrix}}_{\hat{\hat{H}}_j} = \underbrace{\left[\begin{array}{c|cc} +1 & 0 & 0 \\ \hline \Phi_{ji} & \cos \omega t_{ji} & \sin \omega t_{ji} \\ \dfrac{\dot{\Phi}_{ji}}{\omega} & -\sin \omega t_{ji} & \cos \omega t_{ji} \end{array}\right]}_{\hat{\hat{T}}_{ji}} \underbrace{\begin{bmatrix} +1 \\ \hline \Delta_i \\ \dfrac{\dot{\Delta}_i}{\omega} \end{bmatrix}}_{\hat{\hat{H}}_i} \tag{2-12}$$

where $\hat{\hat{H}}_j$, $\hat{\hat{H}}_i$ are the ***dimensionless state vectors*** at j and i respectively, and $\hat{\hat{T}}_{ji}$ is the ***time transport matrix*** in the interval t_{ji}, the coefficients of which are constant for the given interval. For simplicity of writing, the subscript Δ (appearing in ω_Δ and Φ_Δ before) is omitted, but the implication that (*2-12*) constructed by (*2-11*) holds for all three types of Hookean model remains valid.

As its static counterpart, the dynamic transport matrix $\hat{\hat{T}}_{ji}$ gives a clear, systematic and complete record of the dynamic properties of the model during the selected time interval; it is specific for the model, the time interval, and the given load function, and is independent of the end conditions (Problem 2.14).

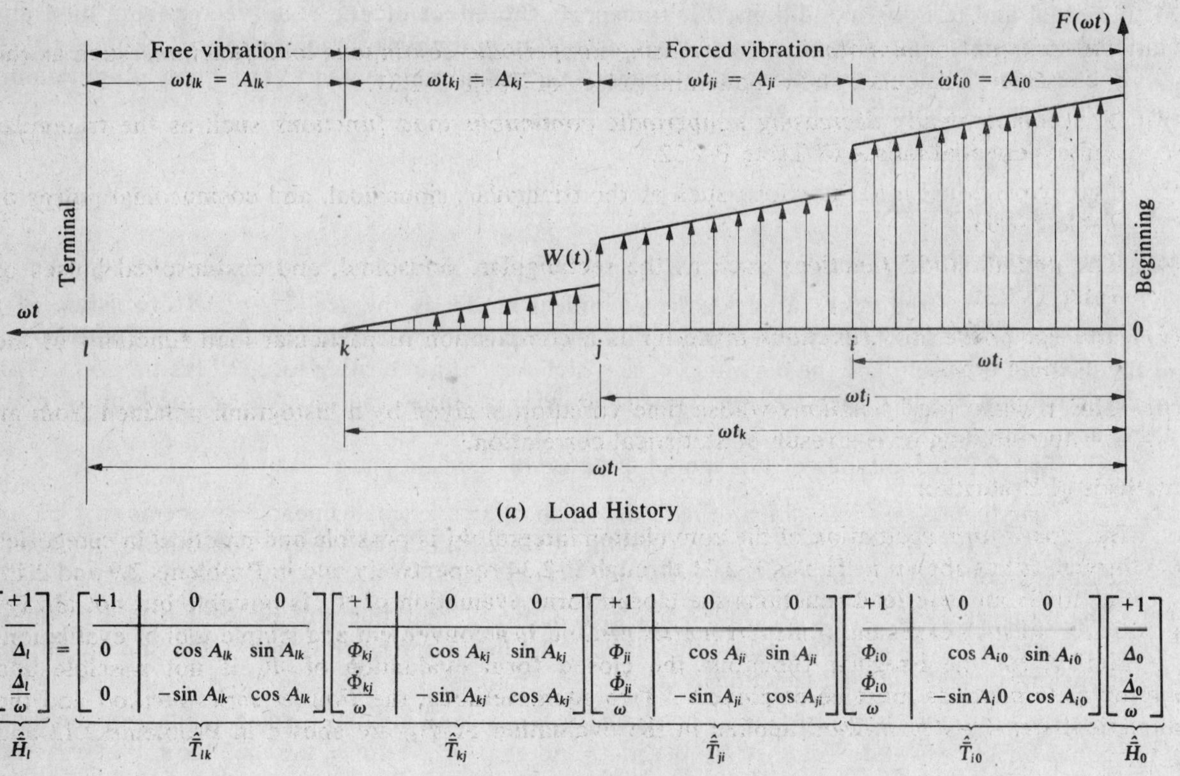

(*a*) Load History

$$\underbrace{\begin{bmatrix} +1 \\ \hline \Delta_l \\ \dfrac{\dot{\Delta}_l}{\omega} \end{bmatrix}}_{\hat{\hat{H}}_l} = \underbrace{\left[\begin{array}{c|cc} +1 & 0 & 0 \\ \hline 0 & \cos A_{lk} & \sin A_{lk} \\ 0 & -\sin A_{lk} & \cos A_{lk} \end{array}\right]}_{\hat{\hat{T}}_{lk}} \underbrace{\left[\begin{array}{c|cc} +1 & 0 & 0 \\ \hline \Phi_{kj} & \cos A_{kj} & \sin A_{kj} \\ \dfrac{\dot{\Phi}_{kj}}{\omega} & -\sin A_{kj} & \cos A_{kj} \end{array}\right]}_{\hat{\hat{T}}_{kj}} \underbrace{\left[\begin{array}{c|cc} +1 & 0 & 0 \\ \hline \Phi_{ji} & \cos A_{ji} & \sin A_{ji} \\ \dfrac{\dot{\Phi}_{ji}}{\omega} & -\sin A_{ji} & \cos A_{ji} \end{array}\right]}_{\hat{\hat{T}}_{ji}} \underbrace{\left[\begin{array}{c|cc} +1 & 0 & 0 \\ \hline \Phi_{i0} & \cos A_{i0} & \sin A_{i0} \\ \dfrac{\dot{\Phi}_{i0}}{\omega} & -\sin A_{i}0 & \cos A_{i0} \end{array}\right]}_{\hat{\hat{T}}_{i0}} \underbrace{\begin{bmatrix} +1 \\ \hline \Delta_0 \\ \dfrac{\dot{\Delta}_0}{\omega} \end{bmatrix}}_{\hat{\hat{H}}_0}$$

(*b*) Transport Matrix Chain

Fig. 2-5. Hookean Model Transport in *l*0 Interval

Transport Chain

Once the general form of the dynamic transport equation is available for a single interval, the extension of this equation to the analysis of motion during the investigated life span of the motion is accomplished by ***matrix multiplication.***

If the ωt axis of the load function in Fig. 2-5 is divided into four time intervals $\omega t_{lk} = A_{lk}$, $\omega t_{kj} = A_{kj}$, $\omega t_{ji} = A_{ji}$, $\omega t_{i0} = A_{i0}$, the state vectors at the respective time stations l, k, j, i become respectively

$$\hat{\hat{H}}_l = \hat{\hat{T}}_{lk}\hat{\hat{H}}_k \qquad \hat{\hat{H}}_k = \hat{\hat{T}}_{kj}\hat{\hat{H}}_j \qquad \hat{\hat{H}}_j = \hat{\hat{T}}_{ji}\hat{\hat{H}}_i \qquad \hat{\hat{H}}_i = \hat{\hat{T}}_{i0}\hat{\hat{H}}_0 \tag{2-13}$$

where $\hat{\hat{T}}_{lk}$, $\hat{\hat{T}}_{kj}$, $\hat{\hat{T}}_{ji}$, $\hat{\hat{T}}_{i0}$ are the transport matrices of the respective intervals the coefficients of which follow the pattern of *(2-12)* and are constants.

Then by successive substitution,

$$\left.\begin{aligned} \hat{\hat{H}}_j &= \hat{\hat{T}}_{ji}\hat{\hat{T}}_{i0}\hat{\hat{H}}_0 = \hat{\hat{T}}_{j0}\hat{\hat{H}}_0 \\ \hat{\hat{H}}_k &= \hat{\hat{T}}_{kj}\hat{\hat{T}}_{ji}\hat{\hat{T}}_{i0}\hat{\hat{H}}_0 = \hat{\hat{T}}_{k0}\hat{\hat{H}}_0 \\ \hat{\hat{H}}_l &= \underbrace{\hat{\hat{T}}_{lk}\hat{\hat{T}}_{kj}\hat{\hat{T}}_{ji}\hat{\hat{T}}_{i0}}_{\hat{\hat{T}}_{l0}}\hat{\hat{H}}_0 = \hat{\hat{T}}_{l0}\hat{\hat{H}}_0 \end{aligned}\right\} \tag{2-14}$$

where the product of $\hat{\hat{T}}$ matrices in each matrix equation is a new matrix. Since the new matrix is a result of a chain product, it is designated as a *transport matrix chain* (Problem 2.15).

Transport Method

This *transport matrix chain* is again characteristic for the given interval ωt_{l0} and is independent of the initial and terminal conditions. It transports the effect of one state vector from one time station 0 to another time station l and it may include any number of subintervals.

Because of its recurrent nature, the transport method is well suited to computer programming and to numerical calculations in general.

2.6 KELVIN MODELS

Physical Characteristics

The next most frequently used ideal mechanical model is the *Kelvin model* consisting of a lumped mass block attached by an elastic spring and a dashpot to a rigid foundation. The function of the dashpot is to simulate the damping force which acts as a retarding force of the motion. This model is again a *single degree freedom system* allowing a small displacement of the lumped mass along or about one axis only.

As in Sec. 2.2, three types of this model are possible and they are:

(*1*) The *longitudinal motion model* (Fig. 2-6) which allows a small linear displacement u of the mass along the X axis and is defined by the mass μ, the linear spring constant k_u, the linear damping constant c_u, and the load function $P_x(t)$.

(*2*) The *transverse motion model* (Fig. 2-7) which allows a small linear displacement v of the mass along the Y axis and is defined by the mass μ, the linear spring constant k_v, the linear damping constant c_v, and the load function $P_y(t)$.

(*3*) The *angular motion model* (Fig. 2-8) which allows a small angular displacement θ of the mass about the Z axis and is defined by the mass moment of inertia ρ, the angular spring constant k_θ, the angular damping constant c_θ, and the load function $Q_z(t)$.

Equations of Motion

The dynamic load acting on the mass of the respective model produces a motion governed by the corresponding differential equation derived from the *conditions of dynamic equilibrium* (Problem 2.16).

Three such differential equations corresponding to the models of Figs. 2-6, 2-7, and 2-8 respectively are:

$$\begin{aligned} \ddot{u} + 2\gamma_u\dot{u} + \omega_u^2 u &= (1/\mu)P_x(t) \\ \ddot{v} + 2\gamma_v\dot{v} + \omega_v^2 v &= (1/\mu)P_y(t) \\ \ddot{\theta} + 2\gamma_\theta\dot{\theta} + \omega_\theta^2\theta &= (1/\rho)Q_z(t) \end{aligned} \tag{2-15}$$

where

$$2\gamma_u = c_u/\mu, \quad 2\gamma_v = c_v/\mu, \quad 2\gamma_\theta = c_\theta/\rho \qquad \text{and} \qquad \omega_u^2 = k_u/\mu, \quad \omega_v^2 = k_v/\mu, \quad \omega_\theta^2 = k_\theta/\rho$$

As in (*2-1a, 2a, 3a*), the displacement coordinates u, v, θ in these equations are related to the initial system X, Y, Z and the load functions are related to the same system.

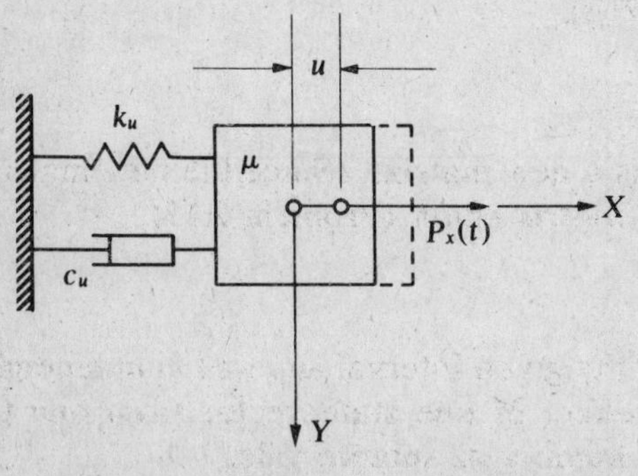

Fig. 2-6. Longitudinal Motion

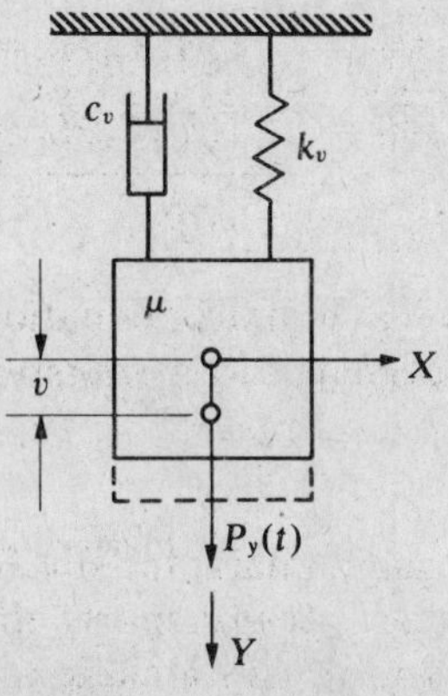

Fig. 2-7. Transverse Motion

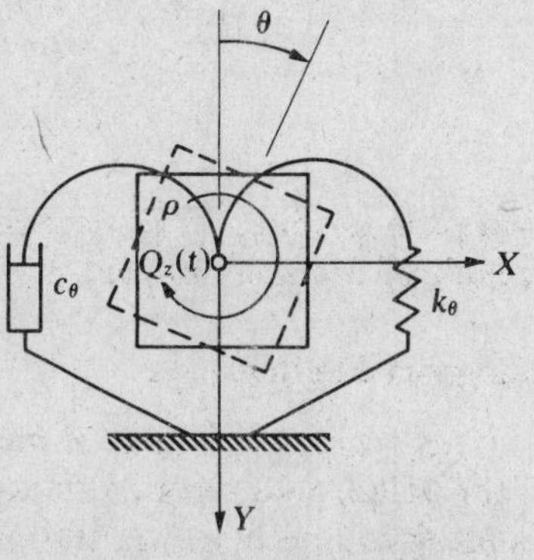

Fig. 2-8. Angular Motion

The *integrals of these equations* obtained by the method of Laplace transforms take on three different forms dependent on the relationship of c to k. Since these solutions again have similar structure, they are written symbolically in terms of Δ, γ_Δ, λ_Δ, μ_Δ, and $W_\Delta(t)$, where Δ is u, v, or θ, γ_Δ is γ_u, γ_v, or γ_θ, λ_Δ is λ_u, λ_v, or λ_θ, μ_Δ is μ or ρ, and $W_\Delta(t)$ is $P_x(t)$, $P_y(t)$, or $Q_z(t)$ as required in the respective equation.

(*a*) *Underdamped motion,* $\gamma_\Delta^2 < \omega_\Delta^2$, $\lambda_\Delta = \sqrt{\omega_\Delta^2 - \gamma_\Delta^2}$:

$$\Delta = e^{-\gamma_\Delta t}\left(\Delta_0 \cos \lambda_\Delta t + \frac{\dot{\Delta}_0 + \gamma_\Delta \Delta_0}{\lambda_\Delta} \sin \lambda_\Delta t\right) + \frac{1}{\mu_\Delta \lambda_\Delta}\int_0^t e^{-\gamma_\Delta(t-\tau)} \sin\left[\lambda_\Delta(t-\tau)\right] W_\Delta(\tau)\, d\tau \tag{2-16a}$$

(*b*) *Critically damped motion,* $\gamma_\Delta^2 = \omega_\Delta^2$, $\lambda_\Delta = 0$:

$$\Delta = e^{-\gamma_\Delta t}[\Delta_0(1 + \gamma_\Delta t) + \dot{\Delta}_0 t] + \frac{1}{\mu_\Delta}\int_0^t (t-\tau)e^{-\gamma_\Delta(t-\tau)} W_\Delta(\tau)\, d\tau \tag{2-16b}$$

(*c*) *Overdamped motion,* $\gamma_\Delta^2 > \omega_\Delta^2$, $\lambda_\Delta = \sqrt{\gamma_\Delta^2 - \omega_\Delta^2}$:

$$\Delta = e^{-\gamma_\Delta t}\left(\Delta_0 \cosh \lambda_\Delta t + \frac{\dot{\Delta}_0 + \gamma_\Delta \Delta_0}{\lambda_\Delta} \sinh \lambda_\Delta t\right) + \frac{1}{\mu_\Delta}\int_0^t e^{-\gamma_\Delta(t-\tau)} \sinh\left[\lambda_\Delta(t-\tau)\right] W_\Delta(\tau)\, d\tau \tag{2-16c}$$

In these equations Δ_0 and $\dot{\Delta}_0$, again called initial conditions, are the *displacements* and *velocities* at $t = 0$, respectively (Problems 2.17, 2.18, 2.19). Equations (*2-16a, b, c*) are the equations of motion where the functions accompanying the initial values are called the *shape functions* and the convolution integrals involving the load functions are designated as the *particular functions.*

2.7 FREE VIBRATION WITH DAMPING

Equations of Motion

As in Sec. 2.3, when the load function in (*2-15*) equals zero and the motion is produced by the initial values only, the resulting motion is called *free vibration with damping.* The equations of motion are (*2-16a, b, c*) with the particular function $\Phi_\Delta = 0$.

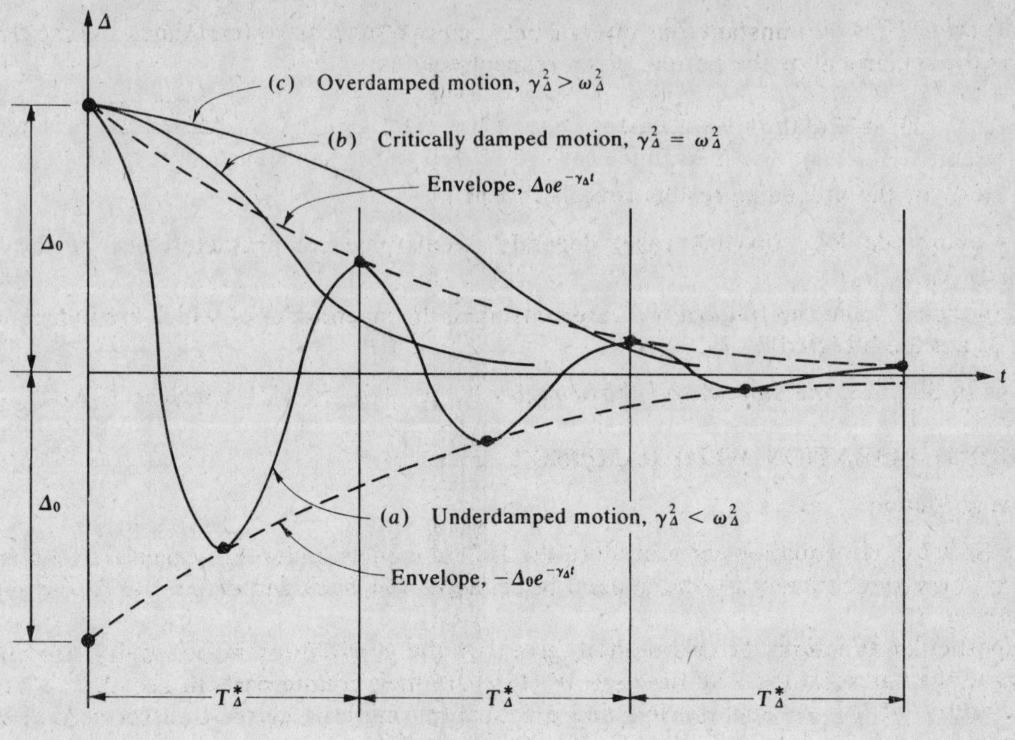

Fig. 2-9. Free Vibration With Damping, $\dot{\Delta}_0 = 0$, Displacement Curves

Graphical Representation

Three curves represent the free vibration with damping (Fig. 2-9), each corresponding to one of the solutions (*2-16a, b, c*) with $\Phi_\Delta = 0$. As in the free vibration without damping, (*2-16a*) may be written as

$$\Delta = R_\Delta^*[\cos(\lambda_\Delta t - \alpha_\Delta)] \qquad \text{or} \qquad \Delta = R_\Delta^*[\sin(\lambda_\Delta t + \beta_\Delta)] \tag{2-17, 18}$$

where the *variable amplitude* is

$$R_\Delta^* = R_\Delta e^{-\gamma_\Delta t} = e^{-\gamma_\Delta t}\sqrt{(\Delta_0)^2 + (\dot{\Delta}_0 + \gamma_\Delta \Delta_0)^2/\lambda_\Delta^2} \tag{2-19}$$

and the *phase angles* representing the *shift of the decaying harmonic curve* on the $\lambda_\Delta t$ axis are

$$\alpha_\Delta = \tan^{-1}\left(\frac{\dot{\Delta}_0 + \gamma_\Delta \Delta_0}{\lambda_\Delta \Delta_0}\right) \qquad \text{and} \qquad \beta_\Delta = \tan^{-1}\left(\frac{\lambda_\Delta \Delta_0}{\dot{\Delta}_0 + \gamma_\Delta \Delta_0}\right) \tag{2-20, 21}$$

The curve (*a*) representing the underdamped motion shows that the motion is oscillatory with diminishing amplitudes, whereas the curves (*b*) and (*c*), representing the critically damped and overdamped motion respectively, show that the magnitude of damping is so large that the motion is no longer oscillatory and decreases exponentially with time (an aperiodic motion). The critically damped motion is the limit of the periodic motion during which the system reaches rest in the shortest time (Problem 2.19).

Parameters of Motion

The *underdamped free vibration* is nonharmonic, nonrepetitive, with decreasing amplitudes spaced on equal intervals (periods). The parameters of this motion are described below.

(*a*) The *amplitude* $R_{\Delta,n}^*$ is a decreasing value given as

$$R_{\Delta,n}^* = R_\Delta e^{-\gamma_\Delta n T_\Delta^*} \tag{2-22}$$

where R_Δ is the initial term in (*2-19*), $n = 0, 1, 2, \ldots,$ and T_Δ^* is defined by (*2-23*).

(*b*) The *period* T_Δ^* is the constant time interval between two successive amplitudes and the *frequency* f_Δ^* is the reciprocal of the period, given respectively as

$$T_\Delta^* = 2\pi/\lambda_\Delta = \frac{2\pi}{\sqrt{\omega_\Delta^2 - \gamma_\Delta^2}} \qquad f_\Delta^* = 1/T_\Delta^* = \lambda_\Delta/2\pi = \frac{\sqrt{\omega_\Delta^2 - \gamma_\Delta^2}}{2\pi} \tag{2-23, 24}$$

The study of the preceding results reveals that:

(*1*) The *amplitude* $R_{\Delta,n}^*$ (in this case) depends on all physical characteristics of the system $(\Delta_0, \dot{\Delta}_0, k_\Delta, c_\Delta, \mu_\Delta)$.

(*2*) The *period* T_Δ^* and the *frequency* f_Δ^* are constants, the magnitudes of which are independent of $\Delta_0, \dot{\Delta}_0$ but are affected by $k_\Delta, c_\Delta, \mu_\Delta$.

Finally, as in Sec. 2.3, the *sum of the phase angles* is $\alpha_\Delta + \beta_\Delta = \pi/2$ which is $T_\Delta^*\lambda_\Delta/4$.

2.8 FORCED VIBRATION WITH DAMPING

Equations of Motion

As in Sec. 2.4, the equations of motion of the Kelvin models under the action of a load function $W_\Delta(t)$ are given symbolically by the general integrals (*2-16a, b, c*) and define the *forced vibration with damping.*

The particular functions of this motion, given by the convolution integrals Φ_Δ^*, are classified according to the nature of the load function $W_\Delta(t)$ as are their counterparts in Sec. 2.4. The *closed form evaluation* of Φ_Δ^* is again possible and practical but more involved in categories (*1*) through (*4*). For the remaining categories the *Fourier Series method,* the *step-by-step integration method* and/or the *transport matrix method* must be employed (Problems 2.23, 2.24, 2.25).

Dynamic Transport Matrix

The extension of the concept of the transport method to the analysis of free or forced vibration with damping meets no particular difficulties and follows the pattern of Sec. 2.5. The *dimensionless state vector* definitions become

$$\hat{\bar{H}}_j = \{+1, \Delta_j, \dot{\Delta}_j/\lambda\} \qquad \text{and} \qquad \hat{\bar{H}}_i = \{+1, \Delta_i, \dot{\Delta}_i/\lambda\}$$

whereas in (*2-12*) the subscript Δ is omitted but implied and in case (*b*) λ is replaced by γ.

The *dimensionless time transport matrix* is constructed by means of the respective equation of motion (*2-16a, b, c*) and its first derivative. This matrix has the same characteristics (but different coefficients) as its counterpart in (*2-12*) and is adaptable to the same manipulations, such as the *construction of transport chains* or computation of Φ_Δ^* for *complex loading* (Problems 2.24 and 2.25).

Special Terminology

Four terms related to the dynamics of single degree freedom systems occur frequently and are defined below.

(*1*) A *periodic motion* is a steady-state oscillatory motion of sustained amplitude and constant period.

(*2*) A *transient motion* is a nonsteady-state oscillatory motion of diminishing amplitude and constant period.

(*3*) An *aperiodic motion* is a nonoscillatory motion exponentially decaying with time.

(*4*) The *resonant motion* (resonance) occurs when the frequency of the load function (excitation) equals the natural frequency of the system.

The transient and aperiodic motions eventually relax to the position of static equilibrium. In contrast, the amplitude of resonant motion is controlled by the amount of damping in the system. If the amount of damping available in the system is not adequate, the amplitude of the resonant motion will increase without limit (runaway motion) causing *destruction* of the system.

Solved Problems

HOOKEAN MODELS

2.1. Derive the governing differential equations (*2-1a, 2a, 3a*) for the Hookean models of Figs. 2-1, 2-2, and 2-3. Use the conditions of dynamic equilibrium of the respective free-body sketch.

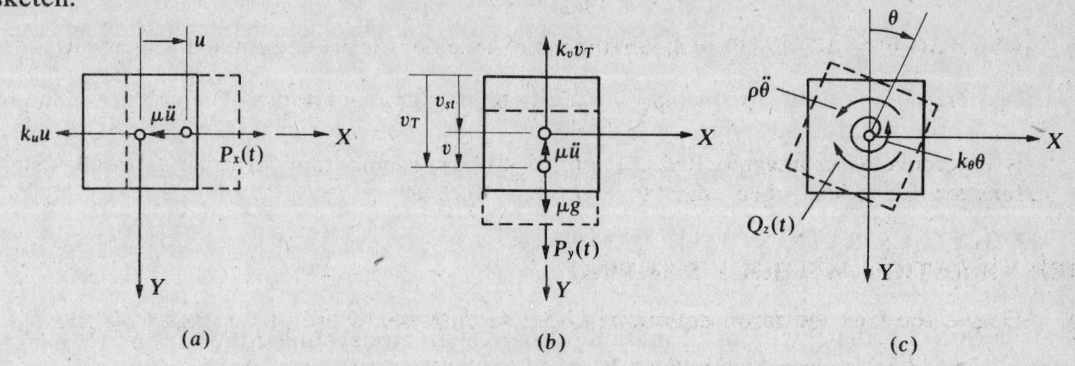

Fig. P-2.1

The free-body diagram shown in Fig. P-2.1*a* corresponds to Fig. 2.1 where the mass μ is displaced a distance u and its neutral position (in this position the spring is neither stretched nor compressed). A displacement u of the mass necessitates stretching the spring, which will exert a pull on the mass by a force $k_u u$. Using Newton's second law (i.e., $\Sigma F_u = \mu\ddot{u}$), we have the inertia force $\mu\ddot{u} = P_x(t) - k_u u$. Thus the condition of equilibrium gives (*2-1a*) as

$$\mu\ddot{u} + k_u u = P_x(t) \tag{1}$$

For Fig. 2.2, let the weight of the mass μg be suspended vertically from the spring and then the mass will be pulled a distance v_{st} to its neutral position as shown in the free-body diagram of Fig. P-2.1*b*. Now let the mass be displaced a distance v and the force in the deformed spring is $k_v v_T$, $v_T = v_{st} + v$. Employing Newton's second law yields

$$\mu\ddot{v}_T = \mu g + P_y(t) - k_v(v_{st} + v_T) \tag{2}$$

since $\mu g = k_v v_{st}$ and $\ddot{v}_T = \ddot{v}$. Upon substitution, we have

$$\mu\ddot{v} + k_v v = P_y(t) \tag{3}$$

Similarly, the free-body diagram corresponding to Fig. 2.3 is shown in Fig. P-2.1*c* in which the restoring force is $k_\theta\theta$ and the mass moment of inertia is $\rho\ddot{\theta}$. The angle θ indicates the angular displacement of the mass inertia from its neutral position. Using the condition of equilibrium of the free-body diagram gives

$$\rho\ddot{\theta} + k_\theta\theta = Q_z(t) \tag{4}$$

2.2. Using Laplace transforms find the general integrals of the differential equations derived in the preceding problem. Assume the displacement and velocity at $t = 0$ to be known.

Let (*1, 3, 4*) of Problem 2.1 be symbolically presented as

$$\ddot{\Delta} + \omega_\Delta^2\Delta = \frac{W_\Delta(t)}{\mu_\Delta} \tag{1}$$

Since $\mathscr{L}(\ddot{\Delta}) = s^2\mathscr{L}(\Delta) - s\Delta(+0) - \Delta'(+0)$, $\mathscr{L}\{\omega_\Delta^2\Delta\} - \omega_\Delta^2\mathscr{L}(\Delta)$, $\mathscr{L}\{W_\Delta(t)/\mu_\Delta\} = 1/\mu_\Delta\mathscr{L}\{W_\Delta(\Delta)\}$, the Laplace transform of (*1*) is

$$s^2 f(s) - s\Delta(+0) - \Delta'(+0) + \omega_\Delta^2 f(s) = \frac{1}{\mu_\Delta}g(s) \tag{2}$$

where $f(s) = \mathscr{L}(\Delta)$ and $g(s) = \mathscr{L}\{W_\Delta(\Delta)\}$.

From (2),

$$f(s) = \Delta(+0)\frac{s}{s^2+\omega_\Delta^2} + \dot{\Delta}(+0)\frac{1}{s^2+\omega_\Delta^2} + \frac{1}{s^2+\omega_\Delta^2}\frac{g(s)}{\mu_\Delta} \tag{3}$$

The inverse of (3) is

$$\Delta = \Delta(+0)\cos\omega_\Delta t + \frac{\dot{\Delta}(+0)}{\omega_\Delta}\sin\omega_\Delta t + \frac{1}{\mu_\Delta\omega_\Delta}\int_0^t W_\Delta(\tau)\sin[\omega_\Delta(t-\tau)]d\tau \tag{4}$$

where $\Delta(+0) = \Delta_0$, $\dot{\Delta}_0(+0) = \dot{\Delta}_0$ are the initial conditions (displacement and velocity at $t = 0$) and

$$\Phi_\Delta = \frac{1}{\mu_\Delta\omega_\Delta}\int_0^t W_\Delta(\tau)\sin[\omega_\Delta(t-\tau)]\,d\tau \tag{5}$$

is the convolution integral (Ref. 2.1, pp. 35–40) representing the particular function (displacement function due to load when $\Delta_0 = 0$ and $\dot{\Delta}_0 = 0$).

FREE VIBRATION WITHOUT DAMPING

2.3. Derive the free vibration equations (2-5*a*, *b*) in terms of the phase angles α_Δ and β_Δ.

Let (2-5*a*) be expanded in the following form without using the subscript Δ:

$$\Delta = R\cos(\omega t - \alpha) = \underbrace{R\cos\alpha}_{\Delta_0}\cos\omega t + \underbrace{R\sin\alpha}_{\dot{\Delta}_0/\omega}\sin\omega t \tag{1}$$

Squaring and adding the constants of (1) yields

$$R = \sqrt{\Delta_0^2 + (\dot{\Delta}_0/\omega)^2} \tag{2}$$

which is (2-6). Taking the ratio of the constants of (1) gives

$$\tan\alpha = \frac{\dot{\Delta}_0}{\omega\Delta_0} \qquad \text{and} \qquad \alpha = \tan^{-1}\frac{\dot{\Delta}_0}{\omega\Delta_0} \tag{3}$$

The inverse of (3) is (2-7*a*).

Similarly, (2-5*b*) expands to

$$\Delta = R\sin(\omega t + \beta) = \underbrace{R\sin\beta}_{\Delta_0}\cos\omega t + \underbrace{R\cos\beta}_{\dot{\Delta}_0/\omega}\sin\omega t$$

where R is the same as shown in (2) and $\tan\beta = \Delta_0\omega/\dot{\Delta}_0$ whose inverse is

$$\beta = \tan^{-1}\frac{\Delta_0\omega}{\dot{\Delta}_0}$$

The equivalence of (2-4) and (2-5*a*) or (2-5*b*) shows that the sum of two harmonic motions at the same frequency (2-4) is harmonic at that frequency with difference phase angles (2-5*a* or 2-5*b*).

2.4. Find (*a*) the lateral displacement equation of the rigid frame shown in Fig. P-2.4*a* due to the initial conditions $u_0 = 1.27$ cm and $\dot{u}_0 = 2.54$ cm/s and (*b*) the moments at B and C at $t = 2$ s. Then (*c*) check the solution by using the dynamic equilibrium condition. Assume the mass to be concentrated on the girder and the columns to be massless spring. Assume $h = 3.05$ m, $L = 6.10$ m, $w = 0.292$ MN/m, $A = 22.86$ cm × 10.16 cm, $I = 1.01$ dm^4, and $E = 206.84$ GN/m^2.

(*a*) Let (2-5*b*) be used for the equation of motion as

$$\Delta = R\sin(\omega_u t + \beta) \tag{1}$$

The spring constant k_u is composed of the shears at B and C due to a unit displacement of the girder (mass center), as

$$k_u = k_B + k_C = \frac{2(3)EI}{h^3} = \frac{2(3)(206.84)(1.01)}{(3.05)^3(10)} = 4.42\text{ MN/m}$$

The mass is

$$\mu = wL/g = (0.292)(6.10)/9.81 = 0.182 \text{ MN-s}^2/\text{m}$$

Thus

$$\omega_u = \sqrt{k_u/\mu} = \sqrt{4.42/0.182} = 4.928 \text{ rad/s}$$
$$R = \sqrt{\Delta_0^2 + (\dot{\Delta}_0/\omega^2)} = \sqrt{(1.27)^2 + (2.54/4.928)^2} = 1.37 \text{ cm}$$
$$\beta = \tan^{-1}(\Delta_0\omega_u/\dot{\Delta}) = \tan^{-1}[(1.27)(4.928)/2.54] = 1.18 \text{ rad}$$

Substituting ω_u, R, and β into (*1*) yields

$$\Delta = 1.37 \sin(4.928t + 1.18) \tag{2}$$

(*b*) Substituting $t = 2$ s into (*2*) gives $\Delta = -1.369$ cm. Consequently the moments are

$$M_B = M_C = \frac{-3EI\Delta}{L^2} = \frac{3(206.84)(1.01)(1.369)}{(6.10)^2} = 23.06 \text{ kN-m (clockwise)}$$

(*c*) The free-body diagram of the girder is shown in Fig. P-2.4*b* where $k_u\Delta = 4.42(-1.369/100) = -60.510$ kN and $\ddot{u}$ is the second derivation of (*2*). When $t = 2$ s, we have

$$\ddot{u} = \frac{d^2\Delta}{dt^2} = -(1.37)(4.928)^2 \sin[(4.928)(2) + 1.18] = 33.27 \text{ cm/s}^2$$

Employing the equilibrium condition $\Sigma F_x = 0$ gives

$$\mu\ddot{u} + k_u\Delta = (0.182)(33.27) - (4.42)(1.369) = 0$$

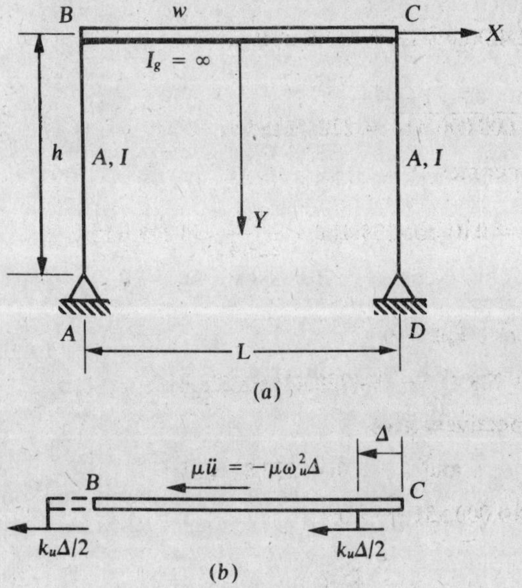

Fig. P-2.4

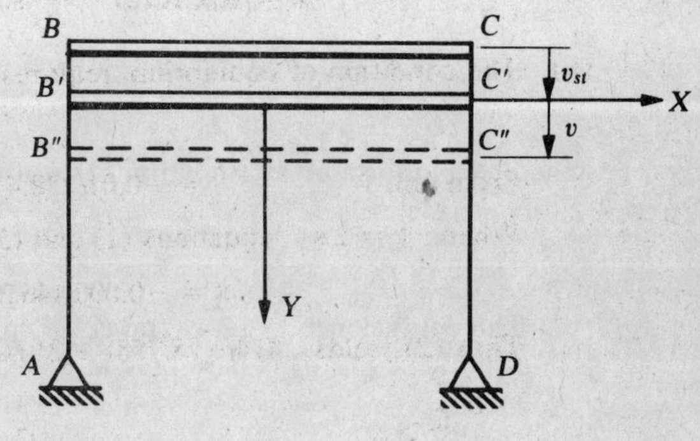

Fig. P-2.5

2.5. Using the rigid frame of the preceding problem find (*a*) the vertical motion equation of the girder for the initial conditions: $v_0 = 0.635$ cm and $\dot{v}_0 = 25.4$ cm/s and (*b*) the axial forces of the columns corresponding to the maximum displacement.

In Fig. P-2.5, v_{st} is the column shortening due to the weight of the girder and v is the vibration displacement measured from the equilibrium position. The spring constant due to a unit downward displacement of the mass is

$$k_v = 2AE/h = 2(232.26)(206.84)/(3.05)(10) = 3150.21 \text{ MN/m}$$

(*a*) Using (*2-5a*) and $\mu = 0.182$ MN-s²/m, we calculate ω_v, R, α, and Δ as follows:

$$\omega_v = \sqrt{k_v/\mu} = \sqrt{3150.21/0.182} = 131.563 \text{ rad/s}$$

$$R = \sqrt{\Delta_0^2 + (\dot{\Delta}_0/\omega_v)^2} = \sqrt{(0.635)^2 + (25.4/131.563)^2} = 0.664 \text{ cm}$$

$$\alpha = \tan^{-1}\frac{\dot{\Delta}_0}{\Delta_0\omega_v} = \tan^{-1}\frac{25.4}{(0.635)(131.563)} = 0.295 \text{ rad}$$

$$\Delta = R\cos(\omega_v t - \alpha) = 0.664\cos(131.563\,t - 0.295)$$

(*b*) When $\cos(131.563\,t - 0.295) = 1$, then $\Delta = \Delta_{max} = 0.664$ cm. Thus the column forces corresponding to downward motion (+) and upward motion (−) are

$$F = \frac{k_v}{2}(v_{st} \pm \Delta_{max}) = \frac{3150.21}{2(100)}\left[\frac{(0.292)(6.10)(3.05)(10)^3}{2(232.26)(206.84)} \pm 0.664\right] = \begin{cases} 11.349 \text{ MN (tension)} \\ 9.568 \text{ MN (compression)} \end{cases}$$

where $v_{st} = wLh/2AE$.

2.6. Using the rigid frame given in Problem 2.4, find (*a*) the equation of rotational motion of the mass due to the initial conditions: $\theta_0 = 1/100$ rad and $\dot{\theta}_0 = 2/5$ rad/s and (*b*) the equilibrium condition at $t = 2$ s. Use U.S. customary units.

In Fig. P-2.6*a*, $B'C'$ is assumed to be the equilibrium position of the mass and v_{st} is the column shortening due to the mass weight. The spring constant may be formulated from Fig. P-2.6*b* as

$$k_\theta = F\left(\frac{L}{2}\right)(2) = \left(\frac{AE}{h}\cdot\frac{L}{2}\right)\left(\frac{L}{2}\right)(2) = \frac{36(30\,000)(20)(20)}{10(2)} = 21\,600\,000.0 \text{ k-ft/rad}$$

The mass is assumed to be uniformly distributed along the girder for which the mass moment of inertia is

$$\rho = \mu L^2/12 = (12.42)(20)^2/12 = 414 \text{ k-s}^2\text{-ft}$$

The circular frequency is

$$\omega_\theta = \sqrt{k_\theta/\rho} = \sqrt{21\,600\,000/414} = 228.41 \text{ rad/s}$$

(*a*) The equation of motion in terms of (*2-4*) becomes

$$\theta = \theta_0\cos\omega_\theta t + \frac{\dot{\theta}_0}{\omega_\theta}\sin\omega_\theta t = 0.01\cos 228.41\,t + \frac{0.4}{228.41}\sin 228.41\,t \qquad (1)$$

(*b*) The condition of equilibrium requires

$$\rho\ddot{\theta} + k_\theta\theta = 0 \qquad (2)$$

From (*1*), $\ddot{\theta} = -0.01(228.41)^2\cos 228.41\,t - 0.4(228.41)\sin 228.41\,t$ (3)

When $t = 2$ s, equations (*1*) and (*3*) respectively give

$$\theta = -0.000\,446\,2 \text{ rad} \qquad \text{and} \qquad \ddot{\theta} = 232.81 \text{ rad/s}^2$$

Thus (*2*) yields $414(-28.788) + 21\,600\,000(0.000\,551\,3) \approx 0$.

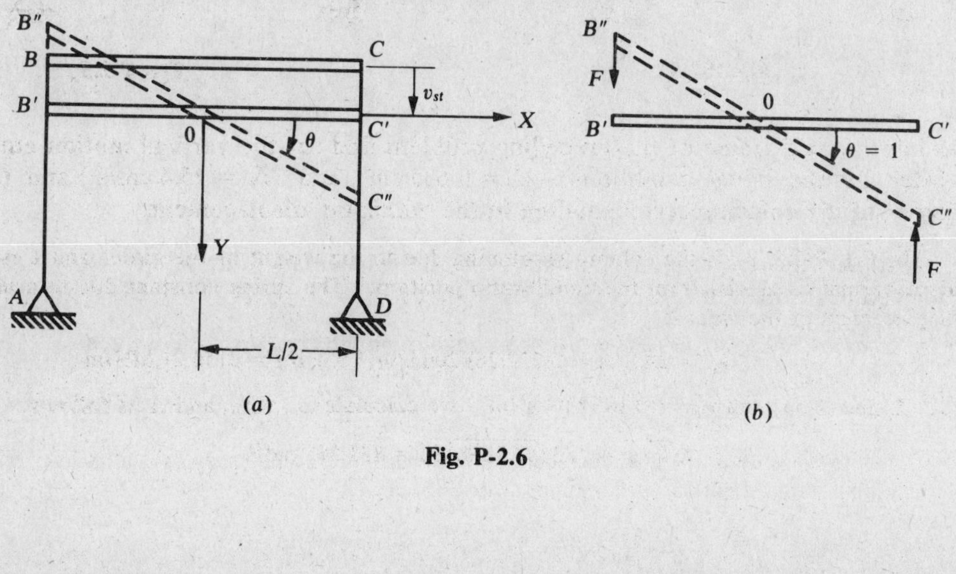

Fig. P-2.6

2.7. Add the harmonic motions: (*a*) $\Delta_1 = 10\cos(\omega t + \pi/4)$ and $\Delta_2 = 12\sin(\omega t - \pi/6)$; (*b*) $\Delta_1 = 10\cos\omega t$ and $\Delta_2 = 6\sin 2\omega t$. Plot the motions.

(*a*) Using the trigonometric transformation,

$$\begin{aligned}\Delta = \Delta_1 + \Delta_2 &= 10\cos(\omega t + \pi/4) + 12\sin(\omega t - \pi/6)\\ &= (10\cos\pi/4 - 12\sin\pi/6)\cos\omega t + (-10\sin\pi/4 + 12\cos\pi/6)\sin\omega t\\ &= 1.070\cos\omega t + 3.322\sin\omega t \end{aligned}\tag{1}$$

Let the amplitude of the desired motion be R and its phase angle with respect to Δ_1 be α. The resultant motion may be expressed as

$$\Delta = R\cos(\omega t - \alpha) = R\cos\alpha\cos\omega t + R\sin\alpha\sin\omega t \tag{2}$$

Equating (*1*) and (*2*) gives $R\cos\alpha = 1.070$ and $R\sin\alpha = 3.322$. Thus

$$R = 3.490, \quad \text{and} \quad \tan\alpha = 3.322/1.070 = 3.105 \quad \text{or} \quad \alpha = 1.259 \text{ rad}$$

From (*2*)

$$\Delta = 3.490\cos(\omega t - 1.259) \tag{3}$$

Δ_1, Δ_2, and Δ are graphically represented by Fig. P-2.7*a*.

(*b*) Since $\Delta_1 = 10\cos\omega t$ and $\Delta_2 = 6\sin 2\omega t$ have different frequencies, the resultant motion will be obtained graphically as shown in Fig. P-2.7*b*.

Note that the resultant motion of (*a*) is harmonic and the resultant motion of (*b*) is not harmonic but periodic.

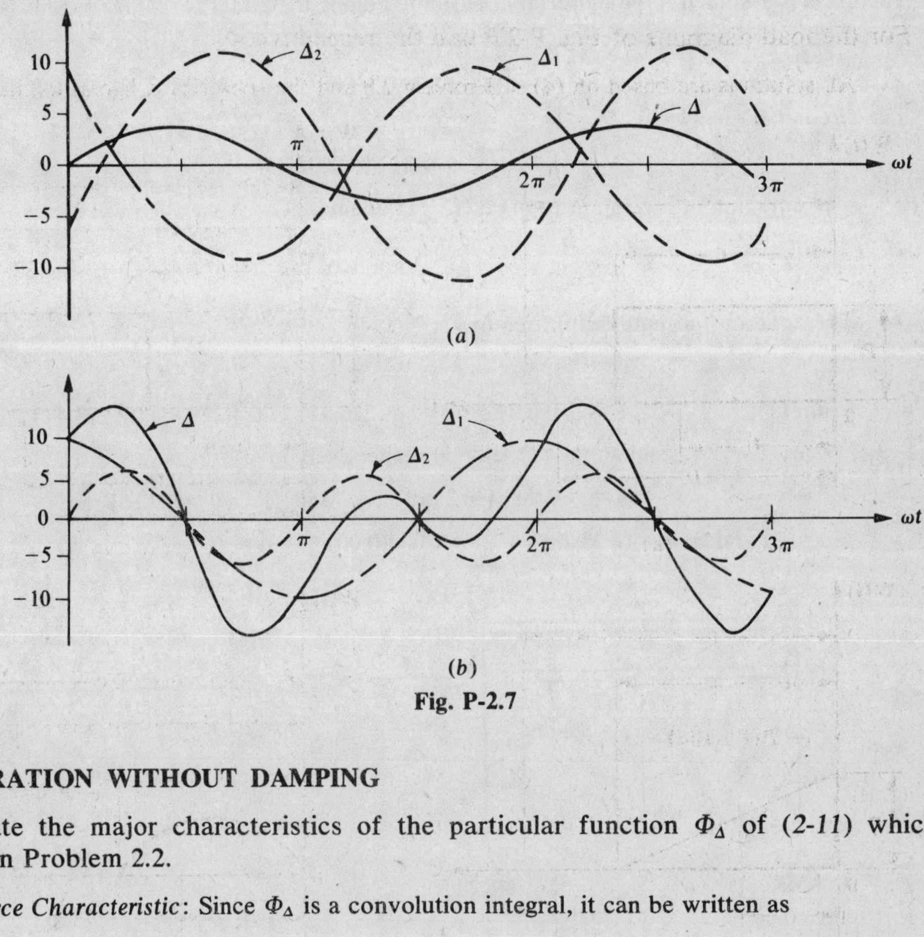

Fig. P-2.7

FORCED VIBRATION WITHOUT DAMPING

2.8. Investigate the major characteristics of the particular function Φ_Δ of (*2-11*) which was derived in Problem 2.2.

(*a*) *Source Characteristic*: Since Φ_Δ is a convolution integral, it can be written as

$$\Phi_\Delta = \frac{1}{\mu_\Delta\omega_\Delta}\int_0^t W_\Delta(\tau)\psi_2(\omega_\Delta t - \omega_\Delta\tau)\,d\tau = \frac{1}{\mu_\Delta\omega_\Delta}\int_0^t W_\Delta(t-\tau)\psi_2(\omega_\Delta\tau)\,d\tau \tag{1}$$

where $\psi_2(\omega_\Delta\tau)$ is the highest shape function of the complementary solution. This property is common to all particular functions (Ref. 2.9, pp. 181, 220).

(*b*) *Superposition*: If $W_\Delta(t) = W_1(t) + W_2(t) + \cdots + W_n(t)$, then

$$\Phi_\Delta = \Phi_1 + \Phi_2 + \cdots + \Phi_n$$

where $\Phi_1, \Phi_2, \ldots, \Phi_n$ are convolution integrals of $W_1(\tau), W_2(\tau), \ldots, W_n(\tau)$ and $\psi_2(\omega_\Delta t - \omega_\Delta t)$, respectively (Problem 2.9).

(*c*) *Differentiation and Integration*: Φ_Δ can be differentiated and/or integrated behind the integral sign with respect to t (Ref. 2.1, pp. 42–44).

(*d*) *Generalized Integration by Parts*: Because of the recurrent nature of multiple integrals of $\sin[\omega_\Delta(t-\tau)]$, Φ_Δ can be easily evaluated by the method of generalized integration by parts (Ref. 2.9, p. 117).

$$\Phi_\Delta = \frac{1}{\mu_\Delta \omega_\Delta}[DI_1 - D'I_2 + D''I_3 - \cdots]_0^t \tag{3}$$

where $$D = W_\Delta(\tau), \qquad D' = \frac{dW_\Delta(\tau)}{d\tau}, \qquad D'' = \frac{d^2W_\Delta(\tau)}{d\tau^2}, \qquad \ldots$$

$$I_1 = \int \sin[\omega_\Delta(t-\tau)]\,d\tau, \qquad I_2 = \int\int \sin[\omega_\Delta(t-\tau)]\,d\tau\,d\tau, \quad \ldots$$

This expansion gives

$$\Phi_\Delta = \frac{1}{\mu_\Delta \omega_\Delta}\left\{\frac{\cos[\omega_\Delta(t-\tau)]}{\omega_\Delta}\left(D - \frac{D''}{\omega_\Delta^2} + \frac{D^{iv}}{\omega_\Delta^4} - \cdots\right) + \frac{\sin[\omega_\Delta(t-\tau)]}{\omega_\Delta}\left(\frac{D'}{\omega_\Delta} - \frac{D'''}{\omega_\Delta^3} + \frac{D^{v}}{\omega_\Delta^5}\cdots\right)\right\}_0^t \tag{4}$$

where the higher derivatives of D are frequently zero.

2.9. For the load diagrams of Fig. P-2.9 find the respective Φ_Δ.

All solutions are based on (*4*) of Problem 2.8 and the subscript Δ is omitted in ω_Δ, W_Δ, and Φ_Δ.

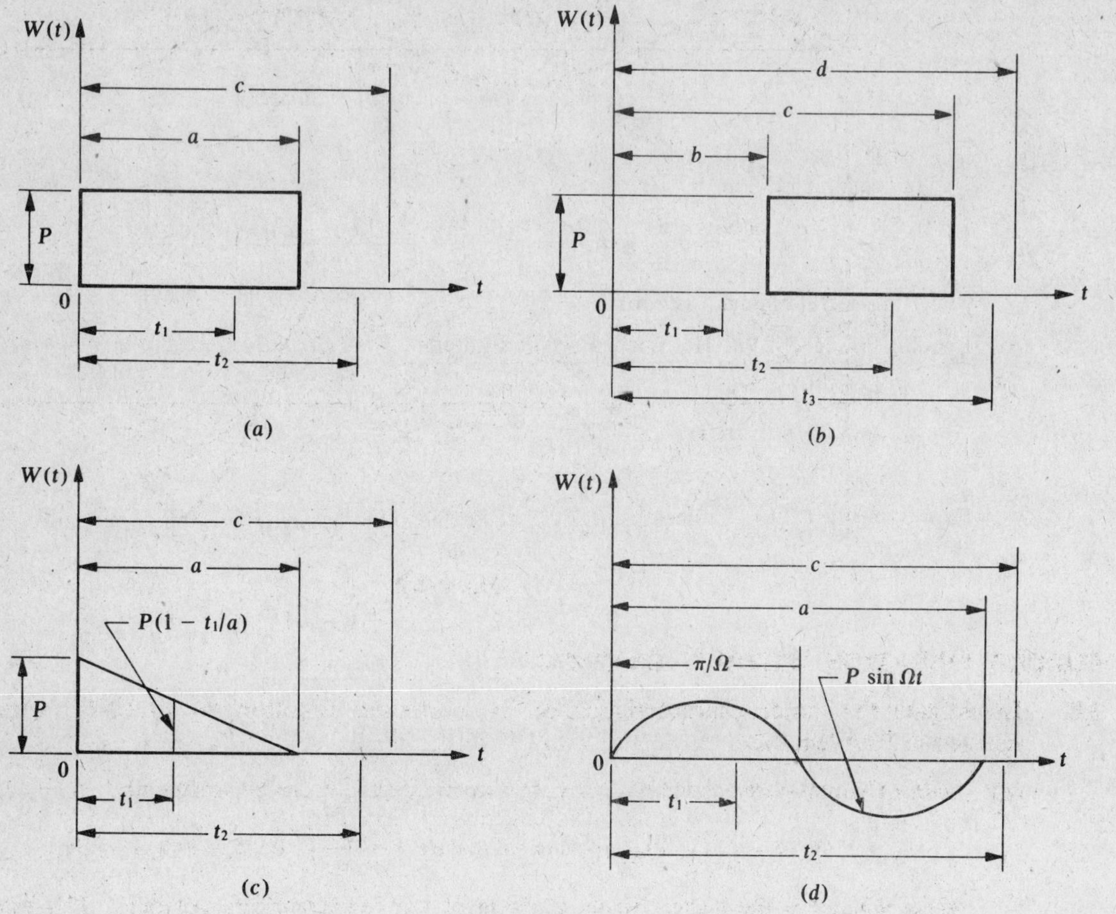

Fig. P-2.9

(*a*) Loading of Fig. P-2.9*a* (Rectangular Pulse, P = constant)

In segment a,

$$\Phi\Big|_0^{t=t_1} = \left\{\frac{P\cos[\omega(t-\tau)]}{\mu\omega^2}\right\}_0^t = \frac{P(1-\cos\omega t)}{\mu\omega^2} \tag{1}$$

where $D = P,\ D' = D'' = \cdots = 0$.

In segment $c - a \leq \infty$,

$$\Phi\Big|_0^{t=t_2} = \Phi\Big|_0^a + \Phi\Big|_a^t = \frac{P\{\cos[\omega(t-a)] - \cos\omega t\}}{\mu\omega^2} \tag{2}$$

where $\Phi|_a^t = 0$, since $D = 0$ in the second integral.

(*b*) Loading of Fig. P-2.9*b* (Rectangular Pulse, P = constant)

In segment b, $\Phi = 0$, and in segment $c - b$,

$$\Phi\Big|_0^{t=t_2} = \Phi\Big|_0^b + \Phi\Big|_b^t = \frac{P\{1 - \cos[\omega(t-b)]\}}{\mu\omega^2} \tag{3}$$

where the first integral is zero.

In segment $d - c \leq \infty$,

$$\Phi\Big|_0^{t=t_3} = \Phi\Big|_0^b + \Phi\Big|_b^c + \Phi\Big|_c^t = \frac{P\{\cos[\omega(t-c)] - \cos[\omega(t-b)]\}}{\mu\omega^2} \tag{4}$$

where the first and third integrals are zero.

(*c*) Loading of Fig. P-2.9*c* (Triangular Pulse, P = constant)

In segment a,

$$\Phi\Big|_0^{t=t_1} = \left\{\frac{P(\omega a - \omega\tau)[\cos\omega(t-\tau)] - P[\sin\omega(t-\tau)]}{a\mu\omega^2}\right\}_0^t$$

$$= \frac{P}{a\mu\omega^2}[\omega(a-t) - \omega a\cos\omega t + \sin\omega t] \tag{5}$$

where $D = P(1 - \tau/a),\ D' = -P/a,\ D'' = D''' = \cdots = 0$.

In segment $c - a \leq \infty$,

$$\Phi\Big|_0^{t=t_2} = \Phi\Big|_0^a + \Phi\Big|_a^t = \frac{P\{\sin\omega t - \sin[\omega(t-a)] - \omega a\cos\omega t]\}}{a\mu\omega^3} \tag{6}$$

where the second integral is zero.

(*d*) Loading of Fig. P-2.9*d* (Harmonic Forcing Function, $P\sin\Omega t$; Ω is constant angular frequency)

In segment a,

$$D = P\sin\Omega\tau, \qquad D'' = -\Omega^2 P\sin\Omega\tau, \qquad D^{iv} = \Omega^4 P\sin\Omega\tau, \qquad \ldots$$

$$D' = \Omega P\cos\Omega\tau, \qquad D''' = -\Omega^3 P\cos\Omega\tau, \qquad D^{v} = \Omega^5 P\cos\Omega\tau, \qquad \ldots$$

Expressing $D, D', D'', \ldots$ in the series (*4*) of Problem 2.8, one may find the convergent geometric series, i.e.,

$$D - \frac{D''}{\omega^2} + \frac{D^{iv}}{\omega^4} - \cdots = \frac{P\sin\Omega\tau}{1 - (\Omega^2/\omega)^2}$$

$$\frac{D'}{\omega} - \frac{D'''}{\omega^3} + \frac{D^{v}}{\omega^5} - \cdots = \frac{P\Omega\cos\Omega\tau}{\omega(1 - \Omega^2/\omega^2)}$$

Then $$\Phi = \frac{P\{\omega\cos[\omega(t-\tau)]\sin\Omega\tau + \Omega\sin[\omega(t-\tau)]\cos\Omega\tau\}_0^t}{\mu\omega(\omega^2 - \Omega^2)}$$

For $t \leq a$, $$\Phi\Big|_0^t = \frac{P(\omega\sin\Omega t - \Omega\sin\omega t)}{\mu\omega(\omega^2 - \Omega^2)} \tag{7}$$

In segment $c - a \leq \infty$,

$$\Phi\Big|_0^{t=t_2} = \Phi\Big|_0^a + \Phi\Big|_a^t = \frac{P\{\omega\cos[\omega(t-a)]\sin\Omega a + \Omega\sin[\omega(t'-a)]\cos\Omega a - \Omega\sin\omega t\}}{\mu\omega(\omega^2-\Omega^2)} \tag{8}$$

where the second integral is zero.

2.10. Using the Dirac delta function $\mathscr{D}$ and the unit step function $\mathscr{U}$ find the particular function Φ for the impulsive load $P\Delta t$ of Fig. P.2-10 where P is constant.

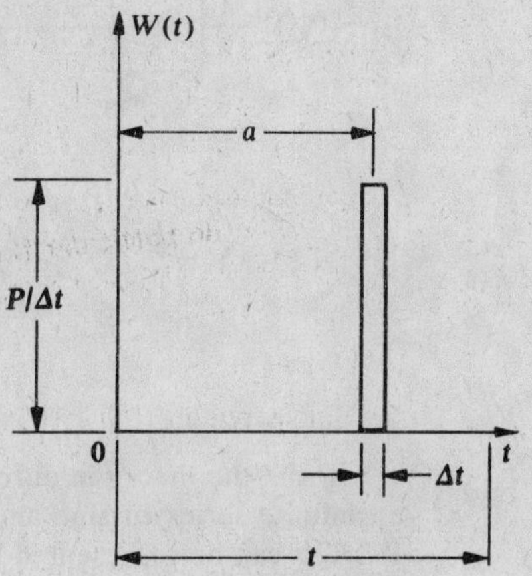

Fig. P-2.10

In terms of the Dirac delta function $\mathscr{D}$,

$$W(t) = P\mathscr{D}(t-a)$$

$$= \lim_{\Delta t\to 0} P\begin{cases} 0 \text{ for } t < (a - \Delta t/2) \\ 1/\Delta t \text{ for } (a-\Delta t/2) < t < (a+\Delta t/2) \\ 0 \text{ for } t > (a+\Delta t/2)\end{cases} \tag{1a}$$

and $$\mathscr{L}\{W(t)\} = Pe^{-as} = g(s) \tag{1b}$$

According to the shifting theorem (Ref. 2.6, p. 3) the inverse of the last product in (3) of Problem 2.2 is

$$\Phi = \mathscr{L}^{-1}\left(\frac{1}{s^2+\omega^2}\cdot\frac{g(s)}{\mu}\right) = \frac{P\sin[\omega(t-a)]}{\mu\omega}\mathscr{U}(t-a) \tag{2}$$

where $\mathscr{U}(t-a)$ is the unit step function (Ref. 2.6, p. 8) defined as

$$\mathscr{U}(t-a) = \begin{cases} 0 \text{ for } 0 \leq t \leq a \\ 1 \text{ for } a < t \leq \infty \end{cases} \tag{3}$$

Equation (2) may be used for the development of the Duhamel's integral which is of course the older name of the convolution integral (5) of Problem 2.2.

2.11. Assume that the rigid frame given in Problem 2.4 is subject to a harmonic force $P(t) = P\sin\Omega t$ at the girder to the right, where $P = 10$ k and $\Omega = 7.417$ rad/s. Find the motion of the mass at $t = 0.4235$ s ($\Omega t = \pi/2$) due to (a) the harmonic force and (b) the harmonic force and the initial displacement and velocity given in Problem 2.4. Use U.S. customary units.

(a) For the given forcing function the particular function was derived in (7) of Problem 2.9. The displacement of the mass may be calculated as

$$\Phi = \frac{P(\omega\sin\Omega t - \Omega\sin\omega t)}{\mu\omega(\omega^2-\Omega^2)} = 0.4104 \text{ in} \tag{1}$$

where $P = 10$ k, $\omega = 4.945$ rad/s, $\Omega = 7.417$ rad/s, $\mu = 12.42$ k-s²/ft, and $t = 0.4235$ s.

(b) For the dynamic response due to harmonic force and initial condition, the equation of motion was derived in (4) of Problem 2.2 and may be written as

$$u = u_0\cos\omega_u t + \frac{\dot{u}_0}{\omega_u}\sin\omega_u t + \Phi \tag{2}$$

where $u_0 = 0.5$ in, $\dot{u}_0 = 1$ in/s, and Φ was shown in (1). From (2),

$$u = 0.5\cos[(4.945)(0.4235)] + (1/4.945)\sin[(4.945)(0.4235)] + 0.4104 = 0.3355 \text{ in} \tag{3}$$

2.12. Using the Fourier series method determine the motion equation of the mass-spring system subject to the periodic forcing function shown in Fig. P-2.12.

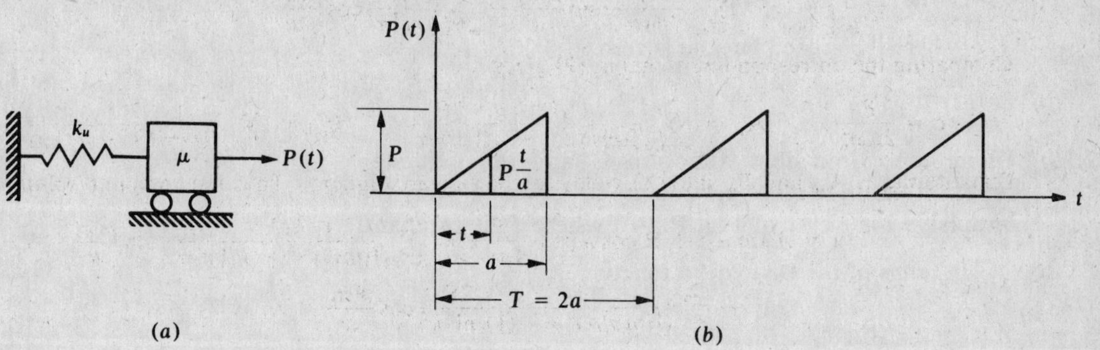

Fig. P-2.12

From the theory in differential equations we know that any periodic function can be expressed by an infinite series of sine and cosine terms. Thus the periodic force $P(t)$ of period T shown in Fig. P-2.12b can be represented by the following Fourier series:

$$P(t) = \frac{c}{2} + \sum_{m=1}^{\infty} A_m \sin\frac{m\pi t}{a} + \sum_{m=1}^{\infty} B_m \cos\frac{m\pi t}{a} \tag{1}$$

where

$$c = \frac{2}{T}\int_0^T P(t)\,dt \tag{2}$$

$$A_m = \frac{2}{T}\int_0^T P(t)\sin\frac{m\pi t}{a}\,dt \tag{3}$$

$$B_m = \frac{2}{T}\int_0^T P(t)\cos\frac{m\pi t}{a}\,dt \tag{4}$$

Substituting $P(t) = Pt/a$ in (2, 3, 4) yields

$$c = \frac{2}{2a}\int_0^a P\frac{t}{a}\,dt = \frac{P}{2} \tag{2a}$$

$$A_m = \frac{2}{2a}\int_0^a P\frac{t}{a}\sin\frac{m\pi t}{a}\,dt = \frac{(-1)^{m+1}P}{m\pi} \qquad m = 1, 2, 3, \ldots, \infty \tag{3a}$$

$$B_m = \frac{2}{2a}\int_0^a P\frac{t}{a}\cos\frac{m\pi t}{a}\,dt = -\frac{2P}{m^2\pi^2} \qquad m = 1, 3, 5, \ldots, \infty \tag{4a}$$

The given forcing function finally becomes

$$P(t) = \frac{P}{4} + \sum_{m=1,2}^{\infty}(-1)^{m+1}\frac{P}{m\pi}\sin\frac{m\pi t}{a} - \sum_{m=1,3}^{\infty}\frac{2P}{m^2\pi^2}\cos\frac{m\pi t}{a} \tag{5}$$

Since the differential equation of motion is $\mu\ddot{u} + k_u u = P(t)$ or

$$\ddot{u} + \omega_u^2 u = P(t)/\mu \tag{6}$$

where $P(t)$ is shown in (5), the complete solution of (6) is composed of the homogeneous solution (u_h) and the particular solution (u_p) as $u = u_h + u_p$ in the following form:

$$u_h = A\sin\omega_u t + B\cos\omega_u t, \qquad u_p = \frac{c}{2} + \sum_{m=1}^{\infty} A_m \sin\frac{m\pi t}{a} + \sum_{m=1}^{\infty} B_m \cos\frac{m\pi t}{a} \tag{7, 8}$$

in which A and B may be determined by using the given initial conditions as usual but c, A_m, and B_m should be determined by substituting $\ddot{u}_p$ and u_p in (6) in the following form:

$$-\sum_{m=1}^{\infty}\left(\frac{m\pi}{a}\right)^2\left(A_m \sin\frac{m\pi t}{a} + B_m \cos\frac{m\pi t}{a}\right) + \omega_u^2\left[\frac{c}{2} + \sum_{m=1}^{\infty} A_m \sin\frac{m\pi t}{a} + \sum_{m=1}^{\infty} B_m \cos\frac{m\pi t}{a}\right]$$

$$= \frac{1}{\mu}\left[\frac{P}{4} + P\sum_{m=1,2}^{\infty}\frac{(-1)^{m+1}}{m\pi}\sin\frac{m\pi t}{a} - \sum_{m=1,3}^{\infty}\frac{2P}{m^2\pi^2}\cos\frac{m\pi t}{a}\right] \tag{9}$$

Comparing the corresponding terms in (9) gives

$$c = \frac{P}{2\mu\omega_u^2} \qquad A_m = \sum_{m=1,2}^{\infty}\frac{(-1)^m}{[(m\pi/a)^2 - \omega_u^2]}\frac{P}{\mu m\pi} \qquad B_m = \sum_{m=1,3}^{\infty}\frac{1}{[(m\pi/a)^2 - \omega_u^2]}\frac{2P}{\mu m^2\pi^2}$$

Substituting c, A_m, and B_m into (8) yields the particular solution. Thus the complete solution is

$$u = A\sin\omega_u t + B\cos\omega_u t + \frac{P}{4\mu\omega_u^2} + \sum_{m=1,2}^{\infty}\frac{(-1)^m}{[(m\pi/a)^2 - \omega_u^2]}\frac{P}{m\pi\mu}\sin\frac{m\pi t}{a}$$

$$+ \sum_{m=1,3}^{\infty}\frac{1}{[(m\pi/a)^2 - \omega_u^2]}\frac{2P}{\mu m^2\pi^2}\cos\frac{m\pi t}{a} \tag{10}$$

Let $u_0 = 0$ and $\dot{u}_0 = 0$ at $t = 0$. Then A and B in (10) may be determined as

$$A = \sum_{m=1,2}^{\infty}\frac{(-1)^{m+1}}{[(m\pi/a)^2 - \omega_u^2]}\frac{P}{a\mu\omega_u} \qquad \text{and} \qquad B = -\frac{P}{4\mu\omega_u^2} - \sum_{m=1,3}^{\infty}\frac{1}{[(m\pi/a)^2 - \omega_u^2]}\frac{2P}{\mu m^2\pi^2}$$

Substituting A and B into (10) and using $a = 1$ s, $t = 0.5$ s, $\omega_u = 4.954$ rad/s, and $\mu = 12.42$ k-s^2/ft, we may find $u = 0.001\,655P$ for $m = 1$, $u = 0.000\,985P$ for $m = 2$, and $u = 0.001\,296P$ for $m = 3$. Comparing with the exact solution $u = 0.001\,23P$ shows that the series is rapidly convergent.

2.13. **Using the structure, forcing function, and initial conditions given in Problem 2.11, find the dynamic response by the step-by-step method.**

For the step-by-step method the forcing function is divided into a number of rectangular pulses as shown in Fig. P-2.13*a*. The response of the second pulse is based on the pulse itself and the displacement and velocity at the end of the first pulse. Thus the final dynamic response is based on the last pulse and its initial conditions (Ref. 2.4, pp. 59–64). Let us find the displacement due to a typical pulse at $t = \tau$ as shown in Fig. P-2.13*b*. The initial conditions at $t = \tau$ are assumed to be u_τ and $\dot{u}_\tau$. From (2-1*b*),

$$u = u_\tau\cos\omega_u(t-\tau) + \frac{\dot{u}_\tau}{\omega_u}\sin\omega_u(t-\tau) + \frac{1}{\mu\omega_u}\int_\tau^t \sin[\omega_u(t-\tau)]P\,d\tau \tag{1}$$

in which the particular function was given in (1) of Problem 2.9. Thus (1) becomes

$$u = \left(u_\tau - \frac{P}{k_u}\right)\cos\omega_u(t-\tau) + \frac{\dot{u}_\tau}{\omega_u}\sin\omega_u(t-\tau) + \frac{P}{k_u} \tag{2}$$

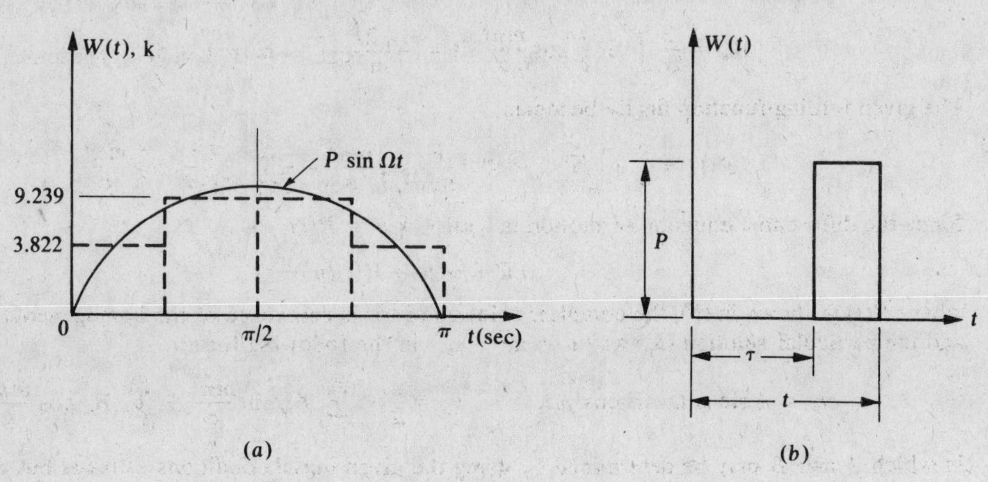

Fig. P-2.13

Let i and $i+1$ be the beginning and the end of an interval of u and $\dot{u}$. For the force P, and the known u_i and $\dot{u}_i$, the displacement and velocity at $i+1$ are

$$u_{i+1} = \left(u_i - \frac{P_i}{k_u}\right)\cos\omega_u(t_{i+1} - t_i) + \frac{\dot{u}_1}{\omega_u}\sin\omega_u(t_{i+1} - t_i) + \frac{P_i}{k_u} \tag{3}$$

$$\frac{\dot{u}_{i+1}}{\omega_u} = -\left(u_i - \frac{P_i}{k_u}\right)\sin\omega_u(t_{i+1} - t_i) + \frac{\dot{u}_i}{\omega_u}\cos\omega_u(t_{i+1} - t_i) \tag{4}$$

Employing Equations 3 and 4, $\omega_u = 4.945$ rad/s, $\Omega = 7.417$ rad/s, $k_u = 25.3125$ k/in, and the forces corresponding to four intervals shown in Fig. P-2.13a gives the displacements and velocities computed in Table P-2.13. At $t = 0.4235$ s, $u = 0.337$ in which checks satisfactorily with $u = 0.3355$ in of (3) in Problem 2.11.

Table P-2.13

t_i	u_i	P_i/k_u	$\left(u_i - \frac{P_i}{k_u}\right)$ cs*	$\dot{u}_i/\omega_u$	$\frac{\dot{u}_i}{\omega_u}$ sn*	u_{i+1}	$-\left(u_i - \frac{P_i}{k_u}\right)$ sn	$\frac{\dot{u}_i}{\omega_u}$ cs	$\dot{u}_{i+1}/\omega_u$
0	0.500	0.151	0.3022	0.2020	0.1010	0.554	−0.1745	0.1750	0.0004
0.106	0.554	0.365	0.1636	0.0004	0.0002	0.529	−0.0945	0.0004	−0.0941
0.212	0.529	0.365	0.1420	−0.0941	−0.0470	0.460	−0.0820	−0.0810	−0.1635
0.318	0.460	0.151	0.2680	−0.1635	−0.0820	0.337	−0.1550	−0.1410	−0.2966

* cs = cos [(4.945)(0.106)] = 0.866 sn = sin [(4.945)(0.106)] = 0.5

TRANSPORT MATRIX

2.14 Construct the time transport matrices $\hat{\hat{T}}_{ji}$ and $\hat{\hat{T}}_{ij}$ for the load given in Fig. P-2.14. Assume the initial conditions at t_i to be Δ_i and $\dot{\Delta}_i$.

From (2-12), the state vector at j may be constructed as

$$\underbrace{\begin{bmatrix} +1 \\ \hline \Delta_j \\ \frac{\dot{\Delta}_j}{\omega} \end{bmatrix}}_{\hat{\hat{H}}_j} = \underbrace{\left[\begin{array}{c|cc} +1 & 0 & 0 \\ \hline \Phi_{ji} & \cos\omega t_{ji} & \sin\omega t_{ji} \\ \frac{\dot{\Phi}_{ji}}{\omega} & -\sin\omega t_{ji} & \cos\omega t_{ji} \end{array}\right]}_{\hat{\hat{T}}_{ji}} \underbrace{\begin{bmatrix} +1 \\ \hline \Delta_i \\ \frac{\dot{\Delta}_i}{\omega} \end{bmatrix}}_{\hat{\hat{H}}_i} \tag{1}$$

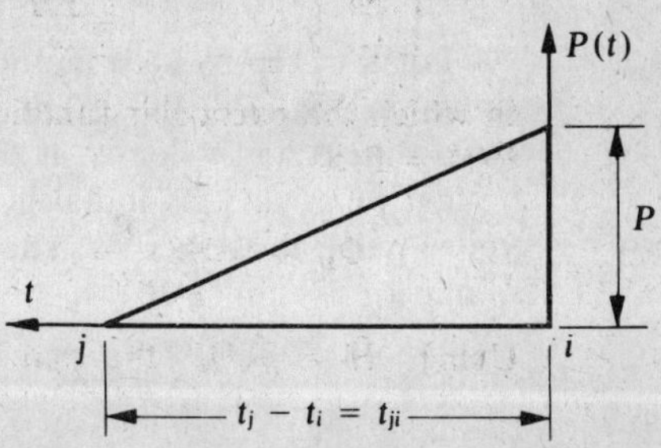

Fig. P-2.14

Using (5) or (6) of Problem 2.9 for $t = a = t_{ji}$ we have

$$\Phi_{ji} = \frac{P}{\mu t_{ji}\omega^3}(\sin\omega t_{ji} - \omega t_{ji}\cos\omega t_{ji}) \qquad \text{and} \qquad \frac{\dot{\Phi}_{ji}}{\omega} = \frac{P}{\mu t_{ji}\omega^3}[\cos\omega t_{ji} + \omega t_{ji}\sin\omega t_{ji} - \omega\cos\omega t_{ji}]$$

Since $\hat{\hat{H}}_j = \hat{\hat{T}}_{ji}\hat{\hat{H}}_i$ and $\hat{\hat{H}}_i = \hat{\hat{T}}_{ij}\hat{\hat{H}}_j$, the relationship between $\hat{\hat{T}}_{ji}$ and $\hat{\hat{T}}_{ij}$ becomes $\hat{\hat{T}}_{ij} = \hat{\hat{T}}_{ji}^{-1}$. Thus

$$\underbrace{\begin{bmatrix} +1 \\ \hline \Delta_i \\ \frac{\dot{\Delta}_i}{\omega} \end{bmatrix}}_{\hat{\hat{H}}_i} = \underbrace{\left[\begin{array}{c|cc} +1 & 0 & 0 \\ \hline \Phi_{ij} & \cos\omega t_{ij} & \sin\omega t_{ij} \\ \frac{\dot{\Phi}_{ij}}{\omega} & -\sin\omega t_{ij} & \cos\omega t_{ij} \end{array}\right]}_{\hat{\hat{T}}_{ij}} \underbrace{\begin{bmatrix} +1 \\ \hline \Delta_j \\ \frac{\dot{\Delta}_j}{\omega} \end{bmatrix}}_{\hat{\hat{H}}_j} \tag{2}$$

where

$$\begin{bmatrix} \Phi_{ij} \\ \frac{\dot{\Phi}_{ij}}{\omega} \end{bmatrix} = \begin{bmatrix} -\cos\omega t_{ij} & \sin\omega t_{ij} \\ \sin\omega t_{ij} & -\cos\omega t_{ij} \end{bmatrix} \begin{bmatrix} \Phi_{ji} \\ \frac{\dot{\Phi}_{ji}}{\omega} \end{bmatrix} \tag{3}$$

In (2, 3) $\cos\omega t_{ij} = \cos\omega t_{ji}$ and $\sin\omega t_{ij} = -\sin\omega t_{ji}$.

2.15. Find $\hat{\hat{H}}_3 = \hat{\hat{T}}_{30}\hat{\hat{H}}_0$ for the given initial conditions Δ_0 and $\dot{\Delta}_0/\omega$ and the forcing function shown in Fig. P-2.15. Use $\hat{\hat{H}}_3$ and then find $\hat{\hat{H}}_0 = \hat{\hat{T}}_{03}\hat{\hat{H}}_3$.

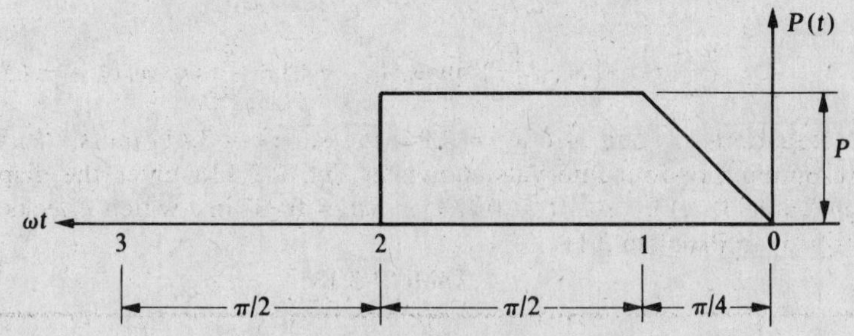

Fig. P-2.15

From (*2-14*) the transport matrix chain is

$$\hat{\hat{H}}_3 = \hat{\hat{T}}_{32}\hat{\hat{T}}_{21}\hat{\hat{T}}_{10}\hat{\hat{H}}_0$$

Substituting the time interval shown in Fig. P-2.15 into (*1*) of the preceding problem gives

$$\hat{\hat{H}}_0 = \begin{bmatrix} +1 \\ \hline \Delta_0 \\ \dfrac{\dot{\Delta}_0}{\omega} \end{bmatrix} \qquad \hat{\hat{T}}_{10} = \left[\begin{array}{c|cc} 1 & 0 & 0 \\ \hline \Phi_{10} & \dfrac{\sqrt{2}}{2} & \dfrac{\sqrt{2}}{2} \\ \dfrac{\dot{\Phi}_{10}}{\omega} & -\dfrac{\sqrt{2}}{2} & \dfrac{\sqrt{2}}{2} \end{array}\right] \qquad \hat{\hat{T}}_{21} = \left[\begin{array}{c|cc} 1 & 0 & 0 \\ \hline \Phi_{21} & 0 & 1 \\ \dfrac{\dot{\Phi}_{21}}{\omega} & -1 & 0 \end{array}\right] \qquad \hat{\hat{T}}_{32} = \left[\begin{array}{c|cc} 1 & 0 & 0 \\ \hline 0 & 0 & 1 \\ 0 & -1 & 0 \end{array}\right]$$

In which the particular functions corresponding to triangular pulse $P(t) = P\tau/a$ and rectangular pulse $P(t) = P$ is

$$\Phi_{10} = 0.098\frac{P}{\mu\omega^2} \qquad \frac{\dot{\Phi}_{10}}{\omega} = 0.372\frac{P}{\mu\omega^2} \qquad \Phi_{21} = 0.292\frac{P}{\mu\omega^2} \qquad \frac{\dot{\Phi}_{21}}{\omega} = 0.708\frac{P}{\mu\omega^2}$$

Using $\hat{\hat{H}}_j = \hat{\hat{T}}_{ji}\hat{\hat{H}}_i$ the state vectors at 1, 2, and 3 respectively are

$$\begin{bmatrix} 1 \\ \hline \Phi_{10} + \dfrac{\sqrt{2}}{2}\left(\Delta_0 + \dfrac{\dot{\Delta}_0}{\omega}\right) \\ \dfrac{\dot{\Phi}_{10}}{\omega} + \dfrac{\sqrt{2}}{2}\left(\Delta_0 + \dfrac{\dot{\Delta}_0}{\omega}\right) \end{bmatrix}, \quad \begin{bmatrix} 1 \\ \hline \Phi_{21} + \dfrac{\dot{\Phi}_{10}}{\omega} + \dfrac{\sqrt{2}}{2}\left(\dfrac{\dot{\Delta}_0}{\omega} - \Delta_0\right) \\ \dfrac{\dot{\Phi}_{21}}{\omega} - \Phi_{10} - \dfrac{\sqrt{2}}{2}\left(\Delta_0 + \dfrac{\dot{\Delta}_0}{\omega}\right) \end{bmatrix}, \quad \begin{bmatrix} 1 \\ \hline \dfrac{\dot{\Phi}_{21}}{\omega} - \Phi_{10} - \dfrac{\sqrt{2}}{2}\left(\Delta_0 + \dfrac{\dot{\Delta}_0}{\omega}\right) \\ -\Phi_{21} - \dfrac{\dot{\Phi}_{10}}{\omega} - \dfrac{\sqrt{2}}{2}\left(\dfrac{\dot{\Delta}_0}{\omega} - \Delta_0\right) \end{bmatrix},$$

The inverse transport chain is

$$\hat{\hat{H}}_0 = \hat{\hat{T}}_{01}\hat{\hat{T}}_{12}\hat{\hat{T}}_{23}\hat{\hat{H}}_3 \tag{1}$$

where $\hat{\hat{H}}_3$ is given above and

$$\hat{\hat{T}}_{23} = \left[\begin{array}{c|cc} 1 & 0 & 0 \\ \hline 0 & 0 & -1 \\ 0 & 1 & 0 \end{array}\right] \qquad \hat{\hat{T}}_{12} = \left[\begin{array}{c|cc} 1 & 0 & 0 \\ \hline -\dfrac{\dot{\Phi}_{21}}{\omega} & 0 & -1 \\ \Phi_{21} & 1 & 0 \end{array}\right] \qquad \hat{\hat{T}}_{01} = \left[\begin{array}{c|cc} 1 & 0 & 0 \\ \hline \dfrac{-\sqrt{2}}{2}\left(\Phi_{10} + \dfrac{\dot{\Phi}_{10}}{\omega}\right) & \dfrac{\sqrt{2}}{2} & \dfrac{-\sqrt{2}}{2} \\ \dfrac{\sqrt{2}}{2}\left(\dfrac{\dot{\Phi}_{10}}{\omega} - \Phi_{10}\right) & \dfrac{\sqrt{2}}{2} & \dfrac{\sqrt{2}}{2} \end{array}\right] \tag{2, 3, 4}$$

Substituting $\hat{\hat{H}}_3$ and (*2, 3, 4*) into (*1*) yields $\hat{\hat{H}}_0$ which must be the same as the given initial conditions at $t = 0$.

KELVIN MODEL

2.16. Derive the governing differential equation (*2-15a, b, c*) for the Kelvin models of Figs. 2-6, 2-7, and 2-8.

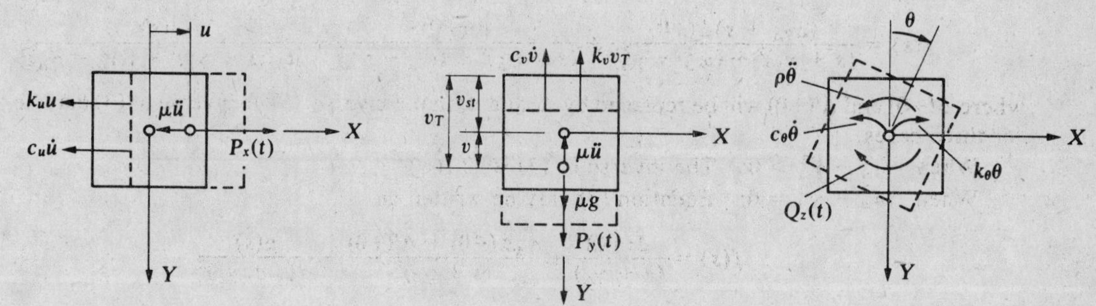

Fig. P-2.16

Figure P-2.16*a* is the free-body diagram of Fig. 2-6 where u is measured from its equilibrium position and $\mu\ddot{u}$, $k_u u$, and $c_u\dot{u}$ represent inertia force, spring force, and damping force, respectively. Applying Newton's second law gives

$$P_x(t) - k_u u - c_u\dot{u} = \mu\ddot{u} \tag{1}$$

Substituting $2\gamma_u = c_u/\mu$ and $\omega_u^2 = k_u/\mu$ into (*1*), we have the governing differential equation

$$\ddot{u} + \omega_u^2 u + 2\gamma_u\dot{u} = \frac{1}{\mu}P_x(t) \tag{2}$$

The free-body diagram shown in Fig. P-2.16*b* is associated with the Kelvin model sketched in Fig. 2-7 where v denotes the displacement from the position of static equilibrium; in this position the spring has a force balancing the weight of the mass: $k_v v_{st} = \mu g$. Measuring v from this position allows us to write the following equation of motion without using $k_v v_{st}$ and μg:

$$P_y(t) - k_v v - c_v\dot{v} = \mu\ddot{v} \tag{3}$$

Substituting $2\gamma_v = c_v/\mu$ and $\omega_v^2 = k_v/\mu$, equation (*3*) becomes

$$\ddot{v} + \omega_v^2 v + 2\gamma_v\dot{v} = \frac{1}{\mu}P_y(t) \tag{4}$$

Figure P-2.16*c* is the free-body diagram corresponding to Fig. 2-8. Applying the condition of equilibrium or Newton's second law, one may similarly obtain the differential equation

$$\ddot{\theta} + \omega_\theta^2\theta + 2\gamma_\theta\dot{v} = \frac{1}{\mu}P_\theta(t) \tag{5}$$

where $\omega_\theta^2 = k_\theta/\mu$, $2\gamma_\theta = c_\theta/\mu$, and θ is measured from its equilibrium position.

2.17. Using Laplace transforms find the general integrals of the differential equation of motion derived in the preceding problem. Assume the displacement and velocity at $t = 0$ to be known.

Let the differential equations of motion (*2, 4, 5*) of the preceding problem be expressed in general form as

$$\ddot{\Delta} + 2\gamma_\Delta\dot{\Delta} + \omega_\Delta^2\Delta = \frac{W_\Delta(t)}{\mu_\Delta} \tag{1}$$

where Δ has the same meaning as in Sec. (2.4).

Referring to Problem 2.2 we may write the Laplace transforms of (*1*) as

$$f(s)(s^2 + \omega_\Delta^2 + 2\gamma_\Delta s) = s\Delta(+0) + \Delta'(+0) + 2\gamma_\Delta\Delta(+0) + \frac{1}{\mu_\Delta}g(s) \tag{2}$$

which may then be expressed as

$$f(s) = \frac{(2\gamma_\Delta + s)\Delta(+0)}{(s+\gamma_\Delta)^2 + (\omega_\Delta^2 - \gamma_\Delta^2)} + \frac{\Delta'(+0)}{(s+\gamma_\Delta)^2 + (\omega_\Delta^2 - \gamma_\Delta^2)} + \frac{g(s)}{\mu_\Delta[(s+\gamma_\Delta)^2 + (\omega_\Delta^2 - \gamma_\Delta)]} \tag{3}$$

where $\Delta(+0)$ and $\Delta'(+0)$ will be replaced by Δ_0 and $\dot{\Delta}_0$ in the inverse. The inverse of (*3*) may be obtained for three cases:

When $\omega_\Delta^2 - \gamma_\Delta^2 > 0$: The inverse of (*3*) is (*2-16a*).

When $\omega_\Delta^2 - \gamma_\Delta^2 = 0$: Equation (*3*) may be written as

$$f(s) = \frac{\Delta(+0)}{(s+\gamma_\Delta)} + \frac{\gamma_\Delta\Delta(+0) + \Delta'(+0)}{(s+\gamma_\Delta)^2} + \frac{g(s)}{\mu(s+\gamma_\Delta)^2} \tag{4}$$

The inverse of (*4*) becomes (*2-16b*).

When $\omega_\Delta^2 - \gamma_\Delta^2 < 0$: We may arrange (*3*) in the following form

$$f(s) = \frac{(s+\gamma_\Delta)\Delta(+0)}{(s+\gamma_\Delta)^2 - (\gamma_\Delta^2 - \omega_\Delta^2)} + \frac{\gamma_\Delta\Delta(+0) + \Delta'(+0)}{(s+\gamma_\Delta)^2 - (\gamma_\Delta^2 - \omega_\Delta^2)} + \frac{g(s)}{\mu[(s+\gamma_\Delta)^2 - (\gamma_\Delta^2 - \omega_\Delta^2)]} \tag{5}$$

of which the inverse is (*2-16c*).

FREE VIBRATION WITH DAMPING

2.18. Using the free vibration (*2-16a*) derive R_Δ^*, α_Δ, and β_Δ, shown in (*2-17*) to (*2-21*).

Expanding (*2-17*) gives

$$\Delta = R_\Delta^* \cos(\lambda_\Delta t - \alpha_\Delta) = R_\Delta^* \cos\alpha_\Delta \cos\lambda_\Delta t + R_\Delta^* \sin\alpha_\Delta \sin\lambda_\Delta t \tag{1}$$

Comparing the corresponding terms in (*1*) with those in the part of free vibration of (*2-16a*) gives

$$R^* \cos\alpha_\Delta = e^{-\gamma_\Delta t}\Delta_0 \qquad \text{and} \qquad R_\Delta^* \sin\alpha_\Delta = e^{-\gamma_\Delta t}\frac{\dot{\Delta}_0 + \gamma_\Delta\Delta_0}{\lambda_\Delta}$$

from which we have

$$R_\Delta^* = e^{-\gamma_\Delta t}\sqrt{\Delta_0^2 + (\dot{\Delta}_0 + \gamma_\Delta\Delta_0)^2/\lambda_\Delta^2} \qquad \text{and} \qquad \alpha_\Delta = \tan^{-1}\frac{\dot{\Delta}_0 + \gamma_\Delta\Delta_0}{\lambda_\Delta\Delta_0} \tag{2, 3}$$

Similarly, (*2-18*) may be written as

$$\Delta = R_\Delta^* \sin(\lambda_\Delta t + \beta_\Delta) = \underbrace{R_\Delta^* \sin\beta_\Delta}_{\Delta_0 e^{-\gamma_\Delta t}} \cos\lambda_\Delta t + \underbrace{R_\Delta^* \cos\beta_\Delta}_{\frac{\dot{\Delta}_0 + \gamma_\Delta\Delta_0}{\lambda_\Delta}} \sin\lambda_\Delta t \tag{4}$$

from which R_Δ^* may be formulated in the same form given in (*2*) and the phase angle is

$$\beta_\Delta = \tan^{-1}\frac{\Delta_0\gamma_\Delta}{\dot{\Delta}_0 + \gamma_\Delta\Delta_0}$$

2.19. Using Problem 2.4 determine the displacement of the mass at $t = 0.5$, 1, and 1.5 s for the damping conditions (*a*) $c = 0.1c_{cr}$, (*b*) $c = c_{cr}$, and (*c*) $c = 3c_{cr}$, where $c_{cr} = 2\mu\omega_\Delta$ is the critical damping constant obtained when $\gamma_\Delta^2 = \omega_\Delta^2$.

(*a*) The equation of underdamped motion is given in (*2-16a*) for $W(\tau) = 0$ as

$$\Delta = e^{-\gamma t}\left(\Delta_0 \cos\lambda t + \frac{\dot{\Delta}_0 + \gamma\Delta_0}{\lambda}\sin\lambda t\right) \tag{1}$$

in which the subscript Δ is omitted. Since $c = (0.1)(2\mu\omega) = 0.2\mu\omega$, $\gamma = c/2\mu = 0.1\omega$, we have

$$\lambda = \sqrt{\omega^2 - \gamma^2} = \sqrt{0.99\omega^2} = 0.995\omega$$

From Problem 2.4, $\omega = 4.945$, $\Delta_0 = 0.5$ in and $\dot{\Delta}_0 = 1$ in/s, Equation (1) becomes

$$\Delta = e^{-0.4945t}\left[0.5\cos 4.92t + \frac{1 + (0.5)(0.4945)}{4.92}\sin 4.92t\right] \tag{2}$$

Substituting $t = 0.5$, 1, and 1.5 s gives $\Delta = -0.1788$ in, -0.1076 in, and 0.216 in, respectively.

(b) For the free vibration with critical damping, Equation (2-16b) is

$$\Delta = e^{-\gamma t}[\Delta_0(1 + \gamma t) + \dot{\Delta}_0 t] \tag{3}$$

where $\gamma = \omega$. Substituting ω, Δ_0, and $\dot{\Delta}_0$ into (3),

$$\Delta = e^{-4.945t}[(0.5)(1 + 4.945t) + t] \tag{4}$$

which gives $\Delta = 0.1885$ in, 0.0282 in, and 0.0034 in, corresponding to $t = 0.5$, 1, and 1.5 s, respectively.

(c) For the free vibration with overdamping, Equation (2-16c) is

$$\Delta = e^{-\gamma t}\left(\Delta_0 \cosh \lambda t + \frac{\dot{\Delta}_0 + \gamma\Delta_0}{\lambda}\sinh \lambda t\right) \tag{5}$$

where $\gamma = 3(2\mu\omega)/2\mu = 3\omega$, $\lambda = \sqrt{\gamma^2 - \omega^2} = \sqrt{(3\omega)^2 - \omega^2} = 2.828\,\omega$. Using ω, Δ_0, and $\dot{\Delta}_0$ given above, we have

$$\Delta = e^{-14.835t}\left[0.5\cosh 13.987t + \frac{1 + (14.835)(0.5)}{13.987}\sinh 13.987t\right]$$

from which the displacements are 0.362 in, 0.236 in, and 0.145 in corresponding to $t = 0.5$, 1, and 1.5 s, respectively.

FORCED VIBRATION WITH DAMPING

2.20. Investigate the major characteristics of the particular function Φ_Δ in (2-16a) which was derived in Problem 2.17.

(a) General Case: Omitting the subscript Δ, the particular function in (2-16a) becomes

$$\Phi = \frac{1}{\mu\lambda}\int_0^t e^{-\gamma(t-\tau)}\sin(\lambda t - \lambda\tau)W(\tau)\,d\tau = -\frac{1}{\mu\lambda}\int_0^t \underbrace{\underbrace{e^{\gamma(\tau-t)}}_{A}\;\underbrace{\sin(\lambda\tau - \lambda t)}_{S}\;\underbrace{W(\tau)}_{W}}_{B}\,d\tau \tag{1}$$

The integration (1) may be (Ref. 2.9, p. 117)

$$\Phi = -\frac{1}{\mu\lambda}[BA_1 - B'A_2 + B''A_3 - B'''A_3 + \cdots]_0^t \tag{2}$$

where

$$A_1 = \int A\,d\tau = \frac{A}{\gamma}, \qquad A_2 = \int A_1\,d\tau = \frac{A}{\gamma^2}, \qquad \ldots \tag{3}$$

and

$$\left.\begin{aligned} B' &= S'W + SW', \qquad B'' = S''W + 2S'W' + SW'', \\ B''' &= S'''W + 3S''W' + 3S'W'' + SW''', \qquad \ldots \end{aligned}\right\} \tag{4}$$

Substituting (3) and (4) into (2) gives

$$\Phi = -\frac{1}{\mu\lambda}A\left[\frac{W}{\gamma}\left(S - \frac{S'}{\gamma} + \frac{S''}{\gamma^2} - \cdots\right) - \frac{W'}{\gamma^2}\left(S - \frac{2S'}{\gamma} + \frac{3S''}{\gamma^2} - \cdots\right)\right.$$
$$\left. + \frac{W'''}{\gamma^3}\left(S - \frac{3S'}{\gamma} + \frac{6S''}{\gamma^2} - \cdots\right) - \frac{W^{iv}}{\gamma^4}\left(S - \frac{4S'}{\gamma} + \frac{10S''}{\gamma^2} - \cdots\right) + \cdots\right]_0^t \tag{5}$$

For $W = a_0 + a_1\tau + a_2\tau^2 + \cdots + a_n\tau^n$, series stops with $d^nW/d\tau^n$ (Problem 2.21).

(b) Special cases: $W(\tau) = P\sin\Omega\tau$ and $W(\tau) = P\cos\Omega\tau$

For $W(\tau) = P\sin\Omega\tau$, Equation (1) becomes

$$\Phi = -\frac{P}{\mu\lambda}\int_0^t e^{\gamma(\tau-t)}\sin(\lambda\tau - \lambda t)\sin\Omega\tau\,d\tau$$
$$= -\frac{P}{2\mu\lambda}\underbrace{\int_0^t e^{\gamma(\tau-t)}\cos(\lambda\tau - \Omega\tau - \lambda t)\,d\tau}_{A} + \frac{P}{2\mu\lambda}\underbrace{\int_0^t e^{\gamma(\tau-t)}\cos(\lambda\tau + \Omega\tau - \lambda t)\,d\tau}_{B} \tag{6}$$

The typical integration of A and B of (6) is

$$\int e^{ax+b}\cos(cx+d)\,dx = \frac{1}{a^2+c^2}e^{ax+b}\,[a\cos(cx+d)+c\sin(cx+d)] \tag{7}$$

Substituting (7) into (6) with $a=\gamma,\ b=-\gamma t,\ c=\lambda-\Omega$ or $c=\lambda+\Omega$ and $d=-\lambda t$ gives the integration of Φ (Problem 2.21).

For $W(\tau)=P\cos\Omega\tau$, Equation (1) becomes

$$\begin{aligned}\Phi &= -\frac{P}{\mu\lambda}\int_0^t e^{\gamma(\tau-t)}\sin(\lambda\tau-\lambda t)\cos\Omega\tau\,d\tau\\ &= -\frac{P}{2\mu\lambda}\int_0^t e^{\gamma(\tau-t)}\sin(\lambda\tau-\Omega\tau-\lambda t)\,d\tau-\frac{P}{2\mu\lambda}\int_0^t e^{\gamma(\tau-t)}\sin(\lambda\tau+\Omega\tau-\lambda t)\,d\tau\end{aligned} \tag{8}$$

Equation (9) given below is a typical integration for (8).

$$\int e^{ax+b}\sin(cx+d)\,dx = \frac{e^{ax+b}}{a^2+c^2}[a\sin(cx+d)-c\cos(cx+d)] \tag{9}$$

where $a=\gamma,\ b=-\gamma t,\ c=\lambda-\Omega$ or $c=\lambda+\Omega$, and $d=-\lambda t$.

2.21. Find the particular function Φ corresponding to each of the load diagrams in Problem 2.9.

(*a*) Loading of Fig. P-2.9*a*, P = constant

From (5) in the preceding problem: $W=P,\quad A=e^{\gamma(\tau-t)},\quad S=\sin(\lambda\tau-\lambda t)$. Thus

$$\Phi = -\frac{1}{\mu\lambda}A\left[\frac{W}{\gamma}\left(S-\frac{S'}{\gamma}+\frac{S''}{\gamma^2}-\cdots\right)\right]_0^t \tag{1a}$$

Since

$$S+\frac{S''}{\gamma^2}+\frac{S^{iv}}{\gamma^4}+\cdots = \frac{\sin\lambda(\tau-t)}{1+(\lambda/\gamma)^2} \quad\text{and}\quad -\frac{S'}{\gamma}-\frac{S'''}{\gamma^3}-\frac{S^{vi}}{\gamma^5}-\cdots = \frac{-\lambda/\gamma\cos\lambda(\tau-t)}{1+(\lambda/\gamma)^2}$$

Equation (1) becomes

$$\Phi = -\frac{P}{\mu\lambda(\gamma^2+\lambda^2)}e^{\gamma(\tau-t)}[\gamma\sin\lambda(\tau-t)-\lambda\cos\lambda(\tau-t)]_0^t \tag{1b}$$

In segment a, (1b) yields

$$\Phi\Big|_0^{t=t_1} - \Phi\Big|_0^t = \frac{P}{\mu\lambda(\gamma^2+\lambda^2)}[\lambda-e^{-\gamma t}(\gamma\sin\lambda t+\lambda\cos\lambda t)] \tag{2}$$

In segment $c-a\le\infty$, (1b) yields

$$\begin{aligned}\Phi\Big|_0^{t=t_2} = \Phi\Big|_0^a = \frac{P}{\mu\lambda(\gamma^2+\lambda^2)}\{&e^{-\gamma(t-a)}[\lambda\cos\lambda(a-1)-\gamma\sin\gamma(a-t)]\\ &-e^{-\gamma t}(\gamma\sin\lambda t+\lambda\cos\lambda t)\}\end{aligned} \tag{3}$$

(*b*) Loading of Fig. P-2.9*b*, P = constant

The particular function of this loading is also (1b).
In segment $c-b$, (1b) becomes

$$\Phi\Big|_0^{t=t_2} = \Phi\Big|_b^t = \frac{P}{\mu\lambda(\gamma^2+\lambda^2)}\{\lambda-e^{-\gamma(t-b)}[\gamma\sin\lambda(t-b)+\lambda\cos\lambda(t-b)]\} \tag{4}$$

In segment $d-c\le\infty$, (1b) becomes

$$\begin{aligned}\Phi\Big|_0^{t=t_3} = \Phi\Big|_b^c = \frac{P}{\mu\lambda(\gamma^2+\lambda^2)}\{&e^{-\gamma(t-c)}[\gamma\sin\lambda(t-c)+\lambda\cos\lambda(t-c)]\\ &-e^{-\gamma(t-b)}[\gamma\sin\lambda(t-b)+\lambda\cos\lambda(t-b)]\}\end{aligned} \tag{5}$$

(c) Loading of Fig. P-2.9c, Triangular Pulse

The particular function is

$$\Phi = \frac{1}{\mu\lambda}\int_0^t e^{-\gamma(t-\tau)} \sin[\lambda(t-\tau)] P\left(1 - \frac{\tau}{a}\right) d\tau$$

$$= \underbrace{-\frac{P}{\mu\lambda}\int_0^t e^{\gamma(\tau-t)} \sin\lambda(\tau - t)\, d\tau}_{\Phi_1} + \underbrace{\frac{P}{\mu\lambda a}\int_0^t e^{\gamma(\tau-t)} \sin[\lambda(\tau-t)]\tau\, d\tau}_{\Phi_2}$$

Let $\Phi = \Phi_1 + \Phi_2$. The solution of Φ_1 was shown in (*1b*). Using (*5*) in the preceding problem we have $W = \tau$, $W' = 1$, and $W'' = 0$. Thus

$$\Phi_2 = \frac{1}{\mu\lambda} A\left[\frac{W}{\gamma}\left(S - \frac{S'}{\gamma} + \frac{S''}{\gamma^2} - \cdots\right) - \frac{W'}{\gamma^2}\left(S - \frac{2S'}{\gamma} + \frac{3S''}{\gamma^2} - \cdots\right)\right]$$

$$= \frac{P}{a\mu\lambda(\gamma^2+\lambda^2)} e^{\gamma(\tau-t)} \left\{\gamma^2\tau\left[\sin\lambda(\tau-t) - \frac{\lambda}{\gamma}\cos\lambda(\tau-t)\right]\right.$$

$$\left. - \frac{1}{\gamma^2+\lambda^2}[(\gamma^2-\lambda^2)\sin\lambda(\tau-t) - 2\lambda\gamma\cos\lambda(\tau-t)]\right\}_0^t \tag{6}$$

In segments a and $c - a \leq \infty$,

$$\Phi\Big|_0^{t=t_1} = \Phi_1\Big|_0^t + \Phi_2\Big|_0^t \qquad \text{and} \qquad \Phi\Big|_0^{t=t_2} = \Phi_1\Big|_0^a + \Phi_2\Big|_0^a \tag{7, 8}$$

which may be obtained by substituting the lower and upper limits into (*1b*) and (*6*).

(d) Loading of Fig. P-2.9d, $P \sin \Omega t$

The solution was shown in (*6*) and (*7*) of the preceding problem. Let $\Phi = \Phi_1 + \Phi_2$ corresponding to A and B in (*6*) of Problem 2.20. Then

$$\Phi_1 = -\frac{P}{2\mu\lambda[\gamma^2 + (\lambda-\Omega)^2]} e^{\gamma(\tau-t)}\{\gamma\cos[(\lambda-\Omega)\tau - \lambda t] + (\lambda-\Omega)\sin[(\lambda-\Omega)\tau - \lambda t]\}_0^t \tag{9}$$

$$\Phi_2 = \frac{P}{2\mu\lambda[\gamma^2 + (\lambda+\Omega)^2]} e^{\gamma(\tau-t)}\{\gamma\cos[(\lambda+\Omega)\tau - \lambda t] + (\lambda+\Omega)\sin[(\lambda+\Omega)\tau - \lambda t]\}_0^t \tag{10}$$

In segments a and $c - a \leq \infty$,

$$\Phi\Big|_0^{t=t_1} = \Phi_1\Big|_0^t + \Phi_2\Big|_0^t \qquad \text{and} \qquad \Phi\Big|_0^{t=t_2} = \Phi_1\Big|_0^a + \Phi_2\Big|_0^a$$

for which the lower and upper limits should be substituted into (*9*) and (*10*).

2.22. Using the Dirac delta function $\mathscr{D}$ and the unit step function $\mathscr{U}$ find the particular function Φ in (*1*) of Problem 2.17 for the impulsive load $P\Delta t$ shown in Fig. P-2.10.

From (*1b*) of Problem 2.10,

$$\mathscr{L}\{W(t)\} = Pe^{-as} = g(s) \tag{1}$$

where Pe^{-as} represents $g(s)$ in the last term of (*3*) of Problem 2.17 for which the inverse of the particular function may be expressed as

$$\Phi = \mathscr{L}^{-1}\left[\frac{1}{(s+\gamma)^2 + (\omega^2 - \gamma^2)}\frac{g(s)}{\mu}\right] \tag{2}$$

Equation (*2*) can be evaluated by using the second translation or shifting property for the following cases:

Case 1: $\omega^2 - \gamma^2 > 0$

$$\Phi = \frac{P}{\mu\lambda} e^{-\gamma(t-a)} \sin[(\lambda(t-a)]\mathscr{U}(t-a) \tag{3}$$

Case 2: $\omega^2 - \gamma^2 = 0$

$$\Phi = \frac{P}{\mu\lambda}(t-a)e^{-\gamma(t-a)}\mathcal{U}(t-a) \tag{4}$$

Case 3: $\omega^2 - \gamma^2 < 0$

$$\Phi = \frac{P}{\mu\lambda}e^{-\gamma(t-a)}\sinh[\lambda(t-a)]\mathcal{U}(t-a) \tag{5}$$

where $\mathcal{U}(t-a)$ is the unit step function discussed in Problem 2.10.

2.23. The rigid frame in Problem 2.4 is subject to a forcing function $P_x(t) = P\sin\Omega t$ and initial conditions $u_0 = 0.5$ in and $\dot{u}_0 = 1$ in/s. Assume $c = 0.1\,c_{cr}$, $P = 10$ k, and $\Omega = 7.417$ rad/s. Find the damped motion at $t = 0.4235$ s.

From (*2-16a*) the motion equation is

$$\Delta = \Delta_1 + \Phi = e^{-\gamma t}\left(\Delta_0\cos\lambda t + \frac{\dot{\Delta}_0 + \gamma\Delta_0}{\lambda}\sin\lambda t\right) + \Phi \tag{1}$$

in which the particular function may be obtained from (*9, 10*) of Problem 2.21 as

$$\Phi = \frac{P}{2\mu\lambda[\gamma^2 + (\lambda-\Omega)^2]}[-\gamma(\cos\Omega t - e^{-\gamma t}\cos\lambda t) + (\lambda-\Omega)(\sin\Omega t - e^{-\gamma t}\sin\lambda t)]$$
$$+\frac{P}{2\mu\lambda[\gamma^2 + (\lambda+\Omega)^2]}[\gamma(\cos\Omega t - e^{-\gamma t}\cos\lambda t) + (\lambda+\Omega)(\sin\Omega t - e^{-\gamma t}\sin\lambda t)] \tag{2}$$

From Problems 2.4 and 2.19 we have $\omega = 4.945$ rad/s, $\gamma = 0.4945$ rad/s, $\lambda = 4.92$ rad/s, and $\mu = 12.42$ k-s²/ft. When $t = 0.4235$ s the displacement is

$$\Delta = \Delta_1 + \Phi = -0.0198 + 0.1109 = 0.0911 \text{ in}$$

TRANSPORT MATRIX

2.24. Construct the time transport matrix $\hat{\hat{T}}_{ji}$ and $\hat{\hat{T}}_{ij}$ for the typical motion of underdamping. Assume the initial conditions at t_i to be Δ_i and $\dot{\Delta}_i$.

From (*2-16a*) the motion equations are

$$\Delta_j = \underbrace{\left(e^{-\gamma t_{ji}}\cos\lambda t_{ji} + \frac{\gamma}{\lambda}\sin\lambda t_{ji}\right)}_{A_{ji}}\Delta_i + \underbrace{e^{-\gamma t_{ji}}\sin\lambda t_{ji}}_{B_{ji}}\frac{\dot{\Delta}_i}{\lambda} + \underbrace{\frac{1}{\mu\lambda}\int_0^t e^{-\gamma(t_{ji}-\tau)}\sin[\lambda(t_{ji}-\tau)]W(\tau)\,d\tau}_{\Phi_{ji}} \tag{1}$$

$$\frac{\dot{\Delta}_j}{\lambda} = \underbrace{\left[(1-e^{-\gamma t_{ji}})\frac{\gamma}{\lambda}\cos\lambda t_{ji} - e^{-\gamma t_{ji}}\sin\lambda t_{ji}\right]}_{C_{ji}}\Delta_i + \underbrace{e^{-\gamma t_{ji}}\left(\cos\lambda t_{ji} - \frac{\gamma}{\lambda}\sin\lambda t_{ji}\right)}_{D_{ji}}\frac{\dot{\Delta}_i}{\lambda} + \frac{\dot{\Phi}_{ji}}{\lambda} \tag{2}$$

Expressing (*1, 2*) in $\hat{\hat{H}}_j = \hat{\hat{T}}_{ji}\hat{\hat{H}}_i$ yields

$$\begin{bmatrix} +1 \\ \hline \Delta_j \\ \dfrac{\dot{\Delta}_j}{\lambda} \end{bmatrix} = \left[\begin{array}{c|cc} 1 & 0 & 0 \\ \hline \Phi_{ji} & A_{ji} & B_{ji} \\ \dfrac{\dot{\Phi}_{ji}}{\lambda} & C_{ji} & D_{ji} \end{array}\right] \begin{bmatrix} +1 \\ \hline \Delta_i \\ \dfrac{\dot{\Delta}_i}{\lambda} \end{bmatrix} \tag{3}$$

Since $\hat{\hat{T}}_{ij} = \hat{\hat{T}}_{ji}^{-1}$, we have

$$\begin{bmatrix} +1 \\ \hline \Delta_i \\ \dfrac{\dot{\Delta}_i}{\lambda} \end{bmatrix} = \left[\begin{array}{c|cc} 1 & 0 & 0 \\ \hline \Phi_{ij} & -D_{ji}/E_{ji} & B_{ji}/E_{ji} \\ \dfrac{\dot{\Phi}_{ij}}{\lambda} & C_{ji}/E_{ji} & -A_{ji}/E_{ji} \end{array}\right] \begin{bmatrix} +1 \\ \hline \Delta_j \\ \dfrac{\dot{\Delta}_j}{\lambda} \end{bmatrix} \tag{4}$$

where
$$\begin{bmatrix} \Phi_{ij} \\ \dfrac{\dot{\Phi}_{ij}}{\lambda} \end{bmatrix} = \frac{-1}{E_{ji}} \begin{bmatrix} -D_{ji} & B_{ji} \\ C_{ji} & -A_{ji} \end{bmatrix} \begin{bmatrix} \Phi_{ji} \\ \dfrac{\dot{\Phi}_{ji}}{\lambda} \end{bmatrix} \qquad E_{ji} = -e^{-2\gamma t_{ji}} + \left(\frac{\gamma}{\lambda}\right)^2 e^{-2\gamma t_{ji}} \sin^2 \lambda t_{ji} \tag{5, 6}$$

2.25. Find $\hat{\hat{H}}_2 = \hat{\hat{T}}_{20}\hat{\hat{H}}_0$ for the forcing function shown in Fig. P-2.25. Assume the initial conditions to be Δ_0 and Δ_0/λ.

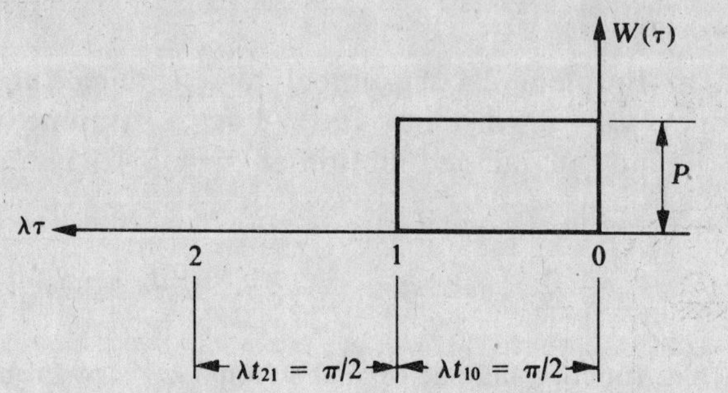

Fig. P-2.25

Using the transport chain gives

$$\hat{\hat{H}}_2 = \hat{\hat{T}}_{21}\hat{\hat{T}}_{10}\hat{\hat{H}}_0 \tag{1}$$

Following (*3*) of the preceding problem, (*1*) becomes

$$\begin{bmatrix} +1 \\ \hline \Delta_2 \\ \dfrac{\dot{\Delta}_2}{\lambda} \end{bmatrix} = \begin{bmatrix} 1 & 0 & 0 \\ \hline 0 & A_{21} & B_{21} \\ 0 & C_{21} & D_{21} \end{bmatrix} \begin{bmatrix} 1 & 0 & 0 \\ \hline \Phi_{10} & A_{10} & B_{10} \\ \dfrac{\dot{\Phi}_{10}}{\lambda} & C_{10} & D_{10} \end{bmatrix} \begin{bmatrix} 1 \\ \hline \Delta_0 \\ \dfrac{\dot{\Delta}_0}{\lambda} \end{bmatrix} \tag{2}$$

Substituting $\lambda t_{10} = \lambda t_{21} = \pi/2$ into (*1, 2*) of the preceding problem gives

$$A_{10} = A_{21} = \gamma/\lambda, \qquad B_{10} = B_{21} = e^{-\pi/2}, \qquad C_{10} = C_{21} = -e^{-\pi/2}, \qquad D_{10} = D_{21} = -\gamma/\lambda e^{-\pi/2} \tag{3}$$

Using (*2*) of Problem 2.21 we have

$$\Phi_{10} = \frac{P}{\mu\lambda(\gamma^2 + \lambda^2)}(\lambda - \gamma e^{-\pi/2}) \qquad \text{and} \qquad \dot{\Phi}_{10}/\lambda = \frac{P}{\mu\lambda^2} e^{-\pi/2} \tag{4}$$

Substituting (*3, 4*) into (*2*) gives $\hat{\hat{H}}_2$.

Supplementary Problems

FREE VIBRATION WITHOUT DAMPING

2.26. A steel structure shown in Fig. P-2.26 has a column (l = 6 ft, d = 6 in) supporting a steel plate (h = 4 in, D = 6 ft). Find the differential equation of (*a*) longitudinal, (*b*) torsional, and (*c*) transverse vibrations and then find the natural circular frequency, natural frequency, and natural period for each of the three motions. Assume that E = 30 000 ksi, G = 11 500 ksi, the column mass is negligible, and the plate mass is lumped at its center.

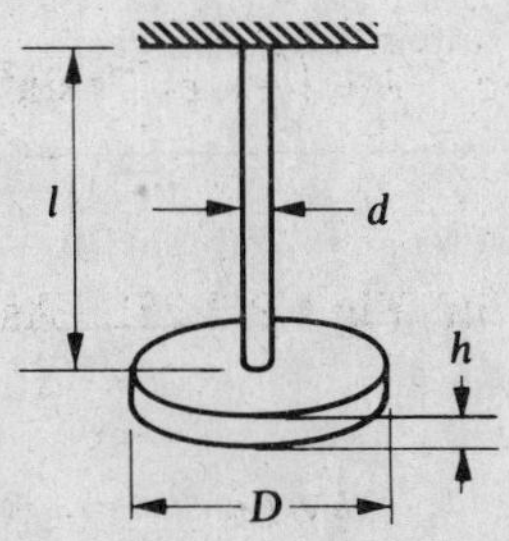

Fig. P-2.26

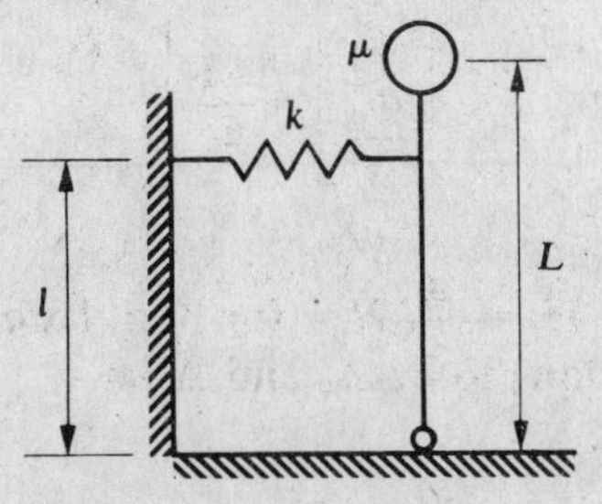

Fig. P-2.27

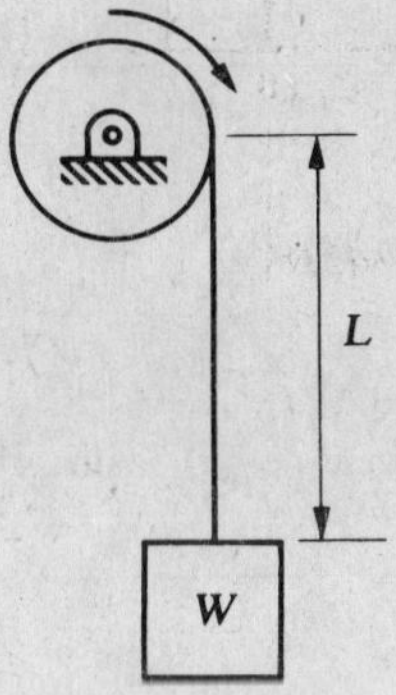

Fig. P-2.29

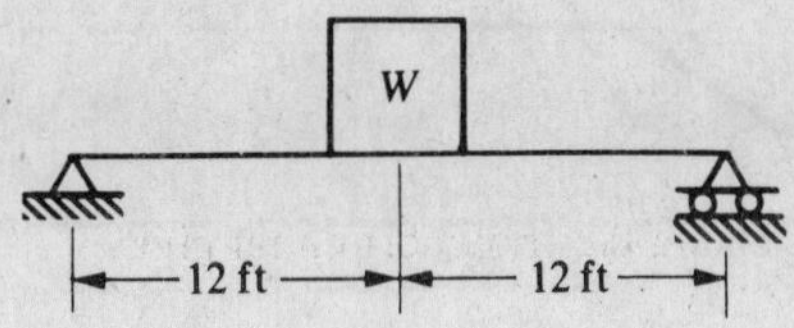

Fig. P-2.30

2.27. Write the differential equation of small vibration of inverted pendulum shown in Fig. P-2.27 and then find its natural frequency. Assume the hinge to be frictionless.

2.28. Rework Problem 2.4 for the modified moment of inertia of the girder as $I_g = 8I$.

2.29. A cage of an elevator shown in Fig. P-2.29 is moving down at a constant velocity $v_0 = 2.5$ ft/s and the spring consists of a steel cable of cross-sectional area = 3.0 in^2 and modulus of elasticity = 16 000 ksi. Assume the cage weight to be $W = 14$ k and the cable length to be $L = 50$ ft. Determine the cable stress if during motion the upper end is suddenly stopped.

2.30. As a piece of heavy machinery is being readied for positioning on its supporting beam (W10 × 45, $I = 248$ in^4, $S = 49.1$ in^2) as shown in Fig. P-2.30, it is *suddenly dropped* by the crane which was lowering it onto the beam. Assume the force of the machine on the beam to be its weight of 12 k; because the distance dropped is small, no increase in force due to acceleration from gravity should be included. If the machine is left on the beam what is the maximum stress it causes? Neglect the beam mass for developing the equation of motion.

FORCED VIBRATION WITHOUT DAMPING

2.31. Find the particular functions for each of the constant and monotonically increasing nonperiodic continuous load functions shown in Table P-2.31.

Table P-2.31 Particular Functions $\Phi(t)$ of the Constant and Monotonically Increasing Nonperiodic Load Functions $W(\tau)$

$W(\tau)$	$\mu\omega\Phi(t) = \int_0^t W(\tau)\sin(\omega t - \omega\tau)\,d\tau$
1. 	$\frac{P}{\omega}(1 - \cos\omega t)$
2. 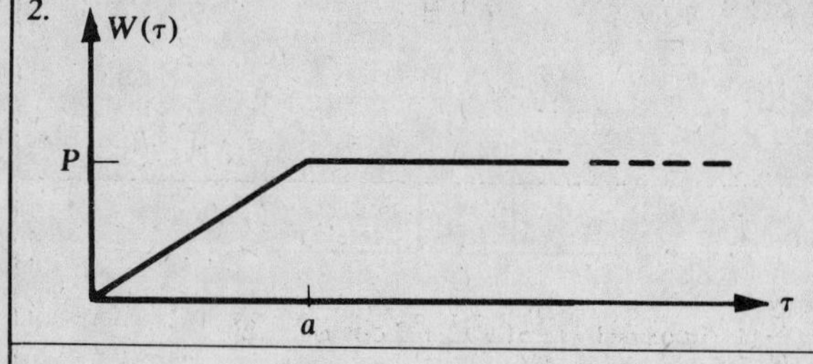	$t < a$: $\frac{P}{\omega^2 a}(\omega t - \sin\omega t)$ $t > a$: $\frac{P}{\omega^2 a}[\omega a + \sin\omega(t - a) - \sin\omega t]$
3. 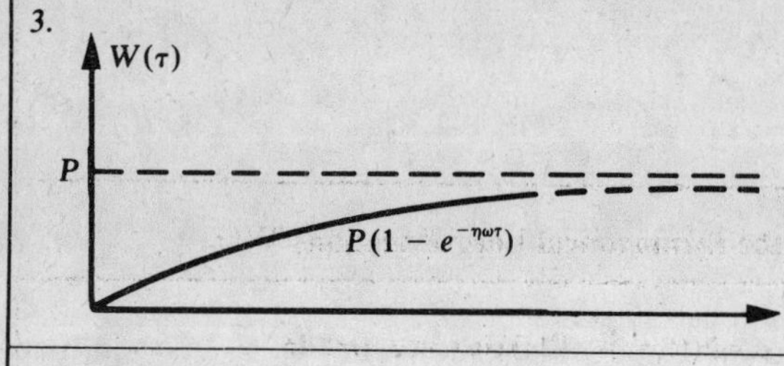	$\frac{P}{\omega}(1 - \cos\omega t) + \frac{P}{\omega(1 + \eta^2)}(-e^{-\eta\omega t} + \cos\omega t - \eta\sin\omega t)$
4. 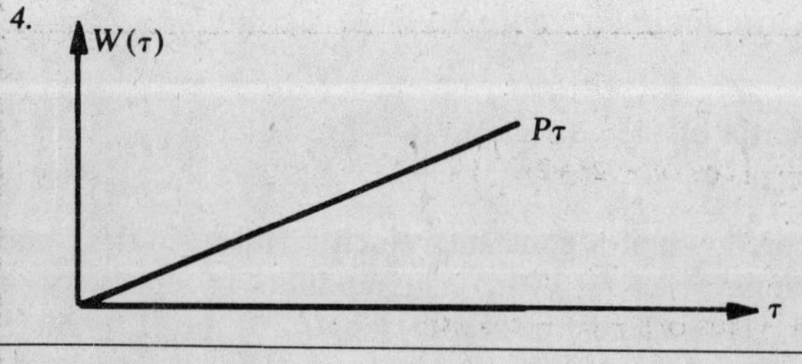	$\frac{P}{\omega^2}(\omega t - \sin\omega t)$
5. 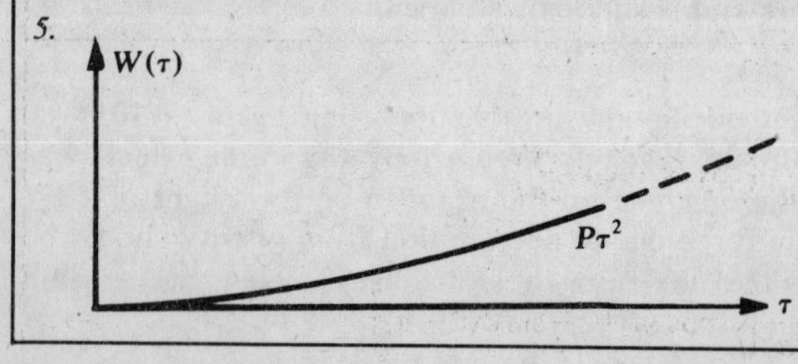	$\frac{P}{\omega^3}(\omega^2 t^2 + 2\cos\omega t - 2)$

2.32. Find the particular functions for each of the monotonically decreasing nonperiodic continuous load functions shown in Table P-2.32.

2.33. Find the particular functions for each of the symmetrical load functions shown in Table P-2.33.

Table P-2.32 Particular Functions $\Phi(t)$ of the Monotonically Decreasing Nonperiodic Continuous Load Functions $W(\tau)$

$W(\tau)$	$\mu\omega\Phi(t) = \int_0^t W(\tau)\sin(\omega t - \omega\tau)\,d\tau$
1. 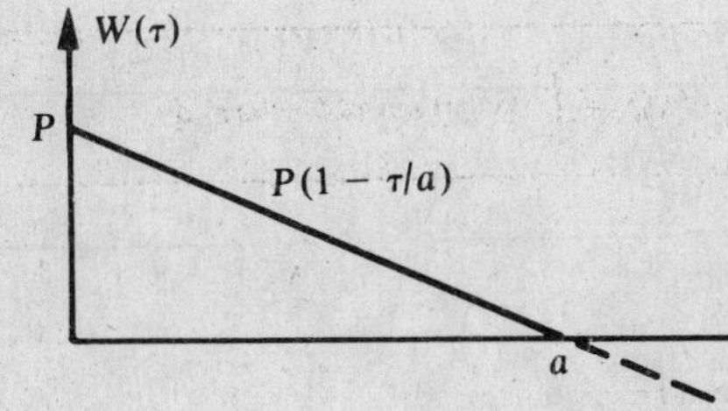	$\frac{P}{a\omega^2}(\sin\omega t - \omega^2 t) + \frac{P}{\omega}(1 - \cos\omega t)$
2. 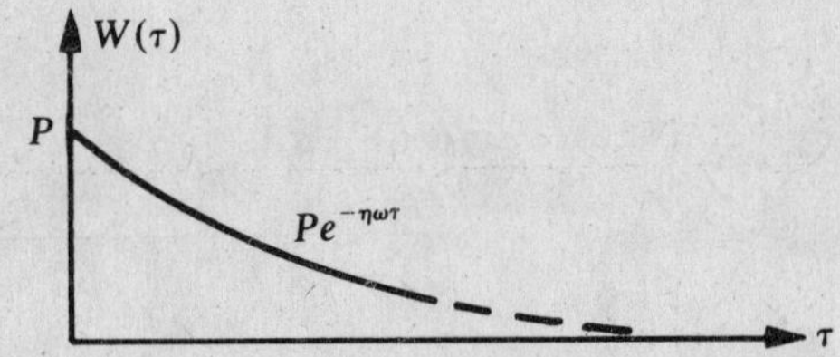	$\frac{P}{\omega(1+\eta^2)}\eta\sin\omega t + \frac{P}{\eta+\omega}(e^{-\eta\omega t} - \cos\omega t)$
3. 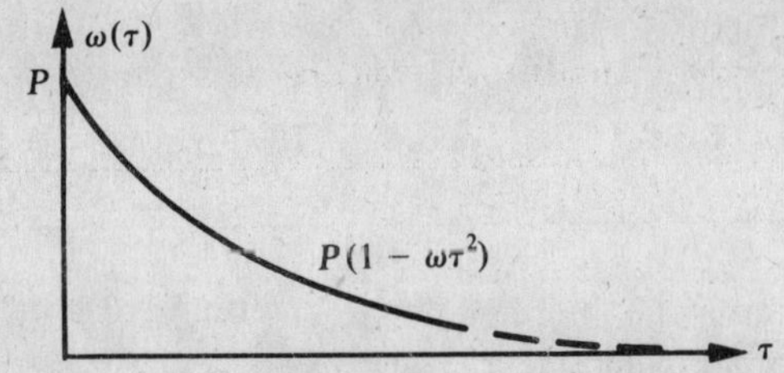	$\frac{P}{\omega}(1 - \cos\omega t) - \frac{P}{\omega^2}(\omega^2 t^2 + 2\cos\omega t - 2)$

Table P-2.33 Particular Functions $\Phi(t)$ of the Symmetrical Load Functions $W(\tau)$

$W(\tau)$	$\mu\omega\Phi(t) = \int_0^t W(\tau)\sin(\omega t - \omega\tau)\,d\tau$
1. 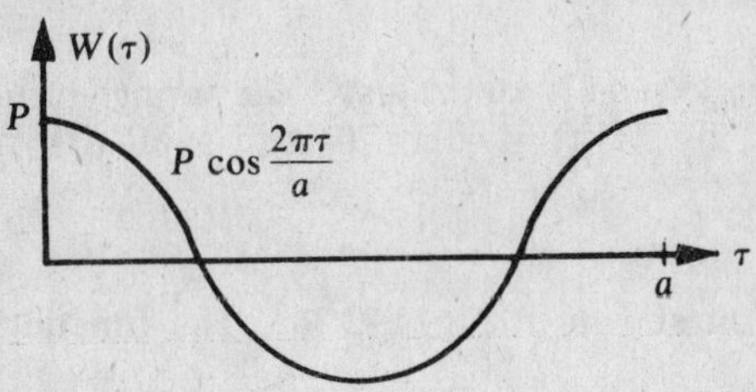	$t < a$: $\frac{a^2\omega P}{4\pi^2 - \omega^2 a^2}\left(\cos\omega t - \cos 2\pi\frac{t}{a}\right)$ $t > a$: $\frac{a^2\omega P}{4\pi^2 - \omega^2 a^2}[\cos\omega(t-a) - \cos\omega t]$
2. 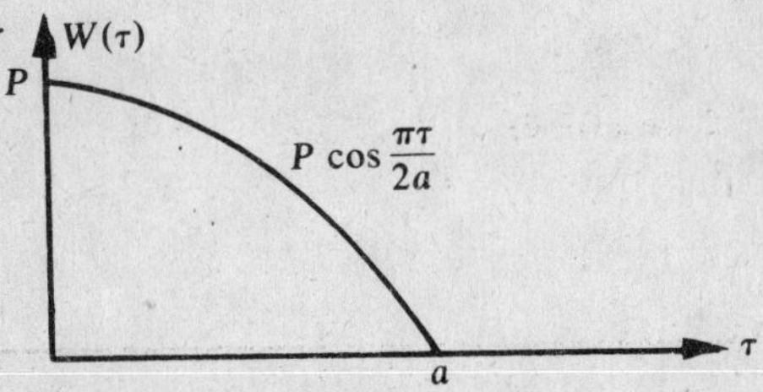	$t < a$: $\frac{4a^2\omega P}{\pi^2 - 4\omega^2 a^2}\left(\cos\omega t - \cos\frac{\pi t}{2a}\right)$ $t > a$: $\frac{4a^2 P}{\pi^2 - 4\omega^2 a^2}\left[\omega\cos\omega t + \frac{\pi}{2a}\sin\omega(t-a)\right]$
3. 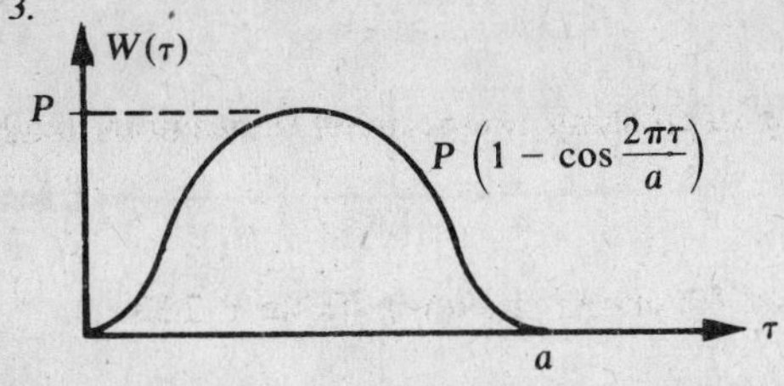	$t < a$: $\frac{a^2\omega P}{4\pi^2 - \omega^2 a^2}\left(\cos\frac{2\pi t}{a} - \cos\omega t\right) - \frac{P}{\omega}(\cos\omega t - 1)$ $t > a$: $\frac{a^2\omega P}{4\pi^2 - \omega^2 a^2}[\cos\omega(t-a) - \cos\omega t] - \frac{P}{\omega}[\cos\omega t - \cos\omega(t-a)]$

2.34. Find the particular functions for each of the periodic load functions shown in Table P-2.34.

Table P-2.34 Particular Functions $\Phi(t)$ of the Periodic Load Function $W(\tau)$

	$W(\tau)$	$\mu\omega\Phi(t) = \int_0^t W(\tau)\sin(\omega t - \omega\tau)\, d\tau$
1.	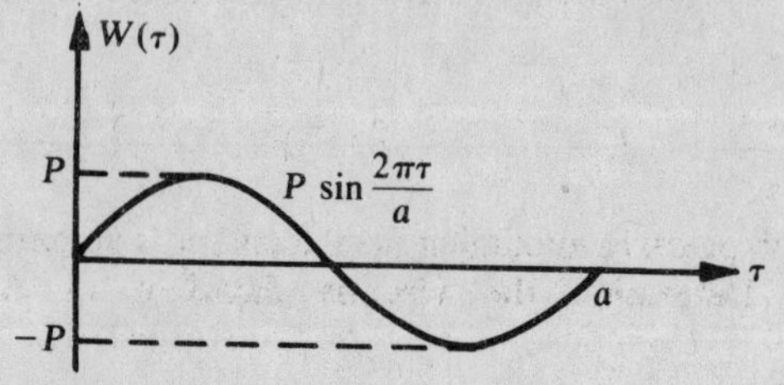	$t < a$: $\dfrac{P\omega a^2}{4\pi^2 - \omega^2 a^2}\left(\dfrac{2\pi}{\omega a}\sin\omega t - \sin\dfrac{2\pi t}{a}\right)$ $t > a$: $\dfrac{2\pi a P}{4\pi^2 - \omega^2 a^2}[\sin\omega t - \sin(\omega t - \omega a)]$
2.	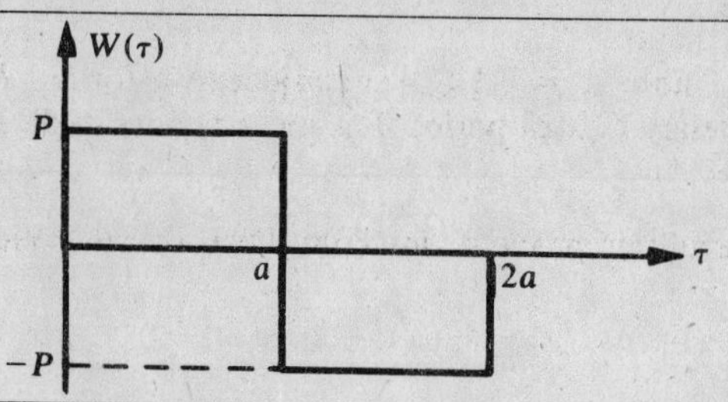	$t < a$: $\dfrac{P}{\omega}(1 - \cos\omega t)$ $a < t < 2a$: $\dfrac{P}{\omega}[2\cos(\omega t - \omega a) - \cos\omega t - 1]$
3.	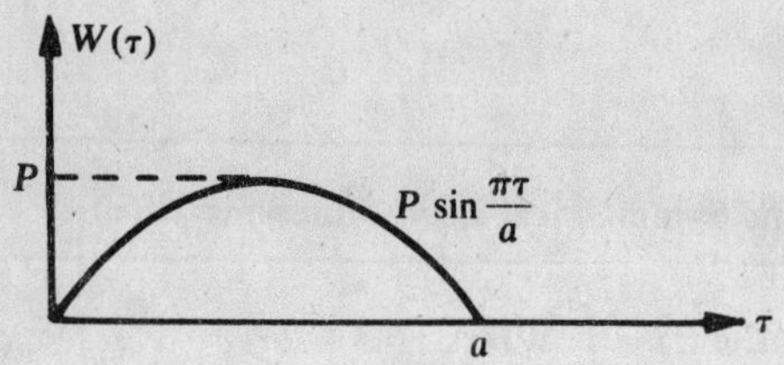	$t < a$: $\dfrac{aP}{\pi^2 - \omega^2 a^2}\left(\pi\sin\omega a - \omega a\sin\pi\dfrac{t}{a}\right)$ $t > a$: $\dfrac{a\pi P}{\pi^2 - \omega^2 a^2}[\sin\omega t + \sin(\omega t - \omega a)]$

2.35. The natural circular frequency of a single degree of freedom system is 16π rad/s. The spring stiffness is 500 lb/ft. Find the amplitude of the vibration caused by $P(t) = 40 \sin 20t$ lbs applied to the mass.

2.36. Find the Fourier series expansion for the periodic function shown in Fig. P-2.36. The function is defined as

$$P(t) = \frac{P}{a}(t + a) \qquad \text{for } -a < t < a$$

$$P(t) = \frac{P}{a}(a - t) \qquad \text{for } 0 < t < a$$

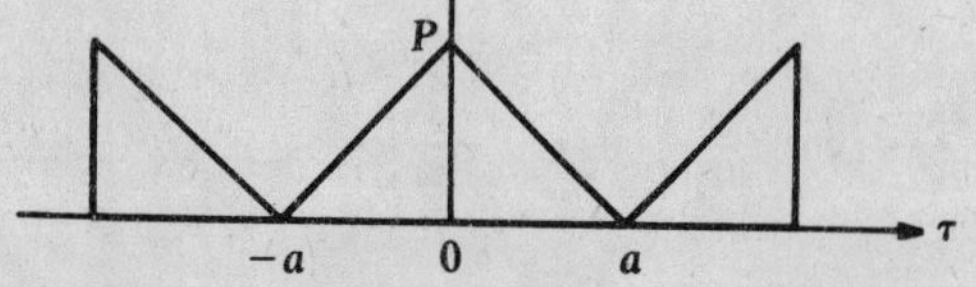

Fig. P-2.36

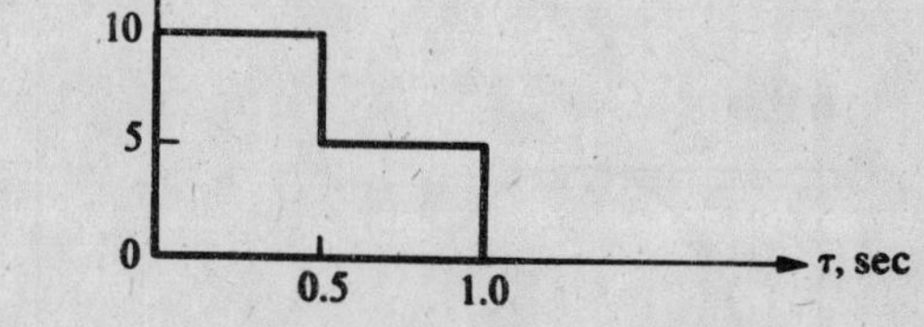

Fig. P-2.37

2.37. Using the step-by-step method find the transverse displacement of Problem 2.26 subject to a forcing function shown in Fig. P-2.37. Use the time interval $\Delta t = 0.5\ \text{s}$ and find the displacement at $t = 1.5\ \text{s}$. Assume $u_0 = 0$, $\dot{u}_0 = 0$.

TRANSPORT MATRIX

2.38. Rework Problem 2.37 by using the transport matrix for which the particular functions may be obtained by convolution integrals.

FREE AND FORCED VIBRATION WITH DAMPING

2.39. The free oscillations of a Kelvin model show that each positive maximum displacement is 90 percent of the preceding positive maximum displacement. Determine the viscous damping c. Assume $\mu g = 50\ \text{k}$ and $k = 3500\ \text{k/in.}$

2.40. A Kelvin model having $\mu g = 2.4\ \text{k}$, $k = 482\ \text{k/ft}$, and $c = 0.1 c_{cr}$ is subject to a force $P(t) = (2 \cos 10t)k$. Find (*a*) critical damping c_{cr}, (*b*) frequency f^*_Δ, (*c*) period T^*_Δ, and (*d*) amplitude R^*_Δ.

2.41. Find the transient motion at $t = 2\ \text{s}$ of the preceding problem including the initial conditions $u_0 = 0.5\ \text{in}$ and $\dot{u}_0 = 2\ \text{in/s}$ at $t = 0$.

Chapter 3

Systems with Several Degrees of Freedom

3.1 SYSTEM

Degrees of Freedom

When the number of degrees of freedom m of a structure exceeds one, the system is said to have *several degrees of freedom* (*multidegree freedom system*), and m independent motions of the system are possible. Although each structural system has an infinite number of degrees of freedom, by lumping the distributed mass at selected points, an *equivalent lumped mass system* with a limited number of degrees of freedom can be introduced for the analysis.

Distributed Mass Equivalents

This lumping of the structural mass may be accomplished in many different ways. For tall structures with relatively stiff girders but flexible columns, the entire system can be replaced by one cantilever beam of identical height with the lumped masses applied at the floor levels (Fig. 3-1*b*). For long, simple, or continuous girders, a number of lumped masses may be introduced and varied in intensity to account for the variable depth (Fig. 3-1*d*). For bent member frames, the lumping of mass at the joints proves to be more desirable (Fig. 3-1*a*). Finally, in frames with curved members, a combination of both concepts yields good results (Fig. 3-1*c*).

Discrete Parameters Model

The lumping of mass at certain nodal points (nodes) of the structure changes the *distributed parameter model* to a *discrete parameter model*, the causes and effects of which are then related to these and only to these nodes. This transformation is of great practical value since it simplifies the problem and makes it suitable to computer programming and numerical analysis in general.

Since in a planar system each lumped mass has three degrees of freedom, the *total number of degrees of freedom m* of a system containing *j lumped masses* and *r constraints*, is

$$m = 3j - r \tag{3-1}$$

where the constraints are real (given by the definition of structure) and/or assumed (introduced as simplifications).

Mechanical Coupling

The *connections of lumped masses* can be again symbolized by springs and dampers, each representing the respective restoring force (moment) and the respective damping force (moment). For a complex system, such diagramming becomes very complicated and is rarely necessary. The centroidal axes of the structural members connecting the lumped masses have assigned deformation characteristics and their *static segmental stiffness matrices* are directly the spring constant matrices of the discrete parameter model. Similar matrices (if experimental data are available) can be constructed for damping.

The utilization of these two concepts is demonstrated in Chapter 5. In the subsequent sections of this chapter, the values of k and c and their positions in the system matrix are assumed to be known.

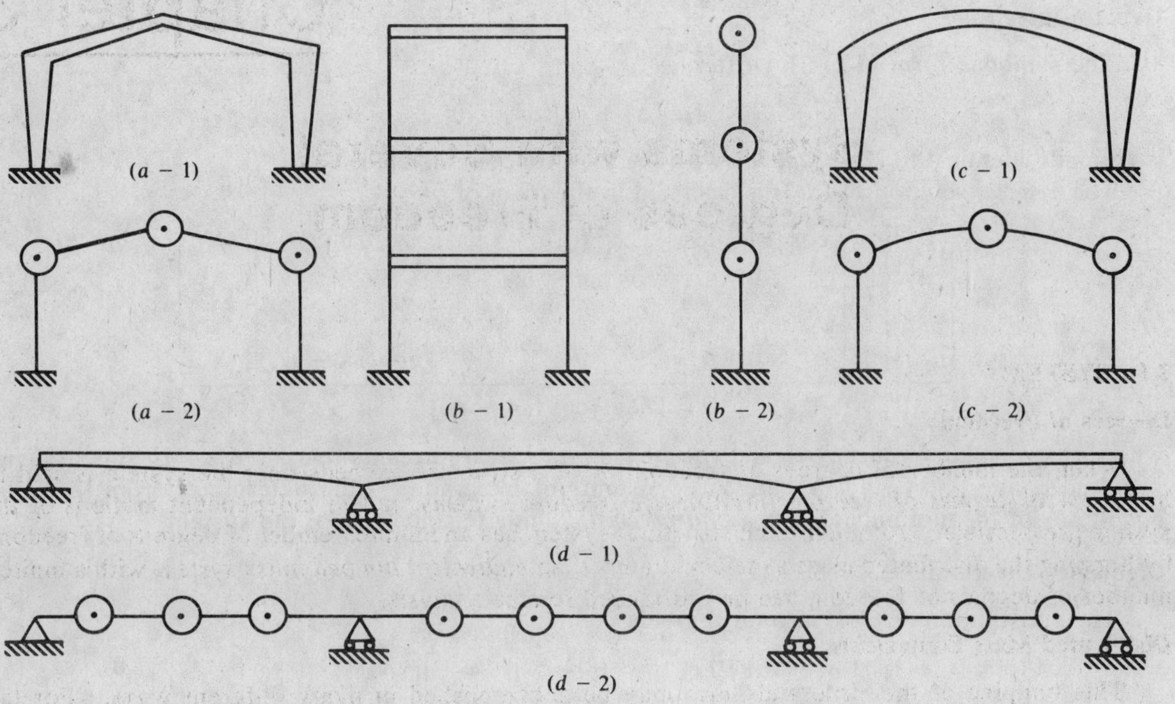

Fig. 3-1. Real Structures (1) and Their Discrete Parameters Models (2)

Dynamic Conditions

As each degree of freedom generates one dynamic condition expressed by the respective *differential equation of motion*, m such equations govern the motion of the structure. Since they are related to the nodes, they are called *nodal equations*. This system of nodal equations may be derived by applying the *principles of dynamic equilibrium* (d'Alambert's principles) to the motion of each mass as shown in this chapter or by using the *concept of work and energy* of the system (Lagrange's equations) as demonstrated in Chapter 4.

Equations of Motion

The *integrals of these differential equations* are then the equations of motion, the constants of which must be determined from the initial conditions. These equations define the *dynamic response of the structure* classified as *periodic*, *transient*, *aperiodic*, or *resonant* and offer the basis for the calculation of stresses and reactions.

3.2 FREE VIBRATION WITHOUT DAMPING

Nodal Equations

For a discrete parameter model with m degrees of freedom in a state of free vibration (of small amplitude) about the configuration of stable equilibrium, the conditions of dynamic equilibrium yield the following *differential matrix equation*:

$$\begin{bmatrix} \mu_{11} & \mu_{12} & \cdots & \mu_{1m} \\ \mu_{21} & \mu_{22} & \cdots & \mu_{2m} \\ \cdots & \cdots & \cdots & \cdots \\ \mu_{m1} & \mu_{m2} & \cdots & \mu_{mm} \end{bmatrix} \begin{bmatrix} \ddot{\Delta}_1 \\ \ddot{\Delta}_2 \\ \cdots \\ \ddot{\Delta}_m \end{bmatrix} + \begin{bmatrix} k_{11} & k_{12} & \cdots & k_{1m} \\ k_{21} & k_{22} & \cdots & k_{2m} \\ \cdots & \cdots & \cdots & \cdots \\ k_{m1} & k_{m2} & \cdots & k_{mm} \end{bmatrix} \begin{bmatrix} \Delta_1 \\ \Delta_2 \\ \cdots \\ \Delta_m \end{bmatrix} = \begin{bmatrix} 0 \\ 0 \\ \cdots \\ 0 \end{bmatrix} \tag{3-2}$$

where μ_{ij} is the mass constant (μ or ρ), k_{ij} is the spring constant (k_u, k_v, or k_θ), and Δ_j and $\ddot{\Delta}_j$ are the displacements and accelerations (u, v, or θ; $\ddot{u}$, $\ddot{v}$, or $\ddot{\theta}$) respectively.

General Integral

The *symbolic form* of (*3-2*), written as

$$[\mu]\{\ddot{\Delta}\} + [k]\{\Delta\} = \{0\} \tag{3-3}$$

is formally identical to (*2-1a, 2a, 3a*) with the load function deleted.

The *general solution* of (*3-3*) is assumed in the form of

$$\underbrace{\begin{bmatrix} \Delta_1 \\ \Delta_2 \\ \cdots \\ \Delta_m \end{bmatrix}}_{\{\Delta\}} = \underbrace{\begin{bmatrix} \cos(\omega t - \alpha) & 0 & \cdots & 0 \\ 0 & \cos(\omega t - \alpha) & \cdots & 0 \\ \cdots & \cdots & \cdots & \cdots \\ 0 & 0 & \cdots & \cos(\omega t - \alpha) \end{bmatrix}}_{\text{Diag. }[\cos(\omega t - \alpha)]} \underbrace{\begin{bmatrix} X_1 \\ X_2 \\ \cdots \\ X_m \end{bmatrix}}_{\{X\}} \tag{3-4}$$

where the *angular frequency* ω and the *phase angle* α are common to all Δ's. In this assumed solution, α and $X_1, X_2, \ldots, X_m$ are the constants of integration to be determined from the initial conditions of the motion and ω is a *characteristic value* (eigenvalue) of the system.

Frequency Determinant

The substitution of (*3-4*) into (*3-2*) yields

$$\underbrace{\begin{bmatrix} (k_{11} - \mu_{11}\omega^2) & (k_{12} - \mu_{12}\omega^2) & \cdots & (k_{1m} - \mu_{1m}\omega^2) \\ (k_{21} - \mu_{21}\omega^2) & (k_{22} - \mu_{22}\omega^2) & \cdots & (k_{2m} - \mu_{2m}\omega^2) \\ \cdots & \cdots & \cdots & \cdots \\ (k_{m1} - \mu_{m1}\omega^2) & (k_{m2} - \mu_{m2}\omega^2) & \cdots & (k_{mm} - \mu_{mm}\omega^2) \end{bmatrix}}_{[[k] - \omega^2[\mu]]} \underbrace{\begin{bmatrix} X_1 \\ X_2 \\ \cdots \\ X_m \end{bmatrix}}_{\{X\}} \cos(\omega t - \alpha) = \underbrace{\begin{bmatrix} 0 \\ 0 \\ \cdots \\ 0 \end{bmatrix}}_{\{0\}} \tag{3-5}$$

where $[k]$ and $[\mu]$ are $(m \times m)$ matrices, ω^2 and $\cos(\omega t - \alpha)$ are scalars, and $\{X\}$ is a $(m \times 1)$ vector of the amplitudes (modal column).

The *consistency* of (*3-5*) for $\{X\}\cos(\omega t - \alpha) \neq 0$ requires that

$$\underbrace{|[k] - \omega^2[\mu]|}_{\text{Det. }(\omega^2)} = 0 \tag{3-6}$$

where Det. (ω^2) is called the *frequency determinant.*

The expansion of this determinant yields a *polynomial of mth degree in* (ω^2), the m roots of which are the *eigenvalues* $\omega_1, \omega_2, \ldots, \omega_m$ (*natural angular frequencies*). The solution of this equation is one of the classical problems of the eigenvalue theory illustrated in Problem 3.1.

Parameters of Motion

The *resultant motion*, in general, is not a harmonic motion, but a *sum of m harmonic motions*, each governed by the respective natural frequency ω.

These natural frequencies, thought to be arranged in order of ascending magnitudes,

$$\omega_1, \omega_2, \ldots, \omega_m \neq 0$$

are called respectively the first, second, ..., mth frequency. They characterize a *definite system.* If one of these values is zero, the system is said to be *semidefinite*, which implies that it can move as a rigid body (Problem 3.2).

The *shape of the polygons* formed by straight lines connecting the extreme positions of the masses in one component motion (amplitudes) is called the *natural* (*characteristic*) *mode.*

Since each mode is governed by one particular frequency, an m degree freedom system vibrates in m natural modes. When the reference amplitude of a given natural mode is unity, the mode is called a *normal mode* and its shape is called the *normal shape.*

Equations of Motion

From the above, m equations define the motion of the system, each containing m components, i.e.,

$$\begin{bmatrix} \Delta_1 \\ \Delta_2 \\ \cdots \\ \Delta_m \end{bmatrix} = \begin{bmatrix} X_{11} \\ X_{21} \\ \cdots \\ X_{m1} \end{bmatrix} \cos(\omega_1 t - \alpha_1) + \begin{bmatrix} X_{12} \\ X_{22} \\ \cdots \\ X_{m2} \end{bmatrix} \cos(\omega_2 t - \alpha_2) + \cdots + \begin{bmatrix} X_{1m} \\ X_{2m} \\ \cdots \\ X_{mm} \end{bmatrix} \cos(\omega_m t - \alpha_m) \tag{3-7}$$

or

$$\{\Delta\} = \sum_{p=1}^{m} \{X_p\} \cos(\omega_p t - \alpha_p)$$

where $\{X_1\}, \{X_2\}, \ldots, \{X_m\}$ are the *eigenvectors* (natural modes, modal columns), the absolute values of which cannot be determined without knowledge of the initial conditions. Their relative values for a known *spectrum* of $\omega_1, \omega_2, \ldots, \omega_m$ are given by (3-5) as shown in Problem 3.1.

Initial Conditions

The *total number of constants of integration* in (3-7) is $m(m+1)$, i.e., the number of unknown amplitudes $(X_{11}, X_{12}, \ldots, X_{m-1,m}, X_{mm})$ is m^2 and the number of unknown phase angles $(\alpha_1, \alpha_2, \ldots, \alpha_m)$ is m.

For their determination *three sets of conditions* are available:

(*a*) The *initial displacements* at $t = 0$, $(\Delta_1(0), \Delta_2(0), \ldots, \Delta_m(0))$; m conditions.

(*b*) The *initial velocities* at $t = 0$, $(\dot{\Delta}_1(0), \dot{\Delta}_2(0), \ldots, \dot{\Delta}_m(0))$; m conditions.

(*c*) The *relationships of amplitudes* given for each ω by (3-5); $m(m-1)$ conditions.

The evaluations of these constants in a particular system are given in Problem 3.2.

3.3 PROPERTIES OF NATURAL MODES

General

The modal columns (eigenvectors, natural modes) in (3-7) reveal special properties which influence considerably the matrix formulation of the analysis. For their study, certain general matrix relationships must be introduced first.

Congruence

A square matrix $[A]$ is said to be *congruent* to another square matrix $[B]$ if there exists a nonsingular matrix $[C]$ such that,

$$[A] = [C]^T[B][C] \tag{3-8}$$

This transformation is called the *congruent transformation* which may take special forms under special conditions.

(*a*) If $[B]$ is a *symmetrical matrix*, then $[A]$ is also a *symmetrical matrix*.

(*b*) If $[B]$ is a *unit matrix* $[I]$ and $[A]$ is also a *unit matrix* $[I]$, then

$$[C]^T = [C]^{-1} \tag{3-9}$$

and $[C]$ is said to be an *orthogonal matrix*.

(*c*) If $[A]$ is a *diagonal matrix*, such that

$$\text{Diag.}\,[A] = [C]^T[B][C] \tag{3-10}$$

then the columns of $[C]$ are said to be *orthogonal*.

(*d*) If $[A]$ is a *diagonal matrix*, the square roots of whose coefficients are real numbers, such that

$$\text{Diag. }[A] = \text{Diag. }[L]\,\text{Diag. }[L]$$

and

$$\underbrace{\text{Diag. }[1/L][C]^T}_{[E]^T}\,[B]\,\underbrace{[C]\,\text{Diag. }[1/L]}_{[E]} = [I] \tag{3-11}$$

then the matrix $[E]$ is said to be *normalized* with respect to $[B]$ and

$$[E]^{-1} = [E]^T[B] \tag{3-12}$$

These properties of the congruent transformation (*3-8*) emerge in the form of *two orthogonality conditions* and *two normalization conditions* in modal analysis of multidegree freedom systems and are described in detail in the subsequent articles of this section.

First Orthogonality Condition

If $\omega_j^2 \neq \omega_k^2$ and the mass matrix $[\mu]$ is symmetrical, two natural modes (eigenvectors), $\{X_j\}$ and $\{X_k\}$, corresponding to these natural angular frequencies are orthogonal with respect to the mass matrix. That is,

$$\{X_j\}^T[\mu]\{X_k\} = 0 \tag{3-13}$$

which is called the *first orthogonality condition* (Problem 3.4).

Second Orthogonality Condition

If $\omega_j^2 \neq \omega_k^2$ and the spring (stiffness) matrix $[k]$ is symmetrical, $\{X_j\}$ and $\{X_k\}$ are also orthogonal with respect to the spring matrix, i.e.,

$$\{X_j\}^T[k]\{X_k\} = 0 \tag{3-14}$$

which is called the *second orthogonality condition* (Problem 3.4).

First Normalization Condition

If $\omega_j^2 = \omega_k^2$ and $\{X_j\} = \{X_k\}$, then

$$\{X_j\}^T[\mu]\{X_j\} = L_j^2 \qquad \text{and} \qquad \{h_j\}^T[\mu]\{h_j\} = 1 \tag{3-15}$$

where $h_j = \{X_{1j}/L_j, X_{2j}/L_j, \ldots, X_{mj}/L_j\}$ is the unit eigenvector corresponding to ω_j and the second form of (*3-15*) is called the first normalization condition (Problem 3.4).

Second Normalization Condition

If $\omega_j^2 = \omega_k^2$ and $\{X_j\} = \{X_k\}$, then

$$\{X_j\}^T[k]\{X_j\} = L_j^2\omega_j^2 \qquad \text{and} \qquad \{h_j\}^T[k]\{h_j\} = \omega_j^2 \tag{3-16}$$

where h_j is the unit eigenvector corresponding to ω_j (same as in *3-15*) and the second form of (*3-16*) is called the second normalization condition (Problem 3.4).

Eigenvector Spectra

The matrix containing all natural eigenvectors, $[X]$, and the matrix containing all normalized eigenvectors, $[h]$, are called the *natural eigenvector spectrum* and the *normalized eigenvector spectrum*, respectively; they are:

$$[X] = \begin{bmatrix} X_{11} & X_{12} & \cdots & X_{1m} \\ X_{21} & X_{22} & \cdots & X_{2m} \\ \cdots & \cdots & \cdots & \cdots \\ X_{m1} & X_{m2} & \cdots & X_{mm} \end{bmatrix} \qquad [h] = \begin{bmatrix} h_{11} & h_{12} & \cdots & h_{1m} \\ h_{21} & h_{22} & \cdots & h_{2m} \\ \cdots & \cdots & \cdots & \cdots \\ h_{m1} & h_{m2} & \cdots & h_{mm} \end{bmatrix} \tag{3-17, 18}$$

By (*3-13*) and (*3-15*), if $[\mu] \neq [I]$,

$$[X]^T[\mu][X] = \begin{bmatrix} L_1^2 & & & \\ & L_2^2 & & \\ & & \ddots & \\ & & & L_m^2 \end{bmatrix} \qquad [h]^T[\mu][h] = \begin{bmatrix} 1 & & & \\ & 1 & & \\ & & \ddots & \\ & & & 1 \end{bmatrix} \tag{3-19, 20}$$

By (*3-14*) and (*3-16*), if $[k] \neq [I]$,

$$[X]^T[k][X] = \begin{bmatrix} L_1^2\omega_1^2 & & & \\ & L_2^2\omega_2^2 & & \\ & & \ddots & \\ & & & L_m^2\omega_m^2 \end{bmatrix} \qquad [h]^T[k][h] = \begin{bmatrix} \omega_1^2 & & & \\ & \omega_2^2 & & \\ & & \ddots & \\ & & & \omega_m^2 \end{bmatrix} \tag{3-21, 22}$$

The result of each congruent transformation in (*3-19, 20*) and (*3-21, 22*) is an $[m \times m]$ *diagonal matrix* (Problem 3.4).

3.4 NORMAL COORDINATES

Concept

The modal equations of motion given by (*3-2*) are frequently called the *coupled equations*, since their solution given by (*3-7*) is a *superposition of m motions*. Although this type of solution causes no special difficulties in the analysis of free vibration, it becomes quite laborious if forced vibration is involved.

In such cases the *concept of uncoupling*, i.e., the transformation of (*3-2*) into a *set of independent* (*uncoupled*) *equations*, is useful. Each of these new equations defines the motion of one mass in terms of one μ, ω, and k. The displacement vectors of this new system (principal system, normal system) called the *normal* (*principal*) *coordinates* are introduced below.

Normalized Nodal Equations

The system of nodal equations (*3-2*) in terms of

$$\underbrace{\begin{bmatrix} \Delta_1 \\ \Delta_2 \\ \cdots \\ \Delta_m \end{bmatrix}}_{\{\Delta\}} = \underbrace{\begin{bmatrix} h_{11} & h_{12} & \cdots & h_{1m} \\ h_{21} & h_{22} & \cdots & h_{2m} \\ \cdots & \cdots & \cdots & \cdots \\ h_{m1} & h_{m2} & \cdots & h_{mm} \end{bmatrix}}_{[h]} \underbrace{\begin{bmatrix} \eta_1 \\ \eta_2 \\ \cdots \\ \eta_m \end{bmatrix}}_{\{\eta\}} \tag{3-23}$$

becomes

$$[\mu][h]\{\ddot{\eta}\} + [k][h]\{\eta\} = \{0\} \tag{3-24}$$

where $\{\Delta\}$ = the *original coordinate* (*displacement*) *vector* introduced at the onset of analysis in (*3-2*).

$\{\eta\}$ = the *normal coordinate* (*displacements*) *vector* introduced in (*3-23*) above.

The premultiplication of (*3-24*) by $[h]^T$ gives $\quad [h]^T[\mu][h]\{\ddot{\eta}\} + [h]^T[k][h][\eta] = 0 \quad$ (*3-25*)

which by (*3-20, 22*) reduces to *m independent nodal equations*.

$$\underbrace{\begin{bmatrix} \ddot{\eta}_1 \\ \ddot{\eta}_2 \\ \cdots \\ \ddot{\eta}_m \end{bmatrix}}_{\{\ddot{\eta}\}} + \underbrace{\begin{bmatrix} \omega_1^2 & & & \\ & \omega_2^2 & & \\ & & \ddots & \\ & & & \omega_m^2 \end{bmatrix}}_{\{\Lambda\}} \underbrace{\begin{bmatrix} \eta_1 \\ \eta_2 \\ \cdots \\ \eta_m \end{bmatrix}}_{\{\eta\}} = \underbrace{\begin{bmatrix} 0 \\ 0 \\ \cdots \\ 0 \end{bmatrix}}_{\{0\}} \tag{3-26}$$

Initial Conditions

The solution of (*3-26*) in symbolic form is

$$\{\eta\} = \text{Diag. }[\cos \omega t]\{\eta_0\} + \text{Diag. }[\sin \omega t]\left\{\frac{\dot{\eta}_0}{\omega}\right\} \tag{3-27}$$

where $\{\eta_0\}$ and $\{\dot{\eta}_0/\omega\}$ are the *initial conditions* stated in vector form (Problem 3.6).

Since the initial conditions are given in the original coordinates, (*3-27*) must be expressed in terms of $\{\Delta\}$, $\{\Delta_0\}$, $\{\dot{\Delta}_0\}$ by means of (*3-23*) and its inverse, i.e.,

$$\{\Delta\} = [h]\text{ Diag. }[\cos \omega t][h]^{-1}\{\Delta_0\} + [h]\text{ Diag. }[\sin \omega t]\text{ Diag. }\left[\frac{1}{\omega}\right][h]^{-1}\{\dot{\Delta}_0\} \tag{3-28}$$

where the diagonal matrices are the same as in (*3-27*).

3.5 FORCED VIBRATION WITHOUT DAMPING

Nodal Equations

The *governing matrix equation* for the forced vibration without damping of a multidegree freedom structure with the forcing function on the right side of the equality sign is

$$[\mu]\{\ddot{\Delta}\} + [k]\{\Delta\} = \{W(t)\} \tag{3-29}$$

where $\{W(t)\} = \{W_1(t), W_2(t), \ldots, W_m(t)\}$ represents the *force and/or moment vector functions* of time.

Normalized Nodal Equations

In terms of the normal coordinates (*3-23*), and after the premultiplication by $[h]^T$ as in (*3-25*), (*3-29*) becomes

$$\underbrace{[h]^T[\mu][h]}_{[I]}[\ddot{\eta}] + \underbrace{[h]^T[k][h]}_{[\Lambda]}\{\eta\} = [h]^T[W(t)] \tag{3-30}$$

which reduces by (*3-20, 22*) to *m independent nodal equations* shown below.

$$\underbrace{\begin{bmatrix} \ddot{\eta}_1 \\ \ddot{\eta}_2 \\ \cdots \\ \ddot{\eta}_m \end{bmatrix}}_{[\ddot{\eta}]} + \underbrace{\begin{bmatrix} \omega_1^2 & & & \\ & \omega_2^2 & & \\ & & \ddots & \\ & & & \omega_m^2 \end{bmatrix}}_{[\Lambda]} \underbrace{\begin{bmatrix} \eta_1 \\ \eta_2 \\ \cdots \\ \eta_m \end{bmatrix}}_{[\eta]} = \underbrace{\begin{bmatrix} h_{11} & h_{21} & \cdots & h_{m1} \\ h_{12} & h_{22} & \cdots & h_{m2} \\ \cdots & \cdots & \cdots & \cdots \\ h_{1m} & h_{2m} & \cdots & h_{mm} \end{bmatrix}}_{[h]^T} \underbrace{\begin{bmatrix} W_1(t) \\ W_2(t) \\ \cdots \\ W_m(t) \end{bmatrix}}_{W(t)} \tag{3-31}$$

Initial Conditions

The solution of (*3-31*) in *symbolic form* is

$$\{\eta\} = \text{Diag. }[\cos \omega t]\{\eta_0\} + \text{Diag. }[\sin \omega t]\left\{\frac{\eta_0}{\omega}\right\} + \{\{h_{1k}\Phi_{1k}\} + \{h_{2k}\Phi_{2k}\} + \cdots + \{h_{mk}\Phi_{mk}\}\} \tag{3-32}$$

where $\{\eta_0\}$, $\{\dot{\eta}_0\}$ are the *initial conditions* and

$$\Phi_{jk} = \frac{1}{\omega_k}\int_0^t \sin[\omega_k(t-\tau)]W_j(\tau)\,d\tau \tag{3-33}$$

with $j = 1, 2, \ldots, m$ and $k = 1, 2, \ldots, m$.

For the *inclusion of the initial conditions* given in the original coordinates, the form of (*3-28*) is retained, i.e.,

$$\{\Delta\} = [h]\,\text{Diag.}\,[\cos\omega t][h]^{-1}\{\Delta_0\} + [h]\,\text{Diag.}\,[\sin\omega t]\,\text{Diag.}\left[\frac{1}{\omega}\right][h]^{-1}\{\dot{\Delta}_0\}$$
$$+ [h]\{\{h_{1k}\Phi_{1k}\} + \{h_{2k}\Phi_{2k}\} + \cdots + \{h_{mk}\Phi_{mk}\}\} \qquad (3\text{-}34)$$

as shown in Problem 3.7.

3.6 FREE AND FORCED VIBRATION WITH DAMPING

Nodal Equations

Finally, when damping is involved, the *nodal equations* assume the form of

$$[\mu]\{\ddot{\Delta}\} + [c]\{\dot{\Delta}\} + [k]\{\Delta\} = \{W(t)\} \qquad (3\text{-}35)$$

where $[c]$ is a symmetrical $(m \times m)$ matrix of damping coefficients and $\{W(t)\}$ is the load vector which is zero in case of free vibration.

Normalized Nodal Equations

In terms of *normal coordinates* (*3-23*) and with transformations similar to those of (*3-30*), (*3-35*) becomes

$$\underbrace{[h]^T[\mu][h]}_{[I]}\{\ddot{\eta}\} + \underbrace{[h]^T[c][h]}_{[2\Gamma]}\{\dot{\eta}\} + \underbrace{[h]^T[k][h]}_{[\Lambda]}\{\eta\} = [h]^T\{W(t)\} \qquad (3\text{-}36)$$

where $[2\Gamma]$ is a diagonal matrix of damping coefficients each corresponding to one node. The numerical values of these coefficients must be assumed or determined by experiments. The formulation of expressions for the coefficients of damping is still in research state and much work remains to be done.

The transformed matrix equation (*3-36*) is again a set of *m independent nodal equations* shown below.

$$\underbrace{\begin{bmatrix}\ddot{\eta}_1\\ \ddot{\eta}_2\\ \cdots\\ \ddot{\eta}_m\end{bmatrix}}_{\{\ddot{\eta}\}} + \underbrace{\begin{bmatrix}2\gamma_1 & & & \\ & 2\gamma_2 & & \\ & & \ddots & \\ & & & 2\gamma_m\end{bmatrix}}_{[2\Gamma]}\underbrace{\begin{bmatrix}\dot{\eta}_1\\ \dot{\eta}_2\\ \cdots\\ \dot{\eta}_m\end{bmatrix}}_{\{\dot{\eta}\}} + \underbrace{\begin{bmatrix}\omega_1^2 & & & \\ & \omega_2^2 & & \\ & & \ddots & \\ & & & \omega_m^2\end{bmatrix}}_{[\Lambda]}\underbrace{\begin{bmatrix}\eta_1\\ \eta_2\\ \cdots\\ \eta_m\end{bmatrix}}_{\{\eta\}}$$
$$= \underbrace{\begin{bmatrix}h_{11} & h_{21} & \cdots & h_{m1}\\ h_{12} & h_{22} & \cdots & h_{m2}\\ \cdots & \cdots & \cdots & \cdots\\ h_{1m} & h_{2m} & \cdots & h_{mm}\end{bmatrix}}_{[h]^T}\underbrace{\begin{bmatrix}W_1(t)\\ W_2(t)\\ \cdots\\ W_m(t)\end{bmatrix}}_{W(t)} \qquad (3\text{-}37)$$

Initial Conditions

The solution of each equation in (*3-37*) takes one of the forms of (*2-16*) depending on γ, ω, and λ. For the *underdamped motion*, $\gamma^2 < \omega^2$, $\lambda = \sqrt{\omega^2 - \gamma^2}$,

$$\{\eta\} = \text{Diag.}\,[(e^{-\gamma t}/\lambda)(\lambda\cos\lambda t + \gamma\sin\lambda t)]\{\eta_0\}$$
$$+ \text{Diag.}\,(e^{-\gamma t})(\sin\lambda t)\left\{\frac{\dot{\eta}_0}{\lambda}\right\} + \{\{h_{1k}\Phi_{1k}\} + \{h_{2k}\Phi_{2k}\} + \cdots + \{h_{mk}\Phi_{mk}\}\} \qquad (3\text{-}38)$$

where again $\{\eta_0\}$ and $\{\dot{\eta}_0\}$ are the *initial conditions* and

$$\Phi_{jk} = \frac{1}{\lambda_k}\int_0^t e^{-\gamma_k(t-\tau)}\sin\lambda_k(t-\tau)W_j(\tau)\,d\tau \qquad (3\text{-}39)$$

with $j = 1, 2, \ldots, m$ and $k = 1, 2, \ldots, m$.

Then as in Sec. 3.5, in terms of the *original coordinates*,

$$\{\Delta\} = [h]\,\text{Diag.}\,[(e^{-\gamma t}/\lambda)(\lambda\cos\lambda t + \gamma\sin\lambda t)][h]^{-1}\{\Delta_0\}$$
$$+ [h]\,\text{Diag.}\,[(e^{-\gamma t})(\sin\lambda t)]\,\text{Diag.}\left[\frac{1}{\lambda}\right][h]^{-1}\{\dot{\Delta}_0\}$$
$$+ [h]\{\{h_{1k}\Phi_{1k}\} + \{h_{2k}\Phi_{2k}\} + \cdots + \{h_{mk}\Phi_{mk}\}\} \tag{3-40}$$

which is similar to (*3-34*) as shown in Problem 3.8.

3.7 TRANSPORT MATRICES

State Vectors

The construction of transport matrices for the systems of this chapter is simplified if it is related to the *normal state vectors*, i.e., if equations (*3-32*) and (*3-38*) are utilized.

The *dimensionless state vector* becomes then

$$\begin{aligned}\hat{\hat{H}}_R &= \{+1, \eta_{1,R}, \eta_{2,R}, \ldots, \eta_{m,R}, \dot{\eta}_{1,R}, \dot{\eta}_{2,R}, \ldots, \dot{\eta}_{m,R}\} \\ \hat{\hat{H}}_L &= \{+1, \eta_{1,L}, \eta_{2,L}, \ldots, \eta_{m,L}, \dot{\eta}_{1,L}, \dot{\eta}_{2,L}, \ldots, \dot{\eta}_{m,L}\}\end{aligned} \tag{3-41}$$

where R and L designate the right and left ends of the time interval and the right end and left end conditions are grouped in column submatrices.

Transport Matrix

The *dimensionless time transport matrix* is constructed by means of the respective equations of motion to conform with the state vectors given by (*3-41*). Since the initial conditions are given in the general system, $[h]^{-1}$ must be used for their inclusion as shown in Problems 3.9 and 3.10.

Solved Problems

FREE VIBRATION WITHOUT DAMPING

3.1. A three-story, one-bay shear building (Ref. 3.2) is analyzed for natural frequencies, natural periods, and natural modes. The structural properties shown in Fig. P-3.1*a* are $l_1 = l_2 = l_3 = 4.27$ m, $h = 6.71$ m, and $w_1 = w_2 = w_3 = 18.009$ kN/m. Assume that the weight on each floor includes the superimposed mass and the mass of the girder. Neglect the masses of columns.

For a shear building the girders are infinitely rigid and the structural joints are not rotatable. If the axial deformations of columns are assumed to be negligible, then each mass will have two constraints along the directions of rotation and vertical displacement. From (*3-1*) the degrees of freedom of the system are

$$m = 3j - r = 3(3) - 2(3) = 3$$

As to the governing differential equations of motion, let us consider a typical girder i having mass μ_i and subject to a displacement u_i. If the stiffnesses of the columns below and above the girder are respectively signified by k_i and k_{i+1}, then the differential equation of motion can be established from Newton's second law as

$$\mu_i\ddot{\Delta}_i = -k_i(\Delta_i - \Delta_{i-1}) + k_{i+1}(\Delta_{i+1} - \Delta_i) \tag{1}$$

Following (*1*) we may write the equations for the present problem as

$$\begin{aligned}\mu_1\ddot{\Delta}_1 &= -k_1(\Delta_1) + k_2(\Delta_2 - \Delta_1) \\ \mu_2\ddot{\Delta}_2 &= -k_2(\Delta_2 - \Delta_1) + k_3(\Delta_3 - \Delta_2) \\ \mu_3\ddot{\Delta}_3 &= -k_3(\Delta_3 - \Delta_2)\end{aligned} \tag{2}$$

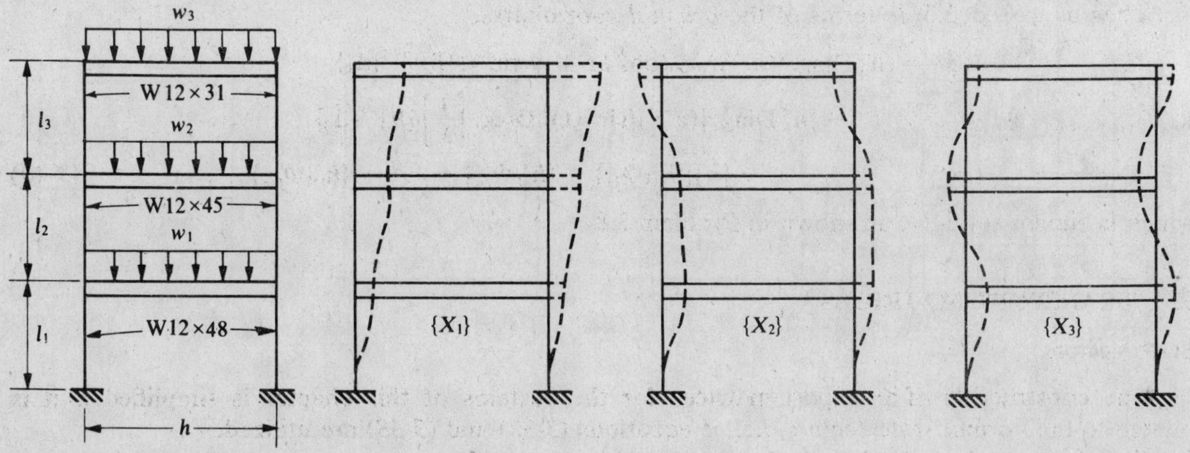

Fig. P-3.1

In matrix notation,

$$\begin{bmatrix} \mu_1 & 0 & 0 \\ 0 & \mu_2 & 0 \\ 0 & 0 & \mu_3 \end{bmatrix} \begin{bmatrix} \ddot{\Delta}_1 \\ \ddot{\Delta}_2 \\ \ddot{\Delta}_3 \end{bmatrix} = \begin{bmatrix} -(k_1 + k_2) & k_2 & 0 \\ k_2 & -(k_2 + k_3) & k_3 \\ 0 & k_3 & -k_3 \end{bmatrix} \begin{bmatrix} \Delta_1 \\ \Delta_2 \\ \Delta_3 \end{bmatrix} \tag{3}$$

As stated in (3-4), the general solution of (3) may be assumed to be $\Delta_i = X_i \cos(\omega t - \alpha)$. Using the solution in (3) yields

$$\begin{bmatrix} (k_1 + k_2) - \mu_1\omega^2 & -k_2 & 0 \\ -k_2 & (k_2 + k_3) - \mu_2\omega^2 & -k_3 \\ 0 & -k_3 & k_3 - \mu_3\omega^2 \end{bmatrix} \begin{bmatrix} X_1 \\ X_2 \\ X_3 \end{bmatrix} \cos(\omega t - \alpha) = \begin{bmatrix} 0 \\ 0 \\ 0 \end{bmatrix} \tag{4}$$

where $\mu_1 = \mu_2 = \mu_3 = (18.009)(6.71)/9.81 = 12.318 \text{ kN-s}^2/\text{m}$, $k_1 = 12EI_1(2)/l_1^3$, $k_2 = 12EI_2(2)/l_2^3$, and $k_3 = 12EI_3(2)l_3^3$. Substituting $I_1 = 1.982 \text{ dm}^4$, $I_2 = 1.460 \text{ dm}^4$, $I_3 = 0.992 \text{ dm}^4$, $E = 206.84 \text{ GN/m}^2$, and $l_1 = l_2 = l_3 = 4.27$ m, we have $k_1 = 12.660$ MN/m, $k_2 = 9.329$ MN/m, and $k_3 = 6.340$ MN/m. For rigid frames with joint rotations the reader may consult standard texts (Ref. 3.1, pp. 130–134) for construction of spring matrix. From (3-6) the frequency determinant yields

$$\omega^6 - 3571.85\,\omega^4 + 3\,005\,746.235\,\omega^2 - 400\,624\,381 = 0 \tag{5}$$

A number of numerical methods are available for solving the characteristic values of polynomial equations as shown in (5), namely Bocher's, Krylov's, and Danilevsky's methods (Refs. 3.3 and 3.4). The roots of (5) yield the natural angular frequencies (ω) and their corresponding natural frequencies (f) and natural periods (T) as

$$\begin{array}{lll} \omega_1^2 = 163.84 & \omega_2^2 = 1030.41 & \omega_3^2 = 2380.464 \\ \omega_1 = 12.80 \text{ rad/s} & \omega_2 = 32.10 \text{ rad/s} & \omega_3 = 48.79 \text{ rad/s} \\ f_1 = 2.037 \text{ cps} & f_2 = 5.109 \text{ cps} & f_3 = 7.765 \text{ cps} \\ T_1 = 0.493 \text{ s} & T_2 = 0.196 \text{ s} & T_3 = 0.129 \text{ s} \end{array}$$

Substituting ω_1^2, ω_2^2, and ω_3^2 into (4) will respectively yield the first, second, and third natural modes as

$$\{X_1\} = \begin{bmatrix} X_{11} \\ 2.140\,X_{11} \\ 3.135\,X_{11} \end{bmatrix} \qquad \{X_2\} = \begin{bmatrix} X_{12} \\ 0.9948\,X_{12} \\ -0.9955\,X_{12} \end{bmatrix} \qquad \{X_3\} = \begin{bmatrix} X_{13} \\ -0.7787\,X_{13} \\ 0.2161\,X_{13} \end{bmatrix}$$

which are graphically presented in Figs. P-3.1b, c, d.

3.2. The shear building given in Problem 3.1 is modified to have roller supports as shown in Fig. P-3.2a for representing a semidefinite system. Determine the natural frequencies, natural periods, and natural modes.

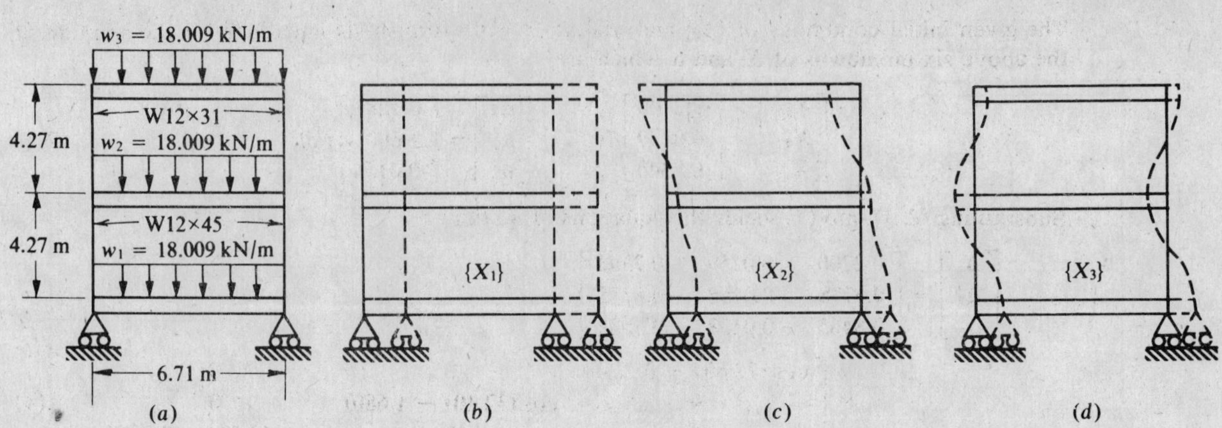

Fig. P-3.2

From (*1, 2*) of Problem 3.1 the governing differential equations may be expressed as

$$\begin{bmatrix} \mu_1 & 0 & 0 \\ 0 & \mu_2 & 0 \\ 0 & 0 & \mu_3 \end{bmatrix} \begin{bmatrix} \ddot{\Delta}_1 \\ \ddot{\Delta}_2 \\ \ddot{\Delta}_3 \end{bmatrix} = \begin{bmatrix} -k_2 & k_2 & 0 \\ k_2 & -(k_2+k_3) & k_3 \\ 0 & k_3 & -k_3 \end{bmatrix} \begin{bmatrix} \Delta_1 \\ \Delta_2 \\ \Delta_3 \end{bmatrix} \tag{1}$$

Using the general solution (*3-4*), we have

$$\begin{bmatrix} k_2 - \mu_1\omega^2 & -k_2 & 0 \\ -k_2 & (k_2+k_3) - \mu_2\omega^2 & -k_3 \\ 0 & -k_3 & k_3 - \mu_3\omega^2 \end{bmatrix} \begin{bmatrix} X_1 \\ X_2 \\ X_3 \end{bmatrix} = 0 \tag{2}$$

where $\mu_1 = \mu_2 = \mu_3 = 12.318\ \text{kN-s}^2/\text{m}$, $k_2 = 9.329\ \text{MN/m}$, and $k_3 = 6.340\ \text{MN/m}$ as shown in Problem 3.1. Employing (*3-6*) gives the following frequency determinant:

$$\det.(\omega^2) = \omega^2(\omega^4 - 2544.084\,\omega^2 + 1\,169\,406.918) = 0 \tag{3}$$

of which the roots are $\omega_1^2 = 0 \qquad \omega_2^2 = 602.203 \qquad \omega_3^2 = 1941.881$

Thus

$$\begin{array}{lll} \omega_1 = 0 \text{ (rigid-body motion)} & \omega_2 = 24.583\ \text{rad/s} & \omega_3 = 44.079\ \text{rad/s} \\ f_1 = 0 & f_2 = 3.912\ \text{cps} & f_3 = 7.015\ \text{cps} \\ T_1 = \infty & T_2 = 0.256\ \text{s} & T_3 = 0.143\ \text{s} \end{array}$$

Substituting ω_1^2, ω_2^2, and ω_3^2 respectively into (*2*) yields

$$\{X_1\} = \begin{bmatrix} X_{11} \\ X_{11} \\ X_{11} \end{bmatrix} \qquad \{X_2\} = \begin{bmatrix} X_{12} \\ 0.203\,X_{12} \\ -1.195\,X_{12} \end{bmatrix} \qquad \{X_3\} = \begin{bmatrix} X_{13} \\ -1.561\,X_{13} \\ 0.565\,X_{13} \end{bmatrix}$$

which are natural modes and shown in Figs. P-3.2*b, c, d* where $\{X_1\}$ is rigid-body motion.

3.3. Find the equations of motion for Problems 3.1 and 3.2 when the initial conditions are $\{\Delta_0\} = \{1, 0, 1\}$ ft and $\{\dot{\Delta}_0\} = \{1.5, 1, 1\}$ ft/s. Determine the displacements at $t = 1/4$ s in U.S. customary units.

(*a*) For Problem 3.1, substituting the values of ω and $\{X\}$ of Problem 3.1 in (*3-7*) yields the general solution

$$\{\Delta\} = \sum_{p=1}^{m} \{X_p\} \cos(\omega_p t - \alpha_p) = \begin{bmatrix} X_{11} & X_{12} & X_{13} \\ 2.140\,X_{11} & 0.9948\,X_{12} & -0.7787\,X_{13} \\ 3.135\,X_{11} & -0.9955\,X_{12} & 0.2161\,X_{13} \end{bmatrix}$$

$$\times \begin{bmatrix} \cos(12.80\,t - \alpha_1) & 0 & 0 \\ 0 & \cos(32.10\,t - \alpha_2) & 0 \\ 0 & 0 & \cos(48.79\,t - \alpha_3) \end{bmatrix} \tag{1}$$

The given initial conditions of $\{\Delta_0\}$ and $\{\dot{\Delta}_0\}$ at $t = 0$ furnish six equations for the solution of the above six unknowns of X and α which are

$$\begin{bmatrix} X_{11} \\ X_{12} \\ X_{13} \end{bmatrix} = \begin{bmatrix} 0.2700 \\ 0.0159 \\ 0.7340 \end{bmatrix} \text{ft} \qquad \begin{bmatrix} \alpha_1 \\ \alpha_2 \\ \alpha_3 \end{bmatrix} = \begin{bmatrix} 0.1286 \\ 1.680 \\ 0.0154 \end{bmatrix} \text{rad} \tag{2, 3}$$

Substituting (2, 3) into (1) yields the equations of motion

$$\begin{bmatrix} \Delta_1 \\ \Delta_2 \\ \Delta_3 \end{bmatrix} = \begin{bmatrix} 0.2700 & 0.0159 & 0.7340 \\ 0.5778 & 0.0158 & -0.5715 \\ 0.8465 & -0.0158 & 0.1586 \end{bmatrix} \times \begin{bmatrix} \cos(12.80t - 0.1286) & 0 & 0 \\ 0 & \cos(32.10t - 1.680) & 0 \\ 0 & 0 & \cos(48.79t - 0.0154) \end{bmatrix} \tag{4}$$

The displacements at $t = 1/4$ s may be obtained from above as

$$\{\Delta\} = \{0.426, -1.096, -0.714\} \text{ ft} \tag{5}$$

(*b*) For Problem 3.2: Using the values of ω and $\{X\}$ of Problem 3.2 in (3-7) we have

$$\begin{bmatrix} \Delta_1 \\ \Delta_2 \\ \Delta_3 \end{bmatrix} = \begin{bmatrix} X_{11} & X_{12} & X_{13} \\ X_{11} & 0.203X_{12} & -1.561X_{13} \\ X_{11} & -1.195X_{12} & 0.565X_{13} \end{bmatrix} \begin{bmatrix} 1 + \dfrac{b_1}{X_{11}}t & 0 & 0 \\ 0 & \cos(24.583t - \alpha_2) & 0 \\ 0 & 0 & \cos(44.079t - \alpha_3) \end{bmatrix} \tag{6}$$

where the solution of Δ_i for the first mode is a polynomial equation of the form $X_{11} + b_1 t$ because ω_1 is zero in $\ddot{\Delta}_1 + \omega_1^2 \Delta_i = 0$. Since $\{\Delta_0\}$ and $\{\dot{\Delta}_0\}$ are known at $t = 0$, the constants of integration in (6) may be determined and are given below.

$$\begin{bmatrix} X_{11} \\ X_{12} \\ X_{13} \end{bmatrix} = \begin{bmatrix} 0.6663 \\ 0.0828 \\ 0.04161 \end{bmatrix} \text{ft} \qquad \begin{bmatrix} b_1 \\ \alpha_2 \\ \alpha_3 \end{bmatrix} = \begin{bmatrix} 1.1666 \text{ ft/s} \\ 3.0424 \text{ rad} \\ 0.00721 \text{ rad} \end{bmatrix} \tag{7, 8}$$

Hence the equations of motion in (6) become

$$\begin{bmatrix} \Delta_1 \\ \Delta_2 \\ \Delta_3 \end{bmatrix} = \begin{bmatrix} 0.6663 & 0.0828 & 0.04161 \\ 0.6663 & 0.0168 & -0.06495 \\ 0.6663 & -0.9895 & 0.02351 \end{bmatrix} \times \begin{bmatrix} 1 + 1.7509t & 0 & 0 \\ & \cos(24.583t - 3.0424) & 0 \\ 0 & 0 & \cos(44.079t - 0.00721) \end{bmatrix} \tag{9}$$

from which the displacements at $t = 1/4$ s are

$$\{\Delta\} = \{0.8823, 0.9301, 1.0608\} \text{ ft} \tag{10}$$

PROPERTIES OF NATURAL MODES

3.4. Derive the orthogonality and normalization conditions given in (3-13, 14, 15, 16).

Consider any two characteristic vectors corresponding to the *i*th and *j*th modes. Then we may write (3-5) as

$$[k]\{X_i\} = \omega_i^2[\mu]\{X_i\} \qquad \text{and} \qquad [k]\{X_j\} = \omega_j^2[\mu]\{X_j\} \tag{1, 2}$$

If we postmultiply the transpose of (1) by $\{X_j\}$ and premultiply (2) by $\{X_i\}^T$, we have

$$([k]\{X_i\})^T\{X_j\} = (\omega_i^2[\mu]\{X_i\})^T\{X_j\} \tag{3}$$

$$\{X_i\}^T([k]\{X_j\}) = \{X_i\}^T(\omega_j^2[\mu]\{X_j\}) \tag{4}$$

and using the definition of matrix transpose we may rewrite (*3, 4*) as

$$\{X_i\}^T[k]^T\{X_j\} = \omega_i^2\{X_i\}^T[\mu]^T\{X_j\} \tag{5}$$

$$\{X_i\}^T[k]\{X_j\} = \omega_j^2\{X_i\}^T[\mu]\{X_j\} \tag{6}$$

Since the mass and stiffness matrices are symmetrical $[\mu] = [\mu]^T$, $[k] = [k]^T$, the left-hand sides of (*5, 6*) are equal. Subtracting (*5*) from (*6*) yields

$$(\omega_i^2 - \omega_j^2)\{X_i\}^T[\mu]\{X_j\} = 0 \tag{7}$$

Since $\omega_i \neq \omega_j$,

$$\{X_i\}^T[\mu]\{X_j\} = 0 \qquad \text{and} \qquad \{X_i\}^T[k]\{X_j\} = 0 \tag{8, 9}$$

which are the first and second orthogonality conditions (*3-13, 14*).

From (*5, 6*) it is apparent that for $i = j$, $\omega_i = \omega_j$ and $\{X_i\} = \{X_j\}$. Subtracting (*5*) from (*6*) yields $\{X_i\}^T[\mu]\{X_i\} \neq 0$, which may be expressed as

$$\{X_i\}^T[\mu]\{X_i\} = L_i^2 \tag{10}$$

Substituting (*10*) into (*5*) or (*6*), we have

$$\{X_i\}^T[k]\{X_i\} = L_i^2\omega_i^2 \tag{11}$$

Letting $\{h_i\} = \{X_i\}/L_i$, (*10* and *11*) become

$$\{h_i\}^T[\mu]\{h_i\} = 1 \qquad \text{and} \qquad \{h_i\}^T[k]\{h_i\} = \omega_i^2 \tag{12, 13}$$

which are the first and second normalization conditions.

3.5. **Using the eigenvalues and eigenvectors obtained in Problem 3.1 find their orthogonality and normalization conditions (*3-19, 20,* and *21, 22*) (derived in Problem 3.4).**

From Problem 3.1 we have

$$[\mu] = \begin{bmatrix} 0.8431 & 0 & 0 \\ 0 & 0.8431 & 0 \\ 0 & 0 & 0.8431 \end{bmatrix} \qquad [k] = \begin{bmatrix} 1506.922 & -639.578 & 0 \\ -639.578 & 1075.074 & -435.496 \\ 0 & -435.496 & 435.496 \end{bmatrix} \tag{1, 2}$$

$$[X] = \begin{bmatrix} X_{11} & X_{12} & X_{13} \\ 2.140X_{11} & 0.9948X_{12} & -0.7787X_{13} \\ 3.135X_{11} & -0.9955X_{12} & 0.2161X_{13} \end{bmatrix} \qquad [\omega^2] = \begin{bmatrix} 163.84 & 0 & 0 \\ 0 & 1030.41 & 0 \\ 0 & 0 & 2380.464 \end{bmatrix} \tag{3, 4}$$

Substituting (*1, 3*) into (*3-19*) results in

$$[X]^T[\mu][X] = \begin{bmatrix} 12.990X_{11}^2 & 0 & 0 \\ 0 & 2.512X_{12}^2 & 0 \\ 0 & 0 & 1.393X_{13}^2 \end{bmatrix} \tag{5}$$

where the zero elements are due to the first orthogonality condition and the diagonal elements yield L_1^2, L_2^2, L_3^2, or $L_1 = 3.604X_{11}$, $L_2 = 1.585X_{12}$, $L_3 = 1.1804X_{13}$ which are then used to divide the eigenvectors in (3) as $\{h_j\} = \{X_j\}/L_j$. Thus

$$[h] = \begin{bmatrix} 0.2775 & 0.6308 & 0.8471 \\ 0.5938 & 0.6276 & -0.6597 \\ 0.8699 & -0.6281 & 0.1831 \end{bmatrix} \tag{6}$$

Using (*1, 6*) for (*3-20*) yields

$$[h]^T[\mu][h] = \begin{bmatrix} 1 & 0 & 0 \\ 0 & 1 & 0 \\ 0 & 0 & 1 \end{bmatrix} \tag{7}$$

which is an identity matrix. The zero off-diagonal elements result from (*5*) and the diagonal elements are due to the first normalization condition.

Substituting (*2, 3*) into (*3-21*) and then performing matrix multiplication we have

$$[X]^T[k][X] = \begin{bmatrix} 2130.057X_{11}^2 & 0 & 0 \\ 0 & 2590.403X_{12}^2 & 0 \\ 0 & 0 & 3320.704X_{13}^2 \end{bmatrix} \tag{8}$$

which shows the second orthogonality condition established in (*3-14*). For the second normalization condition we have

$$[h]^T[k][h] = \begin{bmatrix} 163.90 & 0 & 0 \\ 0 & 1303.80 & 0 \\ 0 & 0 & 2381.00 \end{bmatrix} \tag{9}$$

which approximately conforms with (*4*).

NORMAL COORDINATES

3.6. Use the normalized nodal equations to find the motion equations of Problem 3.1 for which the initial conditions are given in Problem 3.3. Find the displacements at $t = 0.25$ s.

From (*3-28*) the general expression of motion equation is

$$\{\Delta\} = [h]\,\text{Diag.}\,[\cos\omega t][h]^{-1}\{\Delta_0\} + [h]\,\text{Diag.}\,[\sin\omega t][1/\omega][h]^{-1}\{\dot{\Delta}_0\} \tag{1}$$

in which the inverse of $[h]$ is sometimes quite laborious and may not be necessary if we use $[h]^T[\mu]$ instead of $[h]^{-1}$ as shown in (*3-12*). Thus (*1*) may be rewritten as

$$\{\Delta\} = [h]\,\text{Diag.}\,[\cos\omega t][h]^T[\mu]\{\Delta_0\} + [h]\,\text{Diag.}\,[\sin\omega t][1/\omega][h]^T[\mu]\{\dot{\Delta}_0\} \tag{2}$$

From Problems 3.3 and 3.5 we have

$$\{\Delta_0\} = \begin{bmatrix} 1 \\ 0 \\ 1 \end{bmatrix}\text{ ft} \qquad \{\dot{\Delta}_0\} = \begin{bmatrix} 1.5 \\ 1.0 \\ 1.0 \end{bmatrix}\text{ ft/s} \qquad [h] = \begin{bmatrix} 0.2775 & 0.6308 & 0.8471 \\ 0.5938 & 0.6276 & -0.6579 \\ 0.8699 & -0.6281 & 0.1831 \end{bmatrix}(1/\text{s})\sqrt{\text{ft/k}} \tag{3}$$

$$[\mu] = \begin{bmatrix} 0.8431 & 0 & 0 \\ 0 & 0.8431 & 0 \\ 0 & & 0.8431 \end{bmatrix}\text{ k-s}^2/\text{ft} \qquad \{\omega\} = \begin{bmatrix} 12.80 \\ 32.10 \\ 48.79 \end{bmatrix}\text{ rad/s} \tag{4}$$

After substitution and manipulation of (*2*) we have the following motion equations:

$$\begin{bmatrix} \Delta_1 \\ \Delta_2 \\ \Delta_3 \end{bmatrix} = \begin{bmatrix} 0.2688 & 0.0014 & 0.7358 \\ 0.5744 & 0.0014 & -0.5730 \\ 0.8415 & -0.0014 & 0.1590 \end{bmatrix}\begin{bmatrix} \cos 12.80t \\ \cos 32.10t \\ \cos 48.79t \end{bmatrix} + \begin{bmatrix} 0.0344 & 0.0156 & 0.0116 \\ 0.0735 & 0.0156 & -0.0090 \\ 0.1078 & -0.0156 & 0.0025 \end{bmatrix}\begin{bmatrix} \sin 12.80t \\ \sin 32.10t \\ \sin 48.79t \end{bmatrix} \tag{5}$$

from which the displacements at $t = 1/4$ s are

$$\{\Delta\} = \{0.426, -1.096, -0.714\}\text{ ft} \tag{6}$$

which conforms with (*5*) of Problem 3.3.

FORCED VIBRATION WITHOUT DAMPING

3.7. The structure given in Problem 3.1 is subject to time-dependent forces as $W_1(t) = 100(1-2t)$, $W_2(t) = 80(1-2t)$ k, $W_3(t) = 50(1-2t)$ k applied respectively to μ_1, μ_2, and μ_3. Using the normalized nodal matrix formulation find the displacements at $t = 1/4$ s for (*a*) forced vibration only, and (*b*) free and forced vibrations. Assume the initial conditions to be the same as given in Problems 3.3 and 3.6.

(*a*) From (*3-34*), the equation of motion without initial conditions may be written as

$$\{\Delta\} = [h]\{\{h_{1k}\Phi_{1k}\} + \{h_{2k}\Phi_{2k}\} + \cdots + \{h_{mk}\Phi_{mk}\}\} \tag{1}$$

where Φ_{jk} is the convolution integral shown in (*3-33*).

The particular form Φ_{jk} for a triangular pulse was typically given in (5) of Problem 2.9 as

$$\Phi = \frac{P}{a\omega^3}[\omega(a - t) - \omega a \cos \omega t + \sin \omega t] \tag{2}$$

and may be expressed in the following form according to (1):

$$\Phi_{jk} = \frac{P_j}{a\omega_k^3}[\omega_k(a - t) - \omega_k a \cos \omega_k t + \sin \omega_k t] \tag{3}$$

where $j = 1, 2, 3$ $\quad P_1 = 100\text{ k}$ $\quad \omega_1 = 12.80\text{ rad/s}$

$k = 1, 2, 3$ $\quad P_2 = 80\text{ k}$ $\quad \omega_2 = 32.10\text{ rad/s}$

$a = 1/2\text{ s}$ $\quad P_3 = 50\text{ k}$ $\quad \omega_3 = 48.79\text{ rad/s}$

Evaluating (3) for $t = 1/4\text{ s}$ we have

$$\begin{bmatrix} \Phi_{11} & \Phi_{21} & \Phi_{31} \\ \Phi_{12} & \Phi_{22} & \Phi_{32} \\ \Phi_{13} & \Phi_{23} & \Phi_{33} \end{bmatrix} = \begin{bmatrix} 0.9089 & 0.7271 & 0.4544 \\ 0.0710 & 0.0568 & 0.0355 \\ -0.0188 & -0.0150 & -0.0094 \end{bmatrix} \tag{4}$$

Substituting (4) and $[h]$ shown in (3) of Problem 3.6 into (1) yields the following displacements:

$$\{\Delta\} = \{0.3296, 0.6825, 0.9010\}\text{ ft} \tag{5}$$

(b) For free and forced vibrations the equation of motion consists of the motion due to initial conditions and the motion due to forces as shown in (3-34). Since these motions were separately investigated in the preceding problem and above, we may write the total motion as

$$\{\Delta\} = \{\Delta\}(\text{Eq. (6) of Prob. 3.6}) + \{\Delta\}(\text{Eq. (5) above}) = \begin{bmatrix} 0.426 \\ -1.096 \\ -0.714 \end{bmatrix} + \begin{bmatrix} 0.330 \\ 0.683 \\ 0.901 \end{bmatrix} = \begin{bmatrix} 0.756 \\ -0.413 \\ 0.187 \end{bmatrix}\text{ ft} \tag{6}$$

FREE AND FORCED VIBRATION WITH DAMPING

3.8. The structure of Problem 3.1 has forcing functions similar to Fig. P-2.9a as $P_1(t) = 10\text{ k}$, $P_2(t) = 8\text{ k}$, and $P_3(t) = 5\text{ k}$ applied respectively to masses μ_1, μ_2, and μ_3 for $t \ge 0$. Assume the damping constants corresponding to the three masses to be $c_1 = 0.05\, c_{cr}$, $c_2 = 0.1\, c_{cr}$, $c_3 = 0.12\, c_{cr}$. Find the displacements at $t = 1/4\text{ s}$ for (a) forced vibration only and (b) free and forced vibrations. The initial conditions are the same as given in Problems 3.3, 3.6, and 3.7.

(a) For forced vibration the equation of motion may be obtained from (3-40) exclusive of the terms associated with $\{\Delta_0\}$ and $\{\dot{\Delta}_0\}$ as

$$\{\Delta\} = [h]\{\{h_{1k}\Phi_{1k}\} + \{h_{2k}\Phi_{2k}\} + \cdots + \{h_{mk}\}\{\Phi_{mk}\}\} \tag{1}$$

in which the particular function Φ_{jk} for the present forcing function was given in (2) of Problem 2.21 and may be written as

$$\Phi_{jk} = \frac{P_j}{\lambda_k(\gamma_k^2 + \lambda_k^2)}[\lambda_k - e^{-\gamma_k t}(\gamma_k \sin \lambda_k t + \lambda_k \cos \lambda_k t)] \tag{2}$$

where $j = 1, 2, 3$; $k = 1, 2, 3$; $P_1 = 10\text{ k}$; $P_2 = 8\text{ k}$; and $P_3 = 5\text{ k}$. From (2-16)

$$\begin{aligned} \gamma_1 &= c_1/2\mu_1 = 0.05\,\omega_1 = 0.640\text{ rad/s} \\ \gamma_2 &= c_2/2\mu_2 = 0.1\,\omega_2 = 3.2135\text{ rad/s} \\ \gamma_3 &= c_3/2\mu_3 = 0.12\,\omega_3 = 5.852\text{ rad/s} \end{aligned} \tag{3}$$

where the circular frequencies were obtained in Problem 3.1a as $\omega_1 = 12.80\text{ rad/s}$, $\omega_2 = 32.10\text{ rad/s}$, $\omega_3 = 48.79\text{ rad/s}$. Thus $\lambda_i = \sqrt{\omega_i^2 - \gamma_i^2}$ may be calculated as

$$\begin{aligned} \lambda_1 &= \sqrt{\omega_1^2 - (0.05\,\omega_1)^2} = 12.783 \\ \lambda_2 &= \sqrt{\omega_2^2 - (0.1\,\omega_2)^2} = 31.975 \\ \lambda_3 &= \sqrt{\omega_3^2 - (0.12\,\omega_3)^2} = 48.417 \end{aligned} \tag{4}$$

For a typical representation of Φ_{jk}, let us substitute (3, 4) into (2) for Φ_{12}. Thus

$$\Phi_{12} = \frac{10}{31.975[(3.2135)^2 + (31.975)^2]}[31.975 - e^{-3.2135t}(3.2135 \sin 31.975t + 31.975 \cos 31.975t)] \tag{5}$$

At $t = 1/4$ s, Φ_{jk} may be obtained as

$$\begin{bmatrix} \Phi_{11} & \Phi_{21} & \Phi_{31} \\ \Phi_{12} & \Phi_{22} & \Phi_{32} \\ \Phi_{13} & \Phi_{23} & \Phi_{33} \end{bmatrix} = \begin{bmatrix} 0.1127 & 0.0902 & 0.0563 \\ 0.0099 & 0.0079 & 0.0049 \\ 0.0034 & 0.0027 & 0.0017 \end{bmatrix} \tag{6}$$

Substituting (6) and $[h]$ given in (3) of Problem 3.6 into (1) yields

$$\{\Delta\} = \{0.0434, 0.0836, 0.1116\} \text{ ft} \tag{7}$$

(b) The equation associated with initial conditions may be obtained from (3-40) as

$$\{\Delta\} = \underbrace{[h]\,\text{Diag.}\,[(e^{-\gamma t}/\lambda)(\lambda \cos \lambda t + \gamma \sin \lambda t)]}_{[A]}\underbrace{[h]^T[\mu]\{\Delta_0\}}_{[B]} + [h]\underbrace{\text{Diag.}\,[e^{-\gamma t} \sin \lambda t]}_{[C]}\underbrace{\text{Diag.}\,[1/\lambda][h]^T[\mu]\{\dot{\Delta}_0\}}_{[D]} \tag{8}$$

Using the given $[h]$, $[\mu]$, $\{\Delta_0\}$, $\{\dot{\Delta}_0\}$, and (3, 4), we may evaluate (8) for $t = 1/4$ s as follows:

$$[A] = \begin{bmatrix} -0.85306 & 0 & 0 \\ 0 & -0.01785 & 0 \\ 0 & 0 & 0.19479 \end{bmatrix} \qquad \{B\} = \begin{bmatrix} 0.96737 \\ 0.00225 \\ 0.86857 \end{bmatrix} \tag{9}$$

$$[C] = \begin{bmatrix} -0.04610 & 0 & 0 \\ 0 & 0.44343 & 0 \\ 0 & 0 & -0.10320 \end{bmatrix} \qquad \{D\} = \begin{bmatrix} 0.12312 \\ 0.02493 \\ 0.03040 \end{bmatrix} \tag{10}$$

Thus (8) becomes

$$\{\Delta\} = [h]([A]\{B\} + [C]\{D\}) = \begin{bmatrix} -0.08296 \\ -0.59600 \\ -0.69931 \end{bmatrix} \text{ ft} \tag{11}$$

which are the displacements of free vibration. Adding (7) and (11) gives the total displacements

$$\{\Delta\} = \begin{bmatrix} 0.0434 \\ 0.0836 \\ 0.1116 \end{bmatrix} + \begin{bmatrix} -0.08296 \\ -0.59600 \\ -0.69931 \end{bmatrix} = \begin{bmatrix} -0.0395 \\ -0.5124 \\ -0.5877 \end{bmatrix} \text{ ft} \tag{12}$$

TRANSPORT MATRICES

3.9. Construct the transport matrix for Problem 3.7.

Let the last term on the right side of (3-32) be represented as

$$\{\rho\} = \{h_{1k}\Phi_{1k}\} + \{h_{2k}\Phi_{2k}\} + \cdots + \{h_{mk}\Phi_{mk}\} \tag{1}$$

Then (3-32) becomes

$$\underbrace{\begin{bmatrix} 1 \\ \hline \eta \\ \dfrac{\dot{\eta}}{\omega} \end{bmatrix}}_{\hat{\hat{H}}_R} = \underbrace{\left[\begin{array}{c|cc} 1 & 0 & 0 \\ \hline \{\rho\} & [\cos \omega t] & [\sin \omega t] \\ \left\{\dfrac{\dot{\rho}}{\omega}\right\} & [-\sin \omega t] & [\cos \omega t] \end{array}\right]}_{\hat{\hat{T}}_{RL}} \underbrace{\begin{bmatrix} 1 \\ \hline \eta \\ \dfrac{\dot{\eta}}{\omega} \end{bmatrix}}_{\hat{\hat{H}}_L} \tag{2}$$

Substituting (*2*) into (*3-34*) gives the following transport matrix corresponding to the original coordinates

$$\begin{bmatrix} 1 \\ \hline \Delta \\ \dot{\Delta} \end{bmatrix}_R = \left[\begin{array}{c|cc} 1 & 0 & 0 \\ \hline \{\Psi\} & [cs] & [sn] \\ \{\dot{\Psi}\} & [-sn'] & [cs] \end{array}\right] \begin{bmatrix} 1 \\ \hline \Delta \\ \dot{\Delta} \end{bmatrix}_L \tag{3}$$

where
$$\{\Psi\} = [h]\{\rho\} \qquad [cs] = [h]\,\text{Diag.}\,[\cos \omega t][h]^T[\mu]$$
$$\{\dot{\Psi}\} = [h]\{\dot{\rho}\} \qquad [sn] = [h]\,\text{Diag.}\,[\sin \omega t]\,\text{Diag.}\,[1/\omega][h]^T[\mu]$$
$$[sn'] = [h]\,\text{Diag.}\,[\sin \omega t]\,\text{Diag.}\,[\omega][h]^T[\mu]$$

Using Problem 3.7 we may calculate the following:

$$\{\rho\} = \begin{bmatrix} 1.07930 \\ 0.05814 \\ -0.00772 \end{bmatrix} \quad \{\dot{\rho}\} = \begin{bmatrix} -3.43836 \\ 2.32730 \\ -0.30599 \end{bmatrix} \quad \{\Delta_0\} = \begin{bmatrix} 1 \\ 0 \\ 1 \end{bmatrix} \quad \{\dot{\Delta}_0\} = \begin{bmatrix} 1.5 \\ 1.0 \\ 1.0 \end{bmatrix} \tag{4}$$

in which $\{\dot{\rho}\}$ corresponding to $\{h_{1k}\dot{\Phi}_{1k}/\omega_k + h_{2k}\dot{\Phi}_{2k}/\omega_k + \cdots + h_{mk}\dot{\Phi}_{mk}/\omega_k\}$ and $\dot{\Phi} = d\Phi/dt$ is obtained from (*3*) of Problem 3.7. Consequently the other terms are

$$[cs] = \begin{bmatrix} 0.44238 & -0.63498 & -0.02433 \\ -0.63497 & -0.01105 & -0.47318 \\ -0.02403 & -0.47316 & -0.66716 \end{bmatrix} \quad [sn] = \begin{bmatrix} 0.00552 & 0.01309 & -0.01214 \\ 0.01309 & 0.006125 & -0.01143 \\ -0.01214 & -0.01144 & 0.00709 \end{bmatrix} \tag{5}$$

$$\{\Psi\} = \begin{bmatrix} 0.330 \\ 0.683 \\ 0.901 \end{bmatrix} \quad \{\dot{\Psi}\} = \begin{bmatrix} 0.2547 \\ -0.3792 \\ -4.5088 \end{bmatrix} \quad [sn'] = \begin{bmatrix} -0.0744 & 18.738 & -13.018 \\ 18.738 & 3.8299 & -9.046 \\ -13.018 & -9.046 & 9.547 \end{bmatrix} \tag{6}$$

Using (*4, 5, 6*) yields the final solution of (*3*):

$$\{\Delta \,\vdots\, \dot{\Delta}\}_{t=1/4\,s} = \{0.7566 \quad -0.4113 \quad 0.1871 \,\vdots\, 13.351 \quad -11.5067 \quad -2.2142\} \tag{7}$$

3.10. Construct the transport matrix for Problem 3.8.

Let (*3-38*) be represented by $[A_{RL}]$, $[B_{RL}]$, and $\{\rho\}$ as

$$\{\eta\}_R = \underbrace{\text{Diag.}\,[(e^{-\gamma t}/\lambda)(\lambda \cos \lambda t + \gamma \sin \lambda t)]}_{[A_{RL}]}\{\eta\}_L + \underbrace{\text{Diag.}\,[(e^{-\gamma t}/\lambda) \sin \lambda t]}_{[B_{RL}]}\{\dot{\eta}\}_L$$
$$+ \underbrace{\{\{h_{1k}\Phi_{1k}\} + \{h_{2k}\Phi_{2k}\} + \cdots + \{h_{mk}\Phi_{mk}\}\}}_{[\rho]} \tag{1}$$

from which

$$\{\dot{\eta}\}_R = \underbrace{\text{Diag.}\left[\gamma e^{-\gamma t}\left(-\frac{\gamma}{\lambda}\sin \lambda t - \frac{\lambda}{\gamma}\sin \lambda t\right)\right]}_{[C_{RL}]}\{\eta\}_L + \underbrace{\text{Diag.}\left[e^{-\gamma t}\left(\cos \lambda t - \frac{\gamma}{\lambda}\sin \lambda t\right)\right]}_{[D_{RL}]}\{\dot{\eta}\}_L + \{\dot{\rho}\} \tag{2}$$

Expressing (*1, 2*) in transport matrix we have

$$\begin{bmatrix} 1 \\ \hline \eta \\ \dot{\eta} \end{bmatrix}_R = \left[\begin{array}{c|cc} 1 & 0 & 0 \\ \hline \{\rho\} & [A_{RL}] & [B_{RL}] \\ \{\dot{\rho}\} & [C_{RL}] & [D_{RL}] \end{array}\right] \begin{bmatrix} 1 \\ \hline \eta \\ \dot{\eta} \end{bmatrix}_L \tag{3}$$

where $\{\dot{\rho}\} = \{h_{1k}\dot{\Phi}_{1k}\} + \{h_{2k}\dot{\Phi}_{2k}\} + \cdots + \{h_{mk}\dot{\Phi}_{mk}\}$ and $\dot{\Phi} = d\Phi/dt$ from (*2*) of Problem 3.8. Substituting (*3*) into (*3-40*) gives

$$\begin{bmatrix} 1 \\ \hline \Delta \\ \dot{\Delta} \end{bmatrix}_R = \left[\begin{array}{c|cc} 1 & 0 & 0 \\ \hline \{\Psi\} & [E_{RL}] & [F_{RL}] \\ \{\dot{\Psi}\} & [G_{RL}] & [H_{RL}] \end{array}\right] \begin{bmatrix} 1 \\ \hline \Delta \\ \dot{\Delta} \end{bmatrix}_L \tag{4}$$

where $\{\Psi\} = [h]\{\rho\}$ $[E_{RL}] = [h][A_{RL}][h]^T[\mu]$ $[G_{RL}] = [h][C_{RL}][h]^T[\mu]$

$\{\dot{\Psi}\} = [h]\{\dot{\rho}\}$ $[F_{RL}] = [h][B_{RL}][h]^T[\mu]$ $[H_{RL}] = [h][D_{RL}][h]^T[\mu]$

Thus, from Problem 3.8 we may calculate the following:

$$[E_{RL}] = \begin{bmatrix} 0.06698 & -0.25646 & -0.16860 \\ -0.25646 & 0.22304 & -0.45710 \\ -0.16860 & -0.45710 & -0.64602 \end{bmatrix} \qquad [F_{RL}] = \begin{bmatrix} 0.00371 & 0.00608 & -0.00669 \\ 0.00608 & 0.00326 & -0.00707 \\ -0.00669 & -0.00707 & 0.00267 \end{bmatrix} \tag{5}$$

$$[G_{RL}] = \begin{bmatrix} -1.6984 & -7.0869 & 5.5674 \\ -7.0869 & -2.7196 & 4.5010 \\ 5.5675 & 4.5010 & -4.2430 \end{bmatrix} \qquad [H_{RL}] = \begin{bmatrix} 0.03513 & -0.25170 & -0.10960 \\ -0.25170 & -0.21128 & -0.35520 \\ -0.10960 & -0.35520 & -0.57090 \end{bmatrix} \tag{6}$$

$$\{\Psi\} = \begin{bmatrix} 0.04340 \\ 0.08358 \\ 0.11157 \end{bmatrix} \qquad \{\dot{\Psi}\} = \begin{bmatrix} 0.0523 \\ 0.0516 \\ -0.1101 \end{bmatrix} \qquad \{\Delta_0\} = \begin{bmatrix} 1 \\ 1 \\ 1 \end{bmatrix} \qquad \{\dot{\Delta}_0\} = \begin{bmatrix} 1.5 \\ 1.0 \\ 1.0 \end{bmatrix} \tag{7}$$

Substituting (5, 6, 7) into (4) yields

$$\{\Delta \,\vdots\, \dot{\Delta}\}_{t=1/4\,s} = \{-0.03956 \quad -0.5124 \quad -0.5877 \,\vdots\, 3.6127 \quad -3.4793 \quad 0.1239\} \tag{8}$$

Supplementary Problems

3.11. Find the natural frequencies, natural periods, and natural modes of the shear building shown in Fig. P-3.11 where $\mu_1 = \mu_2 = 1.93$ k/ft, $EI = 50\,000$ k-ft^2, $l = 10$ ft, and $h = 20$ ft.

3.12. If two railroad cars vibrate against each other after coupling together, find natural frequency, natural period, and the equation of motion. Assume that the stiffness of the flexible coupling is k and the masses of the cars are μ_1 and μ_2 respectively.

3.13. Write the governing differential equations for the vibrating system shown in Fig. P-3.13 which represents a building having masses μ_1 and ρ_1 and column stiffness k_1 supported by a flexible foundation with spring constants k_θ and k_2. The foundation mate has masses μ_2 and ρ_2 but the column is assumed to be massless.

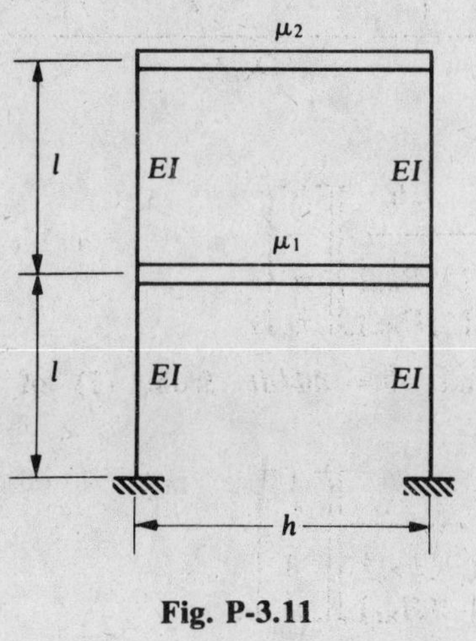

Fig. P-3.11

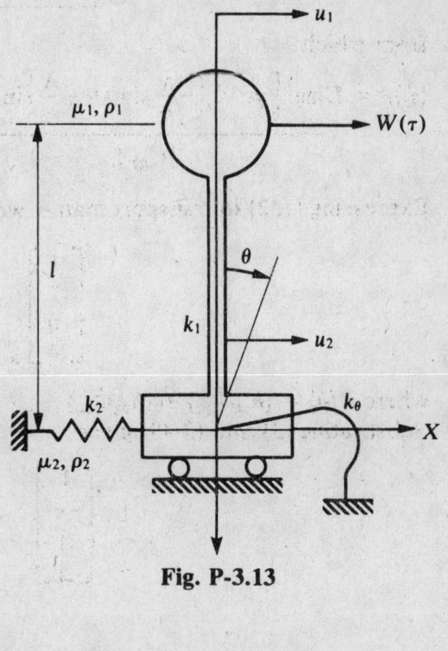

Fig. P-3.13

3.14. Assume that a two-degree-of-freedom system has diagonal mass matrix as $\mu_1 = \mu_2 = 10\ \text{lb-s}^2/\text{in}$, of which the first natural circular frequency and its natural mode are $\omega_1 = 10\ \text{rad/s}$, $\{X\} = \{X_{11}, X_{11}\}$, and $k_{11} = 2000\ \text{lb/in}$. Find the second natural frequency and its natural mode.

3.15. Use the normalized matrix formulation and find the equations of motion of Problem 3.2 with initial conditions as given in Problem 3.3. Determine the displacements at $t = 1/4$ s.

3.16. The structure given in Problem 3.11 is subject to lateral forces $W_1(t) = 100 \sin 12\pi t$ k and $W_2(t) = 80 \sin 12\pi t$ k applied to μ_1 and μ_2 respectively. Using the normalized matrix formulation find the displacements due to the applied forces at $t = 1/4$ s.

3.17. Rework Problem 3.16 assuming damping $c_1 = 0.1\,c_{cr}$ and $c_2 = 0.15\,c_{cr}$ associated with masses μ_1 and μ_2 respectively.

3.18. Construct the transport matrix for Problem 3.16.

3.19. Construct the transport matrix for Problem 3.17.

Chapter 4

Energy Methods

4.1 BASIC CONCEPTS

General

An *alternate approach* to the formulation of nodal equations is based on the *energy principles* (virtual work, kinetic and potential energy, Hamilton's principle, and Lagrange's equations). Although the application of these principles reduces eventually to Newton's laws, these approaches are more general and yet more automatic; they minimize the thought and effort involved in the preparation of the equations of motion.

For the *development of the energy methods* concepts of *generalized coordinates, forces*, and *constraints* must be introduced.

Generalized Coordinates

The position of a planar mechanical system consisting of j lumped masses (or mass particles) is given at any instant t by $3j$ *displacements* (coordinates) measured from the datum and designated as

$$\Delta_i = \{u_i, v_i, \theta_i\}, \qquad i = 1, 2, \ldots, j \tag{4-1}$$

If the motion of the system is subjected to r restrictions called *constraints*, the number of possible and *independent displacements* reduces to

$$m = 3j - r \tag{4-2}$$

where m is the *number of degrees of freedom.*

These independent displacements are called the *generalized coordinates* (displacements) and although their number is fixed by (4-2), many sets of such coordinates are admissible in a given system (Problems 4.1, 4.2, and 4.3). An adroit choice frequently simplifies the analysis, and a particularly important set are the *principal coordinates*, the choice of which decouples the *nodal equations* (Sec. 3.3) and leads to m *independent equations of motion.*

Transformation Equations

The relationships of the generalized coordinates denoted as $q_1, q_2, \ldots, q_m$ to the original coordinates introduced by (4-1) as $u_1, u_2, \ldots, u_j$, $v_1, v_2, \ldots, v_j$, $\theta_1, \theta_2, \ldots, \theta_j$ are given by the *transformation equations* as

$$\begin{aligned} u_i &= u_i(q_1, q_2, \ldots, q_m, t) \\ v_i &= v_i(q_1, q_2, \ldots, q_m, t) \\ \theta_i &= \theta_i(q_1, q_2, \ldots, q_m, t) \end{aligned} \tag{4-3}$$

where t is the time, $i = 1, 2, \ldots, j$, and u_i, v_i, θ_i are *continuous functions*, the derivatives of which are also continuous (Problems 4.4, 4.5, and 4.6).

The *derivatives* of (4-3) written symbolically in terms of Δ_i (which is u_i, v_i, or θ_i) are

$$\frac{d\Delta_i}{dt} = \dot{\Delta}_i = \frac{\partial \Delta_i}{\partial q_1}\dot{q}_1 + \frac{\partial \Delta_i}{\partial q_2}\dot{q}_2 + \cdots + \frac{\partial \Delta_i}{\partial q_m}\dot{q}_m \tag{4-4}$$

$$\frac{\partial \dot{\Delta}_i}{\partial q_k} = \frac{\partial^2 \Delta_i}{\partial q_1 \partial q_k}\dot{q}_1 + \frac{\partial^2 \Delta_i}{\partial q_2 \partial q_k}\dot{q}_2 + \cdots + \frac{\partial^2 \Delta_i}{\partial q_m \partial q_k}\dot{q}_m \tag{4-5}$$

$$\frac{d}{dt}\left(\frac{\partial \Delta_i}{\partial q_k}\right) = \frac{\partial^2 \Delta_i}{\partial q_1 \partial q_k}\dot{q}_1 + \frac{\partial^2 \Delta_i}{\partial q_2 \partial q_k}\dot{q}_2 + \cdots + \frac{\partial^2 \Delta_i}{\partial q_m \partial q_k}\dot{q}_m \tag{4-6}$$

where the derivatives with respect to t are the *total derivatives* and the derivatives with respect to q_k (any generalized coordinate in sequence of $q_1, q_2, \ldots, q_m$) are the *partial derivatives.*

The comparison of (*4-4, 4-5*, and *4-6*) shows that

$$\frac{\partial \Delta_i}{\partial q_k} = \frac{\partial \dot{\Delta}_i}{\partial \dot{q}_k} \qquad \text{and} \qquad \frac{\partial \dot{\Delta}_i}{\partial q_k} = \frac{d}{dt}\left(\frac{\partial \Delta_i}{\partial q_k}\right) \tag{4-7, 8}$$

from which many simplifications may be made hereafter (Problem 4.7).

Classification of Constraints

According to the character of the restrictions placed upon the motion of the system, called the *constraints of motion*, the mechanical systems are classified as scleronomic, rheonomic, holonomic, and nonholonomic.

(*a*) The *scleronomic system* has constraints which are independent of time.

(*b*) The *rheonomic system* has constraints which are functions of time.

(*c*) The *holonomic system* has constraints given by r conditions of the following integrable form:

$$0 = a_s(q_1, q_2, \ldots, q_m, t), \qquad s = 1, 2, \ldots, r \tag{4-9}$$

(*d*) The *nonholonomic system* has constraints given by r conditions of the following nonintegrable form:

$$0 = \sum_{i=1}^{m} b_{si}\, dq_i + b_{si}\, dt = 0, \qquad \begin{Bmatrix} i = 1, 2, \ldots, m \\ s = 1, 2, \ldots, r \end{Bmatrix} \tag{4-10}$$

where $$b_{si} = b_{si}(q_1, q_2, \ldots, q_m, t)$$

Since (*4-10*) cannot be integrated, the coordinates of motion in the direction of constraints cannot be eliminated and the *number of coordinates exceeds the number of degrees of freedom.* As nonholonomic constraints occur rarely in structural dynamics, the problems of this chapter are restricted to systems (*a*), (*b*), and (*c*) (Problem 4.8).

Generalized Causes

The *virtual work* of a system with lumped masses specified by coordinates $\Delta_1, \Delta_2, \ldots, \Delta_p$ with forces $F_1, F_2, \ldots, F_p$ acting at and along the respective coordinates, undergoing a virtual displacement consistent with the constraints, is

$$\delta W = \sum_{k=1}^{p} F_k\, \delta \Delta_k \tag{4-11}$$

In terms of (*4-3*), the *virtual displacement* at k is

$$\delta \Delta_k = \frac{\partial \Delta_k}{\partial q_1}\delta q_1 + \frac{\partial \Delta_k}{\partial q_2}\delta q_2 + \cdots + \frac{\partial \Delta_k}{\partial q_m}\delta q_m \tag{4-12}$$

and (*4-11*) becomes
$$\delta W = \sum_{i=1}^{m} Q_i\, \delta q_i \tag{4-13}$$

where
$$Q_i = \sum_{k=1}^{p} F_k \frac{\partial \Delta_k}{\partial q_i} \tag{4-14}$$

is the *generalized force* associated with the *generalized coordinate* q_i (Problems 4.9, 4.10, and 4.11).

The comparison of (*4-11*) and (*4-13*) shows that F_k is the virtual work due to $\delta\Delta_k = +1$ and Q_i is the virtual work per $\delta q_i = +1$, where the generalized force Q_i is a force or a moment according to the nature of q_i (linear or angular displacement) (Problem 4.12).

Configuration of Equilibrium

The calculation of generalized forces is simplified if the *real forces have potential.* Then,

$$Q_i = -\frac{\partial V(q_1, q_2, \ldots, q_m)}{\partial q_i}, \qquad i = 1, 2, \ldots, m \tag{4-15}$$

If the system is in a *state of equilibrium*,

$$\delta W = \sum_{i=1}^{m} Q_i \, \delta q_i = -\sum_{i=1}^{m} \frac{\partial V(q_1, q_2, \ldots, q_m)}{\partial q_i} \delta q_i = 0$$

from which

$$Q_i = -\frac{\partial V(q_1, q_2, \ldots, q_m)}{\partial q_i} = 0 \tag{4-16}$$

Since the number of (*4-16*) equals the number of generalized coordinates, these equations define the *configuration of equilibrium* (Problem 4.14).

Stability of Equilibrium

Since equations (*4-16*) are the same as those defining the conditions of extremes of a function, these equations also indicate whether the configuration of equilibrium they define is a *stable* or *unstable* one.

A necessary and sufficient condition for the configuration to be in a state of stable equilibrium is that the potential energy in (*4-16*) be a minimum. This statement is generally known as the *Theorem of Minimum Potential Energy.*

4.2. GENERAL METHODS

Components of Generalized Force

The generalized force defined by (*4-14*) consists, in general, of three distinct components:

(*a*) The *spring force* due to the *change of potential energy,*

$$Q_{V,i} = -\frac{\partial V}{\partial q_i} \tag{4-17}$$

where V is the potential energy.

(*b*) The *damping force* due to the *dissipation energy,*

$$Q_{D,i} = -\frac{\partial D}{\partial \dot{q}_i} \tag{4-18}$$

where D is the dissipation energy.

(*c*) The *applied load* component acting *along the respective generalized coordinate,*

$$Q_{W,i} = \frac{\partial W}{\partial q_i} \tag{4-19}$$

where W is the real work of all load components along the generalized coordinates.

Lagrange's Equation

The *generalized force*

$$Q_i = Q_{V,i} + Q_{D,i} + Q_{W,i} \tag{4-20}$$

is related to the kinetic energy T of the system by the equation

$$\frac{d}{dt}\left(\frac{\partial T}{\partial \dot{q}_i}\right) - \frac{\partial T}{\partial q_i} = Q_i, \qquad i = 1, 2, \ldots, m \tag{4-21}$$

which is the *symbolic Lagrange's equation* (Problem 4.13).

With the notation

$$L = T - V$$

called the *Lagrangian function*, and in terms of (*4-17, 4-18,* and *4-19*), (*4-21*) may be written as

$$\frac{d}{dt}\left(\frac{\partial L}{\partial \dot{q}_i}\right) - \frac{\partial L}{\partial q_i} + \frac{\partial D}{\partial \dot{q}_i} = \frac{\partial W}{\partial q_i}, \qquad i = 1, 2, \ldots, m \tag{4-22}$$

which is the *standard Lagrange's equation* (Problem 4.14).

Special Cases.

According to the type of motion, the *standard Lagrange's equation* (*4-22*) takes on one of the following special forms:

(*a*) *Free motion without damping*:

$$\frac{d}{dt}\left(\frac{\partial T}{\partial \dot{q}_i}\right) + \frac{\partial V}{\partial q_i} = 0 \tag{4-23a}$$

(*b*) *Forced motion without damping*:

$$\frac{d}{dt}\left(\frac{\partial T}{\partial \dot{q}_i}\right) + \frac{\partial V}{\partial q_i} = \frac{\partial W}{\partial q_i} \tag{4-23b}$$

(*c*) *Free motion with damping*:

$$\frac{d}{dt}\left(\frac{\partial T}{\partial \dot{q}_i}\right) + \frac{\partial V}{\partial q_i} + \frac{\partial D}{\partial \dot{q}_i} = 0 \tag{4-23c}$$

(*d*) *Forced motion with damping*:

$$\frac{d}{dt}\left(\frac{\partial T}{\partial \dot{q}_i}\right) + \frac{\partial V}{\partial q_i} + \frac{\partial D}{\partial \dot{q}_i} = \frac{\partial W}{\partial q_i} \tag{4-23d}$$

Since $i = 1, 2, \ldots, m$, each of these types (a, b, c, d) yields m equations of motion which are or will reduce to the differential equations of motion derived for the same system by means of Newton's second law.

Hamilton's Equation

A broader and more general method of mechanics known as *Hamilton's method* is based on the relationship

$$\int_{t_1}^{t_2} (\delta T + \delta W)\, dt = \int_{t_1}^{t_2} \sum_{i=1}^{m} \left[\frac{\partial T}{\partial q_i} - \frac{d}{dt}\left(\frac{\partial T}{\partial \dot{q}_i}\right) + Q_i\right] \delta q_i\, dt = 0 \tag{4-24}$$

where the symbols are those of (*4-21*).

Hamilton's Principle

The verbal interpretation of (*4-24*), called *Hamilton's principle*, states that

For the real motion of the system within an arbitrarily selected time interval, $t_1 \to t_2$,

$$\delta I = \int_{t_1}^{t_2} (\delta T + \delta W)\, dt$$

vanishes for any continuously varying virtual displacement if this displacement is zero at t_1 and t_2.

Hamilton's principle may be regarded as the *fundamental law of mechanics*, the special consequences of which are Lagrange's equations and Newton's laws.

The application of (*4-24*) is of limited value in structural dynamics, since for all problems of practical importance (*4-24*) can be reduced to Lagrange's equation of the respective kind (*4-23a, b, c, d*). For this reason, the derivation and applications of (*4-24*) are not given in this book.

4.3. SMALL OSCILLATION IN GENERALIZED COORDINATES

Classification

Lagrange's equations (*4-23a, b, c, d*) find a broad application in the *analysis of linear oscillations* of mechanical systems described in Sec. 3.1. In such cases the analytical form of the *potential energy V*, the *kinetic energy T*, and the *dissipation energy D* of the given system must be determined first. General forms of these functions and their matrix constructions are developed in this section.

Potential Energy

If the *potential energy function V* is expanded in a Taylor series (where the first term is placed equal to zero as a reference value and higher terms beyond the second one are neglected), it reduces to

$$V = \frac{1}{2}\sum_{i=1}^{m}\sum_{k=1}^{m} k_{ik}q_i q_k = \frac{1}{2}\{q\}^T[k]\{q\} \tag{4-25}$$

where $\{q\} = \{q_1, q_2, \ldots, q_m\}$, $\{q\}^T = [q_1, q_2, \ldots, q_m]$, and

$$[k] = \begin{bmatrix} k_{11} & k_{12} & \cdots & k_{1m} \\ k_{21} & k_{22} & \cdots & k_{2m} \\ \cdots & \cdots & \cdots & \cdots \\ k_{m1} & k_{m2} & \cdots & k_{mm} \end{bmatrix} \tag{4-26}$$

is a square symmetrical *matrix of stiffness coefficients*, each of which is defined as

$$k_{ik} = k_{ki} = \frac{\partial^2 V}{\partial q_i \partial q_k} \tag{4-27}$$

The values of coefficients in $[k]$ classify V as:

(*a*) *Positive definite*, if all main diagonal coefficients and all minors of $[k]$ are positive.

(*b*) *Positive semidefinite*, if one or more principal minors of $[k]$ are zero.

(*c*) *Indefinite*, if all main diagonal coefficients are nonzero but one or more are negative.

The conditions (*a*), (*b*), (*c*) also define the datum configuration as *stable, neutral*, and *indifferent* respectively.

Kinetic Energy

If it is assumed that the transformation equations (*4-3*) and the equilibrium configuration are time independent, the kinetic energy is

$$T = \frac{1}{2}\sum_{i=1}^{m}\sum_{k=1}^{m} \mu_{ik}\dot{q}_i\dot{q}_k = \frac{1}{2}\{\dot{q}\}^T[\mu]\{\dot{q}\} \tag{4-28}$$

where $\{\dot{q}\} = \{\dot{q}_1, \dot{q}_2, \ldots, \dot{q}_m\}$, $\{\dot{q}\}^T = [\dot{q}_1, \dot{q}_2, \ldots, \dot{q}_m]$, and

$$[\mu] = \begin{bmatrix} \mu_{11} & \mu_{12} & \cdots & \mu_{1m} \\ \mu_{21} & \mu_{22} & \cdots & \mu_{2m} \\ \cdots & \cdots & \cdots & \cdots \\ \mu_{m1} & \mu_{m2} & \cdots & \mu_{mm} \end{bmatrix} \tag{4-29}$$

is a square symmetrical *matrix of inertia coefficients*, each of which is defined as

$$\mu_{ik} = \mu_{ki} = \sum_{j=1}^{p} \mu_j \left(\frac{\partial \Delta_j}{\partial q_i}\right)\left(\frac{\partial \Delta_j}{\partial q_k}\right) \tag{4-30}$$

In *(4-30)*, μ_j is the *inertia coefficient of the lumped mass* corresponding to the ordinary coordinate Δ_j which is u_j, v_j, or θ_j given by *(4-3)*.

Although the coefficients of $[\mu]$ classify T, for cases studied in this volume, T is always a *positive definite quadratic function.*

Dissipation Energy

The positive semidefinite dissipation energy function is

$$D = \frac{1}{2}\sum_{i=1}^{m}\sum_{k=1}^{m} c_{ik}\dot{q}_i\dot{q}_k = \frac{1}{2}\{\dot{q}\}^T[c]\{\dot{q}\} \tag{4-31}$$

where $\{\dot{q}\}$ and $\{\dot{q}\}^T$ are the same as in *(4-28)*,

and

$$[c] = \begin{bmatrix} c_{11} & c_{12} & \cdots & c_{1m} \\ c_{21} & c_{22} & \cdots & c_{2m} \\ \cdots & \cdots & \cdots & \cdots \\ c_{m1} & c_{m2} & \cdots & c_{mm} \end{bmatrix} \tag{4-32}$$

is a square symmetrical *matrix of damping coefficients*, each of which is defined as

$$c_{ik} = c_{ki} = \sum_{j=1}^{p} c_j\left(\frac{\partial \Delta_j}{\partial q_i}\right)\left(\frac{\partial \Delta_j}{\partial q_k}\right) \tag{4-33}$$

In *(4-33)*, c_j is the *damping coefficient of the lumped mass* corresponding to the ordinary coordinate Δ_j which is u_j, v_j, or θ_j given by *(4-3)*.

Generalized Load Function

The *generalized load function* is by *(4-14)*

$$Q_{W,i} = \frac{\partial W}{\partial q_i} = \sum_{k=1}^{m} F_k \frac{\partial \Delta_k}{\partial q_i} \tag{4-34}$$

and the transformation *(4-3)* in matrix form is

$$\{\Delta\} = [A]\{q\} \tag{4-35}$$

Thus the generalized load vector becomes

$$\{Q_W\} = [A]^T\{W\} \tag{4-36}$$

where $[A]$ is a transformation matrix of given coefficients and $\{W\}$ is a given load vector.

Motion in Generalized Coordinates

The substitution of the participating functions V and T given respectively by *(4-25)* and *(4-28)* in the Lagrange's equation *(4-23a)* yields a system of nodal equations *(3-2)* in generalized coordinates (Problem 4.15).

Similarly, the substitution of V, T, and D given respectively by *(4-25)*, *(4-28)*, and *(4-31)* in the Lagrange's equation *(4-23c)* yields a system of nodal equations *(3-35)* in generalized coordinates (Problem 4.16).

It must be pointed out that the displacement vector $\{\Delta\}$ introduced in *(3-2)* and *(3-35)* includes m components, is a generalized coordinate vector, and equals $\{q\}$, whereas the coordinate vector $\{\Delta\}$ defined by *(4-3)* includes $3j$ components, is not a generalized coordinate vector, and its relationship to $\{q\}$ is given by *(4-35)*.

Motion in Normal Coordinates

Since the normal coordinates $\{\eta\}$ are generalized coordinates of a particular type, the participating functions of *(4-23)* can be written as

$$T = \tfrac{1}{2}\{\dot{\eta}\}^T[h]^T[\mu][h]\{\dot{\eta}\} = \tfrac{1}{2}\{\dot{\eta}\}^T\{\dot{\eta}\} \tag{4-37}$$

$$V = \tfrac{1}{2}\{\eta\}^T[h]^T[k][h]\{\eta\} = \tfrac{1}{2}\{\eta\}^T[\Lambda]\{\eta\} \tag{4-38}$$

$$D = \tfrac{1}{2}\{\dot{\eta}\}^T[h]^T[c][h]\{\dot{\eta}\} = \tfrac{1}{2}\{\dot{\eta}\}^T[2\Gamma]\{\dot{\eta}\} \tag{4-39}$$

$$Q_W = [h]^T\{W\} \tag{4-40}$$

where $[h]$, $[\Lambda]$, $[2\Gamma]$, and $[W]$ have the same meaning as they had in (*3-23*), (*3-31*), and (*3-37*). Their substitution in (*4-23a, b, d*) yields the particular systems of nodal equations identical respectively to (*3-26*), (*3-31*), and (*3-37*) (Problems 4.18 and 4.19).

Solved Problems

BASIC CONCEPTS

4.1. Using (*4-2*) determine the number of degrees of freedom of the frame of Fig. P-4.1 and then show the multiple choice of generalized coordinates.

Let the structural joints 0, 1, 2, 3 shown in Fig. P-4.1 be considered as structural nodes among which nodes 0 and 3 are fixed and furnish 6 constraints. If no axial deformation is considered, members a, b, c will contribute 3 constraints related to longitudinal movements. According to (*4-2*) the nodal system has 3 degrees of freedom:

$$m = 3j - r = 3(4) - 9 = 3 \tag{1}$$

Because the axial deformation is not considered, $v_1 = v_2 = 0$ and the movements of u_1 and u_2 are not independent. Thus the generalized coordinates may be chosen as (a) $\theta_1 = q_1$, $\theta_2 = q_2$, and $u_1 = q_3$ or (b) $\theta_1 = q_1$, $\theta_2 = q_2$, and $u_2 = q_3$.

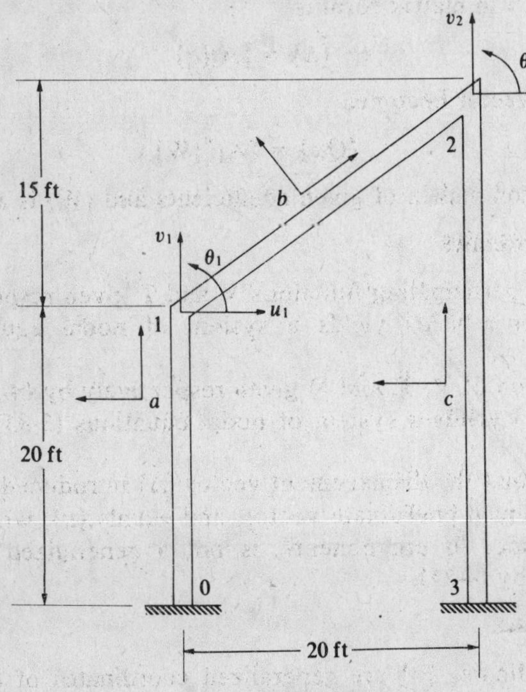

Fig. P-4.1

4.2. Repeat Problem 4.1 for the frame of Fig. P-4.2.

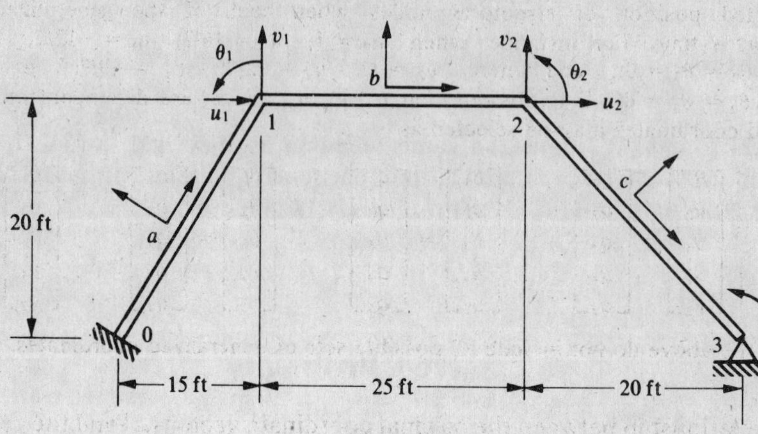

Fig. P-4.2

As in Problem 4.1 the structural joints 0, 1, 2, 3 of Fig. P-4.2 are selected as the reference nodes. Since 0 is fixed (3 constraints) and 3 is hinged (2 constraints), and the lengths of members 01, 12, 23 are assumed to remain constant (3 constraints), the nodal system has 8 constraints. By *(4-2)* the number of degrees of freedom of this system is

$$m = 3j - r = 3(4) - 8 = 4$$

in which the degrees of freedom of joint rotation and that of linear displacement are 3 and 1 respectively. Because the columns are inclined, the linear movements of u_1, v_1, u_2, and v_2 are all possible but not necessarily independent. Thus the generalized coordinates may be chosen as

$$[\theta_1\ \theta_2\ \theta_3\ u_1]^T = [q_1\ q_2\ q_3\ q_4]^T \qquad [\theta_1\ \theta_2\ \theta_3\ v_1]^T = [q_1\ q_2\ q_3\ q_4]^T \tag{1,2}$$

$$[\theta_1\ \theta_2\ \theta_3\ u_2]^T = [q_1\ q_2\ q_3\ q_4]^T \qquad [\theta_1\ \theta_2\ \theta_3\ v_2]^T = [q_1\ q_2\ q_3\ q_4]^T \tag{3,4}$$

4.3. Repeat Problem 4.1 for the frame of Fig. P-4.3.

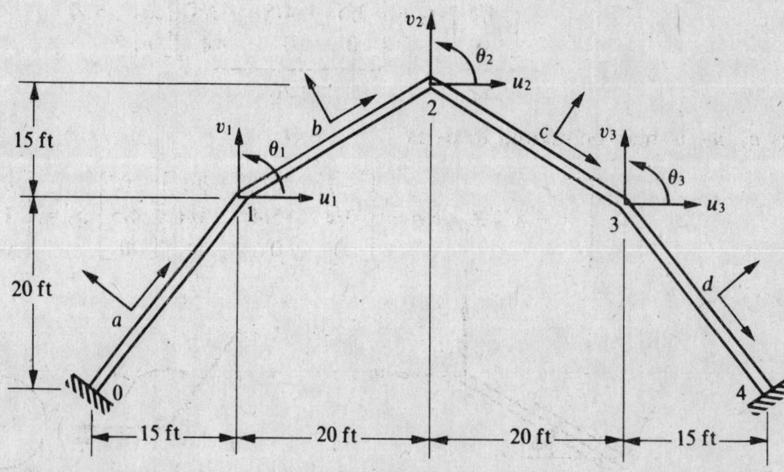

Fig. P-4.3

From Fig. P-4.3 the supports and members furnish $r = 10$, and according to *(4-2)* we have

$$m = 3(5) - 10 = 5$$

in which 2 degrees of freedom are associated with linear displacements. Therefore two displacement coordinates must be carefully selected for generalized coordinates which should be able to define the deflected position of structural nodes when each of the generalized coordinates moves one unit at a time. For instance, when $u_2 = 1$, $u_1 = 16/7$, $v_1 = -12/7$, $u_3 = 16/7$, $v_3 = 12/7$, $v_2 = \theta_1 = \theta_2 = \theta_3 = 0$; and when $v_2 = 1$, $u_1 = 12/7$, $v_1 = -9/7$, $u_3 = -12/7$, $v_3 = -9/7$, $u_2 = \theta_1 = \theta_2 = \theta_3 = 0$. It is obvious that u_1, v_1, u_3, and v_3 are dependent on u_2 and v_2. Thus the generalized coordinates may be selected as

$$\begin{bmatrix}\theta_1\\ \theta_2\\ \theta_3\\ u_2\\ v_2\end{bmatrix} = \begin{bmatrix}q_1\\ q_2\\ q_3\\ q_4\\ q_5\end{bmatrix} \qquad \begin{bmatrix}\theta_1\\ \theta_2\\ \theta_3\\ u_1\\ u_3\end{bmatrix} = \begin{bmatrix}q_1\\ q_2\\ q_3\\ q_4\\ q_5\end{bmatrix} \qquad \begin{bmatrix}\theta_1\\ \theta_2\\ \theta_3\\ v_1\\ v_3\end{bmatrix} = \begin{bmatrix}q_1\\ q_2\\ q_3\\ q_4\\ q_5\end{bmatrix} \qquad \begin{bmatrix}\theta_1\\ \theta_2\\ \theta_3\\ v_1\\ u_3\end{bmatrix} = \begin{bmatrix}q_1\\ q_2\\ q_3\\ q_4\\ q_5\end{bmatrix} \qquad (1, 2, 3, 4)$$

Note that the above do not include all possible sets of generalized coordinates.

4.4. Find the relationship between the original coordinate vector $\{\Delta\}$ and the generalized coordinate vector $\{q\}$ for Problem 4.1 as

$$\{\Delta\} = [A]\{q\}$$

Referring to Fig. P-4.4*a*, the constraint condition of a typical member without axial deformation is expressed as

$$u_i \cos\phi_l + v_i \sin\phi_l - u_j \cos\phi_l - v_j \sin\phi_l = 0 \qquad (1)$$

where ϕ_l is measured from the X axis as shown in Figs. P-4.4*b*, *c*. Employing (*1*) and using the abbreviations $c\phi_l = \cos\phi_l$ and $s\phi_l = \sin\phi_l$, we have

$$\begin{bmatrix}-c\phi_a & -s\phi_a & 0 & 0\\ c\phi_b & s\phi_b & -c\phi_b & -s\phi_b\\ 0 & 0 & -c\phi_c & -s\phi_c\end{bmatrix}\begin{bmatrix}u_1\\ v_1\\ u_2\\ v_2\end{bmatrix} = 0 \quad \text{or} \quad \begin{bmatrix}0 & -1 & 0 & 0\\ 4/5 & 3/5 & -4/5 & -3/5\\ 0 & 0 & 0 & -1\end{bmatrix}\begin{bmatrix}u_1\\ v_1\\ u_2\\ v_2\end{bmatrix} = 0 \qquad (2, 3)$$

(*a*) Using the generalized coordinates chosen in (*a*) of Problem 4.1 as $\theta_1 = q_1$, $\theta_2 = q_2$, and $u_1 = q_3$ we may rewrite (*3*) as

$$\underbrace{\begin{bmatrix}0\\ 4/5\\ 0\end{bmatrix}}_{A_q} q_3 + \underbrace{\begin{bmatrix}-1 & 0 & 0\\ 3/5 & -4/5 & -3/5\\ 0 & 0 & -1\end{bmatrix}}_{A_r}\begin{bmatrix}v_1\\ u_2\\ v_2\end{bmatrix} = 0 \qquad (4)$$

Thus v_1, u_2, v_2 may be related to q_3 as

$$\begin{bmatrix}v_1\\ u_2\\ v_2\end{bmatrix} = -A_r^{-1}A_q q_3 = -\begin{bmatrix}-1 & 0 & 0\\ 3/4 & -5/4 & -3/4\\ 0 & 0 & 1\end{bmatrix}\begin{bmatrix}0\\ 4/5\\ 0\end{bmatrix} q_3 = \begin{bmatrix}0\\ 1\\ 0\end{bmatrix} q_3 \qquad (5)$$

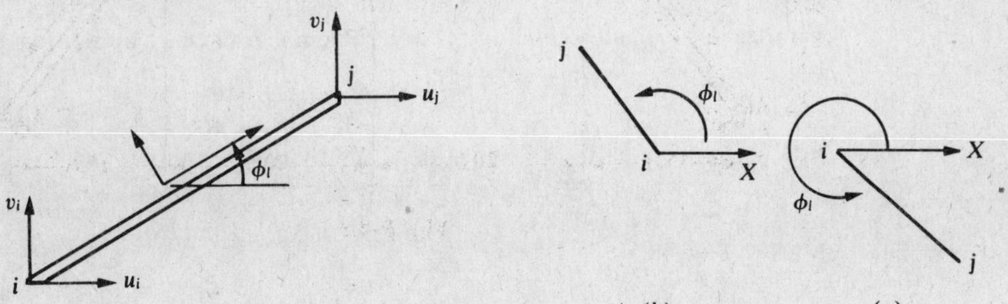

(*a*) (*b*) (*c*)

Fig. P-4.4

which shows that $v_1 = v_2 = 0$ and $u_2 = q_3$ as discussed in Problem 4.1. The relationship between $\{\Delta\}$ and $\{q\}$ becomes

$$\begin{bmatrix} \theta_1 \\ \theta_2 \\ \hline u_1 \\ \hline v_1 \\ u_2 \\ v_2 \end{bmatrix} = \left[\begin{array}{cc|c} 1 & 0 & 0 \\ 0 & 1 & 0 \\ \hline 0 & 0 & 1 \\ \hline 0 & 0 & 0 \\ 0 & 0 & 1 \\ 0 & 0 & 0 \end{array}\right] \begin{bmatrix} q_1 \\ q_2 \\ \hline q_3 \end{bmatrix} \tag{6}$$

(*b*) For the generalized coordinates given in (*b*) of Problem 4.1 as $\theta_1 = q_1$, $\theta_2 = q_2$, and $u_2 = q_3$, we may rewrite (3) as

$$\underbrace{\begin{bmatrix} 0 \\ -4/5 \\ 0 \end{bmatrix}}_{A_q} q_3 + \underbrace{\begin{bmatrix} 0 & 1 & 0 \\ 4/5 & 3/5 & -3/5 \\ 0 & 0 & 1 \end{bmatrix}}_{A_r} \begin{bmatrix} u_1 \\ v_1 \\ v_2 \end{bmatrix} = 0 \tag{7}$$

from which

$$\begin{bmatrix} u_1 \\ v_1 \\ v_2 \end{bmatrix} = -A_r^{-1} A_q q_3 = -\begin{bmatrix} -3/4 & 5/4 & 3/4 \\ 1 & 0 & 0 \\ 0 & 0 & 1 \end{bmatrix} \begin{bmatrix} 0 \\ -4/5 \\ 0 \end{bmatrix} q_3 = \begin{bmatrix} 1 \\ 0 \\ 0 \end{bmatrix} q_3 \tag{8}$$

Finally the transformation matrix is

$$\begin{bmatrix} \theta_1 \\ \theta_2 \\ \hline u_1 \\ \hline v_1 \\ u_2 \\ v_2 \end{bmatrix} = \left[\begin{array}{cc|c} 1 & 0 & 0 \\ 0 & 1 & 0 \\ \hline 0 & 0 & 1 \\ \hline 0 & 0 & 0 \\ 0 & 0 & 1 \\ 0 & 0 & 0 \end{array}\right] \begin{bmatrix} q_1 \\ q_2 \\ \hline q_3 \end{bmatrix} \tag{9}$$

4.5. Find $\{\Delta\} = [A]\{q\}$ for the structure given in Problem 4.2.

Using the typical constraint equation (*1*) of Problem 4.4, the constraint conditions for Fig. P-4.2 are

$$\begin{bmatrix} -c\phi_a & -s\phi_a & 0 & 0 \\ c\phi_b & s\phi_b & -c\phi_b & -s\phi_b \\ 0 & 0 & c\phi_c & s\phi_c \end{bmatrix} \begin{bmatrix} u_1 \\ v_1 \\ u_2 \\ v_2 \end{bmatrix} = 0 \quad \text{or} \quad \begin{bmatrix} -3/5 & -4/5 & 0 & 0 \\ 1 & 0 & -1 & 0 \\ 0 & 0 & 4/5 & -3/5 \end{bmatrix} \begin{bmatrix} u_1 \\ v_1 \\ u_2 \\ v_2 \end{bmatrix} = 0 \tag{1, 2}$$

(*a*) For the generalized coordinates given in (*1*) of Problem 4.2, (*2*) may be written as

$$\underbrace{\begin{bmatrix} -3/5 \\ 1 \\ 0 \end{bmatrix}}_{A_q} q_4 + \underbrace{\begin{bmatrix} -4/5 & 0 & 0 \\ 0 & -1 & 0 \\ 0 & 4/5 & -3/5 \end{bmatrix}}_{A_r} \begin{bmatrix} v_1 \\ u_2 \\ v_2 \end{bmatrix} = 0 \tag{3}$$

from which

$$\begin{bmatrix} v_1 \\ u_2 \\ v_2 \end{bmatrix} = -A_r^{-1} A_q q_4 = \begin{bmatrix} 5/4 & 0 & 0 \\ 0 & 1 & 0 \\ 0 & 4/3 & 5/3 \end{bmatrix} \begin{bmatrix} -3/5 \\ 1 \\ 0 \end{bmatrix} q_4 = \begin{bmatrix} -3/4 \\ 1 \\ 4/3 \end{bmatrix} q_4 \tag{4}$$

Consequently, $\{\Delta\}$ may be related to $\{q\}$ as

$$\begin{bmatrix}\theta_1\\ \theta_2\\ \theta_3\\ \hline u_1\\ \hline v_1\\ u_2\\ v_2\end{bmatrix} = \left[\begin{array}{ccc|c} 1 & 0 & 0 & 0\\ 0 & 1 & 0 & 0\\ 0 & 0 & 1 & 0\\ \hline 0 & 0 & 0 & 1\\ \hline 0 & 0 & 0 & -3/4\\ 0 & 0 & 0 & 1\\ 0 & 0 & 0 & 4/3 \end{array}\right] \begin{bmatrix} q_1\\ q_2\\ q_3\\ \hline q_4\end{bmatrix} \tag{5}$$

(*b*) For the generalized coordinates given in (*2*) of Problem 4.2, from (*2*) we have

$$\underbrace{\begin{bmatrix}-4/5\\ 0\\ 0\end{bmatrix}}_{A_q} q_4 + \underbrace{\begin{bmatrix}-3/5 & 0 & 0\\ 1 & -1 & 0\\ 0 & 4/5 & -3/5\end{bmatrix}}_{A_r} \begin{bmatrix}u_1\\ u_2\\ v_2\end{bmatrix} = 0 \tag{6}$$

from which
$$\begin{bmatrix}u_1\\ u_2\\ v_2\end{bmatrix} = -A_r^{-1}A_q q_4 = -\begin{bmatrix}-5/3 & 0 & 0\\ -5/3 & 1 & 0\\ -20/9 & 4/3 & 5/3\end{bmatrix}\begin{bmatrix}-4/5\\ 0\\ 0\end{bmatrix} q_4 = \begin{bmatrix}-4/3\\ -4/3\\ -16/9\end{bmatrix} q_4 \tag{7}$$

The transformation matrix may be finally expressed as

$$\begin{bmatrix}\theta_1\\ \theta_2\\ \theta_3\\ \hline v_1\\ \hline u_1\\ u_2\\ v_2\end{bmatrix} = \left[\begin{array}{ccc|c} 1 & 0 & 0 & 0\\ 0 & 1 & 0 & 0\\ 0 & 0 & 1 & 0\\ \hline 0 & 0 & 0 & 1\\ \hline 0 & 0 & 0 & -4/3\\ 0 & 0 & 0 & -4/3\\ 0 & 0 & 0 & -16/9 \end{array}\right] \begin{bmatrix} q_1\\ q_2\\ q_3\\ \hline q_4\end{bmatrix} \tag{8}$$

4.6. Find $\{\Delta\} = [A]\{q\}$ for the structure given in Problem 4.3.

The constraint equations for Fig. P-4.3 may be formulated according to (*1*) of Problem 4.4 as

$$\begin{bmatrix} -c\phi_a & -s\phi_a & 0 & 0 & 0 & 0\\ c\phi_b & s\phi_b & -c\phi_b & -s\phi_b & 0 & 0\\ 0 & 0 & c\phi_c & s\phi_c & -c\phi_c & -s\phi_c\\ 0 & 0 & 0 & 0 & c\phi_d & s\phi_d \end{bmatrix} [u_1 \;\; v_1 \;\; u_2 \;\; v_2 \;\; u_3 \;\; v_3]^T = 0 \tag{1}$$

Using the given structural dimensions we write (*1*) as

$$\begin{bmatrix} -3/5 & -4/5 & 0 & 0 & 0 & 0\\ 4/5 & 3/5 & -4/5 & -3/5 & 0 & 0\\ 0 & 0 & 4/5 & -3/5 & -4/5 & 3/5\\ 0 & 0 & 0 & 0 & 3/5 & -4/5 \end{bmatrix} [u_1 \;\; v_1 \;\; u_2 \;\; v_2 \;\; u_3 \;\; v_3]^T = 0 \tag{2}$$

(*a*) For the generalized coordinates given in (*1*) of Problem 4.3, (*2*) may be arranged as

$$\underbrace{\begin{bmatrix} 0 & 0\\ -4/5 & -3/5\\ 4/5 & -3/5\\ 0 & 0\end{bmatrix}}_{A_q} \begin{bmatrix}q_4\\ q_5\end{bmatrix} + \underbrace{\begin{bmatrix} -3/5 & -4/5 & 0 & 0\\ 4/5 & 3/5 & 0 & 0\\ 0 & 0 & -4/5 & 3/5\\ 0 & 0 & 3/5 & -4/5\end{bmatrix}}_{A_r} \begin{bmatrix}u_1\\ v_1\\ u_3\\ v_3\end{bmatrix} = 0 \tag{3}$$

from which

$$\begin{bmatrix} u_1 \\ v_1 \\ u_3 \\ v_3 \end{bmatrix} = -A_r^{)-1} A_q \begin{bmatrix} q_4 \\ q_5 \end{bmatrix}$$

$$= \begin{bmatrix} -115/21 & -20/7 & 0 & 0 \\ -20/7 & 15/7 & 0 & 0 \\ 0 & 0 & 20/7 & 15/7 \\ 0 & 0 & 15/7 & 20/7 \end{bmatrix} \begin{bmatrix} 0 & 0 \\ -4/5 & -3/5 \\ 4/5 & -3/5 \\ 0 & 0 \end{bmatrix} \begin{bmatrix} q_4 \\ q_5 \end{bmatrix} = \begin{bmatrix} 16/7 & 12/7 \\ -12/7 & -9/7 \\ 16/7 & -12/7 \\ 12/7 & -9/7 \end{bmatrix} \begin{bmatrix} q_4 \\ q_5 \end{bmatrix} \tag{4}$$

The relationship between $\{\Delta\}$ and $\{q\}$ becomes

$$\begin{bmatrix} \theta_1 \\ \theta_2 \\ \theta_3 \\ \hline u_2 \\ v_2 \\ \hline u_1 \\ v_1 \\ u_3 \\ v_3 \end{bmatrix} = \left[\begin{array}{ccc|cc} 1 & 0 & 0 & 0 & 0 \\ 0 & 1 & 0 & 0 & 0 \\ 0 & 0 & 1 & 0 & 0 \\ \hline 0 & 0 & 0 & 1 & 0 \\ 0 & 0 & 0 & 0 & 1 \\ \hline 0 & 0 & 0 & 16/7 & 12/7 \\ 0 & 0 & 0 & -12/7 & -9/7 \\ 0 & 0 & 0 & 16/7 & -12/7 \\ 0 & 0 & 0 & 12/7 & -9/7 \end{array}\right] \begin{bmatrix} q_1 \\ q_2 \\ q_3 \\ \hline q_4 \\ q_5 \end{bmatrix} \tag{5}$$

(*b*) Similarly, (*2*) may be rewritten according to the generalized coordinates given in (*2*) of Problem 4.4 as

$$\underbrace{\begin{bmatrix} -3/5 & 0 \\ 4/5 & 0 \\ 0 & -4/5 \\ 0 & 3/5 \end{bmatrix}}_{A_q} \begin{bmatrix} q_4 \\ q_5 \end{bmatrix} + \underbrace{\begin{bmatrix} -4/5 & 0 & 0 & 0 \\ 3/5 & -4/5 & -3/5 & 0 \\ 0 & 4/5 & -3/5 & 3/5 \\ 0 & 0 & 0 & -4/5 \end{bmatrix}}_{A_r} \begin{bmatrix} v_1 \\ u_2 \\ v_2 \\ v_3 \end{bmatrix} = 0 \tag{6}$$

Expressing v_1, u_2, v_2, and v_3 in terms of q_4 and q_5, we have

$$\begin{bmatrix} v_1 \\ u_2 \\ v_2 \\ v_3 \end{bmatrix} = -A_r^{)-1} A_q \begin{bmatrix} q_4 \\ q_5 \end{bmatrix}$$

$$= \begin{bmatrix} 5/4 & 0 & 0 & 0 \\ -15/32 & 5/8 & -5/8 & -15/32 \\ -5/8 & 5/6 & 5/6 & 5/8 \\ 0 & 0 & 0 & 5/4 \end{bmatrix} \begin{bmatrix} -3/5 & 0 \\ 4/5 & 0 \\ 0 & -4/5 \\ 0 & 3/5 \end{bmatrix} \begin{bmatrix} q_4 \\ q_5 \end{bmatrix} = \begin{bmatrix} -3/4 & 0 \\ 7/32 & 7/32 \\ 7/24 & 7/24 \\ 0 & 3/4 \end{bmatrix} \begin{bmatrix} q_4 \\ q_5 \end{bmatrix} \tag{7}$$

Thus, the relationship between $\{\Delta\}$ and $\{q\}$ is

$$\begin{bmatrix} \theta_1 \\ \theta_2 \\ \theta_3 \\ \hline u_1 \\ u_3 \\ \hline v_1 \\ u_2 \\ v_2 \\ v_3 \end{bmatrix} = \left[\begin{array}{ccc|cc} 1 & 0 & 0 & 0 & 0 \\ 0 & 1 & 0 & 0 & 0 \\ 0 & 0 & 1 & 0 & 0 \\ \hline 0 & 0 & 0 & 1 & 0 \\ 0 & 0 & 0 & 0 & 1 \\ \hline 0 & 0 & 0 & -3/4 & 0 \\ 0 & 0 & 0 & 7/32 & 7/32 \\ 0 & 0 & 0 & 7/24 & -7/24 \\ 0 & 0 & 0 & 0 & 3/4 \end{array}\right] \begin{bmatrix} q_1 \\ q_2 \\ q_3 \\ \hline q_4 \\ q_5 \end{bmatrix} \tag{8}$$

4.7. Derive the partial derivative and the time derivative of

$$\Delta_i = \Delta_i(q_1, q_2, \ldots, q_m, t)$$

and then prove (*4-7, 8*).

From the theory of partial derivatives we know that for a given function of two variables $f(x, y)$ we can have two first-partial derivatives that specify the rate of change of the function $f(x, y)$ when x is varied and y is held constant and when y is varied and x is held constant as

$$\frac{\partial f}{\partial x} = \lim_{\Delta x \to 0} \frac{f(x + \Delta x, y) - f(x, y)}{\Delta x} \tag{1}$$

and

$$\frac{\partial f}{\partial y} = \lim_{\Delta y \to 0} \frac{f(x, y + \Delta y) - f(x, y)}{\Delta y} \tag{2}$$

Partial differentiation of (*1* and *2*) yields four second-partial derivatives

$$\frac{\partial^2 f}{\partial x^2} \qquad \frac{\partial^2 f}{\partial y \partial x} \qquad \frac{\partial^2 f}{\partial y^2} \qquad \frac{\partial^2 f}{\partial x \partial y} \tag{3}$$

where $\partial^2 f/\partial x \partial y$ is the derivative of $\partial f/\partial y$ with respect to x.

If the variables x and y in $f(x, y)$ are functions of t, the total derivative of $f(x, y)$ may be obtained as

$$\frac{df}{dt} = \frac{\partial f}{\partial x}\frac{dx}{dt} + \frac{\partial f}{\partial y}\frac{dy}{dt} \tag{4}$$

From (*4-1* and *4-3*) the displacements Δ_i may be expressed in terms of the generalized coordinates q's as

$$\Delta_i = \Delta_i(q_1, q_2, \ldots, q_k, \ldots, q_m, t) = \Delta_i(q\text{'s}, t) \tag{5}$$

where q's are a function of t. Thus the first derivative of (*5*) with respect to q_k is similar to (*1*) or (*2*) as

$$\frac{\partial \Delta_i}{\partial q_k} = \frac{\partial \Delta_i(q\text{'s}, t)}{\partial q_k} \tag{6}$$

Following (*4*) the total derivative of (*5*) with respect to t yields

$$\frac{d\Delta_i}{dt} = \dot{\Delta}_i = \frac{\partial \Delta_i(q\text{'s}, t)}{\partial q_1}\dot{q}_1 + \frac{\partial \Delta_i(q\text{'s}, t)}{\partial q_2}\dot{q}_2 + \cdots + \frac{\partial \Delta_i(q\text{'s}, t)}{\partial q_k}\dot{q}_k + \cdots + \frac{\partial \Delta_i(q\text{'s}, t)}{\partial q_m}\dot{q}_m \tag{7}$$

in which $\partial \Delta_i(q\text{'s}, t)/\partial q$ is independent of velocities $\dot{q}_1, \dot{q}_2, \ldots, \dot{q}_m$, so that the partial derivative of (*7*) with respect to $\dot{q}_k$ gives

$$\frac{\partial \dot{\Delta}_i}{\partial \dot{q}_k} = \frac{\partial \Delta_i(q\text{'s}, t)}{\partial q_k} \tag{8}$$

Note the right side of (*8*) is the same as the right side of (*6*). Thus

$$\frac{\partial \Delta_i}{\partial q_k} = \frac{\partial \dot{\Delta}_i}{\partial \dot{q}_k} \tag{9}$$

which is (*4-7*).

Following (*6*) we may derive (*4-5*) from (*4-4*) in which $\dot{q}_1, \dot{q}_2, \ldots, \dot{q}_m$ are independent of $q_1, q_2, \ldots, q_m$. Applying (*4*) to (*6*) yields (*4-6*) which is the total derivative with respect to t. Comparing the identical terms in (*4-5*) and (*4-6*), we derive (*4-8*).

4.8. Discuss typical examples of scleronomic, rheonomic, and holonomic systems.

(*a*) Since the scleronomic system is defined as a system with constraints independent of time, the constraint conditions may be expressed as

$$a_s(q_1, q_2, q_3, \ldots, q_m) = 0, \qquad s = 1, 2, \ldots, r \tag{1}$$

which represents a smooth surface on which the motion of the mass is confined. A typical case is the vibration without damping of various structural systems.

(*b*) For the rheonomic system the motion of the mass is confined to the surface expressed as

$$a_s(q_1, q_2, \ldots, q_m, t) = 0, \qquad s = 1, 2, \ldots, r \tag{2}$$

which implies that the surface is in motion. For example, constraint equations may be obtained from force-equilibrium considerations in which the forces involved depend upon velocities and displacements. Damped vibration of various structural systems are examples of this situation.

(*c*) Holonomic constraints can be typically represented by (*2*) of integrable form. If the holonomic constraint does not contain the time explicitly and hence is of the form (*1*), the system is scleronomic as defined by (*a*). If the holonomic constraint is of the type (*2*) the system is rheonomic as discussed in (*b*).

GENERALIZED CAUSES

4.9. Using the generalized coordinates given in (*6*) of Problem 4.4 find the generalized causes for the external loads shown in Fig. P-4.9 by employing (*4-14*).

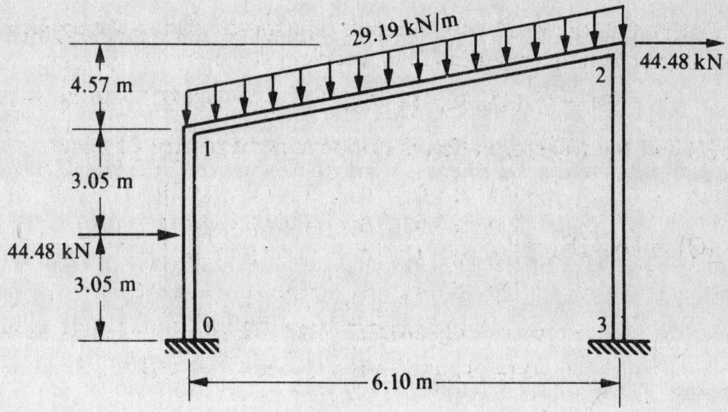

Fig. P-4.9

From (*4-14*),

$$Q_i = \sum_{k=1}^{p} F_k \frac{\partial \Delta_k}{\partial q_i} \tag{1}$$

where $\partial\Delta_k/\partial q_i$ can be determined from $\{\Delta\} = [A]\{q\}$ as $[\partial\Delta/\partial q] = [A]$. Substituting the generalized coordinates chosen in (*6*) of Problem 4.4 yields

$$\left[\frac{\partial \Delta}{\partial q}\right] = [A] = \begin{bmatrix} 1 & 0 & 0 \\ 0 & 1 & 0 \\ 0 & 0 & 1 \\ 0 & 0 & 0 \\ 0 & 0 & 1 \\ 0 & 0 & 0 \end{bmatrix} \tag{2}$$

F_k in (*1*) represents the fixed-end moment or force, due to the applied loads, acting at and along the displacement coordinate Δ_k. From Fig. P-4.9 and the original coordinates arranged in (*2*), we have

$$\{F\} = [-56.50 \quad 90.39 \quad 22.24 \quad -88.96 \quad 44.48 \quad -88.96]^T \tag{3}$$

Substituting (*2*) and (*3*) into (*1*) gives

$$\{Q\} = [A]^T\{F\} = [-56.50 \quad 90.39 \quad 66.72]^T \tag{4}$$

which are the generalized forces corresponding to q_1, q_2, and q_3.

4.10. Using the generalized coordinates given in (*5*) of Problem 4.5 find $\{Q\}$ for the external loads shown in Fig. P-4.10.

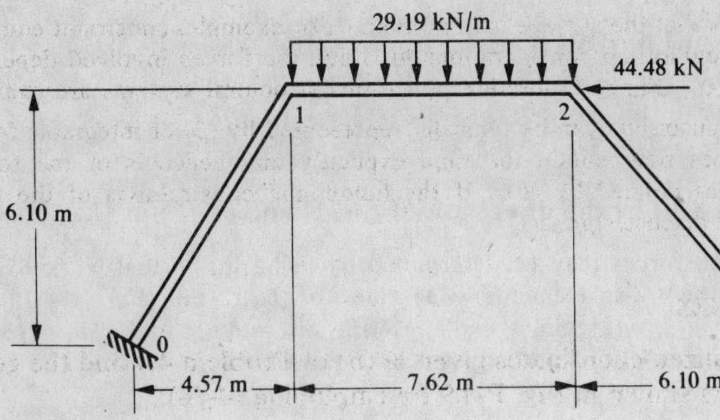

Fig. P-4.10

From Fig. P-4.10, the fixed-end moments and forces, due to the applied loads, acting at and along the original coordinates arranged in (*5*) of Problem 4.5 are

$$\{F\} = [-141.232 \quad 141.232 \quad 0 \quad 0 \quad -111.21 \quad -44.48 \quad -111.21]^T \tag{1}$$

Using (*1*) and the transformation matrix in (*5*) of Problem 4.5 gives

$$\{Q\} = [A]^T\{F\} = \begin{bmatrix} 1 & 0 & 0 & 0 & 0 & 0 & 0 \\ 0 & 1 & 0 & 0 & 0 & 0 & 0 \\ 0 & 0 & 1 & 0 & 0 & 0 & 0 \\ 0 & 0 & 0 & 1 & -3/4 & 1 & 4/3 \end{bmatrix}$$

$$\times [-141.232 \quad 141.232 \quad 0 \quad 0 \quad -111.21 \quad -44.48 \quad -111.21]^T$$

$$= [-141.232 \quad 141.232 \quad 0 \quad -109.35]^T \tag{2}$$

4.11. Using the generalized coordinates given in (*8*) of Problem 4.6 find $\{Q\}$ for the external loads shown in Fig. P-4.11.

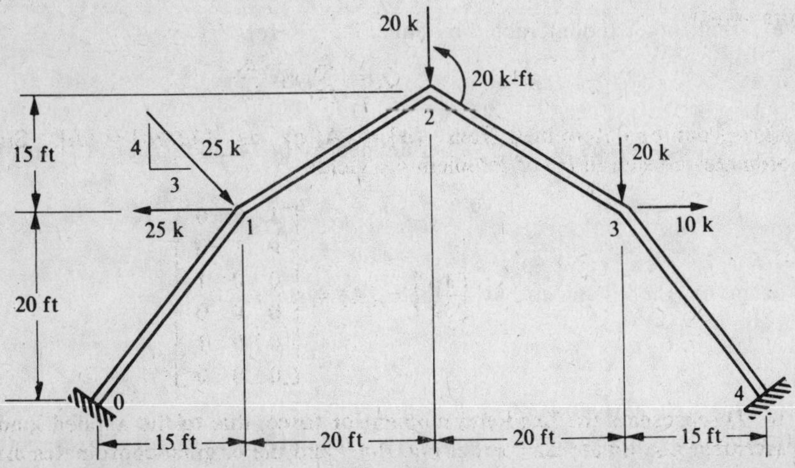

Fig. P-4.11

Since the external forces are acting at the structural nodes, $\{F\}$ can be obtained only by resolving the applied moments and forces along the original coordinates as

$$\{F\} = [0 \quad 20. \quad 0 \quad -10. \quad 10. \quad -20. \quad 0 \quad -20. \quad -20.]^T \tag{1}$$

Using the transformation matrix in (8) of Problem 4.6 we have $\{Q\} = [A]^T\{F\}$ as

$$\{Q\} = \begin{bmatrix} 1 & 0 & 0 & 0 & 0 & 0 & 0 & 0 \\ 0 & 1 & 0 & 0 & 0 & 0 & 0 & 0 \\ 0 & 0 & 1 & 0 & 0 & 0 & 0 & 0 \\ 0 & 0 & 0 & 1 & 0 & -3/4 & 7/32 & 7/24 & 0 \\ 0 & 0 & 0 & 0 & 1 & 0 & 7/32 & -7/24 & 3/4 \end{bmatrix} [0 \quad 20. \quad 0 \quad -10. \quad 10. \quad -20. \quad 0 \quad -20. \quad -20]^T$$

$$= [0 \quad 20. \quad 0 \quad -5/6 \quad 5/6] \tag{2}$$

4.12. Rework Problem 4.10 by the direct use of (4-11) and (4-13) in U.S. customary units.

The generalized forces may be determined by using the virtual work δW done by the applied forces on the virtual displacements $\{\delta\Delta\}$ due to $\{\delta q\}$. For Fig. P-4.10 we have $\delta\theta_1 = \delta q_1$, $\delta\theta_2 = \delta q_2$, $\delta\theta_3 = \delta q_3$, $\delta u_1 = \delta q_4$, $\delta v_1 = -\frac{3}{4}\delta q_4$, $\delta u_2 = \delta q_4$, and $\delta v_2 = \frac{4}{3}\delta q_4$. Thus using $\{F\}$ given in (1) of Problem 4.10 for virtual work (4-11), we have

$$\delta W = \{F\}^T\{\delta\Delta\}$$

$$= [-104.167 \quad 104.167 \quad 0. \quad 0. \quad -25. \quad -10. \quad -25.]\left[\delta q_1 \quad \delta q_2 \quad \delta q_3 \quad \delta q_4 \quad -\frac{3}{4}\delta q_4 \quad \delta q_4 \quad \frac{4}{3}\delta q_4\right]^T$$

$$= \left(-104.167\delta q_1 + 104.167\delta q_2 + 0\delta q_3 - \frac{295}{12}\delta q_4\right) \tag{1}$$

Comparing the above with (4-13) expressed as

$$\delta W = \sum_{i=1}^{4} Q_i \delta q_i \tag{2}$$

we have

$$Q_1 = -104.167 \text{ k-ft} \qquad Q_2 = 104.167 \text{ k-ft} \qquad Q_3 = 0 \qquad Q_4 = -24.583 \text{ k}$$

which checks with (2) of Problem 4.10 in SI units.

LAGRANGE'S EQUATION

4.13. Derive the symbolic Lagrange's equation (4-21).

Consider the lumped mass system of Sec. 4.1. The position of this system at any instant t is defined by

$$\Delta_k = \Delta_k(q_1, q_2, \ldots, q_m) \tag{1}$$

and the external forces (counteracted by equal inertial forces) by

$$F_k = \mu_k \ddot{\Delta}_k \tag{2}$$

where $k = 1, 2, \ldots, p$.

The generalized forces of this system defined by (4-14) in terms of (2) are then

$$Q_i = \sum_{k=1}^{p} F_k \frac{\partial \Delta_k}{\partial q_i} = \sum_{k=1}^{p} \mu_k \ddot{\Delta}_k \frac{\partial \Delta_k}{\partial q_i} \tag{3}$$

where $i = 1, 2, \ldots, m$ and $p \geq m$.

By definition, the kinetic energy of the entire system is

$$T = \frac{1}{2}\sum_{k=1}^{p} \mu_k \dot{\Delta}_k^2 \tag{4}$$

where $\dot{\Delta}_k$ is the time derivative of (1).

The partial derivative and the change in partial derivative of (4) are respectively

$$\frac{\partial T}{\partial q_i} = \sum_{k=1}^{p} \mu_k \dot{\Delta}_k \frac{\partial \dot{\Delta}_k}{\partial q_i} \qquad \text{and} \qquad \frac{\partial T}{\partial \dot{q}_i} = \sum_{k=1}^{p} \mu_k \dot{\Delta}_k \frac{\partial \dot{\Delta}_k}{\partial \dot{q}_i} = \sum_{k=1}^{p} \mu_k \dot{\Delta}_k \frac{\partial \Delta_k}{\partial q_i} \tag{5, 6}$$

where the second form of (6) is based on (4-7).

The total derivative of (6) in terms of (4-8) is then

$$\frac{d}{dt}\left(\frac{\partial T}{\partial \dot{q}_i}\right) = \sum_{k=1}^{p} \mu_k \left(\ddot{\Delta}_k \frac{\partial \Delta_k}{\partial q_i} + \dot{\Delta}_k \frac{d}{dt}\frac{\partial \Delta_k}{\partial q_i}\right)$$

$$= \sum_{k=1}^{p} \mu_k \left(\ddot{\Delta}_k \frac{\partial \Delta_k}{\partial q_i} + \dot{\Delta}_k \frac{\partial \dot{\Delta}_k}{\partial q_i}\right) \tag{7}$$

where the last term in (7) is based on (4-8).

Finally recognizing in (7) the right-side terms of (3) and (5), we may write (7) as

$$\frac{d}{dt}\left(\frac{\partial T}{\partial \dot{q}_i}\right) = \frac{\partial T}{\partial q_i} + Q_i \tag{8}$$

which is the symbolic Lagrange's equation (4-21). For the given system, m such equations may be written and they define the motion of the system.

4.14. Derive the standard Lagrange's equation (4-22) and then using this equation derive the differential equation of motion for the pendulum shown in Fig. P-4.14.

From (4-20),

$$Q_i = Q_{V,i} + Q_{D,i} + Q_{W,i} \tag{1}$$

Substituting (1) into (8) of the preceding problem and then using (4-17), (4-18), and (4-19) we have

$$\frac{d}{dt}\left(\frac{\partial T}{\partial \dot{q}_i}\right) - \frac{\partial T}{\partial q_i} + \frac{\partial V}{\partial q_i} + \frac{\partial D}{\partial \dot{q}_i} = \frac{\partial W}{\partial q_i} \tag{2}$$

Employing the Lagrangian function $L = T - V$ in (2) gives (4-22) as

$$\frac{d}{dt}\left(\frac{\partial L}{\partial \dot{q}_i}\right) - \frac{\partial L}{\partial q_i} + \frac{\partial D}{\partial \dot{q}_i} = \frac{\partial W}{\partial q_i}, \qquad i = 1, 2, \ldots, m \tag{3}$$

For the pendulum shown in Fig. P-4.14, we have $x = l \sin\theta$, $y = l\cos\theta$, $\dot{x} = l\dot{\theta}\cos\theta$, and $\dot{y} = -l\dot{\theta}\sin\theta$. Using θ as generalized coordinate

$$V = \mu g l(1 - \cos\theta) \tag{4}$$

$$T = \tfrac{1}{2}\mu\dot{x}^2 + \tfrac{1}{2}\mu\dot{y}^2 = \tfrac{1}{2}\mu l^2\dot{\theta}^2 \tag{5}$$

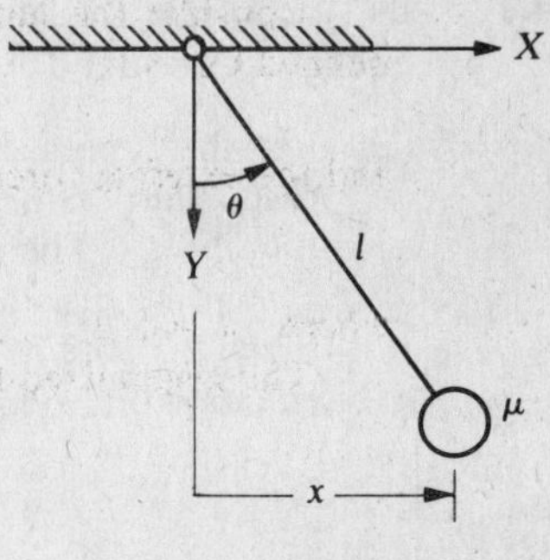

Fig. P-4.14

and substituting (4, 5) into the Lagrangian function gives

$$L = \tfrac{1}{2}\mu l^2\dot{\theta}^2 - \mu g l(1 - \cos\theta) \tag{6}$$

from which

$$\frac{d}{dt}\left(\frac{\partial L}{\partial \dot{\theta}}\right) = \mu l^2 \ddot{\theta} \tag{7}$$

$$\frac{\partial L}{\partial \theta} = -\mu g l \sin\theta \tag{8}$$

Upon the substitution of (7, 8) into (3) with $D = W = 0$ the differential equation becomes

$$\ddot{\theta} + (g/l)\sin\theta = 0 \tag{9}$$

Since the force shown in (4) has potential, the configuration of equilibrium may be obtained by satisfying (4-16). Substituting (4) into (4-16) yields

$$\frac{\partial V}{\partial \theta} = \mu g l \sin\theta = 0 \tag{10}$$

Therefore the configuration of equilibrium corresponds to $\theta = 0$.

SPECIAL CASES

4.15. Using the special Lagrange's equation (*4-23b*) derive the differential equation of motion (*2-1a*).

From (*4-23b*)

$$\frac{d}{dt}\left(\frac{\partial T}{\partial \dot{q}_i}\right) + \frac{\partial V}{\partial q_i} = \frac{\partial W}{\partial q_i} \tag{1}$$

Referring to Fig. 2.1 we write

$$T = \tfrac{1}{2}\mu\dot{u}^2 \qquad V = \tfrac{1}{2}k_u u^2 \qquad \text{and} \qquad \delta W = P_x(t)\delta u$$

from which

$$\frac{d}{dt}\left(\frac{\partial T}{\partial \dot{u}}\right) = \mu\ddot{u} \qquad \frac{\partial V}{\partial u} = k_u u \qquad \text{and} \qquad \frac{\partial W}{\partial u} = P_x(t)$$

Upon substitution, (*1*) becomes (*2-1a*) as

$$\mu\ddot{u} + k_u\mu = P_x(t) \tag{2}$$

4.16. Using (*4-23d*) derive the differential equation of motion (*2-15a*).

Referring to Fig. 2.6, we may write the following energy equations:

$$D = \tfrac{1}{2}c_u\dot{u}^2 \qquad T = \tfrac{1}{2}\mu\dot{u}^2 \qquad V = \tfrac{1}{2}k_u u^2 \qquad \text{and} \qquad \delta W = P_x(t)\delta u$$

where the dissipation function D is defined as one half of the actual dissipation of energy per unit time. Using (*4-23d*), we write

$$\frac{d}{dt}\frac{\partial T}{\partial \dot{u}} = \mu\ddot{u} \qquad \frac{\partial V}{\partial u} = k_u u \qquad \frac{\partial D}{\partial \dot{u}} = c_u\dot{u} \qquad \text{and} \qquad \frac{\partial W}{\partial u} = P_x(t)$$

and then

$$\mu\ddot{u} + c_u\dot{u} + k_u u = P_x(t)$$

which is confirmed by (*2-15a*).

ENERGY FUNCTIONS

4.17. Derive the algebraic and matrix forms of the (*a*) potential, (*b*) kinetic, and (*c*) dissipation energy introduced in (*4-25*), (*4-28*), and (*4-31*) respectively.

(*a*) Let the generalized coordinates $\{q\}$ be chosen for describing the geometric configuration corresponding to a stable equilibrium position at which a system vibrates with small initial disturbances. The generalized coordinates should be chosen to satisfy the condition that, when a system is in its stable equilibrium position, the coordinates are zero, i.e., $q_1 = q_2 = \cdots = q_m = 0$. Since the potential energy of a system is a function of coordinates only, we may write the energy equation in Taylor series form as

$$V(q_1, q_2, \ldots, q_m) = V_e + \sum_{i=1}^{m}\left(\frac{\partial V}{\partial q_i}\right)_e q_i + \sum_{i=1}^{m}\sum_{j=1}^{m}\frac{1}{2}\left(\frac{\partial^2 V}{\partial q_i \partial q_j}\right)_e q_i q_j + \cdots \tag{1}$$

where the subscript e denotes the stable equilibrium position. Note (*1*) is a general form of Taylor series. However, for the present case, the potential energy is measured from the equilibrium position and should be a minimum at equilibrium. Thus we have

$$V_e = 0 \qquad \left(\frac{\partial V}{\partial q_i}\right)_e = 0 \tag{2}$$

Because of small oscillations, all terms higher than the second order shown in (*1*) are negligible. So

$$V = \frac{1}{2}\sum_{i=1}^{m}\sum_{j=1}^{m}\left(\frac{\partial^2 V}{\partial q_i \partial q_j}\right)_e q_i q_j \tag{3}$$

or

$$V = \frac{1}{2}\sum_{i=1}^{m}\sum_{j=1}^{m} k_{ij} q_i q_j = \frac{1}{2}\{q\}^T[k]\{q\} \tag{4}$$

If we take the partial derivative of V with respect to a coordinate q_l, then the terms associated with q_l in (4) are the elements in row l and column l and other elements will not contribute to their derivation. From the reciprocal relationship, $[k]$ is known to be symmetric and thus the sum of the terms of the product of row l and column l is

$$\tfrac{1}{2}[2k_{11}q_lq_1 + 2k_{l2}q_lq_2 + \cdots + k_{ll}q_lq_l + \cdots + 2k_{lm}q_lq_m] \tag{5}$$

Taking the partial derivative with respect to q_l, we have

$$\frac{\partial V}{\partial q_l} = k_{l1}q_1 + k_{l2}q_2 + \cdots + k_{ll}q_l + \cdots + k_{lm}q_m \tag{6}$$

which represents the nodal force at l due to displacements at m coordinates. If we now take a partial derivative with respect to q_k in (6),

$$\frac{\partial^2 V}{\partial q_l \partial q_k} = k_{lk} \tag{7}$$

which is a single stiffness coefficient.

(b) The kinetic energy of a system can be expressed as

$$T = \frac{1}{2}\sum_{j=1}^{p} \mu_j \dot{\Delta}_j^2 \tag{8}$$

where $\Delta_j = \Delta_j(q_1, q_2, \ldots, q_m)$. Thus

$$\dot{\Delta}_j = \sum_{i=1}^{m} \frac{\partial \Delta_j}{\partial q_i}\dot{q}_i \qquad \dot{\Delta}_j^2 = \sum_{i=1}^{m}\sum_{k=1}^{m} \frac{\partial \Delta_j}{\partial q_i}\frac{\partial \Delta_j}{\partial q_k}\dot{q}_i\dot{q}_k \tag{9, 10}$$

Substituting (9, 10) into (8) yields

$$T = \frac{1}{2}\sum_{i=1}^{m}\sum_{k=1}^{m} \dot{q}_i\dot{q}_k \sum_{j=1}^{p} \mu_j \frac{\partial \Delta_j}{\partial q_i}\frac{\partial \Delta_j}{\partial q_k} \tag{11}$$

Let

$$\mu_{ik} = \sum_{j=1}^{p} \mu_j \frac{\partial \Delta_j}{\partial q_i}\frac{\partial \Delta_j}{\partial q_k} \tag{12}$$

Then (11) becomes

$$T = \frac{1}{2}\sum_{i=1}^{m}\sum_{k=1}^{m} \mu_{ik}\dot{q}_i\dot{q}_k = \frac{1}{2}\{\dot{q}\}^T[\mu]\{\dot{q}\} \tag{13}$$

(c) The dissipation energy may be expressed as

$$D = \frac{1}{2}\sum_{j=1}^{p} c_j \dot{\Delta}_j^2 \tag{14}$$

Using (10) in (14),

$$D = \frac{1}{2}\sum_{i=1}^{m}\sum_{k=1}^{m} \dot{q}_i\dot{q}_k \sum_{j=1}^{p} c_j \frac{\partial \Delta_j}{\partial q_i}\frac{\partial \Delta_j}{\partial q_k} \tag{15}$$

Let

$$c_{ik} = \sum_{j=1}^{p} c_j \frac{\partial \Delta_j}{\partial q_i}\frac{\partial \Delta_j}{\partial q_k} \tag{16}$$

The dissipation energy finally becomes

$$D = \frac{1}{2}\sum_{i=1}^{m}\sum_{k=1}^{m} c_{ik}\dot{q}_i\dot{q}_k = \frac{1}{2}\{\dot{q}\}^T[c]\{\dot{q}\} \tag{17}$$

SMALL OSCILLATION IN GENERALIZED COORDINATES

4.18. The system shown in Fig. P-4.18 is subject to initial conditions: $\Delta_{10} = \Delta_{20} = 1$ in, $\dot{\Delta}_{10} = 4$ in/s, $\dot{\Delta}_{20} = 6$ in/s. Use (a) the matrix form (4-37) and (4-38) and (b) the algebraic form (4-25) and (4-28) to find the equations of motion and the displacements at $t = 1/4$ s. Let $\mu_1 = \mu_2 = 1$ k-s^2/in and $k_1 = k_2 = 1$ k/in.

(*a*) For free vibration without damping, (*4-23a*) gives

$$\frac{d}{dt}\left(\frac{\partial T}{\partial \dot{q}_i}\right) + \frac{\partial V}{\partial q_i} = 0 \qquad (1)$$

Substituting (*4-37*) and (*4-38*) into (*1*) gives

$$\{\ddot{\eta}\} + [\Lambda]\{\eta\} = 0 \qquad (2)$$

Note that $\{q\}$ are generalized coordinates of $\{\Delta\}$ and $\{\eta\}$ are normal coordinates of $\{q\}$ which, in this case, are eigenvectors of $\{\Delta\}$. Thus this is identical to modal matrix analysis which gives a general equation of motion (*3-28*) as

$$\{\Delta\} = [h]\,\text{Diag.}\,[\cos \omega t][h]^T[\mu]\{\Delta_0\} + [h]\,\text{Diag.}\,[\sin \omega t]\,\text{Diag.}\left[\frac{1}{\omega}\right][h]^T[\mu]\{\dot{\Delta}_0\} \qquad (3)$$

in which

$$\{\Delta_0\} = \begin{bmatrix} 1 \\ 1 \end{bmatrix} \qquad \{\dot{\Delta}_0\} = \begin{bmatrix} 4 \\ 6 \end{bmatrix} \qquad [\mu] = \begin{bmatrix} 1 & 0 \\ 0 & 1 \end{bmatrix} \qquad [k] = \begin{bmatrix} 1 & -1 \\ -1 & 2 \end{bmatrix} \qquad (4a, b, c, d)$$

Employing modal matrix analysis (Problems 3.5 and 3.6) gives

$$\{X\} = \begin{bmatrix} q_1 & q_2 \\ 0.618\,q_1 & -1.618\,q_2 \end{bmatrix} \qquad \left[\frac{1}{\omega}\right] = \begin{bmatrix} 1/0.618 & 0 \\ 0 & 1/1.618 \end{bmatrix} \qquad [h] = \begin{bmatrix} 0.8507 & 0.5257 \\ 0.5257 & -0.8507 \end{bmatrix} \qquad (5a, b, c)$$

Substituting (*4, 5*) into (*3*) and evaluating the displacements at $t = 1/4$ s, we have

$$\{\Delta\} = \begin{bmatrix} 2.004 \\ 2.448 \end{bmatrix} \text{in} \qquad (6)$$

(*b*) The potential energy may be expressed as

$$V = \tfrac{1}{2}k_2\Delta_2^2 + \tfrac{1}{2}k_1(\Delta_1 - \Delta_2)^2 \qquad (7)$$

Using the eigenvectors (*5a*) as generalized coordinates expressed below,

$$\{\Delta\} = \begin{bmatrix} 1 & 1 \\ 1 & -1.618 \end{bmatrix}\begin{bmatrix} q_1 \\ q_2 \end{bmatrix} = [A]\{q\} \qquad (8)$$

Thus (*7*) becomes

$$V = \frac{1}{2}(0.5278\,q_1^2 + 9.472\,q_2^2) = \frac{1}{2}\begin{bmatrix} q_1 \\ q_2 \end{bmatrix}^T \begin{bmatrix} 0.5278 & 0 \\ 0 & 9.472 \end{bmatrix}\begin{bmatrix} q_1 \\ q_2 \end{bmatrix} \qquad (9)$$

For the kinetic energy (*4-28*),

$$T = \frac{1}{2}\begin{bmatrix} \dot{q}_1 \\ \dot{q}_2 \end{bmatrix}^T [\mu]\begin{bmatrix} \dot{q}_1 \\ \dot{q}_2 \end{bmatrix} = \frac{1}{2}\begin{bmatrix} \dot{q}_1 \\ \dot{q}_2 \end{bmatrix}^T \begin{bmatrix} 1.3819 & 0 \\ 0 & 3.618 \end{bmatrix}\begin{bmatrix} \dot{q}_1 \\ \dot{q}_2 \end{bmatrix} \qquad (10)$$

where μ_{ik} is obtained by using μ_j and $\partial\Delta/\partial q$ in (*4-30*):

$$\begin{aligned} \mu_{11} &= 1(1)(1) + 1(0.618)(0.618) = 1.3819 & \mu_{12} &= 1(1)(1) + 1(0.618)(-1.618) = 0 \\ \mu_{21} &= 1(1)(1) + 1(-1.618)(0.618) = 0 & \mu_{22} &= 1(1)(1) + 1(-1.618)(-1.618) = 3.618 \end{aligned} \qquad (10a)$$

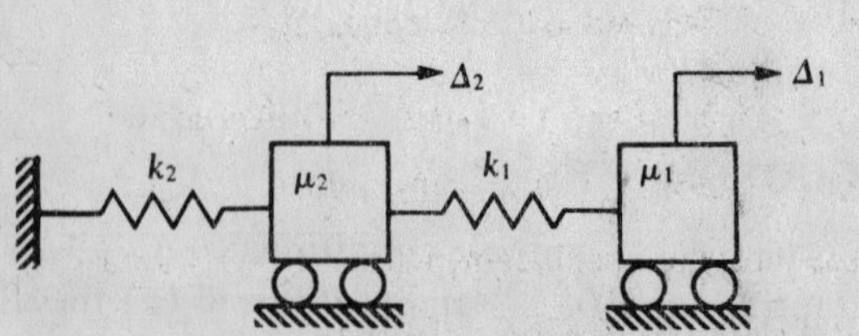

Fig. P-4.18

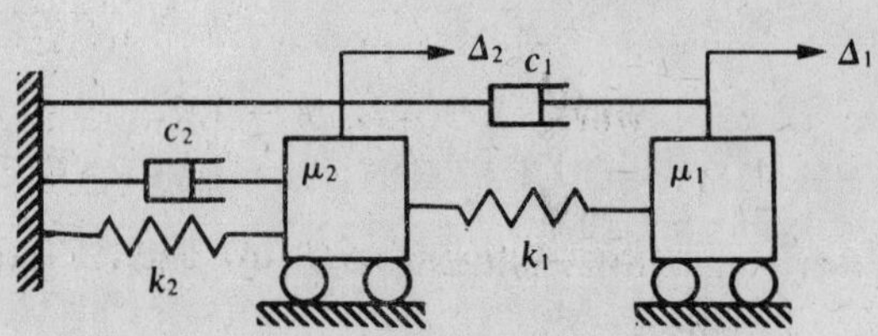

Fig. P-4.19

Applying (*1*) to (*9*) and (*10*) we have

$$\begin{bmatrix}\ddot{q}_1\\ \ddot{q}_2\end{bmatrix} + \begin{bmatrix}0.3879 & 0\\ 0 & 2.618\end{bmatrix}\begin{bmatrix}q_1\\ q_2\end{bmatrix} = 0 \tag{11}$$

of which the general solution is

$$\{q\} = \text{Diag. }[\cos \omega t]\{q_0\} + \text{Diag. }\left([\sin \omega t]\left[\frac{1}{\omega}\right]\right)\{\dot{q}_0\} \tag{12}$$

From (*8*) and (*11*).

$$\{q_0\} = [A]^{-1}\{\Delta_0\} \qquad \{\dot{q}_0\} = [A]^{-1}\{\dot{\Delta}_0\}$$

$$[A]^{-1} = \begin{bmatrix}0.7236 & 0.4472\\ 0.2763 & -0.4472\end{bmatrix} \qquad \left\{\frac{1}{\omega}\right\} = \begin{bmatrix}1/0.618 & 0\\ 0 & 1/1.618\end{bmatrix} \tag{13}$$

Substituting (*8*) and (*13*) into (*12*) and evaluating $\{\Delta\}$ at $t = 1/4$ s, we have the same solution as shown in (*6*).

4.19. The damped-spring-mass system shown in Fig. P-4.19 is subject to a suddenly applied force 100 k at μ_1 with initial conditions of $\Delta_{10} = \Delta_{20} = 1$ in, $\dot{\Delta}_{10} = 4$ in/s, and $\dot{\Delta}_{20} = 6$ in/s. Use (*a*) the matrix form (*4-37 to 4-40*) and (*b*) the algebraic form (*4-25*), (*4-28*), (*4-31*), and (*4-34*) to find the equations of motion and the displacement at 1/4 s. Let $\mu_1 = \mu_2 = 1$ k-s^2/in, $k_1 = k_2 = 1$ k/in, $c_1 = c_2 = 1$ k-s/in.

(*a*) The Lagrange's equation for forced vibration with damping is

$$\frac{d}{dt}\left(\frac{\partial T}{\partial \dot{q}_i}\right) + \frac{\partial V}{\partial q_i} + \frac{\partial D}{\partial \dot{q}_i} = \frac{\partial W}{\partial q_i} \tag{1}$$

Substituting T, V, D, and Q_W (*4-37 to 4-40*) into (*1*) we have

$$\{\ddot{\eta}\} + [2\Gamma]\{\dot{\eta}\} + [\Lambda]\{\eta\} = [h]^T\{W\} \tag{2}$$

which are identical to the differential equations of motion obtained by modal matrix analysis. Thus the solution of (*2*) is

$$\begin{aligned}\{\Delta\} &= \{\Delta\}_h + \{\Delta\}_p\\ &= [h]\Big\{\text{Diag. }[(e^{-\gamma t}/\lambda)(\lambda\cos\lambda t + \gamma\sin\lambda t)]\,[h]^T[\mu]\{\Delta_0\}\\ &\qquad + \text{Diag. }[(e^{-\gamma t})\sin\lambda t]\text{ Diag. }\left[\frac{1}{\lambda}\right][h]^T[\mu]\{\dot{\Delta}_0\}\Big\} + [h]\left\{\{h_{1k}\Phi_{1k}\} + \{h_{2k}\Phi_{2k}\}\right\}\end{aligned} \tag{3}$$

in which $\{\Delta\}_h$ and $\{\Delta\}_p$ are displacements associated with free and forced vibrations respectively; $[h]$, $[\mu]$, $\{\Delta_0\}$, $\{\dot{\Delta}_0\}$, and $\{1/\omega\}$ are the same as given in (*4*) and (*5*) of Problem 4.18; λ, γ, and Φ_{ik} may be calculated by similarly following Problem 3.8. From (*2*),

$$2\gamma_1 = 1 \quad\text{or}\quad \gamma_1 = 1/2 \qquad\text{and}\qquad 2\gamma_2 = 1 \quad\text{or}\quad \gamma_2 = 1/2 \tag{4}$$

Since $\lambda_i = \sqrt{\omega_i^2 - \gamma_i^2}$, we have

$$\lambda_1 = 0.3632 \qquad\text{and}\qquad \lambda_2 = 1.5388 \tag{5}$$

For the given load, the particular function may be obtained from (*2*) of Problem 2.21 as

$$\Phi_{jk} = \frac{P_{jk}}{\lambda_k(\gamma_k^2 + \lambda_k^2)}[\lambda_k - e^{-\gamma_k t}(\gamma_k \sin\lambda_k t + \lambda_k\cos\lambda_k t)] \tag{6}$$

where $j = 1, 2$; $k = 1, 2$; $P_1 = 100$ k; $P_2 = 0$. For $t = 1/4$ s, (*6*) becomes

$$\Phi_{11} = 2.86 \qquad \Phi_{12} = 2.85 \qquad \Phi_{21} = \Phi_{22} = 0 \tag{7}$$

Substituting (*4*), (*5*), (*6*), and (*7*) into (*3*) we have the displacements at $t = 1/4$ s as

$$\{\Delta\} = \{\Delta\}_h + \{\Delta\}_p = \begin{bmatrix}1.889\\ 2.279\end{bmatrix} + \begin{bmatrix}2.856\\ 0.005\end{bmatrix} = \begin{bmatrix}4.745\\ 2.284\end{bmatrix}\text{ in} \tag{8}$$

(*b*) For (*4-25*), (*4-28*), and (*4-34*), the matrices $[A]$, $[k]$, and $[\mu]$ are the same as shown in (*8*), (*9*), and (*10*) of Problem 4.18. The dissipation energy (*4-31*) may be expressed as

$$D = \frac{1}{2}\begin{bmatrix}\dot{q}_1\\ \dot{q}_2\end{bmatrix}^T [c]\begin{bmatrix}\dot{q}_1\\ \dot{q}_2\end{bmatrix} = \frac{1}{2}\begin{bmatrix}\dot{q}_1\\ \dot{q}_2\end{bmatrix}^T \begin{bmatrix}1.3819 & 0\\ 0 & 3.618\end{bmatrix}\begin{bmatrix}\dot{q}_1\\ \dot{q}_2\end{bmatrix} \tag{9}$$

where c_{ik} is obtained by using (*4-33*) in the similar manner shown in (*10a*) of the preceding problem.

The generalized load function may be obtained from (*4-34*) as

$$Q_{W1} = F_1\frac{\partial\Delta_1}{\partial q_1} + F_2\frac{\partial\Delta_2}{\partial q_1} = 100(1) + 0(1) = 100$$

$$Q_{W2} = F_1\frac{\partial\Delta_1}{\partial q_2} + F_2\frac{\partial\Delta_2}{\partial q_2} = 100(1) + 0(-1.618) = 100 \tag{10}$$

Applying Lagrange's equation (*1*), we have

$$\begin{bmatrix}1.3819 & 0\\ 0 & 3.618\end{bmatrix}\begin{bmatrix}\ddot{q}_1\\ \ddot{q}_2\end{bmatrix} + \begin{bmatrix}1.3819 & 0\\ 0 & 3.618\end{bmatrix}\begin{bmatrix}\dot{q}_1\\ \dot{q}_2\end{bmatrix} + \begin{bmatrix}0.5278 & 0\\ 0 & 9.472\end{bmatrix}\begin{bmatrix}q_1\\ q_2\end{bmatrix} = \begin{bmatrix}100.\\ 100.\end{bmatrix} \tag{11}$$

of which the general solution is

$$q_i = e^{-\gamma_i t}\left(q_{i0}\cos\lambda_i t + \frac{\dot{q}_{i0} + \gamma_i q_{i0}}{\lambda_i}\sin\lambda_i t\right) + \frac{Q_{Wi}}{\mu_i\lambda_i(\gamma_i^2 + \lambda_i^2)}[\lambda_i - e^{-\gamma_i t}(\gamma_i\sin\lambda_i t + \lambda_i\cos\lambda_i t)] \tag{12}$$

where $\gamma_1 = \gamma_2 = 1/2$, $\lambda_1 = 0.3632$, $\lambda_2 = 1.5388$, $\mu_1 = 1.3819$, and $\mu_2 = 3.618$. Since the relationship $\{\Delta\}$ and $\{q\}$ was given in (*8*) and (*13*) of Problem 4.18, when $t = 1/4$ s, (*12*) gives

$$\begin{bmatrix}q_1\\ q_2\end{bmatrix} = \begin{bmatrix}4.454\\ 0.290\end{bmatrix} \quad\text{and}\quad \{\Delta\} = [A]\{q\} = \begin{bmatrix}4.745\\ 2.284\end{bmatrix}\text{ in}$$

Supplementary Problems

BASIC CONCEPTS

4.20. Using (*4-2*) determine the number of degrees of freedom of Fig. P-4.20 and then show the multiple choices of generalized coordinates.

4.21. Rework Problem 4.3 and show other possible generalized coordinates.

Fig. P-4.20

4.22. Find $\{\Delta\} = [A]\{q\}$ for the structure given in Problem 4.20.

4.23. Find $\{\Delta\} = [A]\{q\}$ for Problem 4.21.

4.24. Use the generalized coordinates obtained in Problem 4.22 and find the generalized causes for the external load shown in Fig. P-4.20.

4.25. Find the generalized causes by using the transformation matrix obtained in Problem 4.23.

APPLICATIONS

4.26. Rework Problem 4.24 by the direct use of (*4-11*) and (*4-13*).

4.27. Calculate the natural frequencies of the system shown in Fig. P-4.27 by using Lagrange's equation for small amplitudes of motion.

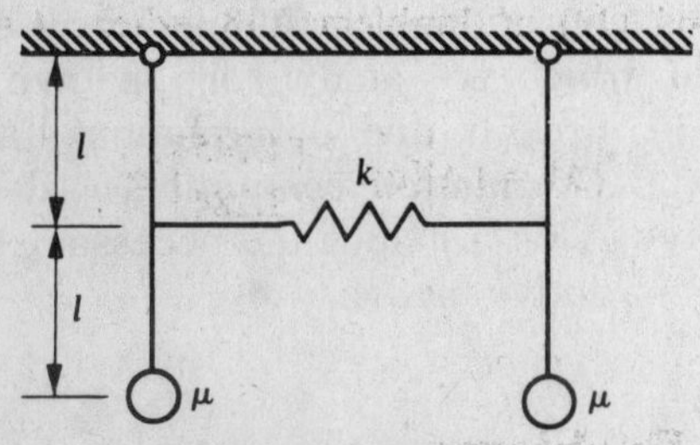

Fig. P-4.27

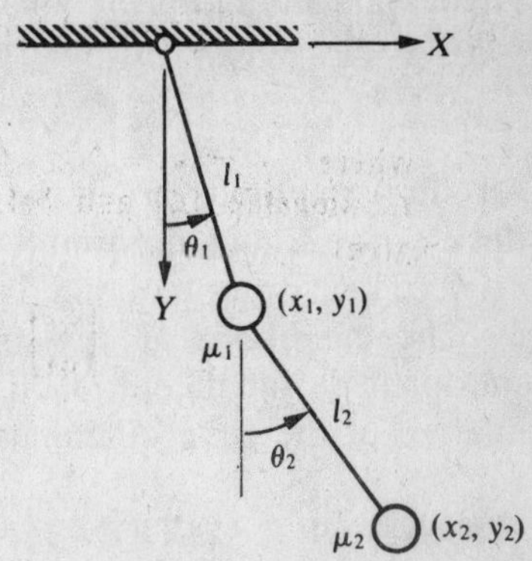

Fig. P-4.28

4.28. Find the natural frequencies of the double pendulum shown in Fig. P-4.28 by means of Lagrange's equation for small amplitudes of motion.

4.29. Rework Problem 4.18 for a suddenly applied force $P = 100$ k at μ_1 without initial conditions. Use the same generalized coordinates given in that problem.

4.30. Use the generalized coordinates $\{\Delta\} = \begin{bmatrix} 1 & 0 \\ 1 & -1 \end{bmatrix}\{q\}$ and rework Problem 4.29 by means of algebraic form.

4.31. Rework (*b*) of Problem 4.19 by using the generalized coordinates $\{\Delta\} = \begin{bmatrix} 1 & 0 \\ 1 & -1 \end{bmatrix}\{q\}$.

Chapter 5

Matrix Methods

5.1 INTRODUCTION

General

The nodal equations (*3-2*), derived from the conditions of dynamic equilibrium or by means of Lagrange's equations in the preceding chapters, can be also developed by methods analogous to those of static structural analysis, designated as

(*1*) The *dynamic transport matrix method* (II, pp. 32–37).

(*2*) The *dynamic flexibility matrix method* (II, pp. 48–60).

(*3*) The *dynamic stiffness matrix method* (II, pp. 188–199).

Their applications in the analysis of lumped parameter models is shown in this chapter; their extension to the analysis of distributed parameter models is developed in Chapters 6, 7, 8, and 9.

Concepts

All three methods treat the structural members connecting the lumped masses as elastic massless springs, the deformation characteristics of which are analytically known from static analysis. Their sole function is the transmission of stresses and deformations from node to node. The *conditions of dynamic equilibrium* and of *deformation compatibility*, defined by the system, must be satisfied at each node simultaneously. This provides the necessary basis for the formulation of the governing matrix equation of the respective method.

5.2 TRANSPORT MATRIX METHOD

Segmental Matrices

The *dynamic transport matrix method* operates (as does its static counterpart) in terms of three types of matrices: the *state vector matrices*, $\hat{H}^L_L$, $\hat{H}^R_R$, [7 × 1]; the *transport matrices*, $\hat{L}^{LR}_{LR}$, $\hat{L}^{RL}_{RL}$, [7 × 7]; and the *nodal (joint) effect matrices*, $\hat{J}^j_{LjR}$, $\hat{J}^j_{RjL}$, [7 × 7].

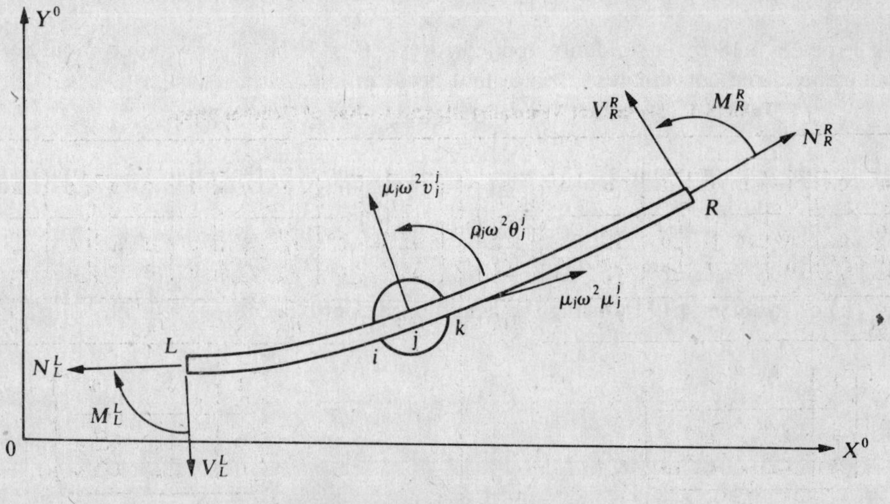

Fig. 5-1. Curved Bar—*LR*

For the isolated segment *LR* of Fig. 5-1, the transport matrix equations are

$$\hat{H}_L^L = \hat{T}_{Li}^{Lj}\hat{I}_{ik}^{j}\hat{T}_{kR}^{jR}\hat{H}_R^R = \hat{L}_{LR}^{LR}\hat{H}_R^R \qquad \hat{H}_R^R = \hat{T}_{Rk}^{Rj}\hat{I}_{ki}^{j}\hat{T}_{iL}^{jL}\hat{H}_L^L = \hat{L}_{RL}^{RL}\hat{H}_L^L \tag{5-1, 2}$$

where $\hat{L}_{LR}^{LR}$ and $\hat{L}_{RL}^{RL}$ are the dynamic transport matrices, $[7 \times 7]$. These matrices are results of the respective triple matrix product consisting of the *static transport matrices*, $\hat{T}_{Li}^{Lj}$, $\hat{T}_{kR}^{jR}$ or $\hat{T}_{Rk}^{Rj}$, $\hat{T}_{iL}^{jL}$, and the respective *inertia matrix*, $\hat{I}_{ik}^{j}$ or $\hat{I}_{ki}^{i}$ shown in Table 5-1.

The inertia matrices, $[7 \times 7]$, contain besides the unity 1 and the unit submatrices I, the *inertia submatrices*

$$j_{ik}^{j} = \text{Diag.}\,[\mu_j\omega^2, \mu_j\omega^2, \rho_j\omega^2] \qquad j_{ki}^{j} = \text{Diag.}\,[-\mu_j\omega^2, -\mu_j\omega^2, -\rho_j\omega^2] \tag{5-3, 4}$$

representing the effect of the respective inertia vector due to unit displacement. The coefficients of these matrices are products of the respective mass constant and the square of natural angular frequency. This formulation assumes a free motion, the components of which are harmonic.

The derivations of $\hat{T}_{LR}^{LR}$, $\hat{T}_{RL}^{RL}$ are already available elsewhere (II, pp. 38–42) and are only summarized here (Problems 5.1 and 5.2). The construction of the inertia submatrices based on the equilibrium of the node is shown in Problem 5.1.

Joint Conditions

Since each center of lumped mass can be considered a *structural joint*, the conditions at these points are called the *joint conditions.* To accommodate the interferences that may occur at joints, such as the intermediate rigid supports, elastic supports, hinges, guides, etc., the *joint effect matrix* developed for static analysis is used for this purpose.

These matrices, designated as $\hat{J}_{ik}^{j}$, $\hat{J}_{ki}^{j}$, are $[7 \times 7]$ matrices, the coefficients of which must satisfy the joint conditions and may include the inertia submatrices [Equations (*5-3, 4*)] as shown in Problem 5.2.

Transport Chain

The *application of the segmental transport matrices* and the joint effect matrices in dynamic analysis follows the pattern of static analysis. First the mass of each member is lumped at selected points and the real structure is represented by the ideal model, some constraints of which may be replaced (if preferable) by linear and/or angular springs.

For the three-span frame of Fig. 5-2*a* the column's deformation effects are characterized by their end stiffness matrices, and the remaining curved girders are treated as a three-span continuous girder on elastic supports (Fig. 5-2*b*).

Table 5-1 Segmental Transport Matrices—Bar of General Shape

$$\underbrace{\begin{bmatrix} 1 \\ \hline \sigma_L^L \\ \Delta_L^L \end{bmatrix}}_{\hat{H}_L^L} = \underbrace{\underbrace{\left[\begin{array}{c|cc} 1 & 0 & 0 \\ \hline 0 & t_{Li}^{Lj} & 0 \\ 0 & d_{Li}^{Lj} & s_{Li}^{Lj} \end{array}\right]}_{\hat{T}_{Li}^{Lj}} \underbrace{\left[\begin{array}{c|cc} 1 & 0 & 0 \\ \hline 0 & I & j_{ik}^{j} \\ 0 & 0 & I \end{array}\right]}_{\hat{I}_{ik}^{-j}} \underbrace{\left[\begin{array}{c|cc} 1 & 0 & 0 \\ \hline 0 & t_{kR}^{jR} & 0 \\ 0 & d_{kR}^{jR} & s_{kR}^{jR} \end{array}\right]}_{\hat{T}_{kR}^{jR}}}_{\hat{L}_{LR}^{LR} = \left[\begin{array}{c|cc} 1 & 0 & 0 \\ \hline 0 & L_{\sigma\sigma, LR}^{LR} & L_{\sigma\Delta, LR}^{LR} \\ 0 & L_{\Delta\sigma, LR}^{LR} & L_{\Delta\Delta, LR}^{LR} \end{array}\right]} \underbrace{\begin{bmatrix} 1 \\ \hline \sigma_R^R \\ \Delta_R^R \end{bmatrix}}_{\hat{H}_R^R}$$

$$\underbrace{\begin{bmatrix} 1 \\ \hline \sigma_R^R \\ \Delta_R^R \end{bmatrix}}_{\hat{H}_R^R} = \underbrace{\underbrace{\left[\begin{array}{c|cc} 1 & 0 & 0 \\ \hline 0 & t_{Rk}^{Rj} & 0 \\ 0 & d_{Rk}^{Rj} & s_{Rk}^{Rj} \end{array}\right]}_{\hat{T}_{Rk}^{Rj}} \underbrace{\left[\begin{array}{c|cc} 1 & 0 & 0 \\ \hline 0 & I & j_{ki}^{j} \\ 0 & 0 & I \end{array}\right]}_{\hat{I}_{ki}^{j}} \underbrace{\left[\begin{array}{c|cc} 1 & 0 & 0 \\ \hline 0 & t_{ij}^{jL} & 0 \\ 0 & d_{iL}^{j} & s_{iL}^{jL} \end{array}\right]}_{\hat{T}_{iL}^{jL}}}_{\hat{L}_{RL}^{RL} = \left[\begin{array}{c|cc} 1 & 0 & 0 \\ \hline 0 & L_{\sigma\sigma, RL}^{RL} & L_{\sigma\Delta, RL}^{RL} \\ 0 & L_{\Delta\sigma, RL}^{RL} & L_{\Delta\Delta, RL}^{RL} \end{array}\right]} \underbrace{\begin{bmatrix} 1 \\ \hline \sigma_L^L \\ \Delta_L^L \end{bmatrix}}_{\hat{H}_L^L}$$

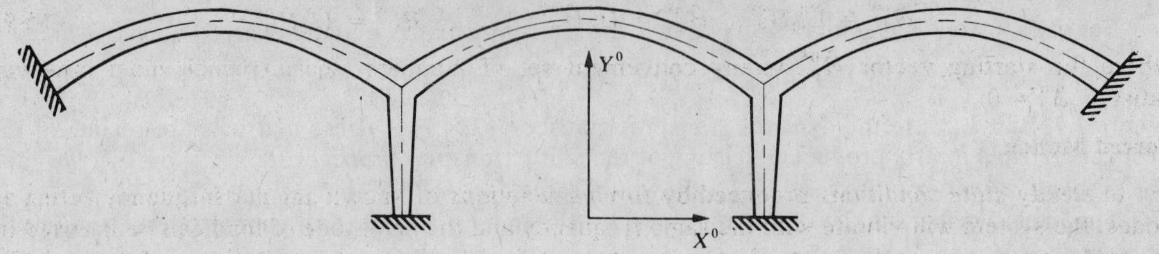

(a) Real Structure

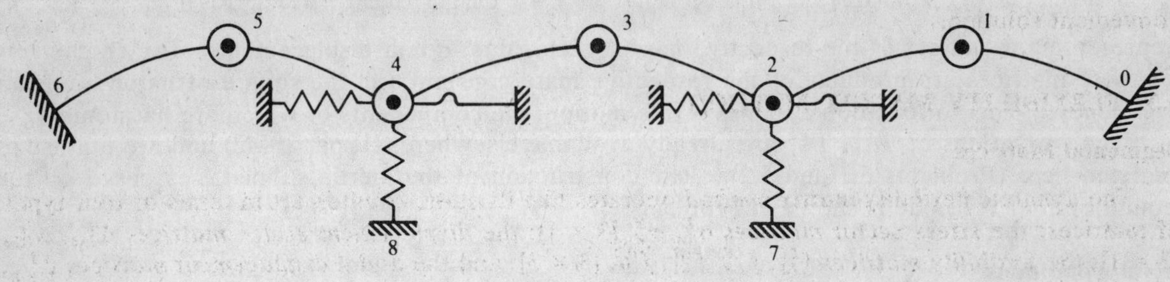

(b) Lumped Parameters Model

Fig. 5-2. Three-Span Rigid Frame with Parabolic Girders

Then, beginning at 0 and with all vector components in the reference system, the *transport chain* becomes

$$\hat{H}_6^0 = \underbrace{\hat{L}_{64}^0 \hat{J}_4^0 \hat{L}_{42}^0 \hat{J}_2^0 \hat{L}_{20}^0}_{\hat{L}_{60}^0} \hat{H}_0^0 \tag{5-5}$$

where $\hat{J}_4^0$ and $\hat{J}_2^0$ represent the spring effects and the inertia effect of the respective columns.

The *alternate approach* to the same problem uses static transport matrices and includes the effect of inertia vectors in the joint effect matrices, i.e.,

$$\hat{H}_6^0 = \underbrace{\hat{J}_6^0 \hat{T}_{65}^0 \hat{J}_5^0 \hat{T}_{54}^0 \hat{J}_4^0 \hat{T}_{43}^0 \hat{J}_3^0 \hat{T}_{32}^0 \hat{J}_2^0 \hat{T}_{21}^0 \hat{J}_1^0 \hat{T}_{10}^0}_{\hat{L}_{60}^0} \hat{H}_0^0 \tag{5-6}$$

where $\hat{L}_{60}^0$ must be the same as in (*5-5*). Since the direction of $\hat{J}_1^0, \hat{J}_2^0, \ldots, \hat{J}_5^0$ is apparent from the chain correspondence of subscripts, the triple subscript is replaced by the single subscript of the node.

Frequency Determinant

The result of the chain product in (*5-5*) is a new transport matrix L_{60}^0, from which the *frequency determinant* can be constructed by selecting the parts of the matrix corresponding to the existing eigenvector.

From (*5-5*) written in submatrix form with $\Delta_6^0 = 0$, $\Delta_0^0 = 0$ (fixed ends) as

$$\begin{bmatrix} 1 \\ \hline \sigma_6^0 \\ 0 \end{bmatrix} = \left[\begin{array}{c|cc} 1 & 0 & 0 \\ \hline 0 & L_{\sigma\sigma,60}^0 & L_{\sigma\Delta,60}^0 \\ 0 & L_{\Delta\sigma,60}^0 & L_{\Delta\Delta,60}^0 \end{array}\right] \begin{bmatrix} 1 \\ \hline \sigma_0^0 \\ 0 \end{bmatrix} \tag{5-7}$$

the *frequency determinant equation* becomes

$$|L_{\Delta\sigma,60}^0| = 0 \tag{5-8}$$

and yields the desired *natural angular frequencies*. With the natural frequencies known, substituting back into the chain (*5-5*) gives the relative values of the *modal columns*, each computed as

$$\hat{H}_1^{0*} = \hat{L}_{10}^{0}\hat{H}_0^{0*}, \quad \hat{H}_2^{0*} = \hat{L}_{20}^{0}\hat{H}_0^{0*} \quad ,\ldots, \quad \hat{H}_6^{0*} = \hat{L}_{60}^{0}\hat{H}_0^{0*} \tag{5-9}$$

where the starting vector $\hat{H}_0^{0*}$ is any convenient set of numbers (scales) which must however satisfy $\Delta_0^0 = 0$.

Forced Motion

In *steady-state conditions* produced by *forcing functions* of known angular frequency acting at nodes, the system will vibrate with the same frequency and the amplitude of load can be inserted in the transport matrix or the joint effect matrix (as required by the system) (Problems 5.9 and 5.10).

In cases of *nonperiodic excitations*, the stiffness methods of Sec. 5.4 and 5.5 offer a more convenient solution.

5.3 FLEXIBILITY MATRIX METHOD

Segmental Matrices

The *dynamic flexibility matrix method* operates like its static counterpart in terms of four types of matrices: the *stress vector matrices* σ_L^L, σ_R^R, [3 × 1]; the *displacement vector matrices* Δ_{LR}^L, Δ_{RL}^R, [3 × 1]; the *flexibility matrices* f_{LL}^{LL}, f_{LR}^{LR}, f_{RL}^{RL}, f_{RR}^{RR}, [3 × 3]; and the *nodal displacement matrices* Δ_{LjR}^j, Δ_{RjL}^j, [3 × 1].

The relations of the flexibility method vectors and those of the transport method (II, p. 5) are

$$\Delta_{LR}^L = -\Delta_L^L \qquad \Delta_{RL}^R = \Delta_R^R \qquad \sigma_L^L = \sigma_L^L \qquad \sigma_R^R = \sigma_R^R \tag{5-10}$$

Then for the isolated segment LR of Fig. 5-1, the *end displacement equations* are

$$\Delta_{LR}^L = f_{LL}^{LL}\sigma_L^L + f_{LR}^{LR}\sigma_R^R \qquad \Delta_{RL}^R = f_{RL}^{RL}\sigma_L^L + f_{RR}^{RR}\sigma_R^R \tag{5-11}$$

where f_{LL}^{LL} and f_{RR}^{RR} are the *near end (direct) dynamic segmental flexibility matrices*, [3 × 3], and f_{LR}^{LR} and f_{RL}^{RL} are the *far end (indirect) dynamic segmental flexibility matrices*, [3 × 3]. The derivation of (*5-11*) based on the results of Table 5-1 is given in Table 5-2*a*.

Table 5-2

(*a*) Segmental Flexibility Matrices	(*b*) Segmental Stiffness Matrices
(1) Transport Relations (Table 5-1)	(1) Transport Relations (Table 5-1)
$\sigma_L^L = L_{\sigma\sigma,LR}^{LR}\sigma_R^R + L_{\sigma\Delta,LR}^{LR}\Delta_R^R$	$\Delta_L^L = L_{\Delta\sigma,LR}^{LR}\sigma_R^R + L_{\Delta\Delta,LR}^{LR}\Delta_R^R$
$\sigma_R^R = L_{\sigma\sigma,RL}^{RL}\sigma_L^L + L_{\sigma\Delta,RL}^{RL}\Delta_L^L$	$\Delta_R^R = L_{\Delta\sigma,RL}^{RL}\sigma_L^L + L_{\Delta\Delta,RL}^{RL}\Delta_L^L$
(2) Change in Signs	(2) Change in Signs
$\Delta_{LR}^L = -\Delta_L^L, \quad \Delta_{RL}^R = \Delta_R^R$	$\sigma_{LR}^L = -\sigma_L^L, \quad \sigma_{RL}^R = \sigma_R^R$
(3) Flexibility Equations (*5-10*) and (*5-11*)	(3) Stiffness Equations (*5-18*) and (*5-19*)
$\Delta_{LR}^L = \underbrace{L_{\sigma\Delta,RL}^{RL)-1}L_{\sigma\sigma,RL}^{RL}}_{f_{LL}^{LL}}\sigma_L^L - \underbrace{L_{\sigma\Delta,RL}^{RL)-1}}_{f_{LR}^{LR}}\sigma_R^R$	$\sigma_{LR}^L = \underbrace{L_{\Delta\sigma,RL}^{RL)-1}L_{\Delta\Delta,RL}^{RL}}_{k_{LL}^{LL}}\Delta_L^L - \underbrace{L_{\Delta\sigma,RL}^{RL)-1}}_{k_{LR}^{LR}}\Delta_R^R$
$\Delta_{RL}^R = \underbrace{L_{\sigma\Delta,LR}^{LR)-1}}_{f_{RL}^{RL}}\sigma_L^L - \underbrace{L_{\sigma\Delta,LR}^{LR)-1}L_{\sigma\sigma,LR}^{LR}}_{f_{RR}^{RR}}\sigma_R^R$	$\sigma_{RL}^R = \underbrace{L_{\Delta\sigma,LR}^{LR)-1}}_{k_{RL}^{RL}}\Delta_L^L - \underbrace{L_{\Delta\sigma,LR}^{LR)-1}L_{\Delta\Delta,LR}^{LR}}_{k_{RR}^{RR}}\Delta_R^R$

Joint Conditions

As in Sec. 5.2, each center of lumped mass can be considered a *structural joint* and the conditions at this point are again called the *joint conditions*. To accommodate the interferences that may occur at this joint, the *joint displacement matrix* developed for static analysis is again introduced here.

This matrix designated as $f_{j\Delta}^j$ is a [3 × 3] matrix, the coefficients of which must satisfy the prescribed joint conditions.

Flexibility Chain

The application of the segmental flexibility matrices and the joint displacement matrices in dynamic analysis follows again the pattern of static analysis. The mass of each member is lumped at selected points and the real structure is replaced by the ideal model as before.

For the three-span frame of Fig. 5-2*a*, the column's deformation effects are characterized by their end flexibility matrices and the remaining curved girders can again be treated as a three-span continuous girder on elastic supports (Fig. 5-2*b*).

In general, the *condition of deformation compatibility* must be satisfied at each joint of the system, i.e., the algebraic sum of displacement vectors at the section of release must either equal zero or equal the prescribed displacement.

Thus at the fixed end 0 in Fig. 5-2*b*,

$$\Delta_{02} = f^0_{00}\sigma^0_0 + f^0_{02}\sigma^0_2 = 0 \tag{5-12}$$

and at joint 2,

$$\begin{cases} \Delta_{20} + \Delta_{27} = f^0_{20}\sigma^0_0 + (f^0_{22,0} + f^0_{2\Delta})\sigma^0_{20} - f^0_{2\Delta}\sigma^0_{24} = 0 & (5\text{-}13) \\ \Delta_{27} + \Delta_{23} = -f^0_{2\Delta}\sigma^0_{20} + (f^0_{2\Delta} + f^0_{22,4})\sigma^0_{24} + f^0_{24}\sigma^0_{42} = 0 & (5\text{-}14) \end{cases}$$

where the spring support at 2 introduces two compatibility conditions and one stress discontinuity $(\sigma^0_{20} \neq \sigma^0_{24})$ (II, pp. 104–108).

Frequency Determinant

The recurrent use of these equations gives the *system flexibility matrix equation* shown in Table 5-3*a*, from which the *frequency determinant equation* becomes

$$|\Phi| = 0 \tag{5-15}$$

and yields the desired natural angular frequencies.

Table 5-3*a* System Flexibility Matrix Equation

$$\underbrace{\left[\begin{array}{c|c|c|c|c|c} f^0_{00} & f^0_{02} & & & & \\ \hline f^0_{20} & f^0_{22,0} + f^0_{2\Delta} & -f^0_{2\Delta} & & & \\ \hline & -f^0_{2\Delta} & f^0_{2\Delta} + f_{22,4} & f^0_{24} & & \\ \hline & & f^0_{42} & f^0_{44,2} + f^0_{4\Delta} & -f^0_{4\Delta} & \\ \hline & & & -f_{4\Delta} & f^0_{4\Delta} + f^0_{44,6} & f^0_{46} \\ \hline & & & & f^0_{64} & f^0_{66} \end{array}\right]}_{[\Phi]} \underbrace{\left[\begin{array}{c} \sigma^0_0 \\ \hline \sigma_{20} \\ \hline \sigma^0_{24} \\ \hline \sigma^0_{42} \\ \hline \sigma^0_{46} \\ \hline \sigma^0_6 \end{array}\right]}_{[\sigma]} = \underbrace{\left[\begin{array}{c} 0 \\ \hline 0 \\ \hline 0 \\ \hline 0 \\ \hline 0 \\ \hline 0 \end{array}\right]}_{[0]}$$

Forced Motion

In the case of steady-state motion induced by joint loads of given angular frequency, the *load amplitude* W^0_j at the joint j (any joint) produces a stress discontinuity at that joint. That is,

$$-\sigma^0_{j,i} + W^0_j + \sigma^0_{j,k} = 0 \tag{5-16}$$

and the compatibility condition at j yields

$$f^0_{ji}\sigma^0_i + (f^0_{jj,i} + f^0_{jj,k})\sigma^0_j + f^0_{jk}\sigma^0_k + f^0_{jj,i}W^0_j = 0 \tag{5-17}$$

where the last term accounts for the presence of the joint load (Problem 5.10).

In cases of nonperiodic excitations, the methods of Sec. 5.5 offer a more convenient solution.

5.4 STIFFNESS MATRIX METHOD

Segmental Matrices

The *dynamic stiffness matrix method* operates like its static counterpart in terms of four types of matrices: the *stress vector matrices* σ_{LR}^L, σ_{RL}^R, [3 × 1]; the *displacement vector matrices* Δ_L^L, Δ_R^R, [3 × 1]; the *stiffness matrices* k_{LL}^{LL}, k_{LR}^{LR}, k_{RL}^{RL}, k_{RR}^{RR}, [3 × 3]; and the *joint load matrices* W_j^j, [3 × 1].

The relations of the stiffness method vectors and those of the transport method (II, p. 5) are given below.

$$\Delta_L^L = \Delta_L^L \qquad \Delta_R^R = \Delta_R^R \qquad \sigma_{LR}^L = -\sigma_L^L \qquad \sigma_{RL}^R = \sigma_R^R \tag{5-18}$$

Then for the isolated segment LR of Fig. 5-1, the *end stress equations* are

$$\sigma_{LR}^L = k_{LL}^{LL}\Delta_L^L + k_{LR}^{LR}\Delta_R^R \qquad \Delta_{RL}^R = k_{RL}^{RL}\Delta_L^L + k_{RR}^{RR}\Delta_R^R \tag{5-19}$$

where k_{LL}^{LL} and k_{RR}^{RR} are the *near end (direct) dynamic segmental stiffness matrices*, [3 × 3], and k_{LR}^{LR} and k_{RL}^{RL} are the *far end (indirect) dynamic segmental stiffness matrices*, [3 × 3]. The derivation of (*5-19*) based on the results of Table 5-1 is given in Table 5-2*b*.

Joint Conditions

The conditions at the center of the lumped mass are again called the *joint conditions*, represented by the *joint load amplitude vector* W_j^j. In free motion, $W_j^j = 0$.

Stiffness Chain

The application of the segmental stiffness matrices and the joint load matrices in dynamic analysis again follows the pattern of static analysis.

In general, the *condition of equilibrium* must be satisfied at each joint of the system, i.e., the algebraic sum of stress vectors at a given joint must either equal zero or equal the joint load vector W_j.

For the three-span frame of Fig. 5-2*a*, the column's deformation effects are characterized by their end stiffness matrices. The remaining curved girders can again be treated as a three-span continuous girder on elastic supports (Fig. 5-2*b*).

Then at joint 2,

$$\sigma_{20}^0 + \sigma_{2\Delta}^0 + \sigma_{24}^0 = (k_{22,0}^0 + k_{2\Delta}^0 + k_{22,4}^0)\Delta_2^0 + k_{24}^0\Delta_4^0 = 0 \tag{5-20}$$

and at joint 4,

$$\sigma_{42}^0 + \sigma_4^0 + \sigma_{46}^0 = k_{42}^0\Delta_2^0 + (k_{44,2}^0 + k_4^0 + k_{22,4}^0)\Delta_4^0 = 0 \tag{5-21}$$

where $k_{2\Delta}^0$ and $k_{4\Delta}^0$ are the near end segmental stiffness matrices [3 × 3] of columns 2 and 4 (II, pp. 217–219).

Table 5-3*b* System Stiffness Matrix Equation

$$\underbrace{\begin{bmatrix} k_{22,0} + k_{2\Delta} + k_{22,4} & k_{24} \\ k_{42} & k_{44,2} + k_{4\Delta} + k_{44,6} \end{bmatrix}}_{[\Omega]} \underbrace{\begin{bmatrix} \Delta_2^0 \\ \Delta_4^0 \end{bmatrix}}_{[\Delta]} = \underbrace{\begin{bmatrix} 0 \\ 0 \end{bmatrix}}_{[0]}$$

Frequency Determinant

The recurrent use of these equations gives the *system stiffness matrix* equation shown in Table 5-3*b*, from which the *frequency determinant equation* becomes

$$|\Omega| = 0 \tag{5-22}$$

and yields the desired natural frequencies.

Forced Motion

In the case of steady-state motion induced by joint forces of given angular frequency, the *load amplitude vector* W_j^0 must be included in the joint equilibrium equation, which in general becomes

$$k_{j1}^0\Delta_1^0 + \cdots + k_{jn}^0\Delta_n^0 + (k_{jj,1}^0 + \cdots + k_{jj,n}^0)\Delta_j^0 = W_j^0 \tag{5-23}$$

and must be written for all joints of the system.

In cases of nonperiodic excitations, the methods of Sec. 5.5 offer a more convenient solution.

5.5 GENERAL MATRIX METHODS

General

The special methods presented in the preceding sections (5.2), (5.3), and (5.4) have one common handicap, namely they are not (readily) applicable in cases of *nonperiodic excitation* and/or *damping*. In such cases, the *general matrix methods* introduced in this section are far superior. They are the general stiffness matrix method and the general flexibility matrix method.

General Stiffness Matrix

The basic idea of this method is to construct for the given ideal system (with selected lumped masses) the *static system stiffness matrix* Ω (II, p. 229) and insert this matrix in the matrix equation (*3-35*). This then becomes

$$[\mu]\{\ddot{\Delta}\} + [c]\{\dot{\Delta}\} + [\Omega]\{\Delta\} = \{W(t)\} \tag{5-24}$$

where $[\mu]$ and $[c]$ are diagonal matrices of given coefficients.

With (*5-23*) constructed, the remaining part of the solution follows the procedure described in Sec. 3.6 (Problems 5.15 to 5.17).

General Flexibility Matrix

The inverse of the previous procedure is achieved by taking

$$\{R\} = [\Omega]\{\Delta\} \tag{5-25}$$

from which

$$\{\Delta\} = [\Phi]\{R\} \tag{5-26}$$

where $[\Phi] = [\Omega]^{-1}$ is the *system flexibility matrix* giving displacements of joints due to unit causes (forces or moments) (Problems 5.11 to 5.14).

If

$$\{R\} = -[\mu]\{\ddot{\Delta}\} - [c]\{\dot{\Delta}\} + \{W(t)\} \tag{5-27}$$

then (*5-26*) becomes the flexibility formulation of this problem, yielding the nodal equations in the form

$$[\Phi][\mu]\{\ddot{\Delta}\} + [\Phi][c]\{\dot{\Delta}\} + \{\Delta\} = [\Phi]\{W(t)\} \tag{5-28}$$

which can be directly obtained from (*5-24*) on premultiplying by $[\Omega]^{-1}$ (Problem 5.18).

Solved Problems

TRANSPORT MATRICES

5.1. Construct the transport matrices $\hat{L}_{LR}^{0}$ and $\hat{L}_{LR}^{0}$ for the segment LR shown in Fig. P-5.1*a*.

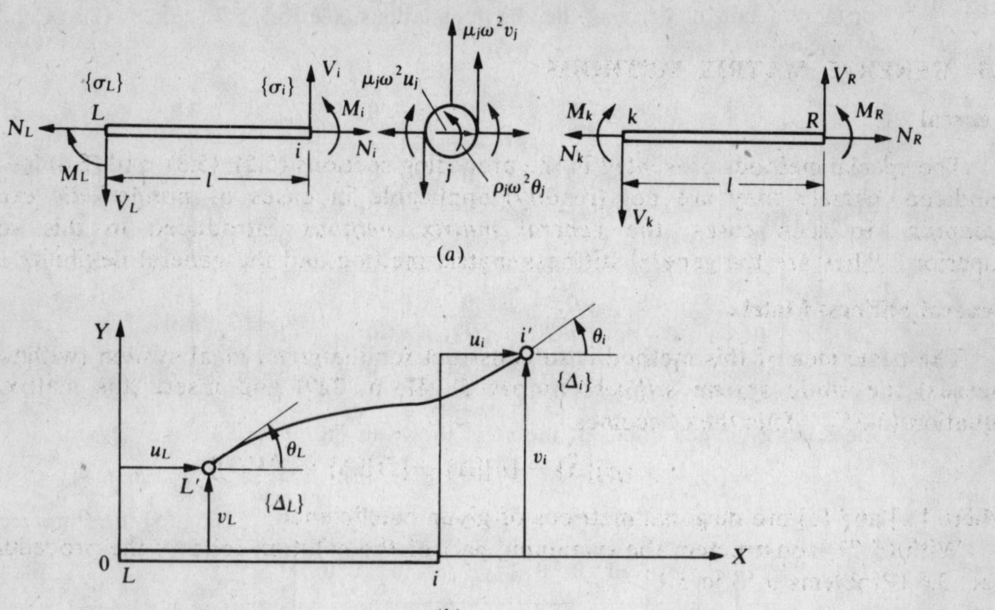

Fig. P-5.1

(*a*) From Table 5-1, the typical transport matrix $\hat{L}_{LR}^{0}$ may be expressed as

$$\left[\begin{array}{c}1\\ \hline \sigma_L\\ \Delta_L\end{array}\right]=\left[\begin{array}{c|cc}1 & 0 & 0\\ \hline 0 & t_{Li} & 0\\ 0 & d_{Li} & s_{Li}\end{array}\right]\left[\begin{array}{c|cc}1 & 0 & 0\\ \hline 0 & I & j_{ik}\\ 0 & 0 & I\end{array}\right]\left[\begin{array}{c|cc}1 & 0 & 0\\ \hline 0 & t_{kR} & 0\\ 0 & d_{kR} & s_{kR}\end{array}\right]\left[\begin{array}{c}1\\ \hline \sigma_R\\ \Delta_R\end{array}\right] \tag{1a}$$

or

$$\hat{H}_L^0=\underbrace{\hat{T}_{Li}^0\hat{I}_{ik}^0\hat{T}_{kR}^0}_{\hat{L}_{LR}^0}\hat{H}_R^0 \tag{1b}$$

in which the elements may be explained by studying Fig. P-5.1*a* and *b*. Let a typical member L_i be shown in Fig. P-5.1 in which the end stresses and deformations, $\{\sigma\}$ and $\{\Delta\}$, are positive as shown. Using equilibrium conditions we have

$$N_L = N_i \qquad V_L = V_i \qquad M_L = V_i l + M_i \tag{2}$$

and in matrix form,

$$\begin{bmatrix}N_L\\ V_L\\ M_L\end{bmatrix}=\begin{bmatrix}1 & 0 & 0\\ 0 & 1 & 0\\ 0 & l & 1\end{bmatrix}\begin{bmatrix}N_i\\ V_i\\ M_i\end{bmatrix} \qquad \text{or} \qquad \{\sigma_L\}=[t_{Li}]\{\sigma_i\} \tag{2a, 2b}$$

For the geometric conditions, we have

$$u_L=-\frac{l}{AE}N_i+u_i \qquad v_L=\frac{l^3}{6EI}V_i+\frac{l^2}{2EI}M_i+v_i-l\theta_i \qquad \theta_L=-\frac{l^2}{2EI}V_i-\frac{l}{EI}M_i+\theta_i \tag{3}$$

In matrix form

$$\begin{bmatrix} u_L \\ v_L \\ \theta_L \end{bmatrix} = \begin{bmatrix} -l/AE & 0 & 0 \\ 0 & l^3/6EI & l^2/2EI \\ 0 & -l^2/2EI & -l/EI \end{bmatrix} \begin{bmatrix} N_i \\ V_i \\ M_i \end{bmatrix} + \begin{bmatrix} 1 & 0 & 0 \\ 0 & 1 & -l \\ 0 & 0 & 1 \end{bmatrix} \begin{bmatrix} u_i \\ v_i \\ \theta_i \end{bmatrix} \tag{3a}$$

or

$$\{\Delta_L\} = [d_{Li}]\{\sigma_i\} + [s_{Li}]\{\Delta_i\} \tag{3b}$$

For more details of deriving the above equations, see Ref. 5.7 (pp. 38–41). Combining (*2a*) and (*3a*) yields

$$\begin{bmatrix} 1 \\ \hline N_L \\ V_L \\ M_L \\ \hline u_L \\ v_L \\ \theta_L \end{bmatrix} = \left[\begin{array}{c|ccc|ccc} 1 & 0 & 0 & 0 & 0 & 0 & 0 \\ \hline 0 & 1 & 0 & 0 & 0 & 0 & 0 \\ 0 & 0 & 1 & 0 & 0 & 0 & 0 \\ 0 & 0 & l & 1 & 0 & 0 & 0 \\ \hline 0 & -l/AE & 0 & 0 & 1 & 0 & 0 \\ 0 & 0 & l^3/6EI & l^2/2EI & 0 & 1 & -l \\ 0 & 0 & -l^2/2EI & -l/EI & 0 & 0 & 1 \end{array}\right] \begin{bmatrix} 1 \\ \hline N_i \\ V_i \\ M_i \\ \hline u_i \\ v_i \\ \theta_i \end{bmatrix} \quad \text{or} \quad \hat{H}^0_L = \hat{T}^0_{Li}\hat{H}^0_i \tag{4a, 4b}$$

The equilibrium conditions of the mass shown in Fig. P-5.1*a* are

$$N_i = \mu_j\omega^2 u_j + N_k \qquad V_i = \mu_j\omega^2 v_j + V_k \qquad M_i = \rho_j\omega^2\theta_j + M_k \tag{5}$$

where $u_i = u_j = u_k$, $v_i = v_j = v_k$, and $\theta_i = \theta_j = \theta_k$. Using (*5*) the state vector matrix is

$$\begin{bmatrix} 1 \\ \hline N_i \\ V_i \\ M_i \\ \hline u_i \\ v_i \\ \theta_i \end{bmatrix} = \left[\begin{array}{c|ccc|ccc} 1 & 0 & 0 & 0 & 0 & 0 & 0 \\ \hline 0 & 1 & 0 & 0 & \mu_j\omega^2 & 0 & 0 \\ 0 & 0 & 1 & 0 & 0 & \mu_j\omega^2 & 0 \\ 0 & 0 & 0 & 1 & 0 & 0 & \rho_j\omega^2 \\ \hline 0 & 0 & 0 & 0 & 1 & 0 & 0 \\ 0 & 0 & 0 & 0 & 0 & 1 & 0 \\ 0 & 0 & 0 & 0 & 0 & 0 & 1 \end{array}\right] \begin{bmatrix} 1 \\ \hline N_k \\ V_k \\ M_k \\ \hline u_k \\ v_k \\ \theta_k \end{bmatrix} \quad \text{or} \quad \hat{H}^0_i = \hat{I}^0_{ik}\hat{H}^0_k \tag{6a, 6b}$$

Since segments Li and kR are identical,

$$\hat{T}^0_{kR} = \hat{T}^0_{Li} \tag{7}$$

Substituting (*4a*), (*6a*), and (*7*) into (*1a*) yields

$$\hat{L}^0_{LR} = \left[\begin{array}{c|c|c|c|c|c|c} 1 & 0 & 0 & 0 & 0 & 0 & 0 \\ \hline 0 & 1 - \dfrac{\mu_j\omega^2 l}{EA} & 0 & 0 & \mu_j\omega^2 & 0 & 0 \\ 0 & 0 & 1 + \dfrac{\mu_j\omega^2 l^3}{6EI} & \dfrac{\mu_j\omega^2 l^2}{2EI} & 0 & \mu_j\omega^2 & -\mu_j\omega^2 l \\ 0 & 0 & 2l + \dfrac{\omega^2 l^2}{2EI}\left(\dfrac{\mu_j l^2}{3} - \rho_j\right) & 1 + \dfrac{\omega^2 l}{EI}\left(\dfrac{\mu_j l^2}{2} - \rho_j\right) & 0 & \mu_j\omega^2 l & (\rho_j - \mu_j l^2)\omega^2 \\ 0 & -\dfrac{l}{EA}\left(2 - \dfrac{\mu_j\omega^2 l}{EA}\right) & 0 & 0 & 1 - \dfrac{\mu_j\omega^2 l}{EA} & 0 & 0 \\ 0 & 0 & \dfrac{4l^3}{3EI} + \dfrac{\omega^2 l^4}{4(EI)^2}\left(\dfrac{\mu_j l^2}{9} - \rho_j\right) & \dfrac{2l^2}{EI} + \dfrac{\omega^2 l^3}{2(EI)^2}\left(\dfrac{\mu_j l^2}{6} - \rho_j\right) & 0 & 1 + \dfrac{\mu_j\omega^2 l^3}{6EI} & -2l + \dfrac{\omega^2 l^2}{2EI}\left(\rho_j - \dfrac{\mu_j l^2}{3}\right) \\ 0 & 0 & \dfrac{-2l^2}{EI} + \dfrac{\omega^2 l^3}{2(EI)^2}\left(\rho_j - \dfrac{\mu_j l^2}{6}\right) & \dfrac{-2l}{EI} + \dfrac{\omega^2 l^2}{(EI)^2}\left(\rho_j - \dfrac{\mu_j l^2}{4}\right) & 0 & -\dfrac{\mu_j\omega^2 l^2}{2EI} & 1 + \dfrac{\omega^2 l}{EI}\left(\dfrac{\mu_j l^2}{2} - \rho_j\right) \end{array}\right] \tag{8}$$

(*b*) For the transport matrix $\hat{L}^0_{RL}$ (Table 5-1) we write

$$\hat{H}^0_R = \hat{T}^0_{Rk}\hat{I}^0_{ki}\hat{T}^0_{iL}\hat{H}^0_L \tag{9}$$

Expressing the equilibrium equations of Fig. P-5.1 from right to left, we have

$$N_R = N_k \qquad V_R = V_k \qquad M_R = M_k - V_R l \tag{10}$$

and in matrix form

$$\begin{bmatrix} N_R \\ V_R \\ M_R \end{bmatrix} = \begin{bmatrix} 1 & 0 & 0 \\ 0 & 1 & 0 \\ 0 & -l & 1 \end{bmatrix}\begin{bmatrix} N_k \\ V_k \\ M_k \end{bmatrix} \qquad \text{or} \qquad \{\sigma_R\} = [t_{Rk}]\{\sigma_R\} \tag{10a, 10b}$$

The geometric equations are

$$\left.\begin{aligned} u_R &= \frac{l}{AE}N_k + u_k \\ v_R &= \frac{-l^3}{6EI}V_k + \frac{l^2}{2EI}M_k + v_k + l\theta_k \\ \theta_R &= \frac{-l^2}{2EI}V_k + \frac{l}{EI}M_k + \theta_k \end{aligned}\right\} \tag{11}$$

and in matrix form

$$\underbrace{\begin{bmatrix} u_R \\ v_R \\ \theta_R \end{bmatrix}}_{\{\Delta_R\}} = \underbrace{\begin{bmatrix} l/AE & 0 & 0 \\ 0 & -l^3/6EI & l^2/2EI \\ 0 & -l^2/2EI & l/EI \end{bmatrix}}_{[d_{Rk}]}\underbrace{\begin{bmatrix} N_k \\ V_k \\ M_k \end{bmatrix}}_{\{\sigma_k\}} + \underbrace{\begin{bmatrix} 1 & 0 & 0 \\ 0 & 1 & l \\ 0 & 0 & 1 \end{bmatrix}}_{[s_{Rk}]}\underbrace{\begin{bmatrix} u_k \\ v_k \\ \theta_k \end{bmatrix}}_{\{\Delta_k\}} \tag{11a}$$

Combining (*10a*) and (*11a*) yields

$$\begin{bmatrix} 1 \\ \hline N_R \\ V_R \\ M_R \\ \hline u_R \\ v_R \\ \theta_R \end{bmatrix} = \left[\begin{array}{c|ccc|ccc} 1 & 0 & 0 & 0 & 0 & 0 & 0 \\ \hline 0 & 1 & 0 & 0 & 0 & 0 & 0 \\ 0 & 0 & 1 & 0 & 0 & 0 & 0 \\ 0 & 0 & -l & 1 & 0 & 0 & 0 \\ \hline 0 & l/AE & 0 & 0 & 1 & 0 & 0 \\ 0 & 0 & -l^3/6EI & l^2/2EI & 0 & 1 & l \\ 0 & 0 & -l^2/2EI & l/EI & 0 & 0 & 1 \end{array}\right]\begin{bmatrix} 1 \\ \hline N_k \\ V_k \\ M_k \\ \hline u_k \\ v_k \\ \theta_k \end{bmatrix} \qquad \text{or} \qquad \hat{H}^0_R = \hat{T}^0_{Rk}\hat{H}^0_k \tag{12}$$

From Fig. P-5.1*a* we have

$$\hat{T}^0_{iL} = \hat{T}^0_{Rk} \tag{13}$$

Now the equilibrium equations of the mass at *j* become

$$N_k = N_i - \mu_j\omega^2 u_j \qquad V_k = V_i - \mu_j\omega^2 v_j \qquad M_k = M_i - \rho_j\omega^2\theta_j \tag{14}$$

In state vector matrix form,

$$\begin{bmatrix} 1 \\ \hline N_k \\ V_k \\ M_k \\ \hline u_k \\ v_k \\ \theta_k \end{bmatrix} = \left[\begin{array}{c|ccc|ccc} 1 & 0 & 0 & 0 & 0 & 0 & 0 \\ \hline 0 & 1 & 0 & 0 & -\mu_j\omega^2 & 0 & 0 \\ 0 & 0 & 1 & 0 & 0 & -\mu_j\omega^2 & 0 \\ 0 & 0 & 0 & 1 & 0 & 0 & -\rho_j\omega^2 \\ \hline 0 & 0 & 0 & 0 & 1 & 0 & 0 \\ 0 & 0 & 0 & 0 & 0 & 1 & 0 \\ 0 & 0 & 0 & 0 & 0 & 0 & 1 \end{array}\right]\begin{bmatrix} 1 \\ \hline N_i \\ V_i \\ M_i \\ \hline u_i \\ v_i \\ \theta_i \end{bmatrix} \qquad \text{or} \qquad \hat{H}^0_k = \hat{I}^0_{ki}\hat{H}^0_i \tag{15}$$

Substituting (*12*), (*13*), and (*15*) into (*9*) gives

$$\hat{L}^0_{RL} = \left[\begin{array}{c|c|c|c|c|c|c} 1 & 0 & 0 & 0 & 0 & 0 & 0 \\ \hline 0 & 1-\dfrac{\mu_j\omega^2 l}{EA} & 0 & 0 & -\mu_j\omega^2 & 0 & 0 \\ 0 & 0 & 1+\dfrac{\mu_j\omega^2 l^3}{6EI} & \dfrac{-\mu_j\omega^2 l^2}{2EI} & 0 & -\mu_j\omega^2 & -\mu_j\omega^2 l \\ 0 & 0 & -2l-\dfrac{\omega^2 l^2}{2EI}\left(\dfrac{\mu_j l^2}{3}-\rho_j\right) & 1+\dfrac{\omega^2 l}{EI}\left(\dfrac{\mu_j l^2}{2}-\rho_j\right) & 0 & \mu_j\omega^2 l & (\mu_j l^2-\rho_j)\omega^2 \\ 0 & \dfrac{l}{EA}\left(2-\dfrac{\mu_j\omega^2 l}{EA}\right) & 0 & 0 & 1-\dfrac{\mu_j\omega^2 l}{AE} & 0 & 0 \\ 0 & 0 & \dfrac{\omega^2 l^4}{4(EI)^2}\left(\rho_j-\dfrac{\mu_j l^2}{9}\right)-\dfrac{4l^3}{3EI} & \dfrac{2l^2}{EI}+\dfrac{\omega^2 l^3}{2(EI)^2}\left(\dfrac{\mu_j l^2}{6}-\rho_j\right) & 0 & 1+\dfrac{\mu_j\omega^2 l^3}{6EI} & 2l+\dfrac{\omega^2 l^2}{2EI}\left(\rho_j-\dfrac{\mu_j l^2}{3}\right) \\ 0 & 0 & \dfrac{-2l^2}{EI}+\dfrac{\omega^2 l^3}{2(EI)^2}\left(\rho_j-\dfrac{\mu_j l^2}{6}\right) & \dfrac{2l}{EI}-\dfrac{\omega^2 l^2}{(EI)^2}\left(\rho_j-\dfrac{\mu_j l^2}{4}\right) & 0 & \dfrac{l^2\mu_j\omega^2}{2EI} & 1-\dfrac{\omega^2 l}{EI}\left(\rho_j-\dfrac{\mu_j l^2}{2}\right) \end{array}\right] \tag{16}$$

5.2. Derive the joint effect matrices $\hat{J}^0_{ik}$ and $\hat{J}^0_{ki}$ to include the inertia forces and the elastic spring constants shown in Fig. P-5.2*a*.

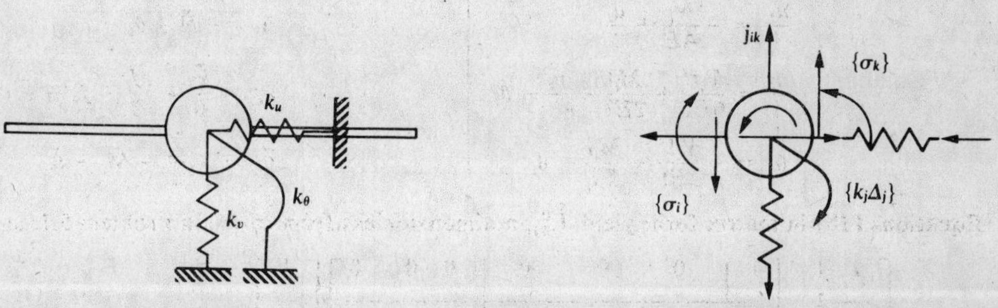

(*a*) (*b*)

Fig. P-5.2

Figure P-5.2*b* shows the stresses, $\{\sigma_i\}$ and $\{\sigma_k\}$; spring forces, $\{k_j\Delta_j\}$; and inertia forces, j_{ik}. From the dynamic equilibrium,

$$N_i = N_k + \mu_j\omega^2\mu_j - k_u u_j \qquad V_i = V_k + \mu_j\omega^2 v_j - k_v v_j \qquad M_i = M_k + \rho_j\omega^2\theta_j - k_\theta\theta_j \tag{1}$$

Since the deformations at *i*, *j*, and *k* are the same, we have

$$u_i = u_j = u_k \qquad v_i = v_j = v_k \qquad \text{and} \quad \theta_i = \theta_j = \theta_k \tag{2}$$

Using (*1*) and (*2*) yields

$$\hat{H}^0_i = \hat{J}^0_{ik}\hat{H}^0_k = \left[\begin{array}{c|ccc|ccc} 1 & 0 & 0 & 0 & 0 & 0 & 0 \\ \hline 0 & 1 & 0 & 0 & \mu_j\omega^2-k_u & 0 & 0 \\ 0 & 0 & 1 & 0 & 0 & \mu_j\omega^2-k_v & 0 \\ 0 & 0 & 0 & 1 & 0 & 0 & \rho_j\omega^2-k_\theta \\ \hline 0 & 0 & 0 & 0 & 1 & 0 & 0 \\ 0 & 0 & 0 & 0 & 0 & 1 & 0 \\ 0 & 0 & 0 & 0 & 0 & 0 & 1 \end{array}\right]\hat{H}^0_k \tag{3}$$

Similarly,

$$\hat{H}^0_k = \hat{J}^0_{ki}\hat{H}^0_i = \left[\begin{array}{c|ccc|ccc} 1 & 0 & 0 & 0 & 0 & 0 & 0 \\ \hline 0 & 1 & 0 & 0 & k_u - \mu_j\omega^2 & 0 & 0 \\ 0 & 0 & 1 & 0 & 0 & k_v - \mu_j\omega^2 & 0 \\ 0 & 0 & 0 & 1 & 0 & 0 & k_\theta - \rho_j\omega^2 \\ \hline 0 & 0 & 0 & 0 & 1 & 0 & 0 \\ 0 & 0 & 0 & 0 & 0 & 1 & 0 \\ 0 & 0 & 0 & 0 & 0 & 0 & 1 \end{array}\right] \hat{H}^0_i \tag{4}$$

5.3. **Derive the dimensionless transport matrices $\hat{\bar{L}}^0_{LR}$ and $\hat{\bar{L}}^0_{RL}$ for the typical segments shown in Fig. P-5.1a.**

Let (2) and (3) of Problem 5.1 be expressed in dimensionless form as

$$\left.\begin{aligned} \frac{N_L}{AE} &= \frac{N_i}{AE} \\ \frac{V_L l^2}{EI} &= \frac{V_i l^2}{EI} \\ \frac{M_L l}{EI} &= \frac{V_i l^2}{EI} + \frac{M_i l}{EI} \\ \frac{u_L}{l} &= -\frac{N_i}{AE} + \frac{u_i}{l} \\ \frac{v_L}{l} &= \frac{V_i l^2}{6EI} + \frac{M_i l}{2EI} + \frac{v_i}{l} - \theta_i \\ \theta_L &= -\frac{V_i l^2}{2EI} - \frac{M_i l}{EI} + \theta_i \end{aligned}\right\} \quad \text{or} \quad \begin{aligned} \bar{N}_L &= \bar{N}_i \\ \bar{V}_L &= \bar{V}_i \\ \bar{M}_L &= \bar{V}_i + \bar{M}_i \\ \bar{u}_L &= -\bar{N}_i + \bar{u}_i \\ \bar{v}_L &= \frac{\bar{V}_i}{6} + \frac{\bar{M}_i}{2} + \bar{v}_i - \bar{\theta}_i \\ \bar{\theta}_L &= -\frac{\bar{V}_i}{2} - \bar{M}_i + \bar{\theta}_i \end{aligned} \tag{1a, 1b}$$

Equations (1b) in matrix form yield $\hat{\bar{T}}^0_{Li}$, a dimensionless form of (4) in Problem 5.1, as

$$\begin{bmatrix} 1 \\ \hline \bar{N}_L \\ \bar{V}_L \\ \bar{M}_L \\ \hline \bar{u}_L \\ \bar{v}_L \\ \bar{\theta}_L \end{bmatrix} = \left[\begin{array}{c|ccc|ccc} 1 & 0 & 0 & 0 & 0 & 0 & 0 \\ \hline 0 & 1 & 0 & 0 & 0 & 0 & 0 \\ 0 & 0 & 1 & 0 & 0 & 0 & 0 \\ 0 & 0 & 1 & 1 & 0 & 0 & 0 \\ \hline 0 & -1 & 0 & 0 & 1 & 0 & 0 \\ 0 & 0 & 1/6 & 1/2 & 0 & 1 & -1 \\ 0 & 0 & -1/2 & -1 & 0 & 0 & 1 \end{array}\right] \begin{bmatrix} 1 \\ \hline \bar{N}_i \\ \bar{V}_i \\ \bar{M}_i \\ \hline \bar{u}_i \\ \bar{v}_i \\ \bar{\theta}_i \end{bmatrix} \quad \text{or} \quad \hat{\bar{H}}^0_L = \hat{\bar{T}}^0_{Li}\hat{\bar{H}}^0_i \tag{2}$$

Following (1b), we write the dimensionless form of (6) in Problem 5.1 as

$$\begin{bmatrix} 1 \\ \hline \bar{N}_i \\ \bar{V}_i \\ \bar{M}_i \\ \hline \bar{u}_i \\ \bar{v}_i \\ \bar{\theta}_i \end{bmatrix} = \left[\begin{array}{c|ccc|ccc} 1 & 0 & 0 & 0 & 0 & 0 & 0 \\ \hline 0 & 1 & 0 & 0 & \bar{a} & 0 & 0 \\ 0 & 0 & 1 & 0 & 0 & \bar{b} & 0 \\ 0 & 0 & 0 & 1 & 0 & 0 & \bar{c} \\ \hline 0 & 0 & 0 & 0 & 1 & 0 & 0 \\ 0 & 0 & 0 & 0 & 0 & 1 & 0 \\ 0 & 0 & 0 & 0 & 0 & 0 & 1 \end{array}\right] \begin{bmatrix} 1 \\ \hline \bar{N}_i \\ \bar{V}_i \\ \bar{M}_i \\ \hline \bar{u}_i \\ \bar{v}_i \\ \bar{\theta}_i \end{bmatrix} \quad \text{or} \quad \hat{\bar{H}}^0_i = \hat{\bar{I}}^0_{ik}\hat{\bar{H}}^0_k \tag{3}$$

where $\bar{a} = \mu_j\omega^2 l/EA$, $\bar{b} = \mu_j\omega^2 l^3/EI$, and $\bar{c} = \rho_j\omega^2 l/EI$. Since $\hat{\bar{T}}^0_{kR} = \hat{\bar{T}}^0_{Li}$, we multiply (2) and (3) according to (1b) of Problem 5.1 for $\hat{\bar{L}}^0_{LR}$ as

$$\hat{L}^0_{LR} = \left[\begin{array}{c|ccc|ccc} 1 & 0 & 0 & 0 & 0 & 0 & 0 \\ \hline 0 & 1-\bar{a} & 0 & 0 & \bar{a} & 0 & 0 \\ 0 & 0 & 1+\bar{b}/6 & \bar{b}/2 & 0 & \bar{b} & -\bar{b} \\ 0 & 0 & 2+\bar{b}/6-\bar{c}/2 & 1+\bar{b}/2-\bar{c} & 0 & \bar{b} & \bar{c}-\bar{b} \\ \hline 0 & -2+\bar{a} & 0 & 0 & 1-\bar{a} & 0 & 0 \\ 0 & 0 & 4/3+\bar{b}/36-\bar{c}/4 & 2+\bar{b}/12-\bar{c}/2 & 0 & 1+\bar{b}/6 & -2-\bar{b}/6+\bar{c}/2 \\ 0 & 0 & -2-\bar{b}/12+\bar{c}/2 & -2-\bar{b}/4+\bar{c} & 0 & -\bar{b}/2 & 1+\bar{b}/2-\bar{c} \end{array}\right] \tag{4}$$

Similarly, applying the dimensionless notations (*1b*) to (*12*), (*13*), and (*15*) of Problem 5.1 will yield the dimensionless form of $\hat{L}^0_{RL}$:

$$\hat{L}^0_{RL} = \left[\begin{array}{c|ccc|ccc} 1 & 0 & 0 & 0 & 0 & 0 & 0 \\ \hline 0 & 1-\bar{a} & 0 & 0 & -\bar{a} & 0 & 0 \\ 0 & 0 & 1+\bar{b}/6 & -\bar{b}/2 & 0 & -\bar{b} & -\bar{b} \\ 0 & 0 & -2-\bar{b}/6+\bar{c}/2 & 1+\bar{b}/2-\bar{c} & 0 & \bar{b} & -\bar{c}+\bar{b} \\ \hline 0 & 2-\bar{a} & 0 & 0 & 1-\bar{a} & 0 & 0 \\ 0 & 0 & -4/3-\bar{b}/36+\bar{c}/4 & 2+\bar{b}/12-\bar{c}/2 & 0 & 1+\bar{b}/6 & 2+\bar{b}/6-\bar{c}/2 \\ 0 & 0 & -2-\bar{b}/12+\bar{c}/2 & 2+\bar{b}/4-\bar{c} & 0 & \bar{b}/2 & 1+\bar{b}/2-\bar{c} \end{array}\right] \tag{5}$$

5.4. Construct the transport matrix $H^0_L = L^0_{LR} H^0_R$ for the structure of Fig. P-5.4.

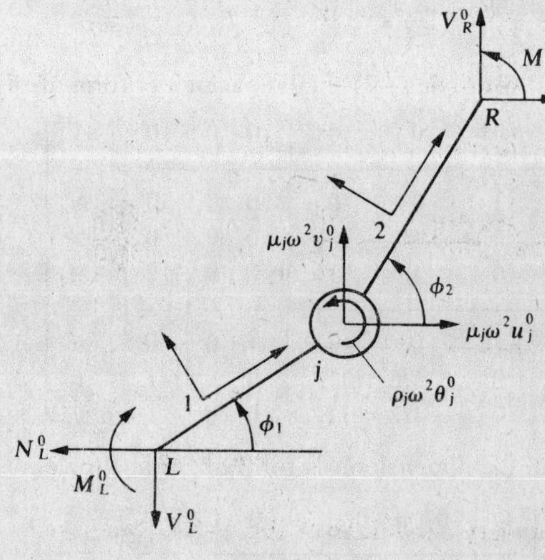

Fig. P-5.4

The transport chain may be written as

$$L^0_{LR} = \Pi^{01} T^1_{Li} \Pi^{10} I^0_{ik} \Pi^{02} T^2_{kR} \Pi^{20} \tag{1}$$

where T^1_{Li} and T^2_{kR} are static transport matrices as shown in (*4*) of Problem 5.1, corresponding to segments 1 and 2 respectively. I^0_{ik} is the inertia matrix of which the typical form is also shown in (*6*) of Problem 5.1. The angular transformation matrix Π^{10} [Equation (*1-4*)] is introduced in (*1*) as Π^{10} and Π^{20} and represents the angular transformations of a state vector from the systems 0 to 1 and 0 to 2 respectively. In full matrix form,

$$\Pi^{20} = \left[\begin{array}{ccc|ccc} \cos\phi_2 & \sin\phi_2 & 0 & 0 & 0 & 0 \\ -\sin\phi_2 & \cos\phi_2 & 0 & 0 & 0 & 0 \\ 0 & 0 & 1 & 0 & 0 & 0 \\ \hline 0 & 0 & 0 & \cos\phi_2 & \sin\phi_2 & 0 \\ 0 & 0 & 0 & -\sin\phi_2 & \cos\phi_2 & 0 \\ 0 & 0 & 0 & 0 & 0 & 1 \end{array}\right] \tag{2}$$

$$\Pi^{10} = \left[\begin{array}{ccc|ccc} \cos\phi_1 & \sin\phi_1 & 0 & 0 & 0 & 0 \\ -\sin\phi_1 & \cos\phi_1 & 0 & 0 & 0 & 0 \\ 0 & 0 & 1 & 0 & 0 & 0 \\ \hline 0 & 0 & 0 & \cos\phi_1 & \sin\phi_1 & 0 \\ 0 & 0 & 0 & -\sin\phi_1 & \cos\phi_1 & 0 \\ 0 & 0 & 0 & 0 & 0 & 1 \end{array}\right], \tag{3}$$

Consequently Π^{01} and Π^{02} represent the angular transformations from system 1 to 0 and 2 to 0 respectively, and have the following orthogonal relationship:

$$\Pi^{20} = \Pi^{02)T} = \Pi^{02)-1} \quad \text{and} \quad \Pi^{10} = \Pi^{01)T} = \Pi^{01)-1} \tag{4}$$

Thus from the member properties and geometric conditions, we obtain $\overset{0}{L}_{LR}$ according to (*1*).

TRANSPORT CHAIN-FREE VIBRATION

5.5. Find the circular frequencies of the single-span beams shown in Fig. P-5.5, in which the lumped masses are due to the beam weight as $\mu = (\gamma A/g)(2l)$ and $\rho = \mu l^2/3$

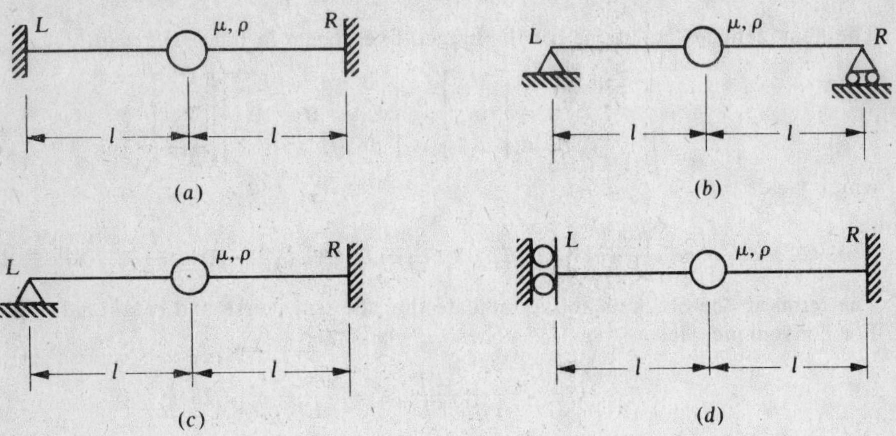

Fig. P-5.5

(*a*) Examining the boundary conditions of Fig. P-5.5*a*, we have

$$\hat{H}_L = [1 \quad N_L \quad V_L \quad M_L \quad 0 \quad 0 \quad 0]^T \tag{1}$$

$$\hat{H}_R = [1 \quad N_R \quad V_R \quad M_R \quad 0 \quad 0 \quad 0]^T \tag{2}$$

Applying (*1*) and (*2*) to the typical transport matrix shown in (*8*) of Problem 5.1, we obtain the following homogeneous equations

$$0 = \begin{bmatrix} u_L \\ v_L \\ \theta_L \end{bmatrix} = \begin{bmatrix} L_{52} & 0 & 0 \\ 0 & L_{63} & L_{64} \\ 0 & L_{73} & L_{74} \end{bmatrix} \begin{bmatrix} N_R \\ V_R \\ M_R \end{bmatrix} \tag{3}$$

where L_{ij} represents the appropriate element at the *i*th row and *j*th column of (*8*) of Problem 5.1.

The determinant of the coefficients of (3) yields

$$\left(2 - \frac{\mu\omega^2 l}{EA}\right)\left(\frac{\mu\omega^2 l^3}{18EI} - \frac{4}{3}\right)\left(\frac{\rho\omega^2 l}{8EI} - 1\right) = 0 \tag{4}$$

which yields

$$\omega_u = \sqrt{\frac{2EA}{\mu l}} \qquad \omega_v = \sqrt{\frac{24EI}{\mu l^3}} \qquad \omega_\theta = \sqrt{\frac{8EI}{\rho l}} \tag{5}$$

where ω_u, ω_v, and ω_θ are uncoupled frequencies corresponding to longitudinal, transverse, and rotational motions respectively. The mode shape associated with ω_θ is antisymmetrical and is not excited by a symmetrical load.

(*b*) The boundary conditions of the simple beam in Fig. P-5.5*b* are

$$\hat{H}_L = [1 \quad N_L \quad V_L \quad 0 \quad 0 \quad 0 \quad \theta_L]^T \tag{6}$$

$$\hat{H}_R = [1 \quad 0 \quad V_R \quad 0 \quad u_R \quad 0 \quad \theta_R]^T \tag{7}$$

Using these conditions in (8) of Problem 5.1 gives the following homogeneous equations

$$0 = \begin{bmatrix} M_L \\ u_L \\ v_L \end{bmatrix} = \begin{bmatrix} L_{43} & 0 & L_{47} \\ 0 & L_{55} & 0 \\ L_{63} & 0 & L_{67} \end{bmatrix} \begin{bmatrix} V_R \\ u_R \\ \theta_R \end{bmatrix} \tag{8}$$

from which the frequency equation is

$$\left(1 - \frac{\mu\omega^2 l}{EA}\right)\left(6 - \frac{\mu\omega^2 l^3}{EI}\right)\left(\frac{\rho\omega^2 l}{EI} - 6\right) = 0 \tag{9}$$

Thus

$$\omega_u = \sqrt{\frac{EA}{\mu l}} \qquad \omega_v = \sqrt{\frac{6EI}{\mu l^3}} \qquad \omega_\theta = \sqrt{\frac{6EI}{\rho l}} \tag{10}$$

(*c*) The homogeneous equations for the hinged-fixed beam in Fig. P-5.5*c* are

$$0 = \begin{bmatrix} M_L \\ u_L \\ v_L \end{bmatrix} = \begin{bmatrix} 0 & L_{43} & L_{44} \\ L_{52} & 0 & 0 \\ 0 & L_{63} & L_{64} \end{bmatrix} \begin{bmatrix} N_R \\ V_R \\ M_R \end{bmatrix} \tag{11}$$

which yields

$$\left(2 - \frac{\mu\omega^2 l}{EA}\right)\left[-\frac{7}{36}\frac{\mu\omega^2 l^6}{(EI)^2} + \frac{8l^3}{3EI} - \rho\omega^2\left(\frac{5}{12}\frac{l^4}{(EI)^2} - \frac{7}{72}\frac{\mu\omega^2 l^7}{(EI)^3}\right)\right] = 0 \tag{12}$$

The terms in the brackets above indicate that the transverse and rotational motions are coupled. For convenience, let $\rho = \mu l^2/3$. Then (12) becomes

$$\left(2 - \frac{\mu\omega^2 l}{EA}\right)\left(\frac{\mu^2 l^6}{(EI)^2}\omega^4 - \frac{36\mu l^3}{EI}\omega^2 + 288\right) = 0 \tag{13}$$

from which

$$\omega_u = \sqrt{\frac{2EA}{\mu l}} \qquad \omega_{v1} = \sqrt{\frac{12EI}{\mu l^3}} \qquad \omega_{v2} = \sqrt{\frac{24EI}{\mu l^3}} \tag{14}$$

If ρ is neglected, $\omega_v = \sqrt{96EI/7\mu l^3}$.

(*d*) The boundary conditions of Fig. P-5.5*d* give

$$0 = \begin{bmatrix} V_L \\ u_L \\ \theta_L \end{bmatrix} = \begin{bmatrix} 0 & L_{33} & L_{34} \\ L_{52} & 0 & 0 \\ 0 & L_{73} & L_{74} \end{bmatrix} \begin{bmatrix} N_R \\ V_R \\ M_R \end{bmatrix} \tag{15}$$

The frequency equation is

$$\left(2 - \frac{\mu\omega^2 l}{EA}\right)\left[\frac{2l}{EI} - \frac{5\mu\omega^2 l^4}{12(EI)^2} + \frac{\mu\rho\omega^4 l^5}{12(EI)^3} - \frac{\rho\omega^2 l^2}{(EI)^2}\right] = 0 \tag{16}$$

Again, letting $\rho = \mu l^2/3$, we have

$$\omega_u = \sqrt{\frac{2EA}{\mu l}} \qquad \omega_{v1} = \sqrt{\frac{3EI}{\mu l^3}} \qquad \omega_{v2} = \sqrt{\frac{24EI}{\mu l^3}} \tag{17}$$

If $\rho = 0$, $\omega_v = \sqrt{24EI/5\mu l^3}$.

5.6. Formulate the symbolic transport chain $\hat{\hat{L}}^0_{LR}$ for the structures shown in Fig. P-5.6.

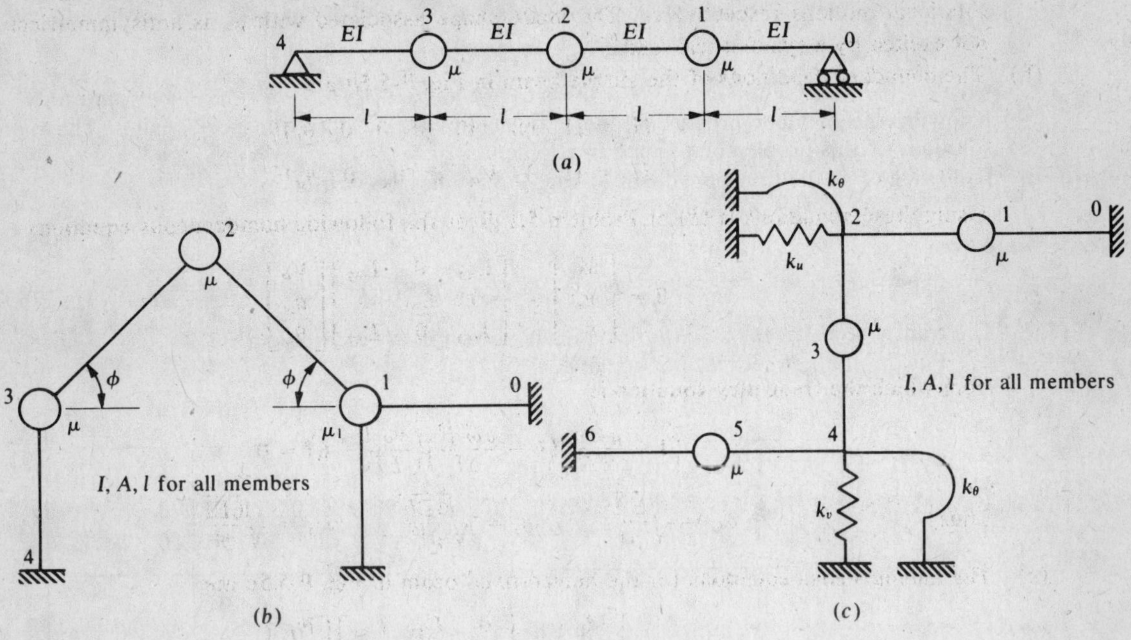

Fig. P-5.6

(*a*) For Fig. P-5.6*a* the transport chain may be written by following (*5-6*) as

$$\hat{\hat{H}}^0_4 = \hat{\hat{T}}^0_{43}\hat{\hat{I}}^0_3\hat{\hat{T}}^0_{32}\hat{\hat{I}}^0_2\hat{\hat{T}}^0_{21}\hat{\hat{I}}^0_{10}\hat{\hat{H}}^0_0 = \hat{\hat{L}}^0_{40}\hat{\hat{H}}^0_0 \tag{1}$$

where the dimensionless static transport matrix $\hat{\hat{T}}^0$ and the dimensionless inertia matrix $\hat{\hat{I}}^0$ are typically shown in (*2*) and (*3*) of Problem 5.3. A general procedure of manipulating (*1*) will be illustrated in Problem 5.7.

(*b*) The transport chain of Fig. P-5.6*b* is

$$\hat{\hat{H}}^0_4 = \hat{\Pi}^{04}\hat{\hat{T}}^4_{43}\hat{\Pi}^{40}\hat{\hat{I}}^0_3\hat{\Pi}^{03}\hat{\hat{T}}^3_{32}\hat{\Pi}^{30}\hat{\hat{I}}^0_2\hat{\Pi}^{02}\hat{\hat{T}}^2_{21}\hat{\Pi}^{20}\hat{\hat{J}}^0_1\hat{\hat{T}}^0_{10}\hat{\hat{H}}^0_0 \tag{2}$$

where the angular transformation matrix Π is similar to (*2*) and (*3*) of Problem 5.4 with the proper substitution of given angles; the dimensionless static transport matrix $\hat{\hat{T}}$ and inertia matrix $\hat{\hat{I}}$ are similar to (*2*) and (*3*) of Problem 5.3 respectively. The joint effect matrix $\hat{\hat{J}}^0_1$, consisting of inertia forces of the mass at node 1 and stiffnesses of column $a - 1$, is

$$\begin{bmatrix} 1 \\ \hline N_{1i} \\ V_{1i} \\ M_{1i} \\ \hline u_{1i} \\ v_{1i} \\ \theta_{1i} \end{bmatrix} = \left[\begin{array}{c|ccc|ccc} 1 & 0 & 0 & 0 & 0 & 0 & 0 \\ \hline 0 & 1 & 0 & 0 & \mu\omega^2 - 12EI/l^3 & 0 & -6EI/l^2 \\ 0 & 0 & 1 & 0 & 0 & \mu\omega^2 - AE/l & 0 \\ 0 & 0 & 0 & 1 & -6EI/l^2 & 0 & -\rho\omega^2 - 4EI/l \\ \hline 0 & 0 & 0 & 0 & 1 & 0 & 0 \\ 0 & 0 & 0 & 0 & 0 & 1 & 0 \\ 0 & 0 & 0 & 0 & 0 & 0 & 1 \end{array}\right] \begin{bmatrix} 1 \\ \hline N_{1k} \\ V_{1k} \\ M_{1k} \\ \hline u_{1k} \\ v_{1k} \\ \theta_{1k} \end{bmatrix} \tag{3}$$

Using the dimensionless notations introduced in (*1*) of Problem 5.3, $\hat{J}_1^0$ in (*3*) becomes

$$\hat{J}_1^0 = \left[\begin{array}{c|ccc|ccc} 1 & 0 & 0 & 0 & 0 & 0 & 0 \\ \hline 0 & 1 & 0 & 0 & \bar{\gamma} & 0 & \bar{s} \\ 0 & 0 & 1 & 0 & 0 & \bar{t} & 0 \\ 0 & 0 & 0 & 1 & -6 & 0 & \bar{u} \\ \hline 0 & 0 & 0 & 0 & 1 & 0 & 0 \\ 0 & 0 & 0 & 0 & 0 & 1 & 0 \\ 0 & 0 & 0 & 0 & 0 & 0 & 1 \end{array}\right] \tag{4}$$

where $\bar{\gamma} = \bar{a} - 12(\gamma/l)^2$, $\bar{s} = -6(\gamma/l)^2$, $\bar{t} = \bar{b} - (l/\gamma)^2$, $\bar{u} = \bar{c} - 4$, and $\gamma^2 = I/A$. Knowing the member properties one may manipulate (*2*) for transport chain. The numerical solutions of this problem are found in Problem 5.20.

(*c*) Following (*5-5*) the transport chain of Fig. P-5.5*c* may be written as

$$\hat{H}_6^0 = \hat{L}_{64}^0\hat{J}_4^0\hat{L}_{42}^0\hat{J}_2^0\hat{L}_{20}^0\hat{H}_0^0 \tag{5}$$

where

$$\hat{L}_{64}^0 = \hat{T}_{65}^0\hat{I}_5^0\hat{T}_{54}^0 \qquad \hat{L}_{20}^0 = \hat{T}_{21}^0\hat{I}_1^0\hat{T}_{10}^0 \qquad \hat{L}_{42}^0 = \hat{\Pi}^{02}\hat{L}_{42}^2\hat{\Pi}^{20} \qquad \hat{L}_4^2 = \hat{T}_{43}^2\hat{I}_3^2\hat{T}_{32}^2 \tag{6, 7, 8, 9}$$

The transport matrices (*6, 7, 8*) are typically shown in (*4*) of Problem 5.3.

For the joint effect matrix, we have

$$\hat{J}_2^0 = \left[\begin{array}{c|ccc|ccc} 1 & 0 & 0 & 0 & 0 & 0 & 0 \\ \hline 0 & 1 & 0 & 0 & \bar{k}_u & 0 & 0 \\ 0 & 0 & 1 & 0 & 0 & 0 & 0 \\ 0 & 0 & 0 & 1 & 0 & 0 & \bar{k}_\theta \\ \hline 0 & 0 & 0 & 0 & 1 & 0 & 0 \\ 0 & 0 & 0 & 0 & 0 & 1 & 0 \\ 0 & 0 & 0 & 0 & 0 & 0 & 1 \end{array}\right] \qquad \hat{J}_4^0 = \left[\begin{array}{c|ccc|ccc} 1 & 0 & 0 & 0 & 0 & 0 & 0 \\ \hline 0 & 1 & 0 & 0 & 0 & 0 & 0 \\ 0 & 0 & 1 & 0 & 0 & \bar{k}_v & 0 \\ 0 & 0 & 0 & 1 & 0 & 0 & \bar{k}_\theta \\ \hline 0 & 0 & 0 & 0 & 1 & 0 & 0 \\ 0 & 0 & 0 & 0 & 0 & 1 & 0 \\ 0 & 0 & 0 & 0 & 0 & 0 & 1 \end{array}\right]$$

where $\bar{k}_u = -k_u l/AE$, $\bar{k}_\theta = -k_\theta l/EI$, and $\bar{k}_v = -k_v l^3/EI$.

Note that the stiffnesses of the intermediate elastic supports should not be too large compared with the stiffnesses of the beam; otherwise numerical difficulties may arise when we carry through the simple transfer-matrix multiplications for determining natural frequencies. A technique called the Delta-matrix method may be used to overcome the difficulties (Ref. 5.3, pp. 193–204).

5.7. Find the circular frequencies of the beam shown in Fig. P-5.6*a* with consideration of longitudinal and transverse vibration only.

Table 5.7 shows the numerical procedures for formulating the transport chain in which the columns 2, 4, and 6 are eliminated because of the boundary conditions $\bar{N}_0 = \bar{M}_0 = \bar{u}_0 = 0$.

From Table 5.7

$$0 = \begin{bmatrix} \bar{M}_4 \\ \bar{u}_4 \\ \bar{v}_4 \end{bmatrix} = \begin{bmatrix} L_{43} & L_{45} & L_{47} \\ L_{53} & L_{55} & L_{57} \\ L_{63} & L_{65} & L_{67} \end{bmatrix} \begin{bmatrix} \bar{M}_0 \\ \bar{u}_0 \\ \bar{v}_0 \end{bmatrix} \tag{1}$$

where $L_{43} = 4 + 23\bar{b}/3 + 18\bar{b}^2/36 + \bar{b}^3/216 \qquad L_{45} = 0 \qquad L_{47} = -10\bar{b} - 2\bar{b}^2 - \bar{b}^3/36$

$L_{53} = 0 \qquad L_{55} = 1 - 6\bar{a} + 5\bar{a}^2 - \bar{a}^3 \qquad L_{57} = 0$

$L_{63} = 64/6 + 59\bar{b}/18 + \bar{b}^2/9 + \bar{b}^3/1296 \qquad L_{65} = 0 \qquad L_{67} = -4 - 23\bar{b}/3 - \bar{b}^2/2 - \bar{b}^3/216$

Substituting the above into (*1*) for determinant yields

$$(1 - 6\bar{a} + 5\bar{a}^2 - \bar{a}^3)(\bar{b}^3 - 33.4285\,\bar{b}^2 + 174.857\,\bar{b} - 61.714) = 0 \tag{2}$$

from which the longitudinal frequencies are

$$\omega_{u1} = \sqrt{\frac{0.1979\,EA}{\mu l}} \qquad \omega_{u2} = \sqrt{\frac{1.555\,EA}{\mu l}} \qquad \omega_{u3} = \sqrt{\frac{3.2468\,EA}{\mu l}}$$

and the transverse frequencies are

$$\omega_{v1} = \sqrt{\frac{0.38\,EI}{\mu l^3}} \qquad \omega_{v2} = \sqrt{\frac{6.003\,EI}{\mu l^3}} \qquad \omega_{v3} = \sqrt{\frac{27.047\,EI}{\mu l^3}}$$

Table 5.7 Transport Chain

Component Matrices (For $\bar{a}$, $\bar{b}$ refer to Problem 5.2)

$$\hat{T}_{43} = \hat{T}_{32} = \hat{T}_{21} = \hat{T}_{10} = \left[\begin{array}{c|ccc} 1 & 0 & 0 & 0 \\ \hline 0 & 0 & 0 & 0 \\ 0 & 1 & 0 & 0 \\ 0 & 1 & 0 & 0 \\ \hline 0 & 0 & 1 & 0 \\ 0 & 1/6 & 0 & -1 \\ 0 & -1/2 & 0 & 1 \end{array}\right] \qquad \hat{I}_3 = \hat{I}_2 = \hat{I}_1 = \left[\begin{array}{c|ccc|ccc} 1 & 0 & 0 & 0 & 0 & 0 & 0 \\ \hline 0 & 1 & 0 & 0 & \bar{a} & 0 & 0 \\ 0 & 0 & 1 & 0 & 0 & \bar{b} & 0 \\ 0 & 0 & 0 & 1 & 0 & 0 & 0 \\ \hline 0 & 0 & 0 & 0 & 1 & 0 & 0 \\ 0 & 0 & 0 & 0 & 0 & 1 & 0 \\ 0 & 0 & 0 & 0 & 0 & 0 & 1 \end{array}\right]$$

Symbolic Matrix Chain

$$\hat{H}_4 = \hat{T}_{43}\hat{I}_3\hat{T}_{32}\hat{I}_2\hat{T}_{21}\hat{I}_1\hat{T}_{10}\hat{H}_0 = \left[\begin{array}{c} 1 \\ \hline \bar{N}_4 \\ \bar{V}_4 \\ \bar{M}_4 \\ \hline \bar{u}_4 \\ \bar{v}_4 \\ \bar{\theta}_4 \end{array}\right] = \left[\begin{array}{cccc} 1 & 0 & 0 & 0 \\ \hline 0 & L_{23} & L_{25} & L_{27} \\ 0 & L_{33} & L_{35} & L_{37} \\ 0 & L_{43} & L_{45} & L_{47} \\ \hline 0 & L_{53} & L_{55} & L_{57} \\ 0 & L_{63} & L_{65} & L_{67} \\ 0 & L_{73} & L_{75} & L_{77} \end{array}\right] \left[\begin{array}{c} 1 \\ \hline \bar{V}_0 \\ \bar{u}_0 \\ \bar{\theta}_0 \end{array}\right]$$

5.8. Find the natural frequencies and natural modes of the shear building shown in Fig. P-5.8, which is the same as that of Fig. P-3.1*a*.

The dimensionless transport chain for this shear building is

$$\underbrace{\begin{bmatrix} \bar{V}_3 \\ \bar{v}_3 \end{bmatrix}}_{\bar{H}_3} = \underbrace{\begin{bmatrix} 1 & a_3 \\ 0 & 1 \end{bmatrix}}_{\bar{I}_3} \underbrace{\begin{bmatrix} 1 & 0 \\ \beta_3^3/\alpha_3 & 1 \end{bmatrix}}_{\bar{T}_{32}} \underbrace{\begin{bmatrix} 1 & a_2 \\ 0 & 1 \end{bmatrix}}_{\bar{I}_2} \underbrace{\begin{bmatrix} 1 & 0 \\ \beta_2^3/\alpha_2 & 1 \end{bmatrix}}_{\bar{T}_{21}} \underbrace{\begin{bmatrix} 1 & a_1 \\ 0 & 1 \end{bmatrix}}_{\bar{I}_1} \underbrace{\begin{bmatrix} 1 & 0 \\ 1 & 1 \end{bmatrix}}_{\bar{T}_{10}} \underbrace{\begin{bmatrix} \bar{V}_0 \\ \bar{v}_0 \end{bmatrix}}_{\bar{H}_0} \tag{1}$$

where $\bar{H}_3$ and $\bar{H}_0$ are the reduced state vectors at the upper and lower boundary of the building; $\bar{I}_3$, $\bar{I}_2$, and $\bar{I}_1$ are the inertia matrices of the respective girders; and $\bar{T}_{32}$, $\bar{T}_{21}$, and $\bar{T}_{10}$ are the static transport matrices relating the state vector of the lower end of the respective columns to the state vector of the upper end of the same columns.

The construction of $\bar{I}_i$ and $\bar{T}_{i,i-1}$ follows the pattern of the preceding problems but involves the restrictions imposed by the definition of the shear building.

The equivalents in (1) are

$$\bar{V}_i = -\frac{l_i^2}{24EI_i}V_i \qquad \bar{v}_i = \frac{v_i}{l_i} \qquad a_i = -\frac{\mu_i\omega^2 l_1^3}{24EI_1}$$

$$\alpha_i = \frac{I_i}{I_{i-1}} \qquad \beta_i = \frac{l_i}{l_{i-1}}$$

where V_i is the ith story shear, v_i is the horizontal displacement of the ith girder, and for the given building,

$$\mu_i = \mu_1 = \mu_2 = \mu$$

$$\alpha_i = \alpha_1 = \alpha_2 = \alpha_3 = 1$$

$$l_i = l_1 = l_2 = l_3 = l$$

$$\beta_i = \beta_1 = \beta_2 = \beta_3 = 1$$

$$I_i = I_1 = I_2 = I_3 = I$$

$$a_i = a_1 = a_2 = a_3 = a$$

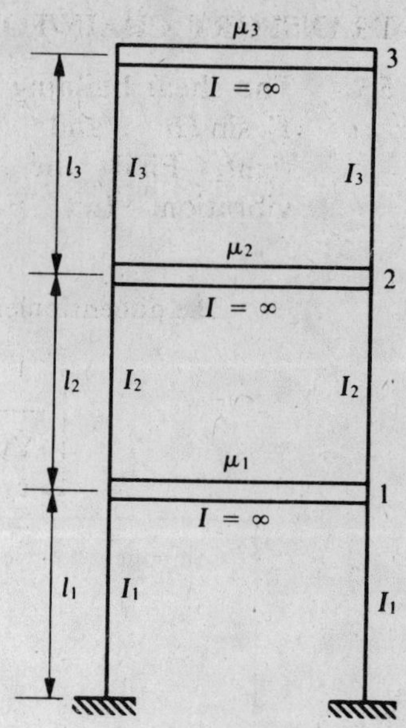

Fig. P-5.8

From Problem 3.1, in U.S. customary units, $I = 476.1\text{ in}^4$, $\mu = 0.8461\text{ k-s}^2/\text{ft}$, $l = 14\text{ ft}$, and $E = 30 \times 10^3\text{ ksi}$. The transport chain equation (1) then becomes

$$\begin{bmatrix} \bar{V}_3 \\ \\ \bar{v}_3 \end{bmatrix} = \begin{bmatrix} \underbrace{1 + 7.7113a + 9.4148a^2 + 2.7101a^3}_{L_{11}} & \underbrace{39 + 6.7083a^2 + 2.7101a^3}_{L_{12}} \\ \underbrace{4.3542 + 8.0613a + 2.7101a^2}_{L_{21}} & \underbrace{1 + 5.3512a + 2.7101a^2}_{L_{22}} \end{bmatrix} \begin{bmatrix} \bar{V}_0 \\ \\ \bar{v}_0 \end{bmatrix} \tag{2}$$

where by inspection $\bar{V}_3 = 0$ and $\bar{v}_0 = 0$.

The frequency determinant abstracted from (2) as $|L_{11}| = 0$ yields

$$2.7101a^3 + 9.4184a^2 + 7.7113a + 1 = 0 \tag{3}$$

from which

$$\begin{aligned} a_{(1)} &= -0.159 &\text{or}\quad \omega_1 &= \sqrt{\frac{24a_{(1)}EI}{\mu l^3}} = 12.79\text{ rad/s} \\ a_{(2)} &= -1.0013 &\text{or}\quad \omega_2 &= \sqrt{\frac{24a_{(2)}EI}{\mu l^3}} = 32.098\text{ rad/s} \\ a_{(3)} &= -2.3148 &\text{or}\quad \omega_3 &= \sqrt{\frac{24a_{(3)}EI}{\mu l^3}} = 48.80\text{ rad/s} \end{aligned} \tag{4}$$

Once ω_1, ω_2, and ω_3 are known, the terms L_{11}, L_{12}, L_{21}, and L_{22} in (2) become distinct constants of the respective frequency and (2) can be used for the calculation of natural modes as shown symbolically below. Since

$$V_3 = 0 \qquad \bar{v}_0 = 0 \qquad \bar{v}_3 = L_{21}\bar{V}_0 \qquad \text{or} \qquad \bar{V}_0 = 1/L_{21}\bar{v}_3$$

then

$$\begin{bmatrix} \bar{V}_1 \\ \bar{v}_1 \end{bmatrix} = \begin{bmatrix} 1 & a \\ 0 & 1 \end{bmatrix} \begin{bmatrix} 1 & 0 \\ 1 & 1 \end{bmatrix} \begin{bmatrix} \bar{v}_3/L_{21} \\ 0 \end{bmatrix} = \begin{bmatrix} (1 + a)\bar{v}_3/L_{21} \\ \bar{v}_3/L_{21} \end{bmatrix} \tag{5}$$

$$\begin{bmatrix} \bar{V}_2 \\ \bar{v}_2 \end{bmatrix} = \begin{bmatrix} 1 & a \\ 0 & 1 \end{bmatrix} \begin{bmatrix} 1 & 0 \\ 1.3591 & 1 \end{bmatrix} \begin{bmatrix} \bar{V}_1 \\ \bar{v}_1 \end{bmatrix} = \begin{bmatrix} (1.3571a + 1)(a + 1)\bar{v}_3/L_{21} \\ 1.3571a(1 + a)\bar{v}_3/L_{21} + \bar{v}_3/L_{21} \end{bmatrix} \tag{6}$$

From (5) and (6),

$$\bar{v}_3 = L_{21}\bar{v}_1 \qquad \bar{v}_2 = 1.371a(1 + a)\bar{v}_1 + \bar{v}_1 \qquad \bar{v}_1 = \bar{v}_1 \tag{7}$$

$$\left.\begin{aligned} &\text{For } a_{(1)} = -0.159, & \bar{v}_1 &= \bar{v}_1, & \bar{v}_2 &= 2.1413\bar{v}_1, & &\text{and} & \bar{v}_3 &= 3.140\bar{v}_1 \\ &\text{For } a_{(2)} = -1.0013, & \bar{v}_1 &= \bar{v}_1, & \bar{v}_2 &= 1.001\,76\bar{v}_1, & &\text{and} & \bar{v}_3 &= -0.9969\bar{v}_1 \\ &\text{For } a_{(3)} = -2.3148, & \bar{v}_1 &= \bar{v}_1, & \bar{v}_2 &= -0.7843\bar{v}_1, & &\text{and} & \bar{v}_3 &= 0.2156\bar{v}_1 \end{aligned}\right\} \tag{8}$$

TRANSPORT CHAIN-FORCED VIBRATION

5.9. The shear building given in the preceding problem is subject to harmonic forces $F_2(t) = F_2 \sin \Omega t$ and $F_3(t) = F_3 \sin \Omega t$ applied at nodes 2 and 3 respectively, to the right. Find the nodal deflections and member end shears for steady state vibration. Let $F_2 = 20\text{ k}$, $F_3 = 10\text{ k}$, and $\Omega = 6.4\text{ rad/s}$.

The dimensionless transport chain for forced vibration of the shear building is

$$\left[\begin{array}{c} 1 \\ \hline \bar{V}_3 \\ \bar{v}_3 \end{array}\right] = \left[\begin{array}{c|cc} 1 & 0 & 0 \\ \hline \bar{F}_3 & 1 & a_3 \\ 0 & 0 & 1 \end{array}\right]\left[\begin{array}{c|cc} 1 & 0 & 0 \\ \hline 0 & 1 & 0 \\ 0 & \beta_3^3/\alpha_3 & 1 \end{array}\right]\left[\begin{array}{c|cc} 1 & 0 & 0 \\ \hline \bar{F}_2 & 1 & a_2 \\ 0 & 0 & 1 \end{array}\right]$$

$$\times\left[\begin{array}{c|cc} 1 & 0 & 0 \\ \hline 0 & 1 & 0 \\ 0 & \beta_2^3/\alpha_2 & 1 \end{array}\right]\left[\begin{array}{c|cc} 1 & 0 & 0 \\ \hline 0 & 1 & a_1 \\ 0 & 0 & 1 \end{array}\right]\left[\begin{array}{c|cc} 1 & 0 & 0 \\ \hline 0 & 1 & 0 \\ 0 & 1 & 1 \end{array}\right]\left[\begin{array}{c} 1 \\ \hline \bar{V}_0 \\ \bar{v}_0 \end{array}\right] \tag{1}$$

where $\quad \bar{F}_i = -\dfrac{F_i l_1^2}{24EI_1} \qquad a_i = \dfrac{-\mu_i \Omega^2 l_1^3}{24EI_1} = a_1 = a_2 = a_3 = a$

After matrix multiplications, (*1*) may be expressed symbolically as

$$\left[\begin{array}{c} 1 \\ \hline \bar{V}_3 \\ \bar{v}_3 \end{array}\right] = \left[\begin{array}{c|cc} 1 & 0 & 0 \\ \hline L_{21} & L_{22} & L_{23} \\ L_{31} & L_{32} & L_{33} \end{array}\right]\left[\begin{array}{c} 1 \\ \hline \bar{V}_0 \\ \bar{v}_0 \end{array}\right] \tag{2}$$

Since $\bar{V}_3 = \bar{v}_0 = 0$, then $\bar{F}_2 = 2\bar{F}_3$. Thus

$$\bar{V}_0 = -L_{21}/L_{22} \qquad \text{and} \qquad \bar{v}_3 = L_{31} - L_{32}L_{21}/L_{22} \tag{3, 4}$$

where $\quad L_{21} = \bar{F}_3 + 2\bar{F}_3(1 + 1.997\,a) \qquad L_{22} = 2.710\,a^3 + 9.4183\,a^2 + 7.7112\,a + 1$

$\quad L_{31} = 3.994\,\bar{F}_3 \qquad L_{32} = 2.710\,a^2 + 8.0612\,a + 4.3541$

Using $I = 476.1\text{ in}^4$, $l = 14\text{ ft}$, $\mu = 0.8461\text{ k-s}^2\text{/ft}$, $\Omega = 6.4\text{ rad/s}$, and $E = 30 \times 10^3\text{ ksi}$, we calculate $a = -\mu\Omega^2 l^3/24EI = -0.039\,81$ and $\bar{F}_3 = -0.000\,822\,5$. Consequently

$$L_{21} = -0.002\,336 \qquad L_{22} = 0.7078 \qquad L_{31} = -0.003\,29 \qquad L_{32} = 3.958$$

Thus $\quad \bar{V}_0 = 0.0033 \qquad \bar{v}_3 = 0.002\,15 \qquad \text{or} \qquad V_0 = -40.08\text{ k} \qquad v_3 = 0.1367\text{ ft} \tag{5}$

By (*1*)

$$\left[\begin{array}{c} 1 \\ \hline \bar{V}_1 \\ \bar{v}_1 \end{array}\right] = \left[\begin{array}{c|cc} 1 & 0 & 0 \\ \hline 0 & 1 & a \\ 0 & 0 & 1 \end{array}\right]\left[\begin{array}{c|cc} 1 & 0 & 0 \\ \hline 0 & 1 & 0 \\ 0 & 1 & 1 \end{array}\right]\left[\begin{array}{c} 1 \\ \hline \bar{V}_0 \\ 0 \end{array}\right] \tag{6}$$

and

$$\left[\begin{array}{c} 1 \\ \hline \bar{V}_2 \\ \bar{v}_3 \end{array}\right] = \left[\begin{array}{c|cc} 1 & 0 & 0 \\ \hline 2\bar{F}_3 & 1 & a \\ 0 & 0 & 0 \end{array}\right]\left[\begin{array}{c|cc} 1 & 0 & 0 \\ \hline 0 & 1 & 0 \\ 0 & 1.3571 & 1 \end{array}\right]\left[\begin{array}{c} 1 \\ \hline \bar{V}_1 \\ \bar{v}_1 \end{array}\right] \tag{7}$$

from which

$$\bar{V}_1 = 0.0462, \qquad \bar{V}_2 = 0.001\,22 \qquad \text{or} \qquad V_1 = -38.48\text{ k}, \qquad V_2 = -14.826\text{ k}$$

$$\bar{v}_1 = 0.0033, \qquad \bar{v}_2 = 0.0076 \qquad \text{or} \qquad v_1 = 0.0462\text{ ft}, \qquad v_2 = 0.1064\text{ ft}$$

5.10. The mass of the beam shown in Fig. P-5.10 is subject to the transverse force $P_v(t) = 8.54 \cos \Omega t$ k and the moment $Q(t) = 17.08 \cos \Omega t$ k-ft. Find the member end forces and deformations for steady state vibration. Let

$$l = 10 \text{ ft} \qquad I = 243 \text{ in}^4 \qquad E = 30 \times 10^3 \text{ ksi}$$
$$\mu = 0.076\,09 \text{ k-s}^2/\text{ft} \qquad \rho = 2.536 \text{ k-s}^2/\text{ft} \qquad \Omega = 39.887 \text{ rad/s}$$

From Problem 5.5, it is known that the given transverse and rotational motions do not induce any longitudinal vibration. Thus the longitudinal deformation and inertia force may not be included in the following.

$$\begin{bmatrix} 1 \\ \hline V_L \\ M_L \\ \hline v_L \\ \theta_L \end{bmatrix} = \left[\begin{array}{c|cc|cc} 1 & 0 & 0 & 0 & 0 \\ \hline 0 & 1 & 0 & 0 & 0 \\ 0 & l & 1 & 0 & 0 \\ \hline 0 & l^3/6EI & l^2/2EI & 1 & -l \\ 0 & -l^2/2EI & -l/EI & 0 & 1 \end{array}\right] \begin{bmatrix} 1 \\ \hline V_i \\ M_i \\ \hline v_i \\ \theta_i \end{bmatrix} \qquad (1)$$

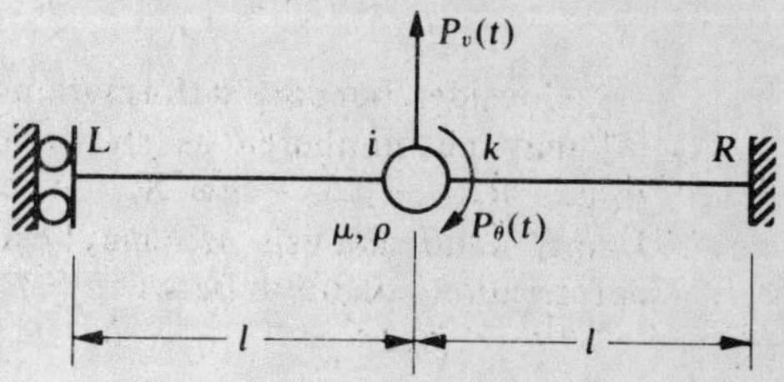

Fig. P-5.10

$$\begin{bmatrix} 1 \\ \hline V_i \\ M_i \\ \hline v_i \\ \theta_i \end{bmatrix} = \left[\begin{array}{c|cc|cc} 1 & 0 & 0 & 0 & 0 \\ \hline 10 & 1 & 0 & \mu\Omega^2 & 0 \\ -20 & 0 & 1 & 0 & \mu\Omega^2 \\ \hline 0 & 0 & 0 & 1 & 0 \\ 0 & 0 & 0 & 0 & 1 \end{array}\right] \begin{bmatrix} 1 \\ \hline V_k \\ M_k \\ \hline v_k \\ \theta_k \end{bmatrix} \qquad (2)$$

Since segments Li and kR are identical, $\hat{T}_{kR} = \hat{T}_{Li}$. Thus the transport matrix may be obtained by using (1) and (2) according to

$$\hat{H}_L = \hat{T}_{Li}\hat{I}_{ik}\hat{T}_{kR}\hat{H}_R \qquad (3)$$

For the given conditions,

$$l^3/6EI = 0.003\,292 \text{ ft/k} \qquad \mu\Omega^2 = 121.052 \text{ k/ft}^2$$
$$l^2/2EI = 0.000\,987\,65 \text{ 1/k} \qquad \rho\Omega^2 = 4035.07 \text{ k/ft}$$
$$l/EI = 0.000\,197\,52 \text{ 1/k-ft,}$$

and (3) becomes numerically

$$\begin{bmatrix} 1 \\ \hline V_L \\ M_L \\ v_L \\ \theta_L \end{bmatrix} = \left[\begin{array}{c|cccc} 1 & 0 & 0 & 0 & 0 \\ \hline 8.5405 & 1.3985 & 0.1196 & 121.0520 & -1210.520 \\ 68.3240 & 20.0000 & 1.3986 & 1210.5200 & -8070.130 \\ 0.0112 & 0.0237 & 0.0036 & 1.3985 & -20.000 \\ -0.0051 & -0.0036 & -0.0004 & -0.1196 & 1.399 \end{array}\right] \begin{bmatrix} 1 \\ \hline V_R \\ M_R \\ v_R \\ \theta_R \end{bmatrix} \qquad (4)$$

Knowing $V_L = \theta_L = v_R = \theta_R = 0$, we have

$$M_L = -153.888 \text{ ft-k} \qquad v_L = -0.357\,57 \text{ ft} \qquad M_R = 320.9546 \text{ ft-k} \qquad V_R = -33.555 \text{ k}$$

GENERAL FLEXIBILITY MATRIX

5.11. Find the natural frequency of the determinate structure shown in Fig. P-5.11*a* for which $I = 1.01 \text{ dm}^4$, $A = 232.26 \text{ cm}^2$, $h = 3.05$ m, $l = 6.10$ m, and $\mu = 181.26 \text{ kN-s}^2/\text{m}$. Assume the columns to be massless.

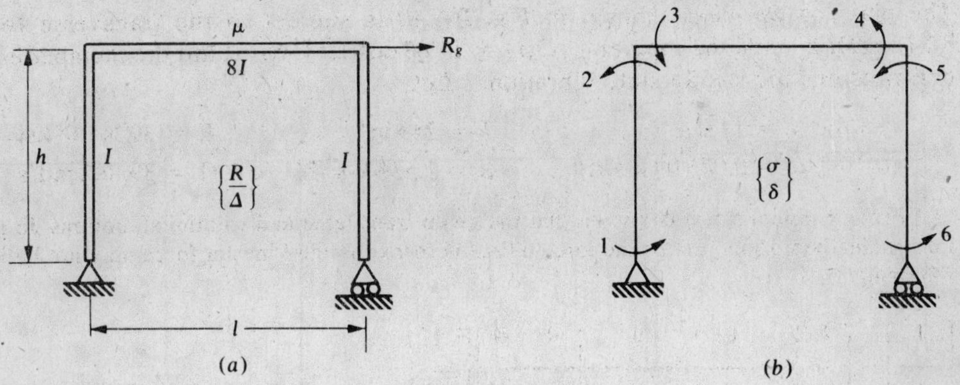

Fig. P-5.11

Consider bending deformations only. The independent moments $\{\sigma\}$ and angular deformations $\{\delta\}$ may be numbered as shown in Fig. P-5.11*b*. The external load vector $\{R\}$ is due to inertia force $R_g = -\mu\ddot{\Delta}_g = \mu\omega^2 X_g$, where $\Delta_g = X_g \cos \omega t$ is the girder's horizontal displacement. Using static analysis we may establish the equilibrium condition between $\{\sigma\}$ and $\{R\}$ and the deformation condition between $\{\delta\}$ and $\{\sigma\}$ as

$$\underbrace{\begin{bmatrix} M_1 \\ M_2 \\ M_3 \\ M_4 \\ M_5 \\ M_6 \end{bmatrix}}_{\{\sigma\}} = \underbrace{\begin{bmatrix} 0 \\ 3.05 \\ -3.05 \\ 0 \\ 0 \\ 0 \end{bmatrix}}_{\{q\}} R_g \tag{1}$$

$$\underbrace{\begin{bmatrix} \theta_1 \\ \theta_2 \\ \theta_3 \\ \theta_4 \\ \theta_5 \\ \theta_6 \end{bmatrix}}_{\{\delta\}} = \underbrace{\begin{bmatrix} h/3EI & -h/6EI & 0 & 0 & 0 & 0 \\ -h/6EI & h/3EI & 0 & 0 & 0 & 0 \\ 0 & 0 & l/24EI & -l/48EI & 0 & 0 \\ 0 & 0 & -l/48EI & l/24EI & 0 & 0 \\ 0 & 0 & 0 & 0 & h/3EI & -h/6EI \\ 0 & 0 & 0 & 0 & -h/6EI & h/3EI \end{bmatrix}}_{[f]} \underbrace{\begin{bmatrix} M_1 \\ M_2 \\ M_3 \\ M_4 \\ M_5 \\ M_6 \end{bmatrix}}_{\{\sigma\}} \tag{2}$$

where $[q]$ = unit load transfer matrix (II, p. 166) and $[f]$ = unit stress segmental flexibility chain (II, p. 165). The signs in $[q]$ and $[f]$ have been adjusted to be consistent with Fig. P-5.11*b*, and to satisfy (*5-25*) and (*5-26*).

In terms of $[q]$ and $[f]$ the system flexibility matrix (Ref. 5.7, p. 167) is

$$[\Phi_R] = [q]^T[f][q] \tag{3}$$

and the dynamic flexibility matrix equation (*5-26*) becomes

$$\{\Delta_R\} = [\Phi_R]\{R\} \tag{4}$$

In this particular case, with

$$\{\Delta_R\} = \Delta_g = X_g \cos \omega t \qquad [\Phi_R] = \frac{35.40}{3EI} \qquad \{R\} = \mu\omega^2 X_g \cos \omega t$$

the governing matrix equation (*4*) reduces to one algebraic equation and is

$$X_g = \frac{35.40}{3EI}\mu\omega^2 X_g \tag{5}$$

from which the natural frequency is

$$\omega = \sqrt{\frac{3EI}{35.40\,\mu}} = 3.1276 \text{ rad/s} \tag{6}$$

5.12. Find the natural frequency of the rigid frame shown in Fig. P-5.12a where $I = 1.01\ \text{dm}^4$, $A = 232.26\ \text{cm}^2$, $h = 3.05\ \text{m}$, $l = 6.10\ \text{m}$, and $\mu = 181.26\ \text{kN-s}^2/\text{m}$. The columns are considered to be massless springs.

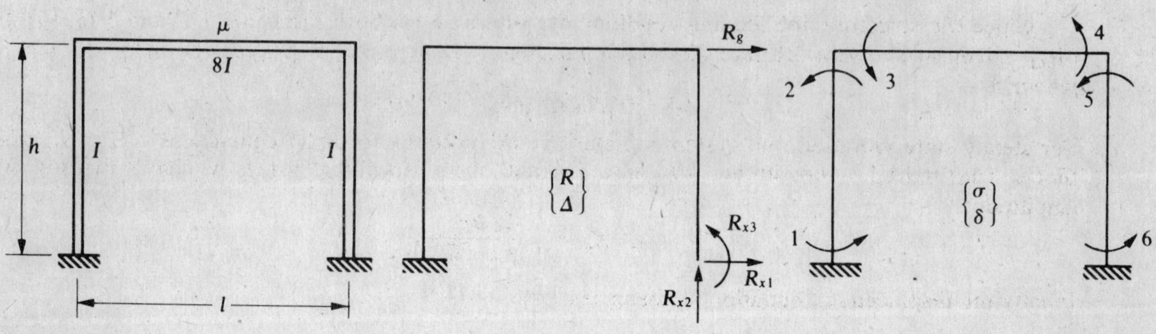

(a) (b) (c)

Fig. P-5.12

Following the general method of flexibility matrix for statics, we change the indeterminate frame into a stable determinate structure (Fig. P-5.12b) in which the inertia force of the girder is $R_g = \mu\omega^2\Delta_g$ and the redundants are the fixed end stresses R_{x1}, R_{x2}, and R_{x3}. Then, as in Problem 5.11,

$$\begin{bmatrix} M_1 \\ M_2 \\ M_3 \\ M_4 \\ M_5 \\ M_6 \end{bmatrix} = \left[\begin{array}{c|ccc} 3.05 & 0 & -6.10 & -1 \\ 0 & 3.05 & 6.10 & 1 \\ 0 & -3.05 & -6.10 & -1 \\ 0 & 3.05 & 0 & 1 \\ 0 & -3.05 & 0 & -1 \\ 0 & 0 & 0 & 1 \end{array}\right] \begin{bmatrix} R_g \\ \hline R_{x1} \\ R_{x2} \\ R_{x3} \end{bmatrix} \tag{1}$$

and

$$\begin{bmatrix} \theta_1 \\ \theta_2 \\ \theta_3 \\ \theta_4 \\ \theta_5 \\ \theta_6 \end{bmatrix} = \frac{1}{(206.84)(1.01)(10)^2} \begin{bmatrix} 3.05/3 & -1.525/3 & 0 & 0 & 0 & 0 \\ -1.525/3 & 3.05/3 & 0 & 0 & 0 & 0 \\ 0 & 0 & 6.10/24 & -6.10/48 & 0 & 0 \\ 0 & 0 & -6.10/48 & 6.10/24 & 0 & 0 \\ 0 & 0 & 0 & 0 & 3.05/3 & -1.525/3 \\ 0 & 0 & 0 & 0 & -1.525/3 & 3.05/3 \end{bmatrix} \begin{bmatrix} M_1 \\ M_2 \\ M_3 \\ M_4 \\ M_5 \\ M_6 \end{bmatrix}$$

or symbolically $\qquad \{\sigma\} = [q_R \mid q_x]\begin{Bmatrix} R_g \\ R_x \end{Bmatrix} \qquad$ and $\qquad \{\delta\} = [f]\{\sigma\}$ $\qquad (2)$

The system flexibility matrix is then

$$[\Phi] = \underbrace{\begin{bmatrix} \Phi_{RR} & \Phi_{Rx} \\ \Phi_{xR} & \Phi_{xx} \end{bmatrix}}_{[q_R \mid q_x]^T[f][q_R \mid q_x]} = 10^{-4} \left[\begin{array}{c|ccc} 4.527 & -2.264 & -13.581 & -2.226 \\ \hline -2.264 & 12.450 & 16.977 & 5.566 \\ -13.581 & 16.977 & 58.853 & 10.019 \\ -2.226 & 5.566 & 10.019 & 3.285 \end{array}\right] \tag{3}$$

and the dynamic flexibility matrix equation (5-26) becomes

$$\begin{bmatrix} \Delta_R \\ \Delta_x \end{bmatrix} = \begin{bmatrix} \Phi_{RR} & \Phi_{Rx} \\ \Phi_{xR} & \Phi_{xx} \end{bmatrix} \begin{bmatrix} R_g \\ R_x \end{bmatrix} \tag{4}$$

where from the condition of compatibility of fixed support, $\{\Delta_x\} = 0$, the second row yields $\{R_x\} = -[\Phi_{xx}^{-1}\Phi_{xR}]\{R_g\}$, and the first row becomes

$$\{\Delta_R\} = [\Phi_{RR} - \Phi_{Rx}\Phi_{xx}^{-1}\Phi_{xR}]\{R_g\} = [\Phi_R]\{R_g\} = [\Phi_R]\mu\omega^2 X_g \cos\omega t \tag{5}$$

from which $[\Phi_R] = 0.000\,063\,2$ and since $\{\Delta_R\} = X_g \cos\omega t$,

$$X_g = 0.000\,063\,2\,\mu\omega^2 X_g \qquad \text{or} \qquad \omega = \sqrt{\frac{15\,833}{181.26}} = 9.346\ \text{rad/s} \tag{6}$$

5.13. Assume the structure shown in the preceding problem is subject to a harmonic force $W(t) = 10 \cos \tau t$ k acting at the girder to the right; $\tau = 21.83$ rad/s. Find the internal moments due to the steady state vibration.

Since the structure and loading condition are the same for both Problems 5.12 and 5.13, $[q]$, $[\Phi]$, $[\Phi_R]$ given in Problem 5.12 are directly applicable. Taking $\{\Delta_R\} = \{X_R\} \cos \tau t$ and $R_g = -\mu \ddot{\Delta} g$ we write

$$\{\Delta_R\} = [\Phi_R]\{-\mu \ddot{\Delta} g + 10 \cos \tau t\} \tag{1}$$

For steady state vibration, the system vibrates according to the forcing frequency as $\Delta_g = X_g \cos \tau t$. Since $\{\Delta_R\} = \{\Delta_g\}$, substituting $\{\Delta_g\}$ into (1) and then cancelling $\cos \tau t$, we have the following amplitude:

$$X_g = \frac{10\Phi_R}{1 - \mu\tau^2\Phi_R} \tag{2}$$

Finally the displacement equation (1) becomes

$$\{\Delta_R\} = [\Phi_R]\left[\frac{10\Phi_R\mu\tau^2}{1 - \mu\tau^2\Phi_R} + 10\right] \cos \tau t \qquad \text{or} \qquad \{\Delta_R\} = [\Phi_R]R \cos \tau t \tag{3}$$

and if $\cos \tau t = 1$, R becomes maximum and $\{\Delta_R\} = \{X_R\}$. Thus $\{R_x\} = -[\Phi_{xx}^{-1}\Phi_{xR}]\{R\}$ and the internal moments are

$$\{\sigma\} = [q_R - q_x\Phi_{xx}^{-1}\Phi_{xR}]\{R\} \tag{4}$$

In (1) and (3) of Problem 5.12 $[q_R]$, $[q_x]$, and $[\Phi_{xR}]$ may be found. Performing (4) yields

$$\{\sigma\} = [34.65 \quad 31.99 \quad -31.99 \quad -31.99 \quad 31.99 \quad 34.65] \text{ ft-k} \tag{5}$$

5.14. Find the natural frequencies of the fixed-end beam shown in Fig. P-5.14*a* assuming three degrees of freedom for each mass. Let $l = 20$ ft, $I = 243$ in^4, $E = 30\,000$ ksi, $\mu = \gamma A/2$, $\rho = \mu l^2/48$, $\gamma = 490$ lb/ft^3, and $A = 36$ in^2.

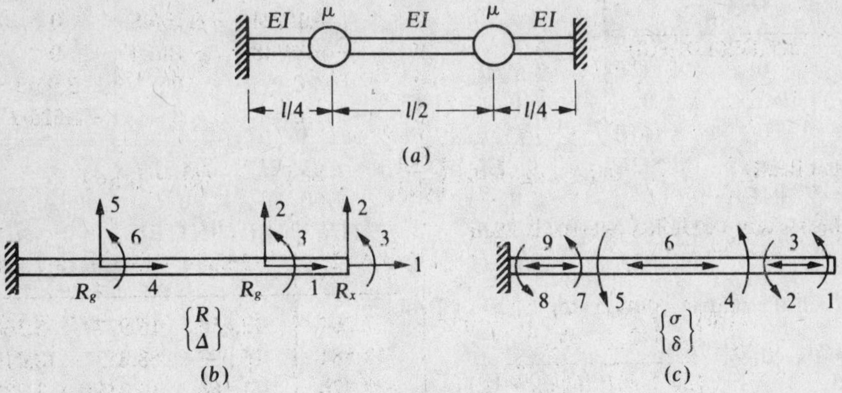

Fig. P-5.14

Let the indeterminate beam be modified to a stable determinate beam as shown in Fig. P-5.14*b*. There the external loads $\{R\}$ are composed of inertia forces $\{R_g\}$ and redundant forces $\{R_x\}$ in the following form:

$$\{R_g\} = \omega^2 \begin{bmatrix} \mu\Delta_{g1} \\ \mu\Delta_{g2} \\ \rho\Delta_{g3} \\ \mu\Delta_{g4} \\ \mu\Delta_{g5} \\ \rho\Delta_{g6} \end{bmatrix} = \omega^2 \text{ Diag. } [\mu]\{\Delta_g\} \qquad \{R_x\} = \begin{bmatrix} R_{x1} \\ R_{x2} \\ R_{x3} \end{bmatrix} \tag{1, 2}$$

The internal forces of moments and axial forces $\{\sigma\}$ are shown in Fig. P-5.14*c*. The relationship between $\{\sigma\}$ and $\{R\}$ yields

$$\underbrace{\begin{bmatrix}\sigma_1\\ \sigma_2\\ \sigma_3\\ \sigma_4\\ \sigma_5\\ \sigma_6\\ \sigma_7\\ \sigma_8\\ \sigma_9\end{bmatrix}}_{\{\sigma\}} = \underbrace{\left[\begin{array}{cccccc|ccc} 0 & 0 & 0 & 0 & 0 & 0 & 0 & 0 & 1\\ 0 & 0 & 0 & 0 & 0 & 0 & 0 & -l/4 & -1\\ 0 & 0 & 0 & 0 & 0 & 0 & 1 & 0 & 0\\ 0 & 0 & 1 & 0 & 0 & 0 & 0 & l/4 & 1\\ 0 & -l/2 & -1 & 0 & 0 & 0 & 0 & -3l/4 & -1\\ 1 & 0 & 0 & 0 & 0 & 0 & 1 & 0 & 0\\ 0 & l/2 & 1 & 0 & 0 & 1 & 0 & 3l/4 & 1\\ 0 & -3/4l & -1 & 0 & -l/4 & -1 & 0 & -l & -1\\ 1 & 0 & 0 & 1 & 0 & 0 & 1 & 0 & 0\end{array}\right]}_{[q_R \quad q_x]} \underbrace{\begin{bmatrix}R_{g1}\\ R_{g2}\\ R_{g3}\\ R_{g4}\\ R_{g5}\\ R_{g6}\\ \hline R_{x1}\\ R_{x2}\\ R_{x3}\end{bmatrix}}_{\begin{Bmatrix}R_g\\ R_x\end{Bmatrix}} \tag{3}$$

The unit stress flexibility chain is

$$\underbrace{\begin{bmatrix}\delta_1\\ \delta_2\\ \delta_3\\ \delta_4\\ \delta_5\\ \delta_6\\ \delta_7\\ \delta_8\\ \delta_9\end{bmatrix}}_{\{\delta\}} = \underbrace{\begin{bmatrix} l/12EI & -l/24EI & 0 & 0 & 0 & 0 & 0 & 0 & 0\\ -l/24EI & l/12EI & 0 & 0 & 0 & 0 & 0 & 0 & 0\\ 0 & 0 & l/4EA & 0 & 0 & 0 & 0 & 0 & 0\\ 0 & 0 & 0 & l/6EI & -l/12EI & 0 & 0 & 0 & 0\\ 0 & 0 & 0 & -l/12EI & l/6EI & 0 & 0 & 0 & 0\\ 0 & 0 & 0 & 0 & 0 & l/2EA & 0 & 0 & 0\\ 0 & 0 & 0 & 0 & 0 & 0 & l/12EI & -l/24EI & 0\\ 0 & 0 & 0 & 0 & 0 & 0 & -l/24EI & l/12EI & 0\\ 0 & 0 & 0 & 0 & 0 & 0 & 0 & 0 & l/4EA\end{bmatrix}}_{[f]} \underbrace{\begin{bmatrix}\sigma_1\\ \sigma_2\\ \sigma_3\\ \sigma_4\\ \sigma_5\\ \sigma_6\\ \sigma_7\\ \sigma_8\\ \sigma_9\end{bmatrix}}_{\{\sigma\}} \tag{4}$$

Manipulating $[q_R \mid q_x]^T[f][q_R \mid q_x]$ yields

$$[\Phi] = \left[\begin{array}{cccccc|ccc} l/4EA & & & & & & & & \\ 0 & 9l^3/64EI & & & \text{Sym.} & & & & \\ 0 & 9l^3/32EI & 3l/4EI & & & & & & \\ l/4EA & 0 & 0 & l/4EA & & & & & \\ 0 & l^3/48EI & l^2/32EI & 0 & l^3/192EI & & & & \\ 0 & 5l^2/32EI & l/4EI & 0 & l^2/32EI & l/4EI & & & \\ 3l/4EA & 0 & 0 & l/4EA & 0 & 0 & l/EA & & \\ 0 & 27l^3/128EI & 15l^2/32EI & 0 & 11l^3/384EI & 7l^2/32EI & 0 & l^3/3EI & \\ 0 & 9l^2/32EI & 3l/4EI & 0 & l^2 & l/4EI & 0 & l^2/2EI & l/EI \end{array}\right] \tag{5}$$

for which the matrix condensation [(*5*) of Problem 5.13] is

$$[\Phi_R] = \begin{bmatrix} -5l/16EA & & & & & \\ 0 & 9l^2/4096EI & & & \text{Sym.} & \\ 0 & 9l^2/1024EI & 21l/256EI & & & \\ l/16EA & 0 & 0 & 3l/16EA & & \\ 0 & 13l^3/12288EI & -7l^2/1024EI & 0 & 27l^3/12288EI & \\ 0 & 7l^2/1024EI & -11l/256EI & 0 & 9l^2/1024EI & 84l/1024EI \end{bmatrix} \tag{6}$$

Consequently, $\{\Delta_R\}$ may be expressed by using (*1*):

$$\{\Delta_R\} = [\Phi_R]\{R_g\} = \omega^2[\Phi_R]\,\text{Diag.}\,[\mu]\{X_g\} \tag{7}$$

Since $\{\Delta_R\} = \{X_g\}\cos\omega t$, performing (*7*) yields

$$\omega_{u1} = 2385.295 \text{ rad/s} \qquad \omega_{u2} = 3373.256 \text{ rad/s} \qquad \omega_{v1} = 226.188 \text{ rad/s} \qquad \omega_{v2} = 383.293 \text{ rad/s} \tag{8}$$

where, for simplicity, ω_{v1} and ω_{v2} are obtained neglecting the mass moment of inertia (i.e., $\rho = 0$).

The longitudinal frequencies (ω_u) and transverse frequencies (ω_v) are not coupled; the characteristics of the uncoupled eigenvalues of this system will be discussed in Problem 5.16

GENERAL STIFFNESS METHOD

5.15. Find the natural frequency of the structure given in Problem 5.12 in U.S. customary units.

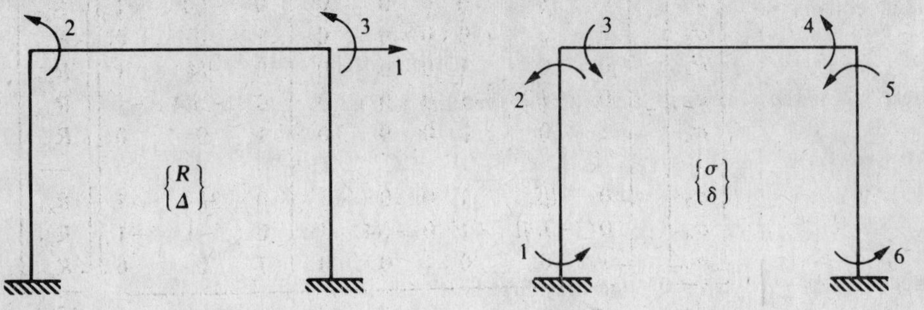

(a) (b)

Let the generalized coordinates $\{\Delta\}$ and generalized force $\{R\}$ be numbered as in Fig. P-5.15*a*. The internal moments $\{\sigma\}$ and their associated bending slopes $\{\delta\}$ are shown in Fig. P-5.15*b*. The equilibrium conditions between the generalized forces and the internal moments are

$$R_1 = \frac{M_1 + M_2}{h} + \frac{M_5 + M_6}{h} = -\mu_1 \ddot{\Delta}_1 = \mu\omega^2 X_1 \cos \omega t$$

$$R_2 = M_2 + M_3 = 0 \tag{1}$$

$$R_3 = M_4 + M_6 - 0$$

where $R_2 = R_3 = 0$ because no rotational inertial forces exist at the joints. Expressing (*1*) in matrix form,

$$\{R\} = [j]\{\sigma\} \tag{2}$$

where the transfer matrix $[j]$ (II, pp. 228–230) is

$$[j] = \begin{bmatrix} 1/10 & 1/10 & 0 & 0 & 1/10 & 1/10 \\ 0 & 1 & 1 & 0 & 0 & 0 \\ 0 & 0 & 0 & 1 & 1 & 0 \end{bmatrix} \tag{3}$$

Using Hooke's law we may write the moment-slope relationships of the system in the following form:

$$\{\sigma\} = [k]\{\delta\} \tag{4}$$

where $[k]$ is composed of stiffness coefficients as

$$[k] = EI \begin{bmatrix} 4/10 & 2/10 & 0 & 0 & 0 & 0 \\ 2/10 & 4/10 & 0 & 0 & 0 & 0 \\ 0 & 0 & 32/20 & 16/20 & 0 & 0 \\ 0 & 0 & 16/20 & 32/20 & 0 & 0 \\ 0 & 0 & 0 & 0 & 4/10 & 2/10 \\ 0 & 0 & 0 & 0 & 2/10 & 4/10 \end{bmatrix} \tag{5}$$

where $EI = 50\,625$ k-ft^2.

The compatibility relationships between the generalized displacements and the internal deformations are

$$\{\delta\} = [b]\{\Delta\}. \tag{6}$$

in which $[b]$ is kinematic matrix obtained by displacing the generalized displacements one at a time for corresponding internal deformations. For our problem, we have

$$[b] = \begin{bmatrix} 1/10 & 0 & 0 \\ 1/10 & 1 & 0 \\ 0 & 1 & 0 \\ 0 & 0 & 1 \\ 1/10 & 0 & 1 \\ 1/10 & 0 & 0 \end{bmatrix} \tag{7}$$

Comparing (3) with (7) yields

$$[j] = [b]^T \tag{8}$$

The above relationship is always true and may be proved by the virtual work method (II, pp. 228–230). Thus, combining (2), (4), (6), and (8) yields the relationship between $\{R\}$ and $\{\Delta\}$ and external displacements as

$$\{R\} = [j][k][j]^T\{\Delta\} = [\Omega]\{\Delta\} \tag{9}$$

where $[\Omega]$ is the stiffness matrix of the system (5-25). By (1), (9) becomes

$$\cos\omega t \left[\begin{array}{c} \mu_1\omega^2 X_1 \\ \hline 0 \\ 0 \end{array}\right] = \left[\begin{array}{c|c} \Omega_{11} & \Omega_{12} \\ \hline \Omega_{21} & \Omega_{22} \end{array}\right] \left[\begin{array}{c} X_1 \\ \hline X_2 \\ X_3 \end{array}\right] \cos\omega t \tag{10}$$

Since $\Omega_{21}\Delta_2 + \Omega_{22}\Delta_3 = 0$, we may first find

$$\begin{bmatrix} X_2 \\ X_3 \end{bmatrix} = -[\Omega_{22}^{-1}\Omega_{21}][X_1] \tag{11}$$

and then

$$\{\mu_1\omega^2 X_1\} = [\Omega_{11} - \Omega_{12}\Omega_{22}^{-1}\Omega_{21}]\{X_1\} = [\Omega_R]\{X_1\} \tag{12}$$

Numerically,

$$[\Omega] = \begin{bmatrix} 1215.0 & 3037.5 & 3037.5 \\ 3037.5 & 101250.0 & 40500.0 \\ 3037.5 & 40500.0 & 101250.0 \end{bmatrix} \quad \text{and} \quad [\Omega_{22}]^{-1} = 10^{-5}\begin{bmatrix} 1.17590 & -0.47034 \\ -0.47034 & 1.17590 \end{bmatrix} \tag{13, 14}$$

Substituting into (12), we have

$$\mu_1\omega^2 X_1 = 1085 X_1 \tag{15}$$

where $\mu_1 = 12.42\ \text{k-s}^2/\text{ft}$. Thus

$$\omega_1 = \sqrt{\frac{1085}{12.42}} = 9.346\ \text{rad/s} \tag{16}$$

5.16. Rework Problem 5.14 using the generalized coordinates and internal forces numbered in Figs. 5.16*a* and *b* respectively.

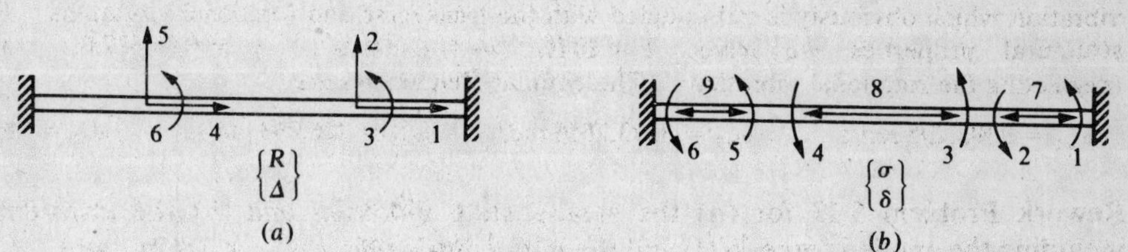

Fig. P-5.16

Employing the equilibrium conditions $\Sigma M = 0$, $\Sigma F_x = 0$, and $\Sigma F_y = 0$ at each nodal point, we have

$$\underbrace{\begin{bmatrix} R_1 \\ R_2 \\ R_3 \\ R_4 \\ R_5 \\ R_6 \end{bmatrix}}_{\{R\}} = \underbrace{\begin{bmatrix} 0 & 0 & -1 & 0 & 0 & 1 & 0 & 0 & 0 \\ 4/l & 4/l & 0 & -2/l & -2/l & 0 & 0 & 0 & 0 \\ 0 & 1 & 0 & 1 & 0 & 0 & 0 & 0 & 0 \\ 0 & 0 & 0 & 0 & 0 & -1 & 0 & 0 & 1 \\ 0 & 0 & 0 & 2/l & 2/l & 0 & -4/l & -4/l & 0 \\ 0 & 0 & 0 & 0 & 1 & 0 & 1 & 0 & 0 \end{bmatrix}}_{[j]} \underbrace{[\sigma_1\sigma_2\sigma_3\sigma_4\sigma_5\sigma_6\sigma_7\sigma_8\sigma_9]^T}_{\{\sigma\}} \tag{1}$$

The stiffness coefficients of the system are

$$\underbrace{\begin{bmatrix}\sigma_1\\ \sigma_2\\ \sigma_3\\ \sigma_4\\ \sigma_5\\ \sigma_6\\ \sigma_7\\ \sigma_8\\ \delta_9\end{bmatrix}}_{\{\sigma\}} = \frac{E}{l}\underbrace{\begin{bmatrix}16I & 8I & 0 & 0 & 0 & 0 & 0 & 0 & 0\\ & 16I & 0 & 0 & 0 & 0 & 0 & 0 & 0\\ & & 4A & 0 & 0 & 0 & 0 & 0 & 0\\ & & & 8I & 4I & 0 & 0 & 0 & 0\\ & & & & 8I & 0 & 0 & 0 & 0\\ & & & & & 2A & 0 & 0 & 0\\ & & & & & & 16I & 8I & 0\\ \text{Sym.} & & & & & & & 16I & 0\\ & & & & & & & & 4A\end{bmatrix}}_{[k]}\underbrace{\begin{bmatrix}\delta_1\\ \delta_2\\ \delta_3\\ \delta_4\\ \delta_5\\ \delta_6\\ \delta_7\\ \delta_8\\ \delta_9\end{bmatrix}}_{\{\delta\}} \tag{2}$$

Using (*1*) and (*2*) we have

$$[\Omega] = [j][k][j]^T = \frac{EI}{l^2}\begin{bmatrix}6l/\gamma^2 & 0 & 0 & -2l/\gamma^2 & 0 & 0\\ & 864/l & 72 & 0 & -96/l & -24\\ & & 24l & 0 & 24 & 4l\\ & & & 6l/\gamma^2 & 0 & 0\\ \text{Sym.} & & & & 864/l & -72\\ & & & & & 24l\end{bmatrix} \tag{3}$$

where $\gamma^2 = I/A$. The inertia forces are

$$\{R\} = \omega^2[\mu]\{X\} \tag{4}$$

in which the diagonal mass matrix $[\mu]$ includes μ and ρ. Since $\{R\} = [\Omega]\{X\} = \omega^2[\mu]\{X\}$,

$$([\Omega] - \omega^2[\mu])\{X\} = 0 \tag{5}$$

The determinant of the coefficients in (*5*) becomes

$$\left[\left(\frac{6l}{\gamma^2} - A\right)^2 - \frac{4l^2}{\gamma^4}\right]\left(151\,077\,888.0 - \frac{24\,772\,608.0\,B}{l} + 71\,424.0\,AB - 663\,552.0\,Al - \frac{755\,712\,B^2}{l^2}\right.$$
$$\left. + 560A^2l^2 - 48A^2Bl + A^2B^2\right) = 0 \tag{6}$$

where $A = \mu\omega^2 l^2/EI$, corresponding to linear (longitudinal and transverse) motion, and $B = \rho\omega^2 l^2/EI$, corresponding to rotational motion. The terms in brackets are associated with longitudinal vibration which obviously is not coupled with the transverse and rotational vibrations. For the given structural properties we have $l = 20$ ft, $A = 0.0003\,\omega^2$, $\gamma^2 = 0.04\,687$ ft^2, and $B = 0$ (neglecting the rotational vibration). The circular frequencies are

$$\omega_{u1} = 2385.295 \text{ rad/s} \qquad \omega_{u2} = 3373.2556 \text{ rad/s} \qquad \omega_{v1} = 226.063 \text{ rad/s} \qquad \omega_{v2} = 383.118 \text{ rad/s}$$

5.17. Rework Problem 5.13 for (*a*) the steady state vibration and (*b*) the transient vibration, including the applied force $W(t)$ and the initial conditions $\Delta_{0(1)} = 1/2$ in and $\dot{\Delta}_{0(1)} = 4$ in/s. The diagrams of generalized coordinates and internal forces are shown in Fig. P-5.15.

(*a*) Using (*5-24*) we write

$$[\mu]\{\ddot{\Delta}\} + [\Omega]\{\Delta\} = \{W(t)\} \tag{1}$$

where $[\Omega]$ is the same as (*10*) or (*13*) of Problem 5.15. Since rotational forces at nodal points 2 and 3 are zero, (*1*) becomes

$$\mu_1\ddot{\Delta}_1 + \Omega_R\Delta_1 = W_1(t) \tag{2}$$

where Ω_R is identical to (*12*) of Problem 5.15. After substitution of $\Delta_1 = X_1 \cos \tau t$ and $W_1(t) = 10 \cos \tau t$, (*2*) gives the amplitude

$$X_1 = \frac{10}{\Omega_R - \mu_1\tau^2} \tag{3}$$

From (2),

$$\Delta_1 = \frac{1}{\Omega_R}\left(10 + \frac{10\mu_1\tau^2}{\Omega_R - \mu_1\tau^2}\right)\cos \tau t \tag{4}$$

Knowing Δ_1, we may calculate Δ_2 and Δ_3 according to (11) of Problem 5.15. Finally, the internal forces may be obtained as

$$\{\sigma\} = [k][j]^T\{\Delta\} \tag{5}$$

Using the numerical values given in Problem 5.15 and $\cos \Omega t = 1$, we have

$$\{\Delta\} = [0.012\,284 \quad -0.000\,263 \quad -0.000\,263]^T \text{ ft} \tag{6}$$

and

$$\{\sigma\} = [34.65 \quad 31.995 \quad -31.995 \quad -31.995 \quad 31.995 \quad 34.65]^T \text{ ft-k} \tag{7}$$

(b) For the analysis with consideration of applied force and initial conditions, the motion equation is

$$\Delta_1 = \Delta_{0(1)}\cos \omega t + \frac{\dot{\Delta}_{0(1)}}{\omega}\sin \omega t + \frac{10(\omega \sin \tau t - \tau \sin \omega t)}{\mu\omega(\omega^2 - \tau^2)} \tag{8}$$

At $t = 1/4$ s,

$$\Delta_1 = 0.0262 \text{ ft} \qquad \Delta_2 = -000\,561 \text{ ft} \qquad \Delta_3 = -0.000\,561 \text{ ft}$$

Consequently the internal moments are

$$\{\sigma\} = [73.90 \quad 68.22 \quad -68.22 \quad -68.22 \quad 68.22 \quad 73.90]^T \text{ ft-k}$$

5.18. The frame shown in Fig. P-5.18a is subject to $W_1(t) = 10(1 - 2t)$ k and $W_2(t) = 20(1 - 2t)$ k. Let $c_1 = c_2 = 0.2\,c_{cr}$, $EI = 15\,000$ k-ft^2, $\mu_1 = \mu_2 = 2$ k-s^2/ft lumped on girders. Find the displacements at $t = 1/4$ s.

The differential equations of motion are

$$\begin{bmatrix} \mu_1 & 0 \\ 0 & \mu_2 \end{bmatrix}\begin{bmatrix} \ddot{\Delta}_1 \\ \ddot{\Delta}_2 \end{bmatrix} + \begin{bmatrix} c_1 & 0 \\ 0 & c_2 \end{bmatrix}\begin{bmatrix} \dot{\Delta}_1 \\ \dot{\Delta}_2 \end{bmatrix} + [\Omega_R]\begin{bmatrix} \Delta_1 \\ \Delta_2 \end{bmatrix} = \begin{bmatrix} W_1(t) \\ W_2(t) \end{bmatrix}$$

Using the numbering systems shown in Fig. P-5.18b and c, formulate

$$[j] = \begin{bmatrix} 1 & 0 & 1/20 & 1/20 & 0 & 0 & 1/20 & 1/20 & 0 & 0 & 0 & 0 \\ 1/20 & 1/20 & -1/20 & -1/20 & 0 & 0 & -1/20 & -1/20 & 1/20 & 1/20 & 0 & 0 \\ 0 & 1 & 1 & 0 & 0 & 0 & 0 & 0 & 0 & 0 & 1 & 0 \\ 0 & 0 & 0 & 1 & 1 & 0 & 0 & 0 & 0 & 0 & 0 & 0 \\ 0 & 0 & 0 & 0 & 0 & 1 & 1 & 0 & 0 & 0 & 0 & 0 \\ 0 & 0 & 0 & 0 & 0 & 0 & 0 & 1 & 1 & 0 & 0 & 1 \end{bmatrix} \tag{1}$$

The stiffness coefficients of a typical member i are

$$[k_i] = \frac{EI_i}{L_i}\begin{bmatrix} 4 & 2 \\ 2 & 4 \end{bmatrix} \tag{2}$$

Consequently,

$$[j][k][j]^T = \frac{EI}{10}\left[\begin{array}{cc|cccc} 12/400 & -12/400 & 3/20 & 3/20 & 3/20 & 3/20 \\ & 24/400 & 0 & -3/20 & -3/20 & 0 \\ \hline & & 16 & 1 & 0 & 6 \\ \text{Sym.} & & & 14 & 6 & 0 \\ & & & & 14 & 1 \\ & & & & & 16 \end{array}\right] = \left[\begin{array}{c|c} \Omega_{11} & \Omega_{12} \\ \hline \Omega_{21} & \Omega_{22} \end{array}\right] \tag{3}$$

from which

$$[\Omega_{22}]^{-1} = \frac{10}{34681EI}\begin{bmatrix} 2546. & -259. & 180. & -966. \\ & 3064. & -1326. & 180. \\ \text{Sym.} & & 3064. & -259. \\ & & & 2546. \end{bmatrix} \tag{4}$$

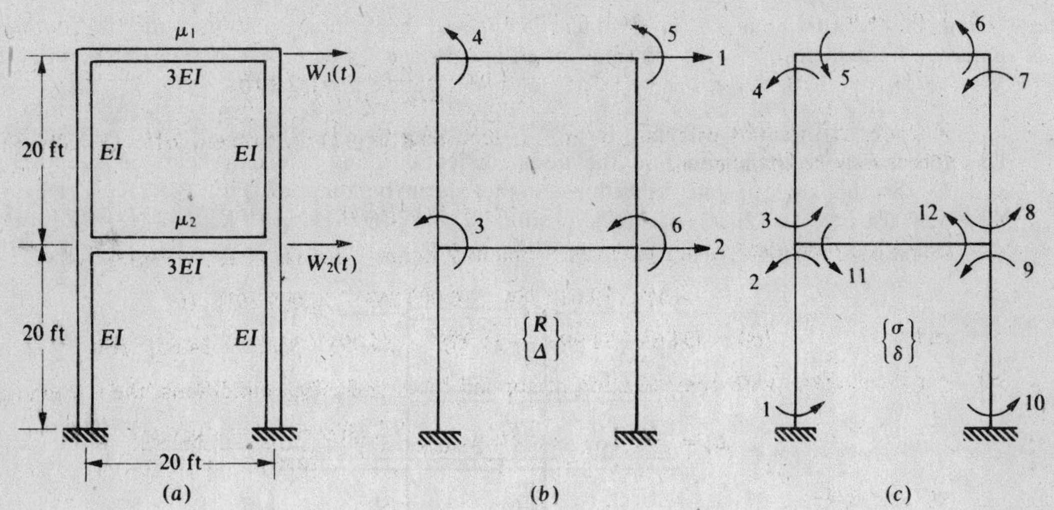

Fig. P-5.18

Thus the condensed stiffness matrix is

$$[\Omega_R] = \begin{bmatrix} 38.85 & -41.70 \\ -41.70 & 86.61 \end{bmatrix} \tag{5}$$

Now we solve (1). Using modal matrix formulation (3-40), the motion equations are

$$\{\Delta\} = [h]\{\{h_{1k}\Phi_{1k}\} + \{h_{2k}\Phi_{2k}\}\} \tag{6}$$

and the particular functions given below are obtained from Problem 2.21c.

$$\Phi_{jk} = \frac{P_j}{\lambda_k(\gamma_k^2 + \lambda_k^2)}\left[\lambda_k - e^{-\gamma_k t}(\gamma_k \sin \lambda_k t + \lambda_k \cos \lambda_k t) + \frac{P_j}{a\lambda_k(\gamma_k^2 + \lambda_k^2)}\right]$$
$$\times \left\{\frac{2\lambda_k\gamma_k}{\gamma_k^2 + \lambda_k^2} - \lambda_k t - \frac{e^{-\lambda_k t}}{\gamma_k^2 + \lambda_k^2}\left[(\gamma_k^2 - \lambda_k^2)\sin \lambda_k t + 2\gamma_k\lambda_k \cos \lambda_k t\right]\right\} \tag{7}$$

Employing modal matrix analysis yields

$$\omega_1 = 2.709 \text{ rad/s} \qquad \omega_2 = 7.338 \text{ rad/s} \qquad \{X_1\} = \begin{bmatrix} X_{11} \\ 0.5797\,X_{11} \end{bmatrix} \qquad \{X_2\} = \begin{bmatrix} X_{12} \\ -1.725\,X_{12} \end{bmatrix} \tag{8}$$

Using the above eigenvalues and eigenvectors, we obtain

$$\left.\begin{aligned} \gamma_1 &= c_1/2\mu_1 = 0.2\,\omega_1 = 0.5418 \\ \gamma_2 &= c_2/2\mu_2 = 0.2\,\omega_2 = 1.4885 \\ \lambda_1 &= \sqrt{\omega_1^2 - \gamma_1^2} = 2.6542 \\ \lambda_2 &= \sqrt{\omega_2^2 - \gamma_2^2} = 7.292 \end{aligned}\right\} \qquad [h] = \begin{bmatrix} 0.612 & 0.3546 \\ 0.3546 & -0.6117 \end{bmatrix} \tag{9, 10}$$

Substituting (9) into (7) at $t = 1/4$ s yields $\Delta_1 = 0.0692$ ft and $\Delta_2 = 0.1282$ ft. Consequently other displacements and internal moments may be calculated.

Supplementary Problems

TRANSPORT MATRICES AND TRANSPORT CHAIN

5.19. Use the dimensionless transport matrix $\hat{L}_{RL}^{0}$ and find the natural frequencies and mode shapes of the beams shown in Figs. P-5.5c and d.

5.20. Using the results obtained in Problem 5.6*b* and neglecting rotational inertia forces, find the natural frequencies of the frame of Fig. P-5.6*b* where $l = 20$ ft, $I = 243$ in^4, $A = 36$ in^2, $\gamma = 490$ lb/ft^3, $\mu = lA\gamma/g$, $\mu_1 = 1.5\mu$, $\phi = 45°$, and $g = 32.2$ ft/s^2.

5.21. Use the transport chain and find the nodal deflections and internal shears at supports *A*, *B*, *C*, and *D* for the steady state vibration of the shear building of Fig. P-5.21. Let $\mu = 2$ k-s^2/ft, $EI = 15\,000$ k-ft^2, $l = 20$ ft, $W_1(t) = 10 \sin \Omega t$ k, $W_2(t) = 15 \sin \Omega t$ k, $W_3(t) = 15 \sin \Omega t$ k, and $\Omega = 10$ rad/s.

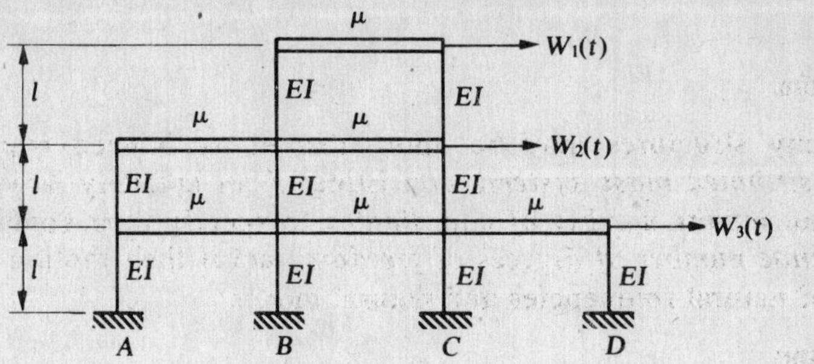

Fig. P-5.21

GENERAL FLEXIBILITY MATRIX

5.22. Find the natural frequencies of the beams shown in Figs. P-5.5*a* and *d* by using the flexibility matrix with consideration of longitudinal, transverse, and rotational motions.

5.23. Find the natural frequencies of the frame shown in Fig. P-5.18 by using the flexibility matrix.

5.24. Consider Problem 5.13 to include the applied force $W(t)$ and the initial conditions of the girder as $\Delta_0 = 1/2$ in and $\dot{\Delta}_0 = 4$ in/s. Find the girder displacement and internal moments at $t = 1/4$ s.

GENERAL STIFFNESS MATRIX

5.25. Using the stiffness method find the natural frequencies and mode shapes of the shear building of Fig. P-5.21. Note that the degrees of freedom of the frame are 3 and the independent internal forces are shears.

5.26. Use the stiffness method and rework Problem 5.21.

5.27. Use the stiffness method and rework Problem 5.18 neglecting damping.

Chapter 6

Distributed Mass Systems: Basic Cases

6.1. SYSTEM

Degrees of Freedom

Although many structures may be approximated by lumped mass systems, in reality all structures are *distributed mass systems* consisting of an infinitely large number of particles and hence requiring an *infinite number of coordinates* to specify their configurations. Consequently, they have an *infinite number of degrees of freedom*, and if their motion is repetitive they have an infinite number of natural frequencies and natural modes.

Definition of Motion

In cases of *small oscillations* about the equilibrium position, the motion of the distributed mass system can be considered as a sum of an infinite number of periodic motions, each of which is simple harmonic and common to all particles of the system, moving in the same frequency and phase. Furthermore, the displacement of any particle is assumed to be in a definite proportion to the displacement of any other particle chosen as a reference.

Equation of Motion

When the system is oscillating in one of these ways, it is said to be *vibrating in normal (principal) modes* and the equation of motion is given in a general form as

$$\Delta(x,t) = \sum_{m=1}^{\infty} X_{\Delta,m}(x) Y_{\Delta,m}(t) \tag{6-1}$$

where $\Delta(x, t)$ is the total displacement at position x and time t, and $X_{\Delta,m}(x)$ and $Y_{\Delta,m}(t)$ are the eigenfunctions and the time functions respectively.

This equation must satisfy both the differential equation of motion and the boundary conditions associated with the problem.

Elastic Bar

The study of *free and forced vibrations* presented in this chapter is confined to the repetitive motion of a finite straight prismatic bar of constant section which is symmetrical with respect to the XY plane, whose X axis coincides with the centroidal axis of the bar. Only *longitudinal* and *transverse motions* are considered.

6.2. FREE LONGITUDINAL VIBRATIONS

Equation of Motion

The *differential equation* of free longitudinal vibrations in the finite, straight bar of Fig. 6-1 is

$$EA\frac{\partial^2 u}{\partial x^2} - \mu\frac{\partial^2 u}{\partial t^2} = 0 \tag{6-2}$$

where the horizontal displacement u is a function of time t and of the horizontal coordinate x measured from L, μ is the mass per unit length, E is the modulus of elasticity, and A is the area of normal section (Problem 6.1).

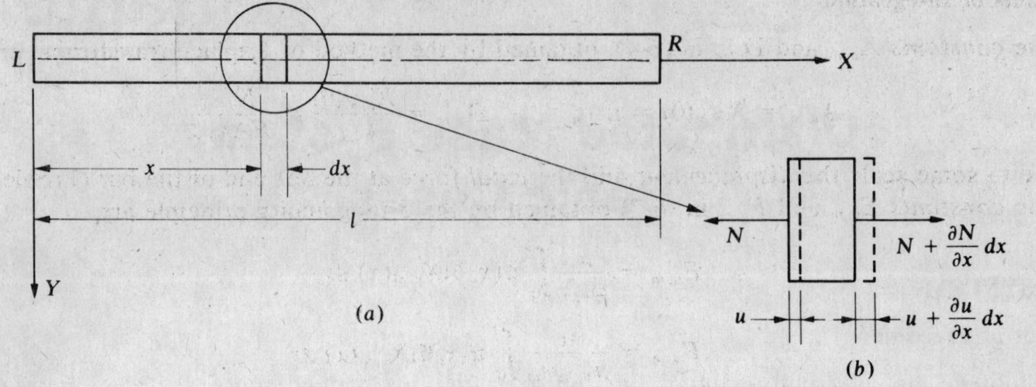

Fig. 6-1. Longitudinal Vibrations of a Straight Bar

The *general solution* of (6-2) is

$$u(x, t) = \sum_{m=1}^{\infty} X_{u,m}(x) Y_{u,m}(t) \tag{6-3}$$

where each member of this series is a product of the *eigenfunction* $X_{u,m}(x)$ and the *time function* $Y_{u,m}(t)$, both corresponding to a given *mode m* and an *angular frequency* ω_m, which is again an *eigenvalue* of the solution.

The *eigenfunction* of the mth mode is

$$X_{u,m}(x) = A_{u,m} a_{u,m}(x) + D_{u,m} d_{u,m}(x) \tag{6-4}$$

where $A_{u,m}$ and $D_{u,m}$ are the integration constants (*6-10*), and

$$a_{u,m}(x) = \cos \lambda_{u,m} x \qquad d_{u,m}(x) = \sin \lambda_{u,m} x \tag{6-5}$$

are the *dynamic parameters* of this motion; they are functions of the *position coordinate* x and the *shape parameter*

$$\lambda_{u,m} = \sqrt{\frac{\mu \omega_m^2}{EA}} \tag{6-6}$$

where ω_m is the *angular frequency*.

The *time function* of the mth mode is

$$Y_{u,m}(t) = E_{u,m} \cos \omega_m t + F_{u,m} \sin \omega_m t \tag{6-7}$$

where $E_{u,m}$ and $F_{u,m}$ are the integration constants (*6-11*) and ω_m is the same angular frequency as in (6-6).

Orthogonality of Eigenfunctions

The *eigenfunctions* $X_{u,m}(x)$ are *orthogonal*, i.e.,

$$\int_0^l X_{u,j}(x) \mu X_{u,k}(x)\, dx = \begin{cases} 0, & j \neq k \\ \mu_{u,m}, & j = k \end{cases} \tag{6-8}$$

$$\int_0^l X'_{u,j}(x) EA X'_{u,k}(x)\, dx = \begin{cases} 0, & j \neq k \\ \omega_m^2 \mu_{u,m}, & j = k \end{cases} \tag{6-9}$$

where

$$\mu_{u,m} = \int_0^l \mu X_{u,m}^2(x)\, dx \tag{6-9a}$$

is called the *generalized mass* in the mth mode (Problem 6.2).

Constants of Integration

The *constants* $A_{u,m}$ and $D_{u,m}$ in (6-4), obtained by the method of Laplace transforms, are

$$A_{u,m} = X_{u,m}(0) = u_{L,m} \qquad D_{u,m} = \frac{X'_{u,m}(0)}{\lambda_{u,m}} = \frac{N_{L,m}}{EA\lambda_{u,m}} \tag{6-10}$$

and are to some scale the *displacement* and the *axial force* at the left end of the bar (Problem 6.3).

The *constants* $E_{u,m}$ and $F_{u,m}$ in (6-7) obtained by the orthogonality principle are

$$E_{u,m} = \frac{\mu}{\mu_{u,m}} \int_0^l u(x,0)X_{u,m}(x)\,dx$$
$$F_{u,m} = \frac{\mu}{\mu_{u,m}\omega_m} \int_0^l \dot{u}(x,0)X_{u,m}(x)\,dx \tag{6-11}$$

where $\mu_{u,m}$ is given by (6-9), and $u(x,0)$ and $\dot{u}(x,0)$ are the *displacement* and *velocity functions* at $t = 0$.

Transport Matrices

The *relationships between the state vectors* of the left and right end and vice versa are given by the following transport equations.

$$\underbrace{\begin{bmatrix} \bar{u}_{R,m} \\ \bar{N}_{R,m} \end{bmatrix}}_{\bar{H}_{Ru,m}} = \underbrace{\begin{bmatrix} a_{u,m} & d_{u,m} \\ -d_{u,m} & a_{u,m} \end{bmatrix}}_{\bar{T}_{RLu,m}} \underbrace{\begin{bmatrix} \bar{u}_{L,m} \\ \bar{N}_{L,m} \end{bmatrix}}_{\bar{H}_{Lu,m}} \qquad \underbrace{\begin{bmatrix} \bar{u}_{L,m} \\ \bar{N}_{L,m} \end{bmatrix}}_{\bar{H}_{Lu,m}} = \underbrace{\begin{bmatrix} a_{u,m} & -d_{u,m} \\ d_{u,m} & a_{u,m} \end{bmatrix}}_{\bar{T}_{LRu,m}} \underbrace{\begin{bmatrix} \bar{u}_{R,m} \\ \bar{N}_{R,m} \end{bmatrix}}_{\bar{H}_{Ru,m}} \tag{6-12, 13}$$

where $\bar{H}_{Lu,m}$, $\bar{H}_{Ru,m}$ are the *dimensionless state vectors* at L and R respectively and $\bar{T}_{RLu,m}$ and $\bar{T}_{LRu,m}$ are the *dimensionless transport matrices* (Problem 6.4).

The *equivalents* in (6-12) and (6-13) are

$$\bar{u}_{R,m} = u_{R,m}/1, \quad \bar{u}_{L,m} = u_{L,m}/1, \quad a_{u,m} = \cos\bar{\lambda}_{u,m},$$
$$\bar{N}_{R,m} = N_{R,m}/EA\lambda_{u,m}, \quad \bar{N}_{L,m} = N_{L,m}/EA\lambda_{u,m}, \quad d_{u,m} = \sin\bar{\lambda}_{u,m} \tag{6-14}$$

and $\bar{\lambda}_{u,m} = l\lambda_{u,m}$ where l = length of the bar.

Frequency Equation

The *angular frequencies* $\omega_1, \omega_2, \ldots$ of this motion are the eigenvalues of the frequency determinant equation constructed from (6-12) or (6-13) by selecting the part of the transport matrix corresponding to the *existing eigenvector* (Problems 6.5 and 6.6).

6.3. FORCED LONGITUDINAL VIBRATIONS

Equation of Motion

The *differential equation* of forced longitudinal vibrations induced in the bar of Fig. 6-1 by a *distributed dynamic load of intensity* $p_u(x,t)$ acting along its centroidal axis is

$$EA\frac{\partial^2 u}{\partial x^2} - \mu\frac{\partial^2 u}{\partial t^2} = -p_u(x,t) \tag{6-15}$$

where the symbols are those of (6-2) (Problem 6.1).

The *general solution* of (6-15), with eigenvalues and eigenfunctions of the free longitudinal vibrations of the bar known, is

$$u(x,t) = \sum_{m=1}^{\infty} X_{u,m}(x)q_{u,m}(t) \tag{6-16}$$

where $q_{u,m}(t)$ is the respective *principal coordinate function* of the mth mode (6-20).

Lagrange's Equation

The *energy functions* in terms of (*6-16*) reduce to

$$T = \frac{1}{2}\int_0^l \mu \left[\frac{\partial u}{\partial t}\right]^2 dx = \frac{1}{2}\sum_{m=1}^{\infty} \mu_{u,m}\left[\frac{dq_{u,m}}{dt}\right]^2$$

$$V = \frac{1}{2}\int_0^l EA\left[\frac{\partial u}{\partial x}\right]^2 dx = \frac{1}{2}\sum_{m=1}^{\infty} \mu_{u,m}\left[\omega_m q_{u,m}\right]^2 \tag{6-17}$$

where T is the *kinetic energy*, V is the *potential energy*, $\mu_{u,m}$ is given by (*6-9a*), and $q_{u,m} = q_{u,m}(t)$.

The *Lagrange's equation of the second kind* (*4-23b*) in terms of (*6-17*) yields

$$\mu_{u,m}\frac{d^2 q_{u,m}}{dt^2} + \mu_{u,m}\omega_m^2 q_{u,m} = Q_{u,m}(t) \tag{6-18}$$

where

$$Q_{u,m}(t) = \int_0^l p_u(x,t)X_{u,m}(x)\,dx \tag{6-19}$$

is the *generalized force* defined by (*4-16*) (Problems 6.7 and 6.8).

The *general solution* of (*6-18*) is then

$$q_{u,m}(t) = E_{u,m}\cos\omega_m t + F_{u,m}\sin\omega_m t + \frac{1}{\mu_{u,m}\omega_m}\int_0^t Q_{u,m}(\tau)\sin\omega_m(t-\tau)\,d\tau \tag{6-20}$$

where $E_{u,m}$ and $F_{u,m}$ are given by (*6-11*).

The *concentrated loads* acting longitudinally on the bar axis can be included in (*6-19*) by simply taking their magnitudes multiplied by the eigenfunction (Problem 6.9). However, the transport matrix approach is more convenient (Problem 6.10).

6.4. FREE TRANSVERSE VIBRATIONS

The *differential equation* of free transverse vibrations in the finite, straight bar of Fig. 6-2 is

$$EI\frac{\partial^4 v}{\partial x^4} + \mu\frac{\partial^2 v}{\partial t^2} = 0 \tag{6-21}$$

where the vertical displacement v is a function of time t and of the position coordinate x measured from L, μ is the mass per unit length, E is the modulus of elasticity, and I is the moment of inertia of the normal section about the Z axis (Problem 6.12).

The *general solution* of (*6-21*) is

$$v(x,t) = \sum_{m=1}^{\infty} X_{v,m}(x)Y_{v,m}(t) \tag{6-22}$$

where each member of this series is again a product of the eigenfunction $X_{v,m}(x)$ and the time function $Y_{v,m}(t)$. Both correspond to a given mode m and to an angular frequency ω_m, which is again an eigenvalue of the solution.

The *eigenfunction* of the mth mode is

$$X_{v,m}(x) = A_{v,m}a_{v,m}(x) + B_{v,m}b_{v,m}(x) + C_{v,m}c_{v,m}(x) + D_{v,m}d_{v,m}(x) \tag{6-23}$$

where $A_{v,m}$, $B_{v,m}$, $C_{v,m}$, and $D_{v,m}$ are the integration constants (*6-29*) and

$$a_{v,m}(x) = \tfrac{1}{2}(\cosh\lambda_{v,m}x + \cos\lambda_{v,m}x) \qquad b_{v,m}(x) = \tfrac{1}{2}(\sinh\lambda_{v,m}x + \sin\lambda_{v,m}x)$$

$$c_{v,m}(x) = \tfrac{1}{2}(\cosh\lambda_{v,m}x - \cos\lambda_{v,m}x) \qquad d_{v,m}(x) = \tfrac{1}{2}(\sinh\lambda_{v,m}x - \sin\lambda_{v,m}x) \tag{6-24}$$

are the *dynamic parameters* of this motion.

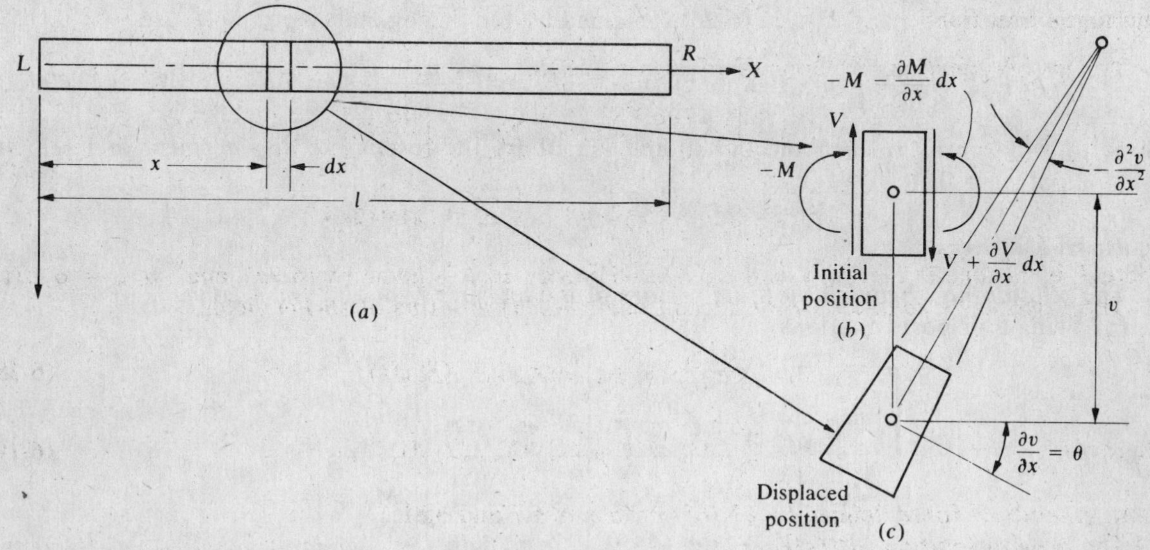

Fig. 6-2. Transverse Vibrations of a Straight Bar

These parameters are functions of the *position coordinate* x and the *shape parameter,*

$$\lambda_{v,m} = \sqrt[4]{\frac{\mu\omega_m^2}{EI}} \tag{6-25}$$

where ω_m is the *angular frequency*.

The *time function* of the mth mode is

$$Y_{v,m}(t) = E_{v,m} \cos \omega_m t + F_{v,m} \sin \omega_m t \tag{6-26}$$

where $E_{v,m}$ and $F_{v,m}$ are the integration constants (*6-30*) and ω_m is the same angular frequency as in (*6-25*).

Orthogonality of Eigenfunctions

The *eigenfunctions are orthogonal*, i.e.,

$$\int_0^l X_{v,j}(x)\mu X_{v,k}(x)\,dx = \begin{cases} 0, & j \neq k \\ \mu_{v,m}, & j = k \end{cases} \tag{6-27}$$

$$\int_0^l X''_{v,j}(x)EIX''_{v,k}(x)\,dx = \begin{cases} 0, & j \neq k \\ \omega_m^2\mu_{v,m}, & j = k \end{cases} \tag{6-28}$$

where

$$\mu_{v,m} = \int_0^l \mu X_{v,m}^2(x)\,dx \tag{6-28a}$$

is again called the *generalized mass* in the mth mode (Problem 6.13).

Constants of Integration

The *constants* $A_{v,m}$, $B_{v,m}$, $C_{v,m}$, and $D_{v,m}$ in (*6-23*), obtained by the method of Laplace transforms, are

$$\begin{aligned} A_{v,m} &= X_{v,m}(0) = v_{L,m} & B_{v,m} &= \frac{X'_{v,m}(0)}{\lambda_{v,m}} = \frac{\theta_{L,m}}{\lambda_{v,m}} \\ C_{v,m} &= \frac{X''_{v,m}(0)}{\lambda_{v,m}^2} = \frac{M_{L,m}}{EI\lambda_{v,m}^2} & D_{v,m} &= \frac{X'''_{v,m}(0)}{\lambda_{v,m}^3} = \frac{-V_{L,m}}{EI\lambda_{v,m}^3} \end{aligned} \tag{6-29}$$

and they are respectively to some scale the *displacement, slope, bending moment,* and *shear* at the left end of the bar.

The *constants* $E_{v,m}$ and $F_{v,m}$ in (6-26), obtained by the orthogonality principle, are

$$E_{v,m} = \frac{\mu}{\mu_{v,m}} \int_0^l v(x,0) X_{v,m}(x)\,dx \qquad F_{v,m} = \frac{\mu}{\mu_{v,m}\omega_m} \int_0^l \dot{v}(x,0) X_{v,m}(x)\,dx \tag{6-30}$$

where μ_m is given by (6-28a) and $v(x,0)$ and $\dot{v}(x,0)$ are the respective *displacement* and *velocity functions* at $t = 0$.

Transport Matrices

The *relationships between the state vectors* of the left and right end and vice versa are given by the following transport equations.

$$\underbrace{\begin{bmatrix} \bar{v}_{R,m} \\ \bar{\theta}_{R,m} \\ \hline \bar{M}_{R,m} \\ \bar{V}_{R,m} \end{bmatrix}}_{\bar{H}_{Rv,m}} = \underbrace{\left[\begin{array}{cc|cc} a_{v,m} & b_{v,m} & c_{v,m} & d_{v,m} \\ d_{v,m} & a_{v,m} & b_{v,m} & c_{v,m} \\ \hline c_{v,m} & d_{v,m} & a_{v,m} & b_{v,m} \\ b_{v,m} & c_{v,m} & d_{v,m} & a_{v,m} \end{array}\right]}_{\bar{T}_{RLv,m}} \underbrace{\begin{bmatrix} \bar{v}_{L,m} \\ \bar{\theta}_{L,m} \\ \hline \bar{M}_{L,m} \\ \bar{V}_{L,m} \end{bmatrix}}_{\bar{H}_{Lv,m}} \tag{6-31}$$

$$\underbrace{\begin{bmatrix} \bar{v}_{L,m} \\ \bar{\theta}_{L,m} \\ \hline \bar{M}_{L,m} \\ \bar{V}_{L,m} \end{bmatrix}}_{\bar{H}_{Lv,m}} = \underbrace{\left[\begin{array}{cc|cc} a_{v,m} & -b_{v,m} & c_{v,m} & -d_{v,m} \\ -d_{v,m} & a_{v,m} & -b_{v,m} & c_{v,m} \\ \hline c_{v,m} & -d_{v,m} & a_{v,m} & -b_{v,m} \\ -b_{v,m} & c_{v,m} & -d_{v,m} & a_{v,m} \end{array}\right]}_{\bar{T}_{LRv,m}} \underbrace{\begin{bmatrix} \bar{v}_{R,m} \\ \bar{\theta}_{R,m} \\ \hline \bar{M}_{R,m} \\ \bar{V}_{R,m} \end{bmatrix}}_{\bar{H}_{Rv,m}} \tag{6-32}$$

where $\bar{H}_{Lv,m}$, $\bar{H}_{Rv,m}$ are the *dimensionless state vectors* at L and R respectively, and $\bar{T}_{LRv,m}$ and $\bar{T}_{RLv,m}$ are the *dimensionless transport matrices* (Problem 6.15).

The *equivalents* in (6-31) and (6-32) are

$$\left.\begin{array}{lll} \bar{v}_{L,m} = v_{L,m}/1 & \bar{v}_{R,m} = v_{R,m}/1 & a_{v,m} = \tfrac{1}{2}(\cosh \bar{\lambda}_{v,m} + \cos \bar{\lambda}_{v,m}) \\ \bar{\theta}_{L,m} = \theta_{L,m}/\lambda_{v,m} & \bar{\theta}_{R,m} = \theta_{R,m}/\lambda_{v,m} & b_{v,m} = \tfrac{1}{2}(\sinh \bar{\lambda}_{v,m} + \sin \bar{\lambda}_{v,m}) \\ \bar{M}_{L,m} = M_{L,m}/EI\lambda_{v,m}^2 & \bar{M}_{R,m} = M_{R,m}/EI\lambda_{v,m}^2 & c_{v,m} = \tfrac{1}{2}(\cosh \bar{\lambda}_{v,m} - \cos \bar{\lambda}_{v,m}) \\ \bar{V}_{L,m} = -V_{L,m}/EI\lambda_{v,m}^3 & \bar{V}_{R,m} = -V_{R,m}/EI\lambda_{v,m}^3 & d_{v,m} = \tfrac{1}{2}(\sinh \bar{\lambda}_{v,m} - \sin \bar{\lambda}_{v,m}) \end{array}\right\} \tag{6-33}$$

where $\bar{\lambda}_{v,m} = l\lambda_{v,m}$ and l = length of the bar.

Frequency Equation

The *angular frequencies* $\omega_1, \omega_2, \ldots,$ of this motion are the eigenvalues of the frequency determinant equation constructed from (6-31) or (6-32) by selecting the part of the transport matrix corresponding to the *existing eigenvector* (Problem 6.16).

6.5. FORCED TRANSVERSE VIBRATION

Equation of Motion

The *differential equation* of forced transverse vibration induced in the bar of Fig. 6-2 by a distributed dynamic load of intensity $p_v(x,t)$, acting normal to the centroidal axis of the bar in its plane of symmetry, is

$$EI\frac{\partial^4 v}{\partial x^4} + \mu\frac{\partial^2 v}{\partial t^2} = p_v(x,t) \tag{6-34}$$

where the symbols are those of (6-21).

The *general solution* of (6-34), with eigenvalues and eigenfunctions of the free transverse vibration of the bar known, is

$$v(x,t) = \sum_{m=1}^{\infty} X_{v,m}(x)q_{v,m}(t) \tag{6-35}$$

where $q_{v,m}(t)$ is the respective *principal coordinate function* of the mth mode (6-39).

Lagrange's Equation

The *energy functions* in terms of (6-35) reduce to

$$\begin{aligned} T &= \frac{1}{2}\int_0^l \mu\left(\frac{\partial v}{\partial t}\right)^2 dx = \frac{1}{2}\sum_{m=1}^{\infty}\mu_{v,m}\left(\frac{dq_{v,m}}{dt}\right)^2 \\ V &= \frac{1}{2}\int_0^l EI\left(\frac{\partial^2 v}{\partial x^2}\right)^2 dx = \frac{1}{2}\sum_{m=1}^{\infty}\mu_{v,m}(\omega_m q_{v,m})^2 \end{aligned} \tag{6-36}$$

where T is the *kinetic energy*, V is the *potential energy*, $\mu_{v,m}$ is given by (6-28), and $q_{v,m} = q_{v,m}(t)$.

The *Lagrange's equation of the second kind*, (4-23b) in terms of (6-36) yields

$$\mu_{v,m}\frac{d^2 q_{v,m}}{dt^2} + \mu_{v,m}\omega_{v,m}^2 q_{v,m} = Q_{v,m}(t) \tag{6-37}$$

where

$$Q_{v,m}(t) = \int_0^l p_v(x,t)X_{v,m}(x)\,dx \tag{6-38}$$

is the *generalized force* defined by (4-16) (Problem 6.20).

The *general solution* of (6-37) is then

$$q_{v,m}(t) = E_{v,m}\cos\omega_m t + F_{v,m}\sin\omega_m t + \frac{1}{\mu_{v,m}\omega_m}\int_0^t Q_{v,m}(\tau)\sin\omega_m(t-\tau)\,d\tau \tag{6-39}$$

where $E_{v,m}$ and $F_{v,m}$ are given by (6-30).

The *concentrated loads* (force or moment) can be included in (6-38) by simply taking their magnitudes multiplied by the eigenfunction (Problem 6.20). Here too the transport matrix approach is more convenient (Problems 6.22, 6.23, and 6.24).

Solved Problems

FREE LONGITUDINAL VIBRATIONS

6.1. Derive the differential equation of longitudinal vibration (6-2) and (6-15) and then find the eigenfunction (6-4) and the time function (6-7).

Referring to Fig. 6-1, x is the position coordinate and u is the displacement of any cross section in the direction of x, measured relative to the static equilibrium position of the cross section. For the element dx shown, the tensile forces are

$$N = AE\frac{\partial u}{\partial x}, \qquad N + \left(\frac{\partial N}{\partial x}\right)dx = AE\left[\frac{\partial u}{\partial x} + \left(\frac{\partial^2 u}{\partial x^2}\right)dx\right]$$

and the inertia force is $\mu\partial^2 u/\partial t^2$. If the distributed dynamic load $p_u(x,t)$ is acting along the x axis, we may write the equilibrium equation of this element as

$$\mu \frac{\partial^2 u}{\partial t^2} dx = -N + N + \frac{\partial N}{\partial x} dx - p_u(x, t)\, dx$$

which yields

$$AE \frac{\partial^2 u}{\partial x^2} - \mu \frac{\partial^2 u}{\partial t^2} = p_u(x, t) \tag{2a}$$

If $p_u(x, t) = 0$ is the condition of free vibration, then (*2a*) becomes

$$a^2 \frac{\partial^2 u}{\partial x^2} - \frac{\partial^2 u}{\partial t^2} = 0 \tag{2b}$$

where $a^2 = AE/\mu$. Equations (*2a*) and (*2b*) correspond to (*6-2*) and (*6-15*) respectively.

For the solution of (*2b*), let

$$u(x, t) = X_u(x) Y_u(t) \tag{3}$$

where X_u is a function of x alone and Y_u is a function of t alone. If $u(x, t)$ satisfies (*2*), then

$$a^2 X_u''(x) Y_u(t) = X_u(x) Y_u''(t) \tag{4}$$

or

$$\frac{a^2 X_u''(x)}{X_u(x)} = \frac{Y_u''(t)}{Y_u(t)} \tag{5}$$

which is the result of *separation of variables*. Both sides of (*5*) must have the same constant value which we call b. Then

$$X_u''(x) - \frac{b}{a^2} X_u(x) = 0 \qquad \text{and} \qquad Y_u''(t) - bY_u(t) = 0 \tag{6, 7}$$

where the constant $b \lesseqgtr 0$. Since we are solving a vibration problem of periodic motion, only the solution corresponding to $b < 0$ is valid. Let $b = -\omega^2$ and use the Laplace transforms method discussed in Chapter 2. Then

$$\mathcal{L}\{X_u''(x)\} = s^2 \mathcal{L}\{X_u(s)\} - sX_u(+0) - X_u'(+0) \qquad \text{and} \qquad \mathcal{L}\left\{\frac{\omega^2}{a^2} X_u(x)\right\} = \frac{\omega^2}{a^2} \mathcal{L}\{X_u(s)\}$$

and the Laplace transforms of (*6*) are

$$\mathcal{L}\{X_u(s)\} = \frac{sX_u(+0)}{s^2 + (\omega^2/a^2)} + \frac{X_u'(+0)}{s^2 + (\omega^2/a^2)} \tag{8}$$

The inverse of (*8*) is

$$X_u(x) = X_u(+0) \cos \frac{\omega}{a} x + X_u'(+0) \frac{a}{\omega} \sin \frac{\omega}{a} x \tag{9}$$

and for the mth mode,

$$X_{u,m}(x) = X_{u,m}(+0) \cos \lambda_{u,m} x + \frac{X_{u,m}'(+0)}{\lambda_{u,m}} \sin \lambda_{u,m} x \tag{10}$$

where $\lambda_{u,m} = \sqrt{\mu \omega_m^2 / AE}$ and m signifies the mode. Equation (*10*) is identical to (*6-4*).

It is obvious that the solution of (*7*) is similar to (*9*) as

$$Y_u(t) = Y_u(+0) \cos \omega t + \frac{Y_u'(+0)}{\omega} \sin \omega t \tag{11}$$

of which the general form for the mth mode is

$$Y_{u,m}(t) = E_{u,m} \cos \omega_m t + F_{u,m} \sin \omega_m t \tag{12}$$

which is the same as (*6-7*).

6.2. Show the orthogonality of eigenfunctions given in (*6-8*) and (*6-9*).

Let the differential equation shown in (*6*) of Problem 6.1 be rewritten for the jth and kth modes in the form

$$AEX_j'' = -\mu\omega_j^2 X_j \quad (1) \qquad AEX_k'' = -\mu\omega_k^2 X_k \quad (2)$$

where the subscript u and the variable x are omitted. Multiplying (*1*) by X_k and (*2*) by X_j and then subtracting (*2*) from (*1*), we have

$$(\omega_k^2 - \omega_j^2)\mu X_k X_j = AE(X_j''X_k - X_jX_k'') \tag{3}$$

Multiply (*3*) by dx and integrate over the length of the beam from $x = 0$ to $x = l$.

$$(\omega_k^2 - \omega_j^2)\int_0^l \mu X_j X_k\,dx = AE\int_0^l (X_j''X_k - X_jX_k'')\,dx \tag{4}$$

Applying integration by parts to the right-hand side of (*4*), we have

$$\int_0^l X_j''X_k\,dx = X_kX_j'\Big|_0^l - \int_0^l X_k'X_j'\,dx \tag{5}$$

$$\int_0^l X_jX_k''\,dx = X_jX_k'\Big|_0^l - \int_0^l X_k'X_j'\,dx \tag{6}$$

Substituting (*5*)and (*6*) into (*4*) yields

$$(\omega_k^2 - \omega_j^2)\int_0^l \mu X_j X_k\,dx = AE(X_kX_j' - X_jX_k')\Big|_0^l \tag{7}$$

The right-hand side of (*7*) is zero for general boundary conditions; for instance, $X = 0$ for a fixed or hinged end and $X' = 0$ for a roller or free end. Consequently,

$$(\omega_k^2 - \omega_j^2)\int_0^l \mu X_j X_k\,dx = 0 \tag{8}$$

Notice that ω_k and ω_j are distinct for different modes. The equation above can be satisfied only if

$$\int_0^l \mu X_j X_k\,dx = 0 \qquad j \neq k \tag{9}$$

and

$$\int_0^l \mu X_j X_k\,dx = \int_0^l \mu X_m^2\,dx \qquad j = k = m \tag{10}$$

Equations (*9*) and (*10*) are identical to (*6-8*).

Now let (*1*) and (*2*) be written as

$$AEX_j''' = -\mu\omega^2 X_j' \quad (11) \qquad AEX_k''' = -\mu\omega^2 X_k' \quad (12)$$

Using the approach shown in (*3*) and (*6*),

$$(\omega_k^2 - \omega_j^2)\int_0^l \mu X_j'X_k'\,dx = AE(X_j''X_k' - X_j'X_k'')\Big|_0^l \tag{13}$$

Again the right-hand side of (*13*) will vanish for general boundary conditions. For example, for a fixed or hinged end, $X'' = 0$. This is due to zero acceleration, $\partial^2u/\partial t^2 = 0$, at that end. For a free or roller support, $\partial u/\partial x = 0$ and $X' = 0$. Thus another form of orthogonality is

$$\int_0^l \mu X_j'X_k'\,dx = 0 \qquad j \neq k \tag{14}$$

For $j = k$, let us first arrange (*5*) as

$$\int_0^l X_j'^2\,dx = -\int_0^l X_j''X_j\,dx + X_jX_j'\Big|_0^l \tag{15}$$

Then using (*1*) and the general boundary conditions of $X = X' = 0$, we have

$$\int_0^l AEX_j'^2\,dx = \int_0^l \mu\omega_j^2 X_j^2\,dx \tag{16}$$

which is identical to (*6-8*) and (*6-9*).

6.3. Derive the constants of integration shown in (*6-10*) and (*6-11*).

From (9) of Problem 6.1, the motion equation obtained by the method of Laplace transforms may be generally expressed as

$$X_{u,m}(x) = X_{u,m}(+0)\cos\lambda_{u,m}x + \frac{X'_{u,m}(+0)}{\lambda_{u,m}}\sin\lambda_{u,m}x \qquad (1)$$

where $X_{u,m}(+0)$ and $X'_{u,m}(+0)$ symbolically represent axial displacement and strain, such as $u_{L,m}$ and $u'_{L,m}$ at the left hand. The relationships between the integration constants (1) and the boundary conditions ($u_{L,m}$ and $u'_{L,m}$) are

$$u_{L,m} = X_{u,m}(+0) \qquad u'_{L,m} = \frac{N_{L,m}}{AE\lambda_{u,m}} = \frac{X'_{u,m}(+0)}{\lambda_{u,m}} \qquad (2, 3)$$

Equations (2) and (3) correspond to (6-10).

Similarly, the general form of (11) in Problem 6.1 is

$$Y_{u,m}(t) = Y_{u,m}(+0)\cos\omega_m t + \frac{Y'_{u,m}(+0)}{\omega_m}\sin\omega_m t \qquad (4)$$

Comparing (12) of Problem 6.1 with (4) yields

$$Y_{u,m}(+0) = E_{u,m} \qquad Y'_{u,m}(+0) = \omega_m F_{u,m} \qquad (5)$$

If the initial conditions of the vibrating member are $u(x,0)$ and $u'(x,0)$, then from (6-3) we may write

$$u(x,0) = X_{u,m}(x)Y_{u,m}(+0) \qquad u'(x,0) = X_{u,m}(x)Y'_{u,m}(+0) \qquad (6, 7)$$

Substituting (5) into (6) and (7), which are then multiplied by $\mu X_{u,m}(x)$, gives

$$u(x,0)\mu X_{u,m}(x) = \mu X^2_{u,m}(x)E_{u,m} \qquad u'(x,0)\mu X_{u,m}(x) = \mu X^2_{u,m}(x)\omega_m F_{u,m} \qquad (8, 9)$$

Integrating both sides of (8) and (9) from $x = 0$ to $x = l$ and using the generalized mass (6-9) we have

$$E_{u,m} = \frac{\mu}{\mu_{u,m}}\int_0^l u(x,0)X_{u,m}(x)\,dx \qquad F_{u,m} = \frac{\mu}{\omega_m\mu_{u,m}}\int_0^l u'(x,0)X_{u,m}(x)\,dx \qquad (10, 11)$$

which are identical to (6-11).

6.4. Derive the transport matrix shown in (6-12) and (6-13).

The transport matrix may be derived directly from the result of Problem 6.15 or by imposing boundary conditions on the shape function as illustrated in this problem. From (1) of the preceding problem we have

$$X_{u,m}(x) = X_{u,m}(+0)\cos\lambda_{u,m}x + \frac{X'_{u,m}(+0)}{\lambda_{u,m}}\sin\lambda_{u,m}x \qquad (1)$$

For the boundary conditions at $L(x = 0)$, given in (2) and (3) of Problem 6.3, the shape function (1) becomes

$$X_{u,m}(x) = u_{L,m}\cos\lambda_{u,m}x + \frac{N_{L,m}}{EA\lambda_{u,m}}\sin\lambda_{u,m}x \qquad (2)$$

and the axial force N may be expressed as

$$N_{u,m}(x) = -u_{L,m}\lambda_{u,m}EA\sin\lambda_{u,m}x + N_{L,m}\cos\lambda_{u,m}x \qquad (3)$$

At $x = l$, coinciding with R of the bar, (2) and (3) become

$$X_{u,m}(l) = u_{R,m} = u_{L,m}\cos\lambda_{u,m}l + \frac{N_{L,m}}{EA\lambda_{u,m}}\sin\lambda_{u,m}l \qquad (4)$$

$$N_{u,m}(l) = N_{R,m} = -u_{L,m}\lambda_{u,m}EA\sin\lambda_{u,m}l + N_{L,m}\cos\lambda_{u,m}l \qquad (5)$$

Expressing (4) and (5) in matrix form and using (6-14) yields the transport matrix shown in (6-12).

Similarly, for the boundary conditions $u_{R,m}$ and $N_{R,m}$ at $R(x = l)$, the integration constants in (1) may be obtained as

$$X_{u,m}(+0) = u_{R,m}\cos\lambda_{u,m}l - \frac{N_{R,m}}{EA\lambda_{u,m}}\sin\lambda_{u,m}l \qquad (6)$$

$$\frac{X'_{u,m}(+0)}{\lambda_{u,m}} = u_{R,m}\sin\lambda_{u,m}l + \frac{N_{R,m}}{EA\lambda_{u,m}}\cos\lambda_{u,m}l \qquad (7)$$

Substituting (6-7) into (1) yields

$$X_{u,m}(x) = u_{R,m}A(x) + \frac{N_{R,m}}{EA\lambda_{u,m}}B(x) \tag{8}$$

where $$A(x) = \cos\lambda_{u,m}l\cos\lambda_{u,m}x + \sin\lambda_{u,m}l\sin\lambda_{u,m}x$$

and $$B(x) = \cos\lambda_{u,m}l\sin\lambda_{u,m}x - \sin\lambda_{u,m}l\cos\lambda_{u,m}x$$

From (8) the expression for N is

$$N_{u,m}(x) = -EAu_{R,m}\lambda_{u,m}B(x) + N_{R,m}A(x) \tag{9}$$

At $x = 0$, coinciding with L of the bar, (8) and (9) become

$$X_{u,m}(0) = u_{L,m} = u_{R,m}\cos\lambda_{u,m}l - \frac{N_{R,m}}{EA\lambda_{u,m}}\sin\lambda_{u,m}l \tag{10}$$

$$N_{u,m}(0) = N_{L,m} = EAu_{R,m}\lambda_{u,m}\sin\lambda_{u,m}l + N_{R,m}\cos\lambda_{u,m}l \tag{11}$$

Expressing (10) and (11) in matrix form and using (6-14) gives the transport matrix shown in (6-13).

6.5. Using the transport matrix (6-12) or (6-13) find the angular frequencies of the beams shown in Fig. P-6.5a and b. Given: $l = 6.10$ m, $A = 232.26\ \text{cm}^2$, $\gamma = 76.973\ \text{kN/m}^3$, $E = 206.84\ \text{GN/m}^2$, $k = EA/2l$.

Fig. P-6.5

(a) Since the boundary conditions of Fig. P-6.5a are $\bar{u}_{L,m} = 0$ and $\bar{N}_{R,m} = 0$, the transport matrix equation (6-12) becomes

$$\begin{bmatrix} \bar{u}_{R,m} \\ 0 \end{bmatrix} = \begin{bmatrix} \cos\bar{\lambda}_{u,m} & \sin\bar{\lambda}_{u,m} \\ -\sin\bar{\lambda}_{u,m} & \cos\bar{\lambda}_{u,m} \end{bmatrix} \begin{bmatrix} 0 \\ \bar{N}_{R,m} \end{bmatrix} \tag{1}$$

from which the frequency equation is

$$\cos\bar{\lambda}_{u,m} = 0 \qquad \text{or} \qquad \bar{\lambda} = \left(\frac{2m-1}{2}\right)\pi, \quad m = 1, 2, \ldots, \infty \tag{2}$$

Thus $$\omega_m = \left(\frac{2m-1}{2}\right)\frac{\pi}{l}\sqrt{\frac{EA}{\mu}}$$

Using the given member properties, $(\pi/l)\sqrt{EA/\mu} = 2646.713$ rad/s, gives the first three frequencies as

$$\omega_1 = 1323.356\ \text{rad/s} \qquad \omega_2 = 3970.070\ \text{rad/s} \qquad \omega_3 = 6616.782\ \text{rad/s} \tag{3}$$

(b) For the boundary conditions of Fig. P-6.5b, $u_{L,m} = 0$ and $N_{R,m} = -ku_{R,m}$, we write the following transport matrix equation

$$\begin{bmatrix} \bar{u}_{R,m} \\ \dfrac{-k\bar{u}_{R,m}}{\lambda_{u,m}EA} \end{bmatrix} = \begin{bmatrix} \cos\bar{\lambda}_{u,m} & \sin\bar{\lambda}_{u,m} \\ -\sin\bar{\lambda}_{u,m} & \cos\bar{\lambda}_{u,m} \end{bmatrix} \begin{bmatrix} 0 \\ \bar{N}_{L,m} \end{bmatrix} \tag{4}$$

from which

$$\bar{u}_{R,m} = \sin\bar{\lambda}_{u,m}\bar{N}_{L,m} \qquad \bar{N}_{R,m} = -\frac{k\bar{u}_{R,m}}{EA\lambda_{u,m}} = \cos\bar{\lambda}_{u,m}\bar{N}_{L,m} \tag{5, 6}$$

Consequently, $$\tan \bar{\lambda}_{u,m} = -2\bar{\lambda}_{u,m} \tag{7}$$

of which the first three solutions are found by trial as $\bar{\lambda}_{u,1} = 1.835$, $\bar{\lambda}_{u,2} = 4.815$, and $\bar{\lambda}_{u,3} = 7.917$.

The angular frequencies in terms of $\omega_m = (\bar{\lambda}_{u,m}/l)\sqrt{EA/\mu}$ are

$$\omega_1 = 1545.958 \text{ rad/s}, \quad \omega_2 = 4056.560 \text{ rad/s}, \quad \omega_3 = 6669.945 \text{ rad/s}, \quad \ldots \tag{8}$$

Substituting (5) and (6) into (8) of the preceding problem yields the following shape function:

$$X_{u,m}(x) = \bar{N}_{L,m} \sin(\bar{\lambda}_{u,m} x/l) \tag{9}$$

Substituting $\bar{\lambda}_{u,m}$ of various modes [equation (7)] into (9) yields the corresponding mode shape. $\bar{N}_L$ is the amplitude.

6.6. Consider a prismatic bar compressed by an axial force P applied at its ends. Find the displacement equation of vibration produced by the sudden removal of these forces.

Due to the symmetry of the loading, the middle section of the bar remains stationary. Thus the initial conditions are

$$u(x,0) = \frac{P}{AE}\left(\frac{l}{2} - x\right) \quad \text{and} \quad u'(x,0) = 0 \tag{1}$$

Applying the boundary conditions to (6-12) yields

$$\begin{bmatrix} \bar{u}_{R,m} \\ 0 \end{bmatrix} = \begin{bmatrix} \cos \bar{\lambda}_{u,m} & \sin \bar{\lambda}_{u,m} \\ -\sin \bar{\lambda}_{u,m} & \cos \bar{\lambda}_{u,m} \end{bmatrix} \begin{bmatrix} \bar{u}_{L,m} \\ 0 \end{bmatrix} \tag{2}$$

Thus

$$\sin \bar{\lambda}_{u,m} = 0 \quad \text{from which} \quad \bar{\lambda}_{u,m} = m\pi, \quad m = 1, 2, 3, \ldots \quad \text{or} \quad \omega_m = \frac{m\pi}{l}\sqrt{\frac{EA}{\mu}} \tag{3, 4}$$

Consequently the shape function is

$$X_{u,m}(x) = X_{u,m}(+0) \cos \lambda_{u,m} x \tag{5}$$

From (6-9) the generalized mass is

$$\mu_{u,m} = X^2_{u,m}(+0)\mu \int_0^l \cos^2(\lambda_{u,m} x)\, dx = \frac{1}{2}\mu l X^2_{u,m}(+0) \tag{6}$$

From (6-11) the constants of time function are

$$F_{u,m} = 0$$

$$E_{u,m} = \frac{2\mu}{\mu l X_{u,m}(+0)} \int_0^l \frac{P}{AE}\left(\frac{l}{2} - x\right)(\cos \lambda_{u,m} x)\, dx$$

$$= \frac{2P}{AElX_{u,m}(+0)}\left[\frac{1 - \cos \lambda_{u,m} l}{\lambda^2_{u,m}}\right] = \frac{4P}{AElX_{u,m}(+0)\lambda^2_{u,m}}, \quad m = \text{odd} \tag{7}$$

Thus the time function is $Y_{u,m}(t) = E_{u,m} \cos \omega_m t$ which combined with (5) gives

$$u(x,t) = \frac{4PL}{AE\pi^2} \sum_{m=1,3,5} \frac{1}{m^2} \cos \frac{m\pi x}{l} \cos\left(m\pi \sqrt{\frac{EA}{\mu}}\frac{t}{l}\right) \tag{8}$$

FORCED LONGITUDINAL VIBRATIONS

6.7. Using the energy method derive the solution of time function (6-20).

The potential energy may be generally expressed as $V = \frac{1}{2}kx^2$, for a bar of an incremental length dx and $k = AE/dx$. Thus the strain energy stored in the incremental length is

$$dV = \frac{1}{2}\frac{AE}{dx}\left(\frac{\partial u}{\partial x}\right)^2 (dx)^2 \tag{1}$$

Therefore $$V = \frac{1}{2}\int_0^l AE\left(\frac{\partial u}{\partial x}\right)^2 dx \tag{2}$$

From (*6-16*),

$$\frac{\partial u}{\partial x} = \sum_{m=1}^{\infty} \frac{dX_{u,m}(x)}{dx} q_{u,m} \qquad \frac{\partial u}{\partial t} = \sum_{m=1}^{\infty} X_{u,m}(x) \frac{dq_{u,m}(t)}{dt} \tag{3a, 3b}$$

Substituting (*3a*) into (*2*), integrating, and then using the orthogonality of eigenfunctions (*6-8*) and (*6-9*) gives

$$V = \frac{1}{2} \sum_{m=1}^{\infty} \mu_{u,m}(\omega_m q_{u,m})^2 \tag{4}$$

The kinetic energy for an increment of length is $dT = \frac{1}{2}\mu \dot{u}^2\, dx$ which, for total length of a member, becomes

$$T = \frac{1}{2} \int_0^l \mu \left(\frac{\partial u}{\partial t}\right)^2 dx \tag{5}$$

Substituting (*3b*) and (*6-8*) and (*6-9*) into (*5*) yields

$$T = \frac{1}{2} \sum_{m=1}^{\infty} \mu_{u,m} \left(\frac{dq_{u,m}}{dt}\right)^2 \tag{6}$$

The work done by the external dynamic force $p_u(x, t)$ during an arbitrary distortion is

$$W = \int_0^l p_u(x,t)u(x,t)\, dx = \int_0^l p_u(x,t)[\Sigma X_{u,m}(x) q_{u,m}(t)]\, dx \tag{7}$$

The rate of change of work with respect to $q_{u,m}$ is

$$\frac{\partial W}{\partial q_{u,m}} = \int_0^l p_u(x,t) X_{u,m}(x)\, dx \tag{8}$$

Equations (*4*), (*6*), and (*8*) correspond to (*6-17*) and to (*6-19*).

Applying the Lagrange's equation (*4-23b*) to (*4*) (*6*) and (*8*) yields

$$q''_{u,m}(t) + \omega_m^2 q_{u,m}(t) = \frac{Q_{u,m}(t)}{\mu_{u,m}} \tag{9}$$

Following Problem 6.1 we may have the Laplace transforms of (*9*) as

$$\mathcal{L}\{q_{u,m}(s)\} = q_{u,m}(+0)\frac{s}{s^2+\omega_m^2} + \dot{q}_{u,m}(+0)\frac{1}{s^2+\omega_m^2} + \frac{1}{s^2+\omega_m^2}\,\frac{\mathcal{L}\{Q_{u,m}(s)\}}{\mu_{u,m}} \tag{10}$$

The inverse of (*10*) is

$$q_{u,m}(t) = q_{u,m}(+0)\cos\omega_m t + \frac{\dot{q}_{u,m}(+0)}{\omega_m} + \frac{1}{\mu_{u,m}\omega_m}\int_0^t Q_{u,m}(\tau)\sin\omega_m(t-\tau)\, d\tau \tag{11}$$

With $E_{u,m} = q_{u,m}(+0)$ and $F_{u,m} = \dot{q}_{u,m}(+0)/\omega_m$, (*11*) becomes (*6-20*).

6.8. **Find the displacement equation of the beam subject to $p(x,t) = p(xt/l)$ as shown in Fig. P-6.8. Assume the initial conditions to be zero.**

From Problem 6.5*a*, we have

$$X_{u,m}(x) = \frac{X'_{u,m}(+0)}{\lambda_{u,m}} \sin \lambda_{u,m} x \qquad \lambda_{u,m} = \left(\frac{2m-1}{2}\right)\frac{\pi}{l}, \quad m = 1, 2, \ldots, \infty \tag{1}$$

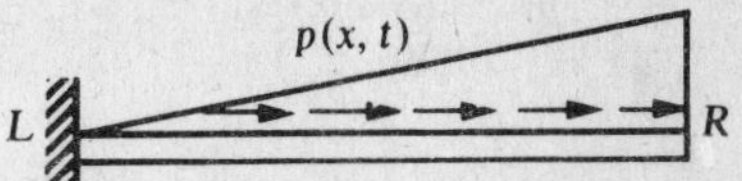

Fig. P-6.8

From (*6-9*) the generalized mass is

$$\mu_{u,m} = \mu \int_0^l \left[\frac{X'_{u,m}(+0)}{\lambda_{u,m}}\right]^2 \sin^2\left(\frac{2m-1}{2}\right)\frac{\pi}{l} x\, dx = \frac{1}{2}\mu l \left[\frac{X'_{u,m}(+0)}{\lambda_{u,m}}\right]^2 \tag{2}$$

The generalized force obtained from (*6-19*) is

$$Q_{u,m}(t) = \int_0^l p\frac{xt}{l}\frac{X'_{u,m}(+0)}{\lambda_{u,m}} \sin\left(\frac{2m-1}{2}\right)\frac{\pi}{l} x\, dx = \frac{pt}{l}\frac{X'_{u,m}(+0)}{\lambda_{u,m}}\frac{\sin(2m-1)\pi/2}{[(2m-1)^2/4](\pi/l)^2} \tag{3}$$

Since the initial conditions are zero, i.e., $E_{u,m} = F_{u,m} = 0$, (6-20) becomes

$$q_{u,m}(t) = \frac{2\lambda_{u,m}}{\mu l X'_{u,m}(+0)\omega_m}\,\frac{p}{l}\,\frac{\sin(2m-1)\pi/2}{[(2m-1)^2/4](\pi/l)^2}\int_0^t \tau \sin\omega_m(t-\tau)\,d\tau$$

$$= \frac{2p\lambda_{u,m}\sin(2m-1)\pi/2}{\mu X'_{u,m}(+0)[(2m-1)^2/4]\pi^2\omega_m}[\omega_m t + \sin\omega_m t] \qquad (4)$$

where $\omega_m = \left(\frac{2m-1}{2}\right)\frac{\pi}{l}\sqrt{\frac{EA}{\mu}}$. Finally, the displacement equation (6-16) is

$$u(x,t) = \sum_{m=1}^{\infty} q_{u,m}(t)X_{u,m}(x) \qquad (5)$$

where $X_{u,m}(x)$ and $q_{u,m}(t)$ correspond to (1) and (4) respectively.

6.9. Find the displacement equation of the beam shown in Fig. P-6.8 which is now subject to a longitudinal force P suddenly applied at the free end. Assume the initial conditions to be zero.

The shape function, eigenvalues, and the generalized mass of this problem are shown in (1) and (2) of Problem 6.8. Since P is a concentrated load, the generalized force obtained by the method of virtual work from

$$\delta W_m = Q_{u,m}\delta q_{u,m} = P\frac{\partial u_{u,m}}{\partial q_{u,m}}\delta q_{u,m} = P\frac{X'_{u,m}(+0)}{\lambda_{u,m}}\sin\left(\frac{2m-1}{2}\right)\pi\,\delta q_{u,m}\bigg|_{x=l} \qquad (1)$$

is

$$Q_{u,m} = P\frac{X'_{u,m}(+0)}{\lambda_{u,m}}\sin\left(\frac{2m-1}{2}\right)\pi \qquad \text{or} \qquad Q_{u,m} = PX_{u,m}(l) \qquad (2)$$

Using the zero initial conditions in (6-20), we may write

$$q_{u,m}(t) = \frac{2P\lambda_{u,m}}{\mu l X'_{u,m}(+0)\omega_m}\sin\left(\frac{2m-1}{2}\right)\pi\int_0^t \sin\omega_m(t-\tau)\,d\tau$$

$$= \frac{2P\lambda_{u,m}}{\mu l X'_{u,m}(+0)\omega_m^2}\sin\left(\frac{2m-1}{2}\right)\pi(1-\cos\omega_m t) \qquad (3)$$

Consequently the displacement equation (6-16) is

$$u(x,t) = \sum_{m=1}^{\infty}\frac{2P}{\mu l\omega_m^2}\sin\left(\frac{2m-1}{2}\right)\pi\sin\left(\frac{2m-1}{2}\right)\pi\frac{x}{l}(1-\cos\omega_m t) \qquad (4)$$

where $\omega_m = \left(\frac{2m-1}{2}\right)\frac{\pi}{l}\sqrt{\frac{EA}{\mu}}$.

6.10. Derive the transport matrix for steady state longitudinal vibration of a straight bar. Using the convolution integral find the load function for the beam shown in Fig. P-6.10.

Consider the straight bar shown in Fig. 6.1 subject to a steady state vibration due to the distributed harmonic force $P(x,t) = P(x)\sin\Omega t$ acting to the right. Then the differential equation of motion (Problem 6.1) is

$$AE\frac{\partial^2 u(x,t)}{\partial x^2} - \mu\frac{\partial^2 u(x,t)}{\partial t^2} = -p(x)\sin\Omega t \qquad (1)$$

For steady state vibration the member vibrates according to the forcing frequency Ω. Thus we may separate t and x as

$$u(x,t) = u(x)\sin\Omega t \qquad (2)$$

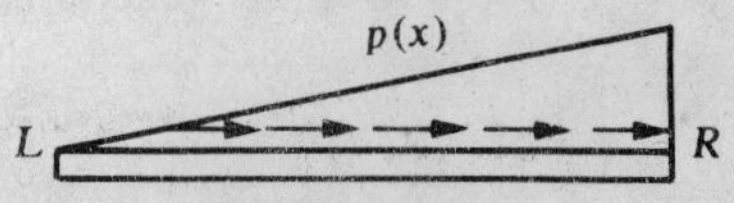

Fig. P-6.10

Substitution of (2) into (1) yields

$$\frac{d^2u(x)}{dx^2} + \lambda_u^2 u(x) = -\frac{p(x)}{AE} \qquad (3)$$

where $\lambda_u^2 = \mu\Omega^2/AE$. The Laplace transforms of (3), similar to (8) of Problem 6.1, are

$$\mathscr{L}\{u(s)\} = \frac{1}{s^2+\lambda_u^2}\left[su(+0) + u'(+0) - \frac{1}{EA}\mathscr{L}\{p(s)\}\right] \tag{4}$$

of which the inverse is

$$u(x) = u(+0)\cos\lambda_u x + \frac{u'(+0)}{\lambda_u}\sin\lambda_u x - \frac{1}{EA\lambda_u}\int_0^x \sin(\lambda_u x - \lambda_u\tau)p(\tau)\,d\tau \tag{5}$$

The first two terms on the right side of (5) are similar to the shape function and may be represented by (6-4). Let

$$\Phi_u(x) = -\frac{1}{EA\lambda_u}\int_0^x \sin(\lambda_u x - \lambda_u\tau)p(\tau)\,d\tau \tag{6}$$

which is the convolution integral representing the load function.

Using the notations of the transport matrix of free vibration (6-12), we write the transport matrix of forced vibration as

$$\begin{bmatrix} 1 \\ \hline \bar{u}_R \\ \bar{N}_R \end{bmatrix} = \left[\begin{array}{c|cc} 1 & 0 & 0 \\ \hline \bar{\Phi}_{RLu} & a_u & d_u \\ \bar{\Phi}'_{RLu} & -d_u & a_u \end{array}\right]\begin{bmatrix} 1 \\ \hline \bar{u}_L \\ \bar{N}_L \end{bmatrix} \tag{7}$$

where $\bar{\Phi}_{RLu} = \Phi(\lambda_u l)$, $\bar{\Phi}'_{RLu} = \Phi'(\lambda_u l)/\lambda_u$, and λ_u is dependent on the given frequency Ω.

To illustrate the application of $\Phi_u(x)$, let us consider the member subject to distributed load $p(x) = px/l$ shown in Fig. P-6.10. Using (5) we have

$$\Phi_u(x) = -\frac{1}{EA\lambda_u}\int_0^x \frac{p\tau}{l}\sin(\lambda_u x - \lambda_u\tau)\,d\tau = \frac{p}{EA\lambda_u^3}(\sin\lambda_u x - \lambda_u x) \tag{8}$$

6.11. The beam shown in Fig. P-6.11 is subject to a concentrated harmonic force $P(x,t) = P\sin\Omega t$, where $P = 177.93$ kN, $\Omega = \omega_1/2$, and ω_1 = fundamental angular frequency. Find the displacement and stress responses at both ends of the beam by using the transport matrix derived in the preceding problem. The member properties are $A = 232.26\text{ cm}^2$, $E = 206.84\text{ GN/m}^2$, $\gamma = 76.973\text{ kN/m}^3$, and $l = 12.19$ m.

The convolution integral derived in (6) of Problem 6.10 applies for distributed load only. For the present case of concentrated load, the Dirac delta function discussed in Problem 2.10 should be used for finding the load function as

$$\frac{d^2u(x)}{dx^2} + \lambda_u^2 u(x) = -\frac{P}{EA}\mathscr{D}(x-l) \tag{1}$$

where $\lambda_u^2 = \mu\Omega^2/AE$. The Laplace transforms of (1) may be expressed as

$$\mathscr{L}\{u(s)\} = \frac{1}{s^2+\lambda_u^2}\left[su(+0) + u'(+0) - \frac{P}{EA}e^{-ls}\right] \tag{2}$$

The inverse of the last term of the right side (2) gives

$$\Phi_u(x) = \frac{-P}{AE\lambda_u}\sin\lambda_u(x-l) \tag{3}$$

Fig. P-6.11

Thus

$$\Phi'_u(x) = \frac{-P}{AE}\cos\lambda_u(x-l) \tag{4}$$

Substitution of $x = l$ into (3) and (4) yields

$$\bar{\Phi}_{RLu} = \Phi(\lambda_u l) = 0 \qquad \bar{\Phi}'_{RLu} = \frac{\Phi'(\lambda_u l)}{\lambda_u} = \frac{-P}{AE\lambda_u} \tag{5}$$

Using (7) of the preceding problem and the boundary conditions of the given beam, we write

$$\begin{bmatrix} 1 \\ \hline \bar{u}_R \\ 0 \end{bmatrix} = \left[\begin{array}{c|cc} 1 & 0 & 0 \\ \hline 0 & \cos\lambda_u l & \sin\lambda_u l \\ -P/AE\lambda_u & -\sin\lambda_u l & \cos\lambda_u l \end{array}\right]\begin{bmatrix} 1 \\ \hline 0 \\ \bar{N}_L \end{bmatrix} \tag{6}$$

from which

$$\bar{u}_R = \sin \lambda_u l \bar{N}_L \qquad \text{and} \qquad 0 = \frac{-P}{AE\lambda_u} + \cos \lambda_u l \bar{N}_L \tag{7}$$

Since

$$\omega_1 = 661.644 \text{ rad/s}, \qquad l = 12.19 \text{ m},$$

$$\lambda_u = \Omega\sqrt{\frac{\mu}{EA}} = \frac{661.644}{2}\sqrt{\frac{182.24}{(232.26)(206.84)10^5}} = 0.064\,43, \qquad \cos \lambda_u l = 0.7071, \quad \sin \lambda_u l = 0.7071$$

we find the dynamic responses from (7) as

$$N_L = 250.88 \text{ kN} \quad \text{and} \quad u_R = 0.0573 \text{ cm} \tag{8}$$

FREE TRANSVERSE VIBRATIONS

6.12. Derive the differential equation of transverse vibration (*6-21*) and then find the general expressions of motion equation (*6-23*) and (*6-26*).

From the free-body diagram shown in Fig. 6-2*b*, the equation of equilibrium of the vertical forces, according to Newton's second law, is

$$\mu\, dx \frac{\partial^2 v}{\partial t^2} = V + \frac{\partial V}{\partial x} dx - V \qquad \text{or} \qquad \mu \frac{\partial^2 v}{\partial t^2} = \frac{\partial V}{\partial x} \tag{1, 2}$$

The sum of moments about any point of the element yields

$$V\, dx - M + M + \frac{\partial M}{\partial x} dx = 0 \qquad \text{or} \qquad \frac{\partial M}{\partial x} = -V \tag{3, 4}$$

Substituting (*4*) into (*2*) gives

$$-\frac{\partial^2 M}{\partial x^2} = \mu \frac{\partial^2 v}{\partial t^2} \tag{5}$$

Since the bar is assumed to have a plane of symmetry and the vibrations occur in that plane, the known differential equation of the deflected curve is

$$EI \frac{\partial^2 v}{\partial^2 x} = M \tag{6}$$

Combining (*5*) and (*6*) yields (*6-21*) as

$$EI \frac{\partial^4 v}{\partial x^4} + \mu \frac{\partial^2 v}{\partial t^2} = 0 \tag{7}$$

Using the method of separation of variables, let

$$v(x, t) = X_v(x) Y_v(t) \tag{8}$$

Substituting (*8*) into (*7*) yields

$$\frac{EIX_v^{iv}(x)}{\mu X_v(x)} = -\frac{Y_v''(t)}{Y_v(t)} \tag{9}$$

in which the term on the left side is dependent only on x and the term on the right side is dependent only on t. They can be equal to each other only if they are equal to the same constant, such as $+\omega^2$. (Note: 0 and $-\omega^2$ are not the solution of the vibration problem). Then (*9*) can be written as

$$X_v^{iv}(x) - \lambda_v^4 X_v(x) = 0 \qquad \text{and} \qquad Y_v''(t) + \omega^2 Y_v(t) = 0 \tag{10, 11}$$

where $\lambda_v^4 = \omega^2 \mu / EI$.

Laplace transforms of (*10*) are

$$\mathscr{L}\{X_v^{iv}(x)\} - \mathscr{L}\{\lambda_v^4 X_v(x)\} = s^4 \mathscr{L}\{X_v(s)\} - s^3 X_v(+0) - s^2 X_v'(+0) - sX_v''(+0) - X_v'''(+0) - \lambda_v^4 \mathscr{L}\{X_v(s)\} = 0 \tag{12}$$

from which

$$\mathscr{L}\{X_v(s)\} = \frac{1}{s^4 - \lambda_v^4}[s^3X_v(+0) + s^2X_v'(+0) + sX_v''(+0) + X_v'''(+0)] \tag{13}$$

The inverse of (13) is

$$X_v(x) = \frac{1}{2}(\cosh \lambda_v x + \cos \lambda_v x)X_v(+0) + \frac{1}{2\lambda_v}(\sinh \lambda_v x + \sin \lambda_v x)X_v'(+0)$$
$$+ \frac{1}{2\lambda_v^2}(\cosh \lambda_v x - \cos \lambda_v x)X_v''(+0) + \frac{1}{2\lambda_v^3}(\sinh \lambda_v x - \sin \lambda_v x)X_v'''(+0) \tag{14}$$

Letting

$$A_v = X_v(+0) \qquad B_v = \frac{X_v'(+0)}{\lambda_v} \qquad C_v = \frac{X_v''(+0)}{\lambda_v^2} \qquad D_v = \frac{X_v'''(+0)}{\lambda_v^3} \tag{15}$$

and using (6-24) we write the shape function (14) for the mth mode as

$$X_{v,m}(x) = A_{v,m}a_{v,m}(x) + B_{v,m}b_{v,m}(x) + C_{v,m}c_{v,m}(x) + D_{v,m}d_{v,m}(x) \tag{16}$$

which is identical to (6-23).

The solution of (11) given in (6-26) is the same as (6-20) which was derived in Problem 6.1.

6.13. Derive the orthogonality of eigenfunctions (6-27) and (6-28).

The governing differential equation shown in (10) of Problem 6.12 may be expressed for the jth and kth modes as

$$EIX_j^{iv} - \omega_j^2\mu X_j = 0 \quad (1) \qquad EIX_k^{iv} - \omega_k^2\mu X_k = 0 \quad (2)$$

where the subscript v and variable x are not shown. Multiplying (1) by X_k and (2) by X_j, where $j \neq k$, and then subtracting (2) from (1), we have

$$(\omega_k^2 - \omega_j^2)\mu X_j X_k = EI(X_j X_k^{iv} - X_k X_j^{iv}) \tag{3}$$

Multiplying both sides of (3) by dx and integrating from $x = 0$ to $x = l$,

$$(\omega_k^2 - \omega_j^2)\int_0^l \mu X_j X_k\,dx = EI\int_0^l (X_j X_k^{iv} - X_k X_j^{iv})\,dx \tag{4}$$

Integrating by parts on the right side of (4) gives

$$\int_0^l X_j X_k^{iv}\,dx = X_j X_k'''\Big|_0^l - X_j' X_k''\Big|_0^l + \int_0^l X_j'' X_k''\,dx \tag{5}$$

$$\int_0^l X_k X_j^{iv}\,dx = X_k X_j'''\Big|_0^l - X_k' X_j''\Big|_0^l + \int_0^l X_j'' X_k''\,dx \tag{6}$$

Substituting (5) and (6) into (4) yields

$$(\omega_k^2 - \omega_j^2)\int_0^l \mu X_j X_k\,dx = EI(X_j X_k''' - X_j' X_k'' - X_k X_j''' + X_k' X_j'')\Big|_0^l \tag{7}$$

Let us examine the following typical boundary conditions:

$$\begin{array}{llll} \text{simple support} & X = X'' = 0, & \text{fixed support} & X = X' = 0 \\ \text{free end} & X'' = X''' = 0, & \text{fixed but guided} & X' = X''' = 0 \end{array} \tag{8}$$

This means that the right side of (7) always vanishes for conventional boundary conditions. Thus (7) becomes

$$(\omega_k^2 - \omega_j^2)\int_0^l \mu X_j X_k\,dx = 0 \tag{9}$$

Since $\omega_k \neq \omega_j$,

$$\int_0^l \mu X_j X_k\,dx = 0, \quad j \neq k \qquad \int_0^l \mu X_j X_k\,dx = \int_0^l \mu X_m^2\,dx, \quad j = k = m \tag{10, 11}$$

Equations (10) and (11) are given in (6-27).

Note that the terms on the right side of (*5*) and (*6*) have the dimensions of energy as the product of a displacement and a shear force or a moment and a rotation. When there is no energy leakage at the supports, these terms are zero. Using (*10*) in (*1*) shows that the left side of (*5*) or (*6*) is also zero. Thus

$$\int_0^l EIX_j''X_k''\,dx = 0 \tag{12}$$

which shows that the second derivatives of shape functions are orthogonal over the interval $(0, l)$ with respect to EI (Eq. 6-*27*).

Using (*1*) in (*5*) or (*6*), for $j = k$ we have

$$\int_0^l X_j''\,dx = \frac{\omega_j^2\mu}{EI}\int_0^l X_j^2\,dx - (X_jX_j''' - X_j'X_j'')\Big|_0^l = \frac{\omega_j^2\mu}{EI}\int_0^l X_j^2\,dx \tag{13}$$

in which the last term vanishes for zero energy leakage. Thus (*13*) becomes

$$\int_0^l X_j''\,dx = \frac{\omega_j^2\mu}{EI}\int_0^l X_j^2\,dx \tag{14}$$

which corresponds to (6-*27*) and (6-*28*).

6.14. Find (*a*) the relationships between the integration constants and the boundary conditions (6-*29*) and (*b*) the constants of time function (6-*30*).

(*a*) The relationships between $A_{v,m}$, $B_{v,m}$, $C_{v,m}$, $D_{v,m}$, and $X_{v,m}(+0)$, $X'_{v,m}(+0)$, $X''_{v,m}(+0)$, $X'''_{v,m}(+0)$ are shown in (*15*) of Problem 6.12 where $X_{v,m}(+0)$, $X'_{v,m}(+0)$, $X''_{v,m}(+0)$, and $X'''_{v,m}(+0)$, obtained by Laplace transforms, respectively represent the boundary conditions of displacement, slope, bending moment, and shear at the left end $(x = 0)$. If the boundary conditions at that end are assumed to be $v_{L,m}$, $\theta_{L,m}$, $M_{L,m}$, and $V_{L,m}$ with positive signs as shown in Fig. 6.2, then the relationships between the integration constants and the boundary conditions are exactly the same as (6-*29*).

(*b*) The constants of time function (6-*30*) are similar to (6-*11*). See Prob. 6.3 for the derivation of (6-*11*).

6.15. Derive the transport matrices shown in (6-*31*) and (6-*32*).

Using (6-*29*) and (6-*33*) we express the shape function as

$$X_{v,m}(x) = \bar{v}_{L,m}a_{v,m}(x) + \bar{\theta}_{L,m}b_{v,m}(x) + \bar{M}_{L,m}c_{v,m}(x) + \bar{V}_{L,m}d_{v,m}(x) \tag{1}$$

The successive derivatives of (*1*) are

$$X'_{v,m}(x) = \lambda_{v,m}[\bar{v}_{L,m}d_{v,m}(x) + \bar{\theta}_{L,m}a_{v,m}(x) + \bar{M}_{L,m}b_{v,m}(x) + \bar{V}_{L,m}c_{v,m}(x)] \tag{2}$$

$$X''_{v,m}(x) = \lambda_{v,m}^2[\bar{v}_{L,m}c_{v,m}(x) + \bar{\theta}_{L,m}d_{v,m}(x) + \bar{M}_{L,m}a_{v,m}(x) + \bar{V}_{L,m}b_{v,m}(x)] \tag{3}$$

$$X'''_{v,m}(x) = \lambda_{v,m}^3[\bar{v}_{L,m}b_{v,m}(x) + \bar{\theta}_{L,m}c_{v,m}(x) + \bar{M}_{L,m}d_{v,m}(x) + \bar{V}_{L,m}a_{v,m}(x)] \tag{4}$$

Let the boundary conditions at the right end $(x = l)$ be

$$v_{R,m} = X_{v,m}(l) \qquad \theta_{R,m} = X'_{v,m}(l) \qquad M_{R,m} = X''_{v,m}(l)/EI \qquad V_{R,m} = -X'''_{v,m}(l)/EI \tag{5}$$

If (*5*) is substituted into (*1*), (*2*), (*3*), and (*4*), which are then expressed in matrix form, we have (6-*31*).

Let the right end (R) of a member correspond to $x = 0$ and the left end (L) correspond to $x = -l$. Then (*1*) may be written as

$$X_{v,m}(x) = \bar{v}_{R,m}a_{v,m}(x) + \bar{\theta}_{R,m}b_{v,m}(x) + \bar{M}_{R,m}c_{v,m}(x) + \bar{V}_{R,m}d_{v,m}(x) \tag{6}$$

of which the derivatives are similar to (*2*), (*3*), and (*4*). Using the boundary conditions at L in (*2*), (*3*), (*4*), and (*6*) we have

$$\begin{bmatrix} V_{L,m} \\ \theta_{L,m} \\ EIM_{L,m} \\ -EIV_{L,m} \end{bmatrix} = \begin{bmatrix} X_{v,m} \\ X'_{v,m} \\ X''_{v,m} \\ X'''_{v,m} \end{bmatrix}_{x=-l} = \begin{bmatrix} a_{v,m} & b_{v,m} & c_{v,m} & d_{v,m} \\ \lambda_{v,m}d_{v,m} & \lambda_{v,m}a_{v,m} & \lambda_{v,m}b_{v,m} & \lambda_{v,m}c_{v,m} \\ \lambda_{v,m}^2c_{v,m} & \lambda_{v,m}^2d_{v,m} & \lambda_{v,m}^2a_{v,m} & \lambda_{v,m}^2b_{v,m} \\ \lambda_{v,m}^3b_{v,m} & \lambda_{v,m}^3c_{v,m} & \lambda_{v,m}^3d_{v,m} & \lambda_{v,m}^3a_{v,m} \end{bmatrix}_{x=-l} \begin{bmatrix} \bar{v}_{R,m} \\ \bar{\theta}_{R,m} \\ \bar{M}_{R,m} \\ \bar{V}_{R,m} \end{bmatrix} \tag{7}$$

When $x = -l$, then $-b_{v,m}(l) = b_{v,m}(-l)$, $-d_{v,m}(l) = d_{v,m}(-l)$, and the other elements remain the same. Using the notations (6-*33*) in (*7*) gives (6-*32*).

6.16. Using the transport matrix (*6-31*) or (*6-32*) find the angular frequencies and shape function of a simply supported beam. Assume $l = 40$ ft, $A = 36\text{ in}^2$, $\gamma = 490\text{ lb/ft}^3$, and $E = 30\,000$ ksi.

Using the boundary conditions, $\bar{v}_{L,m} = \bar{v}_{R,m} = \bar{M}_{L,m} = \bar{M}_{R,m} = 0$, in (*6-31*) we have

$$\begin{bmatrix} 0 \\ \bar{\theta}_{R,m} \\ 0 \\ \bar{V}_{R,m} \end{bmatrix} = \begin{bmatrix} a_{v,m} & b_{v,m} & c_{v,m} & d_{v,m} \\ d_{v,m} & a_{v,m} & b_{v,m} & c_{v,m} \\ c_{v,m} & d_{v,m} & a_{v,m} & b_{v,m} \\ b_{v,m} & c_{v,m} & d_{v,m} & a_{v,m} \end{bmatrix} \begin{bmatrix} 0 \\ \bar{\theta}_{L,m} \\ 0 \\ \bar{V}_{L,m} \end{bmatrix} \tag{1}$$

from which

$$\text{Det.}\begin{bmatrix} b_{v,m} & d_{v,m} \\ d_{v,m} & b_{v,m} \end{bmatrix} = \tfrac{1}{4}[(\sinh \bar{\lambda}_{v,m} + \sin \bar{\lambda}_{v,m})^2 - (\sinh \bar{\lambda}_{v,m} - \sin \lambda_{v,m})^2] = \sinh \bar{\lambda}_{v,m} \sin \bar{\lambda}_{v,m} = 0 \tag{2}$$

Since $\sinh \bar{\lambda}_{v,m} \neq 0$,

$$\sin \bar{\lambda}_{v,m} = 0 \qquad \bar{\lambda}_{v,m} = m\pi \qquad \text{and} \qquad \omega_m = (m\pi)^2 \sqrt{\frac{EI}{\mu l^4}}, \qquad m = 1, 2, \ldots \tag{3}$$

Using the given member properties, $(\pi/l)^2\sqrt{EI/\mu} = 22.50$, we have

$$\omega_1 = 22.50\text{ rad/s}, \qquad \omega_2 = 45.0\text{ rad/s}, \qquad \omega_3 = 67.50\text{ rad/s}, \qquad \ldots \tag{4}$$

The shape function of the beam may be determined either by substituting (*1*) into (*6*) of Problem 6.15 as shown in Problem 5.5 or by solving (*6-23*) directly. For the latter, using the boundary conditions $X = X'' = 0$ at $x = 0$ and $x = l$ in (*6-23*) and its second derivative, we have

$$\begin{aligned} X_{v,m}(0) &= A_{v,m} = 0 & X_{v,m}(l) &= B_{v,m}b_{v,m} + D_{v,m}d_{v,m} = 0 \\ X''_{v,m}(0) &= C_{v,m} = 0 & X''_{v,m}(l) &= B_{v,m}d_{v,m} + D_{v,m}b_{v,m} = 0 \end{aligned} \tag{5}$$

from which $B_{v,m} = -D_{v,m} \tanh \bar{\lambda}_{v,m}$. Thus (*6-23*) becomes

$$X_{v,m}(x) = B_{v,m}b_{v,m}(x) - B_{v,m}d_{v,m}(x) = B_{v,m} \sin \lambda_{v,m}x = B_{v,m} \sin \frac{m\pi x}{l} \tag{6}$$

6.17. Using the transport matrix (*6-31*) or (*6-32*) find the angular frequencies of the beam shown in Fig. P-6.17. Find the shape function of the beam.

The boundary conditions of the fixed and fixed-guided beam are $\bar{v}_{L,m} = \bar{\theta}_{L,m} = \bar{\theta}_{R,m} = \bar{V}_{R,m} = 0$. Substituting these into (*6-31*) yields

$$\text{Det.}\begin{bmatrix} b_{v,m} & c_{v,m} \\ d_{v,m} & a_{v,m} \end{bmatrix} = \cosh \bar{\lambda}_{v,m} \sin \bar{\lambda}_{v,m} + \cos \bar{\lambda}_{v,m} \sinh \bar{\lambda}_{v,m} = 0 \tag{1}$$

or

$$\tan \bar{\lambda}_{v,m} = -\tanh \bar{\lambda}_{v,m}.$$

By trial we found

$$\bar{\lambda}_{v,m} = \frac{4m-1}{4}\pi$$
$$\omega_m = \left(\frac{4m-1}{4}\right)^2 \left(\frac{\pi}{l}\right)^2 \sqrt{\frac{EI}{\mu}} \tag{2}$$

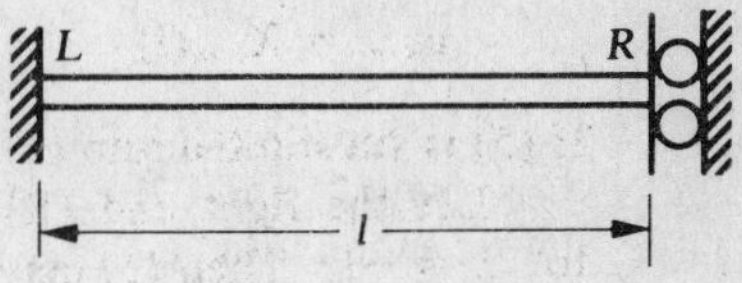

Fig. P-6.17

Substituting the boundary conditions of beam into (*6-23*) yields

$$X_{v,m}(0) = A_{v,m} = 0$$
$$X'_{v,m}(0) = B_{v,m} = 0$$
$$X'_{v,m}(l) = C_{v,m}b_{v,m} + D_{v,m}c_{v,m} = 0$$
$$X'''_{v,m} = C_{v,m}d_{v,m} + D_{v,m}a_{v,m} = 0$$

from which $D_{v,m} = -C_{v,m} \tanh \bar{\lambda}_{v,m}$. Thus (*6-23*) becomes

$$X_{v,m}(x) = C_{v,m}[c_{v,m}(x) - \tanh \bar{\lambda}_{v,m}d_{v,m}(x)] \tag{3}$$

6.18. Find the displacement equation of a simply supported beam subject to a static uniform load p which is suddenly removed at $t = 0$.

The initial conditions of the free vibration are

$$v(x,0) = v_{st} \qquad \text{and} \qquad \dot{v}(x,0) = 0 \tag{1}$$

where v_{st} is the static deflection due to the uniform load p,

$$v_{st} = \frac{p}{24EI}(l^3x - 2lx^3 + x^4) \tag{2}$$

Using the shape function given in (6) of Problem 6.16, we may write the generalized mass (6-28) as

$$\mu_{v,m} = \mu \int_0^l B_{v,m}^2 \sin^2 \frac{m\pi x}{l}\, dx = \frac{1}{2}\mu l B_{v,m}^2 \tag{3}$$

Thus the constants of time function (6-30) are $F_{v,m} = 0$ and

$$E_{v,m} = \frac{\mu}{\mu_{u,m}} \int_0^l v_{st} B_{v,m} \sin \frac{m\pi x}{l}\, dx = \frac{p}{12EIl\mu_{u,m}} \int_0^l (l^3x - 2lx^3 + x^4) \sin \frac{m\pi x}{l}\, dx \tag{4}$$

Upon integration,

$$E_{v,m} = \frac{4pl^4}{EI(m\pi)^5}, \qquad m = 1, 3, 5, \ldots \tag{5}$$

From (6-30),

$$q_{v,m}(t) = E_{v,m} \cos \omega_m t = \frac{4pl^4}{EI(m\pi)^5} \cos \omega_m t, \qquad m = 1, 3, 5 \tag{6}$$

Finally the displacement equation (6-22) becomes

$$v(x,t) = q_{v,m}(t) X_{v,m}(x) = \frac{4pl^4}{EI\pi^5} \sum_{m=1,3}^{\infty} \frac{1}{m^5} \sin \frac{m\pi x}{l} \cos \omega_m t \tag{7}$$

FORCED TRANSVERSE VIBRATIONS

6.19. Using the energy method derive the solution of time function of transverse vibration (6-39).

From (6-35), the general solution of transverse vibration is

$$v(x,t) = \sum_{m=1}^{\infty} X_{v,m}(x) q_{v,m}(t) \tag{1}$$

Utilizing the beam theory we may write the potential energy as

$$V = \frac{1}{2}\int_0^l EI \left(\frac{\partial v}{\partial x}\right)^2 dx \tag{2}$$

and the kinetic energy as

$$T = \frac{1}{2}\int_0^l \mu \left(\frac{\partial v}{\partial t}\right)^2 dx \tag{3}$$

Note that (2) and (3) are analogous to (2) and (5) of Problem 6.7. Thus the derivation of (6-39) is almost exactly the same as that of (6-20) in Problem 6.7.

6.20. Find the displacement equation of a simple beam subject to a concentrated moment, $Q(t) = Q(1 - t/a)$, applied at the mid-span as shown in Fig. P-6.20. Assume the initial conditions to be zero.

From Problem 6.16,

$$X_{v,m}(x) = B_{v,m} \sin \frac{m\pi x}{l}$$

and

$$\omega_m = (m\pi)^2 \sqrt{\frac{EI}{\mu l^4}} \tag{1}$$

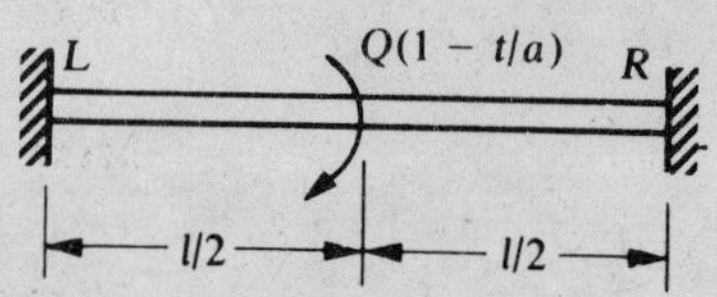

Fig. P-6.20

The work done that is expressed below is the product of the couple and the angle through which it rotates.

$$\delta W_m = Q_{v,m}\delta q_{v,m} = Q(t)\frac{\partial v'_{v,m}}{\partial q_{v,m}}\delta q_{v,m} = Q(t)\frac{m\pi}{l}\cos\frac{m\pi}{2}\delta q_{v,m}\Big|_{x=l/2} \tag{2}$$

from which the generalized force is

$$Q_{v,m} = Q(t)\frac{m\pi}{l}(-1)^{m/2}, \qquad m = \text{even} \tag{3}$$

Using the given initial conditions in (*6-39*) we have $E_{v,m} = F_{v,m} = 0$ and

$$\begin{aligned} q_{v,m}(t) &= \frac{1}{\mu_{v,m}\omega_m}\int_0^t Q_{v,m}(\tau)\sin\omega_m(t-\tau)\,d\tau = \frac{2Qm\pi(-1)^{m/2}}{\mu\omega_m l^2 B_{v,m}}\int_0^t (1-\tau/a)\sin\omega_m(t-\tau)\,d\tau \\ &= \frac{2Qm\pi(-1)^{m/2}}{\mu B_{v,m}(\omega_m l)^2}\left[1-\cos\omega_m t - \frac{1}{a}\left(t - \frac{1}{\omega_m}\sin\omega_m t\right)\right] \end{aligned} \tag{4}$$

where $m = \text{even}$, and $\mu_{v,m} = \mu l B^2_{v,m}/2$. The final displacement equation is

$$v_{v,m}(x,t) = q_{v,m}(t)X_{v,m}(x) \tag{5}$$

in which $m = \text{even number}$ and $X_{v,m}(x)$ and $q_{v,m}(t)$ are given in (*1*) and (*4*) respectively.

6.21. Consider a simply supported beam subject to a distributed load

$$p_v(x,t) = p\sin(\pi x/2l)\sin\Omega t \qquad p = 40\text{ lb/ft}$$

and find the displacement and stress responses for the steady state vibration. The member properties are given in Problem 6.16. The forcing frequency is half of the angular frequency, $\Omega = \omega_1/2$.

From (*6-38*),

$$\begin{aligned} Q_{v,m}(t) &= \int_0^l p\sin\frac{\pi x}{2l}(\sin\Omega t)B_{v,m}\sin\frac{m\pi x}{l}\,dx \\ &= p(\sin\Omega t)B_{v,m}\left[\frac{l}{(2m+1)\pi}\sin\frac{(2m+1)\pi}{2} - \frac{l}{(2m-1)\pi}\sin\frac{(2m-1)\pi}{2}\right] \end{aligned} \tag{1}$$

where $B_{v,m}\sin m\pi x/l$ is the shape function $X_{v,m}$. Using the zero initial conditions for $E_{v,m} = F_{v,m} = 0$ and the generalized mass $\mu_{v,m} = B^2_{v,m}\mu l/2$, we write (*6-39*) as

$$\begin{aligned} q_{v,m}(t) &= \frac{1}{\mu_{v,m}\omega_m}\int_0^t Q_{v,m}(\tau)\sin\omega_m(t-\tau)\,d\tau \\ &= \frac{2p}{\mu}\frac{1}{B_{v,m}}\left[\frac{(-1)^m}{(2m+1)\pi} + \frac{(-1)^m}{(2m-1)\pi}\right]\left[\frac{\sin\Omega t - (\Omega/\omega_m)\sin\omega_m t}{\omega_m^2 - \Omega^2}\right] \end{aligned} \tag{2}$$

Thus the displacement equation (*6-35*) is

$$v(x,t) = \sum_{m=1}^{\infty} X_{v,m}(x)q_{v,m}(t) = \frac{2p}{\mu}\sum_{m=1}^{\infty}\left[\frac{4m(-1)^m}{(4m^2-1)\pi}\right]\left[\frac{\sin\Omega t - (\Omega/\omega_m)\sin\omega_m t}{\omega_m^2-\Omega^2}\right] \tag{3}$$

Note that in (*3*) the terms associated with $\sin\Omega t$ are forced vibration having the same period as the disturbing force and those associated with $\sin\omega_m t$ are free vibration produced by application of the force. The latter vibration due to various kinds of resistance will be gradually damped out. Thus the part of forced vibration called steady state vibration is of practical importance. Now let us consider the steady state vibration of (*3*) by dropping $\sin\omega_m t$ and letting $\sin\Omega t = 1$; then the responses of displacement $v(x)$, moment $M(x)$, and shear $V(x)$ may be obtained from (*3*) as

$$v(x) = \frac{2p}{\mu}\sum_{m=1}^{\infty}\left[\frac{4m(-1)^m}{(4m^2-1)\pi}\right]\left[\frac{\sin m\pi x/l}{\omega_m^2-\Omega^2}\right] \tag{4}$$

$$M(x) = -EI\frac{d^2v(x)}{dx^2} = \frac{2pEI}{\mu}\sum_{m=1}^{\infty}\left(\frac{m\pi}{l}\right)^2\left[\frac{4m(-1)^m}{(4m^2-1)\pi}\right]\left[\frac{\sin m\pi x/l}{\omega_m^2-\Omega^2}\right] \tag{5}$$

$$V(x) = -EI\frac{d^3v(x)}{dx^3} = \frac{2pEI}{\mu}\sum_{m=1}^{\infty}\left(\frac{m\pi}{l}\right)^3\left[\frac{4m(-1)^m}{(4m^2-1)\pi}\right]\left[\frac{\cos m\pi x/l}{\omega_m^2-\Omega^2}\right] \tag{6}$$

Knowing $\omega_m = 22.5$ rad/s, $\Omega = \omega_1/2 = 11.25$ rad/s, and the given member properties, we calculate the following for $m = 1, 2, 3, 4$ as

$$\begin{aligned} M(l/2) &= 7.34 + 0 - 0.1578 + 0 = 7.1822 \text{ ft-k} \\ V(0) &= 0.5765 - 0.08789 + 0.0371 - 0.0206 = 0.5051 \text{ k} \\ V(l) &= 0.5765 - 0.08789 - 0.0371 - 0.0206 = -0.7221 \text{ k} \end{aligned} \tag{7}$$

6.22. Derive the transport matrix for steady state transverse vibration of a straight bar.

Let the straight bar of Fig. 6-2 be subject to a steady state vibration due to a distributed harmonic force $p_v(x, t) = p_v(x) \sin \Omega t$ acting downward. The differential equations of motion are

$$\frac{\partial v(x,t)}{\partial x} = \theta \qquad \frac{\partial^2 v(x,t)}{\partial x^2} = \frac{M}{EI} \qquad \frac{\partial^3 v(x,t)}{\partial x^3} = -\frac{V}{EI} \qquad EI\frac{\partial^4 v(x,t)}{\partial x^4} = -\mu\frac{\partial^2 v(x,t)}{\partial t^2} + p_v(x)\sin \Omega t \tag{1}$$

Using separation of t and x gives

$$v(x,t) = v(x)\sin \Omega t \qquad \theta(x,t) = \theta(x)\sin \Omega t \qquad M(x,t) = M(x)\sin \Omega t \qquad V(x,t) = V(x)\sin \Omega t \tag{2}$$

Consequently (*1*) becomes

$$\frac{dv(x)}{dx} = \theta \qquad \frac{d^2 v(x)}{dx^2} = \frac{M}{EI} \qquad \frac{d^3 v(x)}{dx^3} = \frac{V}{EI} \qquad EI\frac{d^4 v(x)}{dx^4} = \mu\Omega^2 v(x) + p_v(x) \tag{3}$$

Thus

$$\frac{d^4 v(x)}{dx^4} - \lambda_v^4 v(x) = \frac{p_v(x)}{EI} \tag{4}$$

where $\lambda_v^4 = \mu\Omega^2/EI$. Following (*12*) and (*13*) of Problem 6.12, Laplace transforms of (*4*) are

$$\mathcal{L}\{v(s)\} = \frac{1}{s^4 - \lambda_v^4}\left[s^3 v(+0) + s^2 v'(+0) + s v''(+0) + v'''(+0) + \frac{1}{EI}\mathcal{L}\{p_v(s)\}\right] \tag{5}$$

The inverse of this is

$$v(x) = X_v(x) + \Phi_v(x) \tag{6}$$

where $X_v(x)$ is similar to the shape function as shown in (*14*) of Problem 6.12 and $\Phi_v(x)$ is the convolution integral

$$\Phi_v(x) = \frac{1}{2EI\lambda_v^3}\int_0^x [\sinh(\lambda_v x - \lambda_v \tau) - \sin(\lambda_v x - \lambda_v \tau)]p_v(\tau)\,d\tau \tag{7}$$

whose characteristics were discussed in Chapter 2.

Using the notations of free vibrations shown in (*6-31*) and (*6-33*) we may express the transport matrix of forced vibration as

$$\begin{bmatrix} 1 \\ \hline \bar{v}_R \\ \bar{\theta}_R \\ \bar{M}_R \\ \bar{V}_R \end{bmatrix} = \left[\begin{array}{c|cccc} 1 & 0 & 0 & 0 & 0 \\ \hline \bar{\Phi}_{RLv} & a_v & b_v & c_v & d_v \\ \bar{\Phi}'_{RLv} & d_v & a_v & b_v & c_v \\ \bar{\Phi}''_{RLv} & c_v & d_v & a_v & b_v \\ \bar{\Phi}'''_{RLv} & b_v & c_v & d_v & a_v \end{array}\right] \begin{bmatrix} 1 \\ \hline \bar{v}_L \\ \bar{\theta}_L \\ \bar{M}_L \\ \bar{V}_L \end{bmatrix} \tag{8}$$

where $\bar{\Phi}_{RLv} = \Phi_v(l)$, $\bar{\Phi}'_{RLv} = \Phi'_v(l)/\lambda_v$, $\bar{\Phi}''_{RLv} = \Phi''(l)/\lambda_v^2$, and $\bar{\Phi}'''_{RLv} = \Phi'''/\lambda_v^3$. Note that λ_v is dependent on the given forcing frequency Ω.

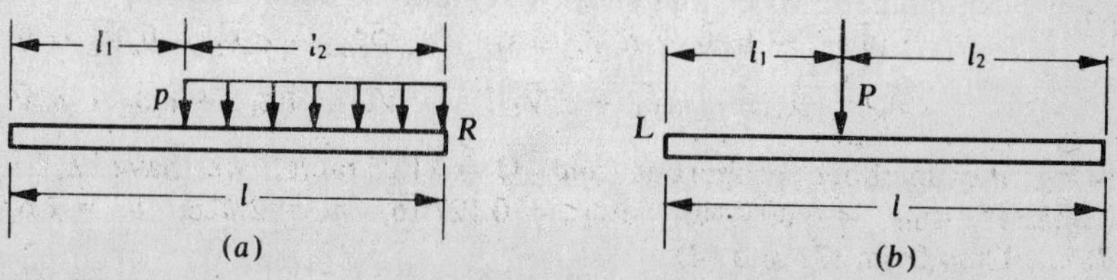

Fig. P-6.23

6.23. Derive the load functions for the beams shown in Fig. P-6.23.

(*a*) For the beam of Fig. P-6.23*a*, $p_v(\tau) = p$, using (*7*) of the preceding problem yields

$$\Phi_v(x) = \frac{p}{2EI\lambda_v^3}\int_{l_1}^{x} [\sinh \lambda_v(x-\tau) - \sin \lambda_v(x-\tau)]\, d\tau$$

$$= \frac{p}{2EI\lambda_v^4}[\cosh \lambda_v(x-l_1) + \cos \lambda_v(x-l_1) - 2], \qquad l_1 < x < l \tag{1}$$

$$\Phi_v(x) = 0, \qquad 0 \le x \le l_1 \tag{2}$$

$$\Phi_v(x) = \frac{p}{2EI\lambda_v^4}[\cosh \lambda_v l_2 + \cos \lambda_v l_2 - 2], \qquad x = l \tag{3}$$

(*b*) For the beam of Fig. P-6.23*b* with concentrated load, using the Dirac delta function gives the following differential equation:

$$\frac{d^4 v(x)}{dx^4} - \lambda_v^4 v(x) = \frac{P}{EI}\mathscr{D}(x - l_1) \tag{4}$$

The solution of the left-hand side (*4*) was shown in (*5*) and (*6*) of Problem 6.22. The Laplace transforms of the right-hand side (*4*) may be written as

$$\mathscr{L}\{v(s)\} = \frac{1}{s^4 - \lambda_v^4}\frac{P}{EI}e^{-l_1 s} \tag{5}$$

of which the inverse is the load function as

$$\Phi_v(x) = \frac{P}{2EI\lambda_v^3}[\sinh \lambda_v(x-l_1) - \sin \lambda_v(x-l_1)], \qquad l_1 < x < l \tag{6}$$

$$\Phi_v(x) = 0, \qquad 0 \le x \le l_1 \tag{7}$$

$$\Phi_v(x) = \frac{P}{2EI\lambda_v^3}(\sinh \lambda_v l_2 - \sin \lambda_v l_2), \qquad x = l \tag{8}$$

6.24. Use the transport matrix to solve Problem 6.21.

The load function of this problem is

$$\Phi_v(x) = \frac{p}{2EI\lambda_v^3}\int_0^x [\sinh \lambda_v(x-\tau) - \sin \lambda_v(x-\tau)] \sin\frac{\pi\tau}{2l}\, d\tau$$

$$= \frac{p}{2EI\lambda_v^4}\left[\frac{1}{1+q^2}\left(q \sinh \lambda_v x - \sin\frac{\pi x}{2l}\right) - \frac{1}{1-q^2}\left(q \sin\frac{\pi x}{2l} - \sin \lambda_v x\right)\right] \tag{1}$$

where $q = \pi/2l\lambda_v$ and $\lambda_v = \sqrt[4]{\mu\Omega^2/EI}$. From (*1*) we find $\Phi'(x)$, $\Phi''(x)$, and $\Phi'''(x)$; therefore $\bar{\Phi}_{RLv}$, $\bar{\Phi}'_{RLv}$, $\bar{\Phi}''_{RLv}$, and $\bar{\Phi}'''_{RLv}$ can also be calculated.

Using the boundary conditions we write the transport matrix as

$$\begin{bmatrix} 1 \\ \hline 0 \\ \bar{\theta}_R \\ 0 \\ \bar{V}_R \end{bmatrix} = \left[\begin{array}{c|cccc} 1 & 0 & 0 & 0 & 0 \\ \hline \bar{\Phi}_{RLv} & a_v & b_v & c_v & d_v \\ \bar{\Phi}'_{RLv} & d_v & a_v & b_v & c_v \\ \bar{\Phi}''_{RLv} & c_v & d_v & a_v & b_v \\ \bar{\Phi}'''_{RLv} & b_v & c_v & d_v & a_v \end{array}\right] \begin{bmatrix} 1 \\ \hline 0 \\ \bar{\theta}_L \\ 0 \\ \bar{V}_L \end{bmatrix} \tag{2}$$

from which

$$\bar{\Phi}_{RLv} + b_v\bar{\theta}_L + d_v\bar{V}_L = 0 \qquad \bar{\Phi}''_{RLv} + d_v\bar{\theta}_L + b_v\bar{V}_L = 0 \tag{3}$$

$$\bar{\theta}_R = \bar{\Phi}'_{RLv} + a_v\bar{\theta}_L + c_v\bar{V}_L \qquad \bar{V}_R = \bar{\Phi}'''_{RLv} + c_v\bar{\theta}_L + a_v\bar{V}_L \tag{4}$$

Knowing the member properties and $\Omega = 11.25$ rad/s, we have $\lambda_v = \sqrt[4]{\mu\Omega^2/EI} = 0.0555$ ft $\bar{\Phi}_{RLv} = 0.025\,31$, $\Phi''_{RLv} = 0.097\,946$, $\bar{\Phi}'''_{RLv} = 0.127\,16$, $a_v = 2.026$, $b_v = 2.6731$, $c_v = 2.630$, and $d_v = 1.8762$. Thus from (*3*) and (*4*)

$$V_L = 0.506 \text{ k}, \qquad V_R = -0.730 \text{ k}, \qquad \theta_L = 0.000\,177 \text{ rad}$$

Supplementary Problems

FREE LONGITUDINAL VIBRATIONS

6.25. Find the expressions of angular frequencies and mode shapes of a beam with both ends fixed by (*a*) applying the boundary conditions directly to the shape function and (*b*) using the transport matrix. Find also the first three natural frequencies. Given: $A = 36\text{ in}^2$, $l = 40\text{ ft}$, $\gamma = 490\text{ lbs/ft}^3$, $I = 243\text{ in}^3$.

6.26. Use the transport matrix to find the expressions of angular frequencies of the beams shown in Fig. P-6.26. Given: l, μ, A, E, and $k = EA/2l$.

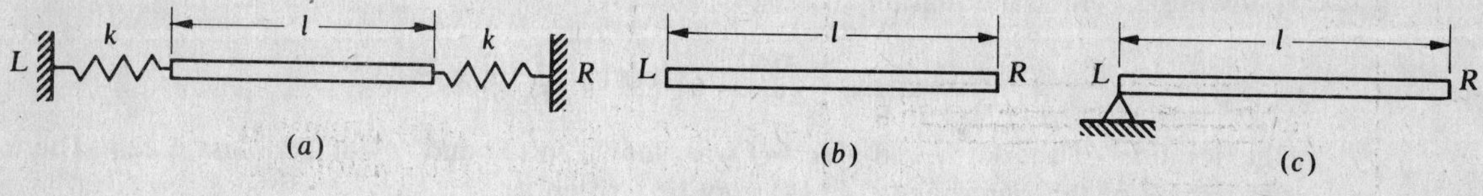

Fig. P-6.26

6.27. Find the displacement equation of a cantilever beam subject to a tensile force P, applied at the free end, which is suddenly removed at $t = 0$.

FORCED LONGITUDINAL VIBRATIONS

6.28. Use the energy method to find the dynamic response of Problem 6.11.

6.29. Find the load functions of the bars shown in Table P-6.29.

6.30. Using the transport matrix method find the displacement and stress responses at $x = l/2$ of the beam given in Problem 6.11.

FREE TRANSVERSE VIBRATIONS

6.31. Find the expressions of angular frequencies and mode shapes of a cantilever beam by (*a*) imposing the boundary conditions directly on the shape function and (*b*) using the transport matrix. Use the member properties given in Problem 6.25 and find the first three natural frequencies of this beam.

6.32. Use the transport matrix to find the expressions of angular frequencies of beams shown in Fig. P-6.32. Given: l, μ, E, I, and k.

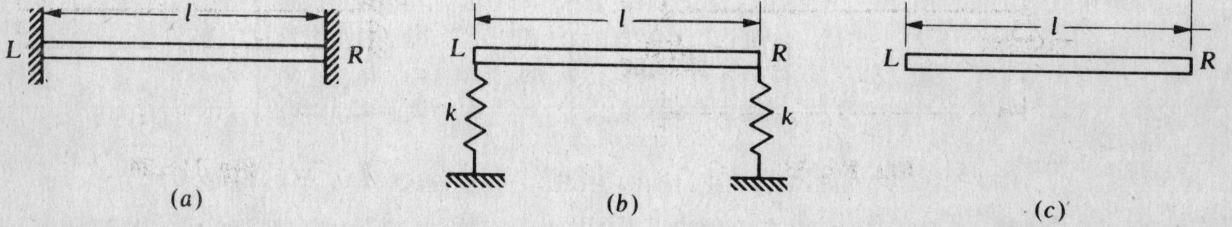

Fig. P-6.32

6.33. Find the displacement equation of the beam of Fig. P-6.33 in which the applied force P is suddenly removed at $t = 0$. Given: l, μ, E, and I.

Table P-6.29 Load Functions $\Phi_u(\lambda_u x)$

Given: $\lambda_u = \sqrt{\mu\Omega^2/EA}$, Ω = longitudinal forcing frequency		
1. 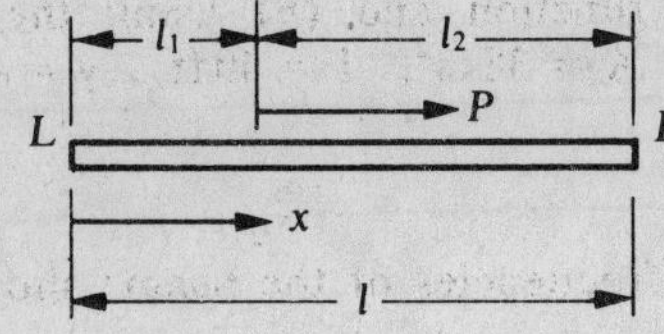	$0 \le x \le l_1$ $l_1 < x < l$ $x = l$	$\Phi(\lambda x) = 0$ $\Phi(\lambda x) = -Pd(\lambda x - \lambda l_1)/\lambda EA$ $\Phi(\lambda l) = -Pd(\lambda l_2)/\lambda EA$
2. 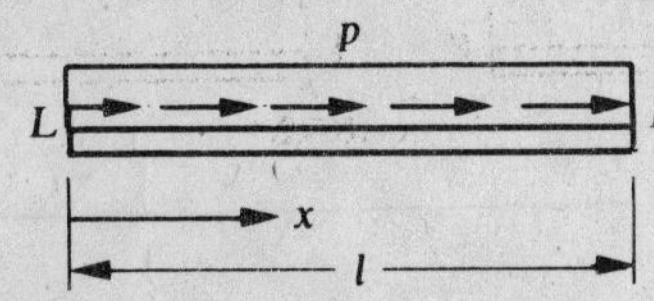	$x = 0$ $0 < x < l$ $x = l$	$\Phi(\lambda x) = 0$ $\Phi(\lambda x) = -p[1 - a(\lambda x)]/EA\lambda^2$ $\Phi(\lambda l) = -p[1 - a(\lambda l)]/EA\lambda^2$
3. 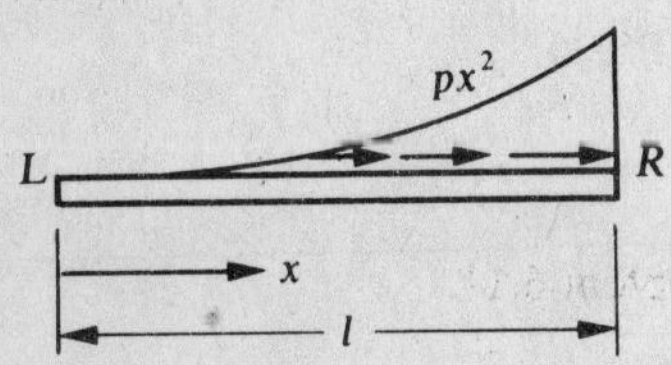	$0 \le x \le l_1$ $0 < x < l$ $x = l$	$\Phi(\lambda x) = 0$ $\Phi(\lambda x) = -p\{x^2 - 2[1 - a(\lambda x)]/\lambda^2\}/EA\lambda^2$ $\Phi(\lambda l) = -p\{l^2 - 2[1 - a(\lambda l)]/\lambda^2\}/EA\lambda^2$
4. 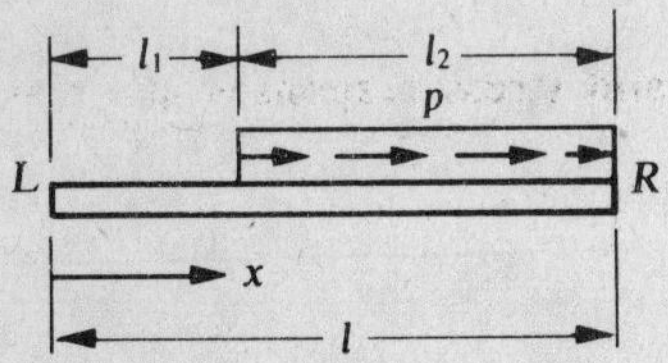	$0 \le x \le l_1$ $l_1 < x < l$ $x = l$	$\Phi(\lambda x) = 0$ $\Phi(\lambda x) = -p[1 - a(\lambda x - \lambda l_1)]/EA\lambda^2$ $\Phi(\lambda l) = -p[1 - a(\lambda l_2)]/EA\lambda^2$

Note: The shape functions a and d appearing in the table above are given analytically in (6-5).

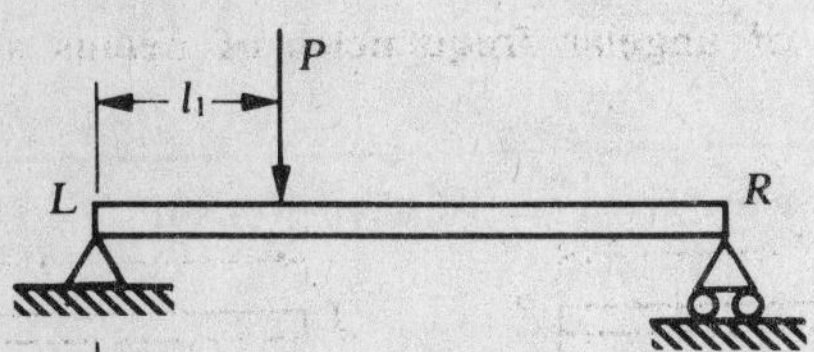

Fig. P-6.33

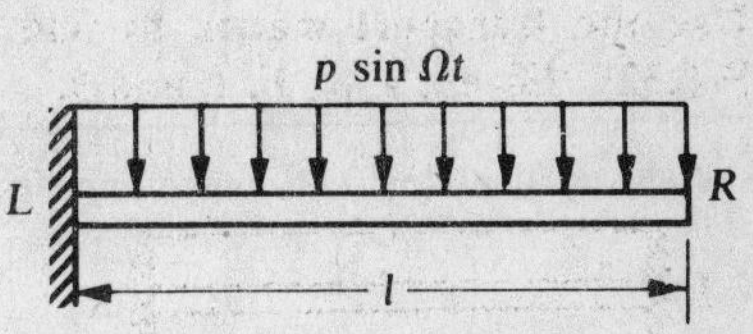

Fig. P-6.34

FORCED TRANSVERSE VIBRATIONS

6.34. Use the energy method and find the displacement and stress responses at both ends of the beam of Fig. P-6.34. Assume $p = 40$ k/ft, $\Omega = \omega_1/2$, and ω_1 = first angular frequencies. The member properties are given in Problem 6.25.

6.35. Find the load functions of the beams shown in Table P-6.35.

6.36. Use the transport matrix to solve Problem 3.34.

Table P-6.35 Load Functions $\Phi_v(\lambda_v x)$

$\lambda_v = \sqrt[4]{\mu\Omega^2/EI}$, Ω = transverse forcing frequency		
1. 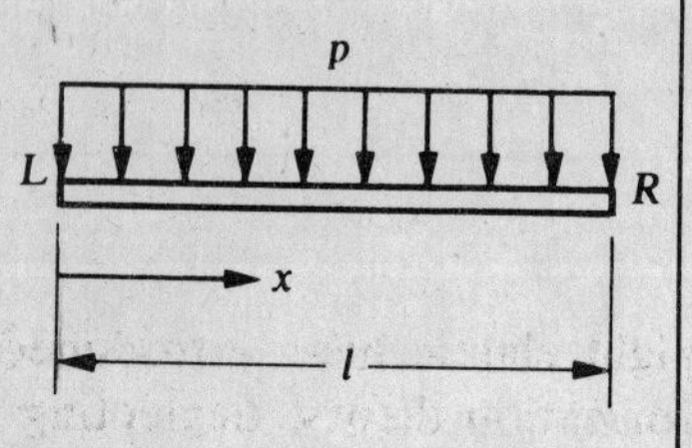	$x = 0$ $0 < x < l$ $x = l$	$\Phi(\lambda x) = 0$ $\Phi(\lambda x) = -p[1 - a(\lambda x)]/EI\lambda^4$ $\Phi(\lambda l) = -p[1 - a(\lambda l)/EI\lambda^4$
2. 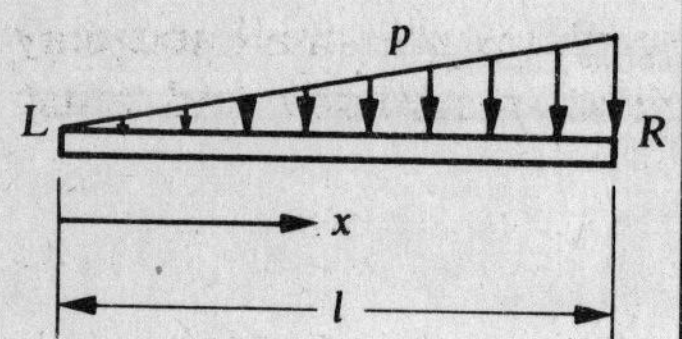	$x = 0$ $0 < x < l$ $x = l$	$\Phi(\lambda x) = 0$ $\Phi(\lambda x) = -p[\lambda x - b(\lambda x)]/EIl\lambda^5$ $\Phi(\lambda l) = -p[\lambda l - b(\lambda l)]/EIl\lambda^5$
3. 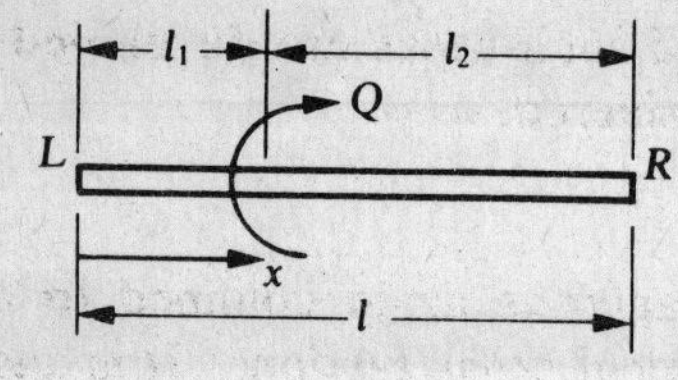	$0 \le x \le l_1$ $l_1 < x < l$ $x = l$	$\Phi(\lambda x) = 0$ $\Phi(\lambda x)' = -Qc(\lambda x - \lambda l_1)/EI\lambda^2$ $\Phi(\lambda l) = -Qc(\lambda l_2)/EI\lambda^2$
4. 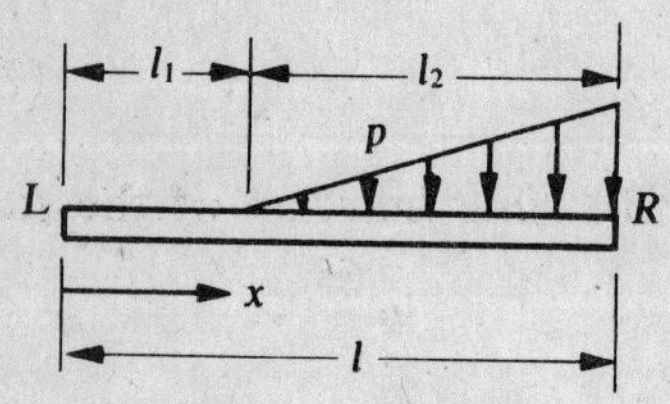	$0 \le x < l_1$ $l_1 < x < l$ $x = l$	$\Phi(\lambda x) = 0$ $\Phi(\lambda x) = \dfrac{-p[(x - l_1)a(\lambda l_1) - l_1 a(\lambda x - \lambda l_1) + b(\lambda l_1)/\lambda - b(\lambda x - \lambda l_1)/\lambda]}{EIl_2\lambda^4}$ $\Phi(\lambda l) = \dfrac{-p[l_2 a(\lambda l_1) - l_1 a(\lambda l_2) + b(\lambda l_1)/\lambda - b(\lambda l_2)/\lambda]}{EIl_2\lambda^4}$

Note: The shape functions a, b, and c appearing in the table above are given analytically in (6-24).

Chapter 7

Distributed Mass Systems: Special Cases

7.1 INTRODUCTION

Constant Secondary Effects

The *differential equations* of free and forced vibrations in straight elastic bars, introduced in Chapter 6, were derived under the assumption of the first order deformation theory, neglecting the effect of resisting medium in longitudinal vibrations, and the effect of normal forces, shearing forces, rotatory inertia moments, and resisting medium in transverse vibrations.

Although in the majority of engineering systems these *secondary effects* are small and may be disregarded, there are cases where these effects reach an appreciable magnitude and must be included in the analysis.

Variable Secondary Effects

In general, the secondary effects mentioned above are *not constant* but vary with position in the span and/or with time. Their inclusion in the analysis leads to differential equations of motion, the coefficients of which are not constants; the direct solution of these equations is not possible. Differential, integral, and energy methods for the solution of such cases are developed in Chapter 10; in this chapter, only constant secondary effects are considered.

Elastic Bar

The study of free and forced vibrations presented in this chapter is again confined to the repetitive motion of a *finite straight elastic bar of constant symmetrical cross section.* It is again assumed that the *equation of motion* can be represented by a series as

$$\Delta(x, t) = \sum_{m=1}^{\infty} X_{\Delta,m}(x) Y_{\Delta,m}(t) \tag{7-1}$$

where $X_{\Delta,m}(x)$ is the *eigenfunction* of the mth mode and $Y_{\Delta,m}(t)$ is the *time function* of the same mode.

7.2 RESISTING MEDIUM

Longitudinal Vibrations

The *differential equation* of free longitudinal vibrations in the finite straight elastic bar of Fig. 7-1 encased along its entire length l in an elastic foundation of resistance $k_u u$ is

$$EA\frac{\partial^2 u}{\partial x^2} - k_u u - \mu\frac{\partial^2 u}{\partial t^2} = 0 \tag{7-2}$$

where u, x, t, E, A, and μ are as defined for (6-2) and k_u is the foundation stiffness defined as the resistance per unit length of the bar due to a unit horizontal displacement.

The *general solution* of (7-2) in the form of (7-1) consists of $X_{u,m}(x)$ and $Y_{u,m}(t)$, both having the same analytical form as (6-4) and (6-7) respectively.

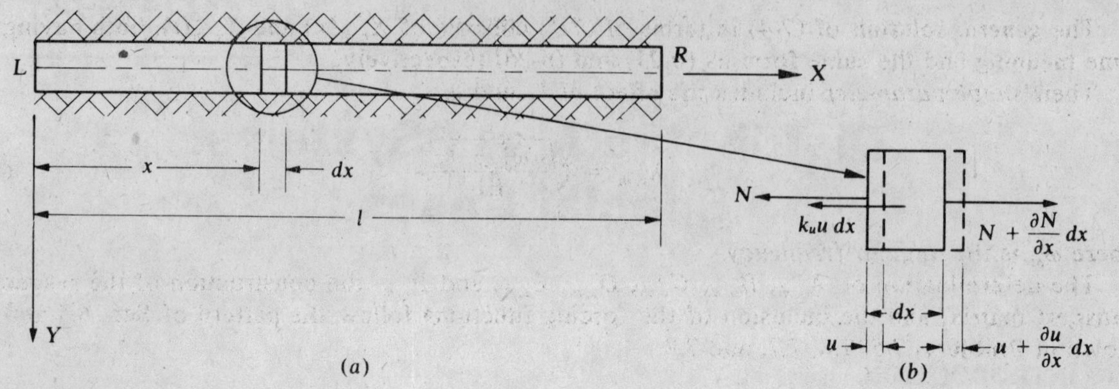

Fig. 7-1. Longitudinal Vibrations of a Straight Bar Encased in Elastic Medium of Resistance, $k_u u$

The *shape parameter*, however, includes the effect of k_u and is

$$\lambda_{u,m} = \sqrt{\frac{\mu\omega_m^2 - k_u}{EA}} \tag{7-3}$$

where ω_m is the *angular frequency*.

The determination of $A_{u,m}$, $D_{u,m}$, $E_{u,m}$, and $F_{u,m}$, the construction of the respective transport matrix, and the inclusion of the forcing function follow the pattern of Sec. 6.2 and Sec. 6.3 and are shown in Problems 7.1, 7.2, 7.3, and 7.4.

Transverse Vibrations

The *differential equation* of free transverse vibrations in the finite straight elastic bar of Fig. 7-2 supported along its entire length l on an elastic foundation of resistance $k_v v$ is

$$EI\frac{\partial^4 v}{\partial x^4} + kv_v + \mu\frac{\partial^2 v}{\partial t^2} = 0 \tag{7-4}$$

where v, x, t, E, I, and μ are as defined for (*6-21*) and k_v is the foundation stiffness defined as the resistance per unit length of the bar due to a unit vertical displacement.

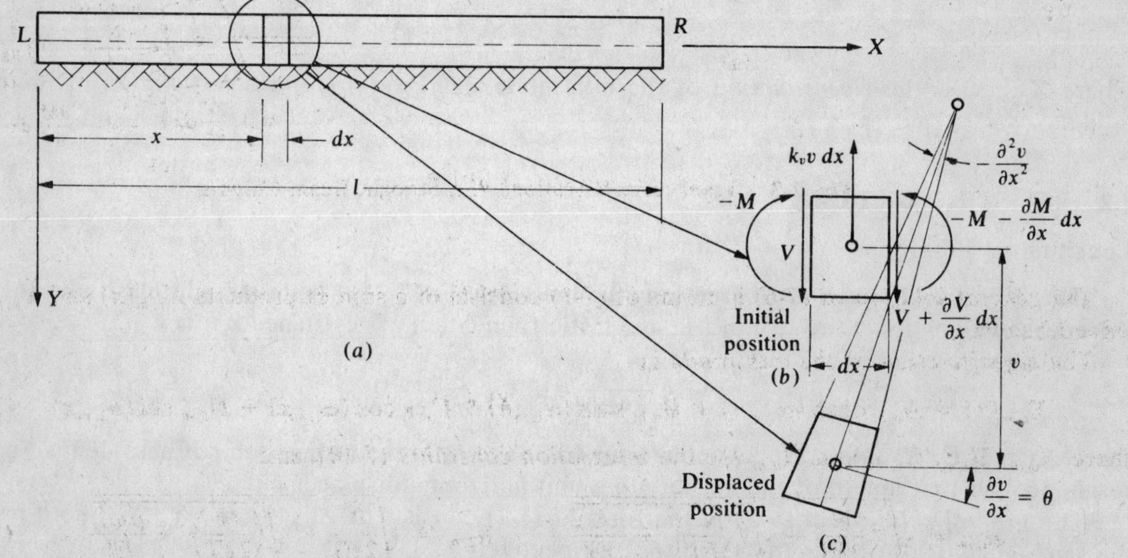

Fig. 7-2. Transverse Vibrations of a Straight Bar on Elastic Foundation of Resistance, $k_v v$

The *general solution* of (*7-4*) in terms of (*7-1*) consists of $X_{v,m}(x)$ and $Y_{v,m}(t)$, both having the same meaning and the same form as (*6-23*) and (*6-26*) respectively.

Their *shape parameter* includes the effect of k_v and is

$$\lambda_{v,m} = \sqrt[4]{\frac{\mu\omega_m^2 - k_v}{EI}} \tag{7-5}$$

where ω_m is the *angular frequency*.

The determination of $A_{v,m}$, $B_{v,m}$, $C_{v,m}$, $D_{v,m}$, $E_{v,m}$, and $F_{v,m}$, the construction of the respective transport matrix, and the inclusion of the forcing functions follow the pattern of Sec. 6.5 and are shown in Problems 7.5, 7.6, 7.7, and 7.8.

7.3 BEAM-COLUMN

Transverse Vibrations

The *differential equation* of free transverse vibrations in the finite straight elastic bar of Fig. 7-3 acted upon by an axial compressive force N is

$$EI\frac{\partial^4 v}{\partial x^4} + N\frac{\partial^2 v}{\partial x^2} + \mu\frac{\partial^2 v}{\partial t^2} = 0 \tag{7-6}$$

where v, x, t, E, I, and μ are those of (*6-21*).

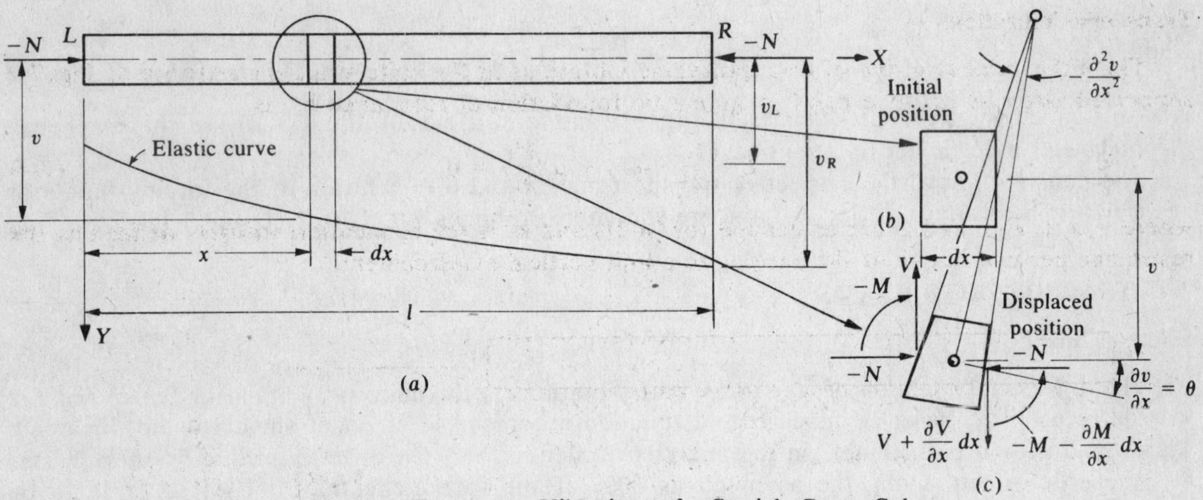

Fig. 7-3. Transverse Vibrations of a Straight Beam-Column

The *general solution* of (*7-6*) in terms of (*7-1*) consists of a sum of products $X_{v,m}(x)$ and $Y_{v,m}(t)$ defined below.

The *eigenfunction of the mth mode* is

$$X_{v,m}(x) = A_{v,m}\cosh(\alpha_{1,m}x) + B_{v,m}\sinh(\alpha_{1,m}x) + C_{v,m}\cos(\alpha_{2,m}x) + D_{v,m}\sin(\alpha_{2,m}x) \tag{7-7}$$

where $A_{v,m}$, $B_{v,m}$, $C_{v,m}$, and $D_{v,m}$ are the *integration constants* (*7-10*), and

$$\alpha_{1,m} = \sqrt{-\frac{N}{2EI} + \sqrt{\left(\frac{N}{2EI}\right)^2 + \frac{\mu\omega_m^2}{EI}}} \qquad \alpha_{2,m} = \sqrt{\frac{N}{2EI} + \sqrt{\left(\frac{N}{2EI}\right)^2 + \frac{\mu\omega_m^2}{EI}}} \tag{7-8}$$

are the *shape parameters* of this motion where ω_m is the *angular frequency*.

The *time function of the mth node* is

$$Y_{v,m}(t) = E_{v,m} \cos \omega_m t + F_{v,m} \sin \omega_m t \tag{7-9}$$

where $E_{v,m}$ and $F_{v,m}$ are the *integration constants* (7-12) and ω_m is the *same angular frequency* as in (7-8).

Constants of Integration

The constants of integration obtained by the method of *Laplace transforms* are

$$\left.\begin{aligned} A_{v,m} &= \frac{\alpha_{2,m}^2 X_{v,m}(+0) + X''_{v,m}(+0)}{\alpha_{1,m}^2 + \alpha_{2,m}^2} & B_{v,m} &= \frac{\alpha_{2,m}^2 X'_{v,m}(+0) + X'''_{v,m}(+0)}{\alpha_{1,m}(\alpha_{1,m}^2 + \alpha_{2,m}^2)} \\ C_{v,m} &= \frac{\alpha_{1,m}^2 X_{v,m}(+0) - X''_{v,m}(+0)}{\alpha_{1,m}^2 + \alpha_{2,m}^2} & D_{v,m} &= \frac{\alpha_{1,m}^2 X'_{v,m}(+0) - X'''_{v,m}(+0)}{\alpha_{2,m}(\alpha_{1,m}^2 + \alpha_{2,m}^2)} \end{aligned}\right\} \tag{7-10}$$

where $\qquad X_{v,m}(+0) = v_{L,m} \qquad X'_{v,m}(+0) = \theta_{L,m} \qquad X''_{v,m}(+0) = M_{L,m}/EI \qquad$ (7-11)

but

$$X'''(+0) = [-V_L + N\theta_{L,m}]/EI$$

which shows that the *integration constants* are to some scale the *displacement, slope, bending moment,* and *shear* of the left end.

The *constants* $E_{v,m}$ and $F_{v,m}$ in (7-9) obtained by the *orthogonality principle* are, as in (6-30),

$$E_{v,m} = \frac{\mu}{\mu_{v,m}} \int_0^l v(x,0) X_{v,m}(x)\,dx \qquad F_{v,m} = \frac{\mu}{\mu_{v,m}\omega_m} \int_0^l \dot{v}(x,0) X_{v,m}(x)\,dx \tag{7-12}$$

where

$$\mu_{v,m} = \int_0^l X_{v,m}^2(x)\,dx \tag{7-13}$$

is again called the *generalized mass in the mth mode* and $v(x,0)$ and $\dot{v}(x,0)$ are the respective *displacement* and *velocity functions* at $t = 0$.

The construction of the respective transport matrix and the inclusion of the forcing functions again follow the pattern of Sec. 6.5 and are shown in Problems 7.9, 7.10, 7.11, and 7.12.

7.4 TIMOSHENKO'S BEAMS

Effect of Shear

The *differential equation* of free transverse vibrations in the finite straight elastic bar of Fig. 6-2 was derived on the basis of displacement function in which the effect of shear was not included. This simplification is justifiable in the analysis of slender bars. For *deep girders* or *short beams* the shear deformation may be as much as 30% of the total deflection and is too large to be neglected. In such cases,

$$EI \frac{\partial^4 v}{\partial x^4} - \frac{\mu EI}{GA^*} \frac{\partial^4 v}{\partial x^2 \partial t^2} + \mu \frac{\partial^2 v}{\partial t^2} = 0 \tag{7-14}$$

where v, x, t, E, I, and μ are as defined for (6-21), G is the modulus of rigidity, and A^* is the modified area of the right section (Problem 7.13).

The *eigenfunction of the mth mode*, $X_{v,m}(x)$, is formally identical to (7-7), the dynamic parameters of which include the *shape parameters*

$$\alpha_{1,m} = \sqrt{-\frac{\mu\omega_m^2}{2GA^*} + \sqrt{\frac{\mu\omega_m^2}{EI} + \left(\frac{\mu\omega_m^2}{2GA^*}\right)^2}} \qquad \alpha_{2,m} = \sqrt{\frac{\mu\omega_m^2}{2GA^*} + \sqrt{\frac{\mu\omega_m^2}{EI} + \left(\frac{\mu\omega_m^2}{2GA^*}\right)^2}} \tag{7-15}$$

The *time function of the same mode*, $Y_{v,m}(t)$, is formally identical to (7-9), and ω_m is the *angular frequency*, identical for both functions.

Effect of Rotatory Inertia

Another effect, which has been disregarded in the preceding chapter, is the *effect of the mass moment of inertia* on the transverse vibrations in straight elastic bars. This effect generates an additional term in (*6-21*) which then becomes

$$EI\frac{\partial^4 v}{\partial x^4} - \mu r^2 \frac{\partial^4 v}{\partial x^2 t^2} + \mu\frac{\partial^2 v}{\partial t^2} = 0 \tag{7-16}$$

and is formally similar to (*7-14*).

Because of this similarity, all preceding functions are applicable here if the shear effect term $\mu\omega_m^2/2GA^*$ is replaced by $\mu r^2\omega_m^2/2EI$, where $r = \sqrt{I/A}$ (Problem 7.14).

General Case

Finally, a completely general case arises when both the *effect of shear* and the *rotatory inertia* are included in (*6-21*). The free vibration of such a bar, called the *Timoshenko's beam*, is governed by

$$EI\frac{\partial^4 v}{\partial x^4} - \left(\mu r^2 + \frac{\mu EI}{GA^*}\right)\frac{\partial^4 v}{\partial x^2 \partial t^2} + \mu\frac{\partial^2 v}{\partial t^2} + \frac{\mu^2 r^2}{GA^*}\frac{\partial^4 v}{\partial t^4} = 0 \tag{7-17}$$

where all terms are those used in (*7-14*) and/or (*7-16*).

The *general solution* of (*7-17*) in terms of (*7-1*) consists again of a sum of products of $X_{v,m}(x)$ and $Y_{v,m}(t)$ given below.

The *eigenfunction of the mth mode*, $X_{v,m}(x)$, is formally identical to (*7-7*), the dynamic parameters of which include the *shape parameters*,

$$\left.\begin{aligned}\alpha_{1,m} &= \sqrt{-\frac{\beta_{1,m}+\beta_{2,m}}{2} + \sqrt{\left(\frac{\beta_{1,m}-\beta_{2,m}}{2}\right)^2 + \beta_{3,m}}}\\ \alpha_{2,m} &= \sqrt{\frac{\beta_{1,m}+\beta_{2,m}}{2} + \sqrt{\left(\frac{\beta_{1,m}-\beta_{2,m}}{2}\right)^2 + \beta_{3,m}}}\end{aligned}\right\} \tag{7-18}$$

where $\beta_{1,m} = \mu\omega_m^2/GA^*$, $\beta_{2,m} = \mu\omega_m^2 r^2/EI$, $\beta_{3,m} = \mu\omega_m^2/EI$

$r = \sqrt{I/A}$ = *radius of gyration* of the right section, ω_m = *angular frequency*

(See Problem 7.15.)

The *time function of the mth mode*, $Y_{v,m}(t)$, is formally identical to (*7-9*), the integration constants of which are those of (*7-12*) and for which ω_m is the same angular frequency as in (*7-18*).

Constants of Integration

The constants of integration of $X_{v,m}(x)$ obtained by the method of *Laplace transforms* are the same as in (*7-10*):

$$\begin{aligned} X_{v,m}(+0) &= v_{L,m} & X''_{v,m}(+0) &= M_{L,m}/EI - \beta_{1,m}v_{L,m}\\ X'_{v,m}(+0) &= \psi_{L,m} + V_{L,m}/GA^* & X_{v,m}(+0) &= -V_{L,m}/EI - \beta_{1,m}v_{L,m} + \beta_{2,m}\psi_{L,m}\end{aligned} \tag{7-19}$$

where $\psi_{L,m}$ is the left end slope of the flexural elastic curve and $v_{L,m}$, $M_{L,m}$, and $V_{L,m}$ are the respective displacement, bending moment, and shear of the left end (Problem 7.16).

Four *typical end conditions* are given below:

(*a*) *Fixed End:* $V_{L,m} = 0$, $\psi_{L,m} = 0$ (*c*) *Hinged End:* $v_{L,m} = 0$, $M_{L,m} = 0$

(*b*) *Fixed Axis:* $v_{L,m} = 0$, $(\partial v/\partial x)_{L,m} = 0$ (*d*) *Free End:* $M_{L,m} = 0$, $V_{L,m} = 0$

The *interpretations* of (*a*) and (*b*) are illustrated by Fig. P-7.16 in Problem 7.16.

Solved Problems

RESISTING MEDIUM

7.1. Derive the differential equation of longitudinal vibration (7-2) of a finite straight prismatic bar encased in elastic medium of resistance $k_u u$ (Fig. 7-1). Then using the method of Laplace transforms find the general integrals of this differential equation.

Applying Newton's second law to the free-body diagram of Fig. 7-1 yields

$$\mu \frac{\partial^2 u}{\partial t^2} dx = -N - k_u u \, dx + N + \frac{\partial N}{\partial x} dx \tag{1}$$

Since $N = AE \, \partial u/\partial x$,

$$AE \frac{\partial^2 u}{\partial x^2} - k_u u - \mu \frac{\partial^2 u}{\partial t^2} = 0 \tag{2}$$

Let $u(x,t) = X_u(x)Y_u(t)$. Then (2) may be rewritten as

$$\frac{AE}{\mu}\frac{X''(x)}{X_u(x)} - \frac{k_u}{\mu} = \frac{Y''_u(t)}{Y_u(t)} = \text{constant} \tag{3}$$

where the constant can be positive or negative and the sign will determine the nature of the solution. For a harmonic motion, only the negative constant will yield the physical solution. Consequently, let the constant be $-\omega^2$ and $\lambda_u = \sqrt{(\mu\omega^2 - k_u)/AE}$; then (3) becomes

$$X''_u(x) + \lambda_u^2 X_u(x) = 0 \tag{4}$$

$$Y''_u(t) + \omega^2 Y_u(t) = 0 \tag{5}$$

If $k_u < \mu\omega^2$, Laplace transform solutions of (4, 5) are

$$X_u(x) = X_u(+0)\cos\lambda_u x + [X'_u(+0)/\lambda_u]\sin\lambda_u x \tag{6}$$

and

$$Y_u(t) = Y_u(+0)\cos\omega_u t + [Y'_u(+0)/\omega_u]\sin\omega_u t \tag{7}$$

where $X_u(+0) = A_u$, $X'_u(+0)/\lambda_u = D_u$, $Y_u(+0) = E_u$, $Y'_u(+0)/\omega_u = F_u$ are the constants of integration of the given mode. Then (6) and (7) in the mth mode are

$$X_{u,m}(x) = A_{u,m}\cos\lambda_{u,m}x + D_{u,m}\sin\lambda_{u,m}x \tag{8}$$

and

$$Y_{u,m}(t) = E_{u,m}\cos\omega_{u,m}t + F_{u,m}\sin\omega_{u,m}t \tag{9}$$

If $k_u > \mu\omega^2$, Equation (4) is

$$X''_u(x) - \lambda_u'^2 X_u(x) = 0 \qquad \lambda'_u = \sqrt{\frac{k_u - \mu\omega^2}{AE}} \tag{10}$$

and (6) becomes

$$X_u(x) = X_u(+0)\cosh\lambda'_{u,m}x + [X'_u(+0)/\lambda'_u]\sinh\lambda'_{u,m}x \tag{11}$$

If $k_u = \mu\omega^2$, then $\lambda_u = 0$; (4) reduces to $X''_u(x) = 0$, the solution of which is

$$X_u(x) = X_u(+0) + X'_u(+0)x \tag{12}$$

Since our definition of the problem involves periodic motion, (11) and (12) do not apply.

7.2. Derive the transport matrix from (8) of the preceding problem.

Since (8) derived in Problem 7.1 is identical to (1) of Problem 6.4, the transport matrix derived in that problem may be directly used here. However, the shape parameter is different and should be expressed as $\lambda_{u,m} = \sqrt{(\mu\omega^2 - k_u)/AE}$ to take account of the elastic medium k_u.

7.3. Use the transport matrix discussed in the preceding problem to find the angular frequencies of the beam given in Fig. P-7.3. Let $l = 6.10$ m, $A = 232.26\ \text{cm}^2$, $\gamma = 76.973\ \text{kN/m}^3$, $E = 206.84\ \text{GN/m}^2$, and $k_u = AE\pi^2/8l^2$.

For the boundary conditions of Fig. P-7.3, the transport matrix may be written as

$$\begin{bmatrix} \bar{u}_{R,m} \\ 0 \end{bmatrix} = \begin{bmatrix} \cos\bar{\lambda}_{u,m} & \sin\bar{\lambda}_{u,m} \\ -\sin\bar{\lambda}_{u,m} & \cos\bar{\lambda}_{u,m} \end{bmatrix} \begin{bmatrix} 0 \\ \bar{N}_{L,m} \end{bmatrix} \qquad (1)$$

from which

$$\cos\bar{\lambda}_{u,m} = 0 \qquad \text{or} \qquad \bar{\lambda}_{u,m} = \frac{2m-1}{2}\pi, \quad m = 1, 2, 3, \ldots \qquad (2)$$

Thus

$$\omega_m = \sqrt{\frac{AE\bar{\lambda}_{u,m}^2}{\mu l^2} + \frac{k_u}{2\mu}} = \frac{\pi}{l}\sqrt{\frac{AE}{\mu}}(2m-1)\sqrt{\frac{3}{8}} \qquad (3)$$

For the given member properties

$$\omega_1 = 1622.435\ \text{rad/s}, \qquad \omega_2 = 4867.305\ \text{rad/s}, \qquad \omega_3 = 8112.375\ \text{rad/s}, \qquad \ldots$$

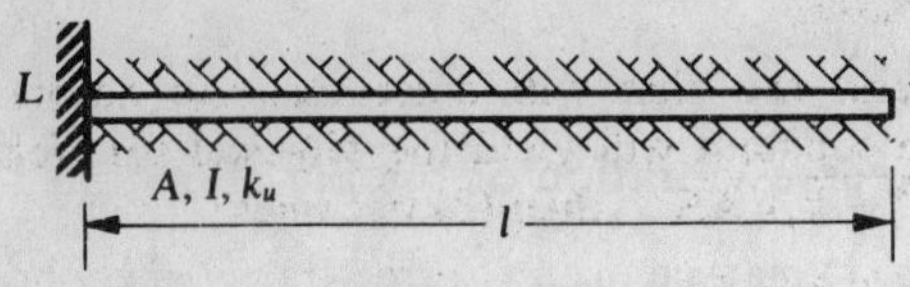

Fig. P-7.3

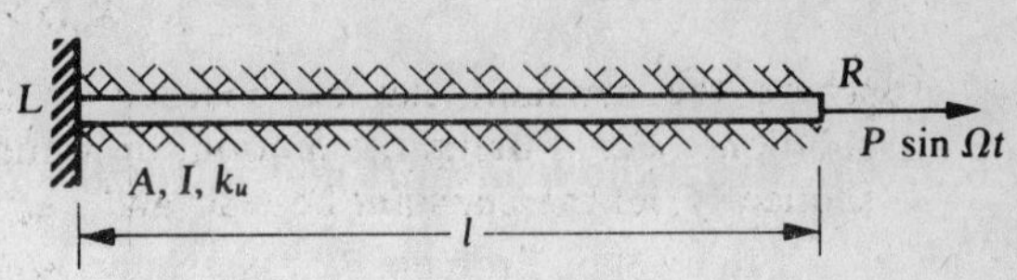

Fig. P-7.4

7.4. Find the displacement and stress responses for the steady state vibration of the beam shown in Fig. P-7.4. Let $k_u = \mu\Omega^2/2$, $\Omega = 330.822$ rad/s, $A = 232.26\ \text{cm}^2$, $E = 206.84\ \text{GN/m}^2$, $\gamma = 76.973\ \text{kN/m}^3$, $l = 12.19$ m, and $P = 177.93$ kN.

Comparing Problem 7.1 with Problem 6.1 shows that the motion equation of longitudinal vibration, whether the vibrating number is in an elastic medium or not, takes the same form with different shape parameters λ_u. Consequently, the transport matrix of forced vibration derived in Problem 6.10 may be directly used here. For the given case of concentrated load (similar to Problem 6.11) we write

$$\begin{bmatrix} 1 \\ \hline \bar{u}_R \\ 0 \end{bmatrix} = \left[\begin{array}{c|cc} 1 & 0 & 0 \\ \hline 0 & \cos\bar{\lambda}_u & \sin\bar{\lambda}_u \\ \bar{\Phi}'_{RLu} & -\sin\bar{\lambda}_u & \cos\bar{\lambda}_u \end{array}\right] \begin{bmatrix} 1 \\ \hline 0 \\ \bar{N}_L \end{bmatrix} \qquad (1)$$

where $\bar{\Phi}'_{RLu} = -P/AE\lambda_u$ and $\bar{\lambda}_u = \lambda_u l$, $\lambda_u = \sqrt{(\mu\Omega^2 - k_u)/AE}$. From (1),

$$\bar{u}_R = \bar{N}_L \sin\bar{\lambda}_u \qquad \text{and} \qquad \bar{\Phi}'_{RLu} + \bar{N}_L\cos\bar{\lambda}_u = 0 \qquad (2)$$

Since $\bar{\lambda}_u = 0.5536$, $\bar{\Phi}'_{RLu} = -8.156 \times 10^{-4}$ m, $\cos\bar{\lambda}_u = 0.8506$, and $\sin\bar{\lambda}_u = 0.5258$, we have

$$N_L = 209.16\ \text{kN} \qquad \text{and} \qquad u_R = 0.005\,041\ \text{m} \qquad (3)$$

7.5. Derive the differential equation of transverse vibration (7-4) of a finite straight prismatic bar on an elastic foundation of resistance $k_v v$ (Fig. 7-2). Find the solution of the differential equation by using Laplace transforms.

Referring to Fig. 7-4, we write the equilibrium equation of vertical forces and the sum of moments as

$$\mu\frac{\partial^2 v}{\partial t^2} = \frac{\partial V}{\partial x} - k_v v \qquad \frac{\partial M}{\partial x} = -V \qquad (1, 2)$$

Substituting (*2*) into (*1*) and using $EI(\partial^2 v/\partial x^2) = M$ gives

$$EI\frac{\partial^4 v}{\partial x^4} + k_v v + \mu\frac{\partial^2 u}{\partial t^2} = 0 \tag{3}$$

which is (*7-4*). Now using $v(x,t) = X_v(x)Y_v(t)$ in (*3*) and rearranging,

$$\frac{EI}{\mu}\frac{X^{iv}(x)}{X_v(x)} + \frac{k_v}{\mu} = -\frac{Y_v''(t)}{Y_v(t)} \tag{4}$$

which must be equal to a constant, say ω^2. Let $\lambda_v^4 = (\mu\omega^2 - k_v)/EI$; then (*4*) becomes

$$X_v^{iv}(x) - \lambda_v^4 X_v(x) = 0 \qquad \text{and} \qquad Y_v''(t) + \omega^2 Y_v(t) = 0 \tag{5, 6}$$

Note that (*5, 6*) are exactly the same as (*10, 11*) of Problem 6.12 of free transverse vibration of a bar without elastic foundation. Consequently the solutions of (*5, 6*) corresponding to the *m*th mode may be similarly obtained as

$$X_{v,m}(x) = A_{v,m}a_{v,m}(x) + B_{v,m}b_{v,m}(x) + C_{v,m}c_{v,m}(x) + D_{v,m}d_{v,m}(x) \tag{7}$$

$$Y_{v,m}(t) = E_{v,m}\cos\omega_m t + F_{v,m}\sin\omega_m t \tag{8}$$

in which $\lambda_{v,m} = \sqrt[4]{(\mu\omega_m^2 - k_v)/EI}$ and the other notations are those of (*6-24*).

7.6. Construct the transport matrix for the free transverse vibration of the bar shown in Fig. 7-2.

From (*7*) of Problem 7.5, the shape function $X_{v,m}(x)$ may be used for free transverse vibration by using the parameter $\lambda_v = \sqrt[4]{(\mu\omega^2 - k_v)/EI}$ which takes account of the foundation stiffness. Thus the derivation of the transport matrix is exactly the same as shown in Problem 6.15 and summarized in (*6-31*) and (*6-32*).

7.7. Using the transport matrix (*6-31*) find the angular frequencies of the beam shown in Fig. P-7.7. Let $l = 40$ ft, $I = 243\ \text{in}^4$, $A = 36\ \text{in}^2$, $\gamma = 490\ \text{lb/ft}^3$, $k_v = 250.273\,EI/l^4$, and $E = 30\,000$ ksi.

Using the boundary conditions of the beam of Fig. P-7.7 in (*6-31*) gives

$$\begin{bmatrix} 0 \\ 0 \\ \bar{M}_{R,m} \\ \bar{V}_{R,m} \end{bmatrix} = \begin{bmatrix} a_{v,m} & b_{v,m} & c_{v,m} & d_{v,m} \\ d_{v,m} & a_{v,m} & b_{v,m} & c_{v,m} \\ c_{v,m} & d_{v,m} & a_{v,m} & b_{v,m} \\ b_{v,m} & c_{v,m} & d_{v,m} & a_{v,m} \end{bmatrix} \begin{bmatrix} 0 \\ 0 \\ \bar{M}_{L,m} \\ \bar{V}_{L,m} \end{bmatrix} \tag{1}$$

from which the frequency equation may be formulated as

$$\text{Det.}\begin{bmatrix} c_{v,m} & d_{v,m} \\ b_{v,m} & c_{v,m} \end{bmatrix} = -\cosh\bar{\lambda}_{v,m}\cos\bar{\lambda}_{v,m} + 1 = 0 \tag{2}$$

By trials, (*2*) gives $\bar{\lambda}_{v,1} = 4.730$, $\bar{\lambda}_{v,2} = 7.853$, $\bar{\lambda}_{v,3} = 10.996$, $\bar{\lambda}_{v,4} = 14.137$. Since

$$\omega_m = \sqrt{\frac{\bar{\lambda}_{v,m}^4 EI}{l^4\mu} + \frac{k_v}{\mu}}$$

the first three angular frequencies (rad/s) are $\omega_1 = 73.06$, $\omega_2 = 149.80$, $\omega_3 = 280.48$.

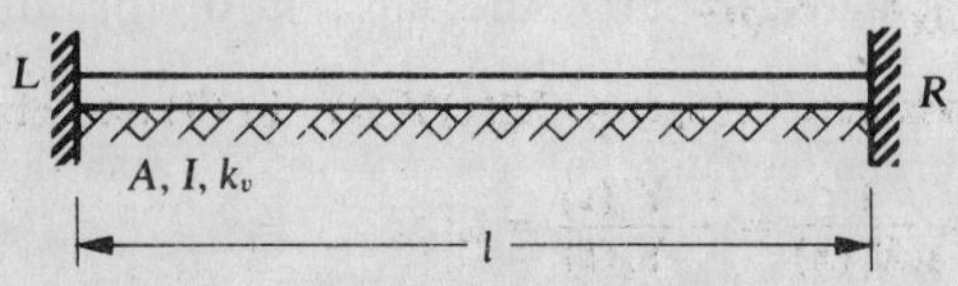

Fig. P-7.7

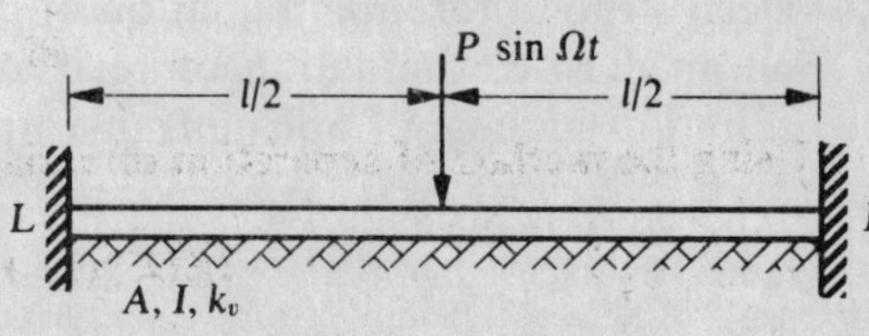

Fig. P-7.8

7.8. Find the moments and shears at both ends of the beam shown in Fig. P-7.8 subject to a steady state vibration caused by the applied force $P \sin \Omega t$. Assume $P = 40$ k, $\Omega = 36.53$ rad/s, and the member properties to be the same as those given in Problem 7.7.

It is evident from Problems 7.6 and 7.7 that the transport matrix derived in Problem 6.22 may be directly used here by changing the parameter λ_v to include the elastic media. Using the Dirac delta function, we write

$$\frac{d^4 v(x)}{dx^4} - \lambda_v^4 v(x) = \frac{P}{EI} \mathscr{D}\left(x - \frac{l}{2}\right) \tag{1}$$

Following Problem 6.23 yields this load function:

$$\Phi_v(x) = \frac{P}{2EI\lambda_v^3}\left[\sinh \lambda_v \left(x - \frac{l}{2}\right) - \sin \lambda_v \left(x - \frac{l}{2}\right)\right] \tag{2}$$

from which

$$\bar{\Phi}_{RLv} = \frac{P}{2EI\lambda_v^3}\left[\sinh\left(\frac{\lambda_v l}{2}\right) - \sin\left(\frac{\lambda_v l}{2}\right)\right] \qquad \bar{\Phi}'_{RLv} = \frac{P}{2EI\lambda_v^3}\left[\cosh\left(\frac{\lambda_v l}{2}\right) - \cos\left(\frac{\lambda_v l}{2}\right)\right]$$

$$\bar{\Phi}''_{RLv} = \frac{P}{2EI\lambda_v^3}\left[\sinh\left(\frac{\lambda_v l}{2}\right) + \sin\left(\frac{\lambda_v l}{2}\right)\right] \qquad \bar{\Phi}'''_{RLv} = \frac{P}{2EI\lambda_v^3}\left[\cosh\left(\frac{\lambda_v l}{2}\right) + \cos\left(\frac{\lambda_v l}{2}\right)\right]$$

The transport matrix is

$$\begin{bmatrix} 1 \\ \hline 0 \\ 0 \\ \bar{M}_R \\ \bar{V}_R \end{bmatrix} = \left[\begin{array}{c|cccc} 0 & 0 & 0 & 0 & 0 \\ \hline \bar{\Phi}_{RLv} & a_v & b_v & c_v & d_v \\ \bar{\Phi}'_{RLv} & d_v & a_v & b_v & c_v \\ \bar{\Phi}''_{RLv} & c_v & d_v & a_v & b_v \\ \bar{\Phi}'''_{RLv} & b_v & c_v & d_v & a_v \end{array}\right] \begin{bmatrix} 1 \\ \hline 0 \\ 0 \\ \bar{M}_L \\ \bar{V}_L \end{bmatrix} \tag{3}$$

Using the given data, we calculate

$\lambda_v = 0.08415$ $\quad a_v = 6.761$ $\quad b_v = 7.120$ $\quad c_v = 7.736$ $\quad d_v = 7.343$

$\bar{\Phi}_{RLv} = 1.0635$ $\quad \bar{\Phi}'_{RLv} = 1.920$ $\quad \bar{\Phi}''_{RLv} = 2.3811$ $\quad \bar{\Phi}'''_{RLv} = 1.7713$

From (3), $\bar{M}_L = 0.7763$, $\bar{M}_R = 0.7763$, $\bar{V}_L = -0.963$, and $\bar{V}_R = 0.963$. Thus the responses are $M_L = M_R = 278.28$ ft-k, $V_L = 29.04$ k, and $V_R = -29.04$ k.

BEAM-COLUMNS

7.9. Derive the differential equation of transverse vibration (7-6) of a finite straight prismatic beam-column (Fig. 7-3). Then find the solution of the differential equation.

Using the free-body diagram of Fig. 7-3,

$$\mu \frac{\partial^2 v}{\partial t^2} dx = -V + V + \frac{\partial v}{\partial x} dx \qquad \text{or} \qquad \frac{\partial v}{\partial x} = \mu \frac{\partial^2 v}{\partial t^2} \tag{1}$$

$$-M + \mu \frac{\partial^2 v}{\partial t^2} dx \left(\frac{dx}{2}\right) + V\,dx + M + \frac{\partial M}{\partial x} dx + N \frac{\partial v}{\partial x} dx = 0 \qquad \text{or} \qquad V = -\frac{\partial M}{\partial x} - N \frac{\partial v}{\partial x} \tag{2}$$

where in (2) the second order term is neglected. Using $\partial^2 v/\partial x^2 = M/EI$ in (2), which is then substituted into (1), gives (7-6) as

$$EI \frac{\partial^4 v}{\partial x^4} + N \frac{\partial^2 v}{\partial x^2} + \mu \frac{\partial^2 v}{\partial t^2} = 0 \tag{3}$$

Using the method of separation of variables, we assume $v(x, t) = X_v(x) Y_v(t)$ in (3) and obtain

$$[EIX_v^{\text{iv}}(x) + NX''_v(x)] \frac{1}{\mu X_v(x)} = -\frac{Y''_v(t)}{Y_v(t)} = \omega^2 \tag{4}$$

where ω is a constant.

From (*4*),

$$X_v^{iv}(x) + \frac{N}{EI}X_v''(x) - \frac{\mu\omega^2}{EI}X_v(x) = 0 \qquad Y_v''(t) + \omega_v^2(t) = 0 \tag{5, 6}$$

Applying Laplace transforms to (*5*), we write

$$\mathscr{L}\{X_v(s)\}\left(s^4 + \frac{N}{EI}s^2 - \frac{\mu\omega^2}{EI}\right) - \left(s^3 + \frac{N}{EI}s\right)X_v(+0) - \left(s^2 + \frac{N}{EI}\right)X_v'(+0) - s^2X_v''(+0) - X_v'''(+0) = 0 \tag{7}$$

Let $\alpha_1 = \sqrt{-\frac{N}{2EI} + \sqrt{\left(\frac{N}{2EI}\right)^2 + \frac{\mu\omega^2}{EI}}} \qquad \alpha_2 = \sqrt{\frac{N}{2EI} + \sqrt{\left(\frac{N}{2EI}\right)^2 + \frac{\mu\omega^2}{EI}}}$ (*8, 9*)

Then (*7*) becomes

$$\mathscr{L}\{X_v(s)\} = \frac{1}{\alpha_1^2 + \alpha_2^2}\left(\frac{1}{s^2 - \alpha_1^2} - \frac{1}{s^2 + \alpha_2^2}\right)$$
$$\times\left[s\left(s^2 + \frac{N}{EI}\right)X_v(+0) + \left(s^2 + \frac{N}{EI}\right)X_v'(+0) + sX_v''(+0) + X_v'''(+0)\right] \tag{10}$$

from which

$$\begin{aligned} X_v(x) = \frac{1}{\alpha_1^2 + \alpha_2^2}\Bigg\{&\left[\left(\alpha_1^2 + \frac{N}{EI}\right)X_v(+0) + X_v''(+0)\right]\cosh\alpha_1 x \\ &+ \left[\left(\alpha_1 + \frac{N}{EI\alpha_1}\right)X_v'(+0) + \frac{1}{\alpha_1}X_v'''(+0)\right]\sinh\alpha_1 x \\ &+ \left[\left(\alpha_2^2 - \frac{N}{EI}\right)X_v(+0) - X_v''(+0)\right]\cos\alpha_2 x \\ &+ \left[\left(\alpha_2 - \frac{N}{\alpha_2 EI}\right)X_v'(+0) - \frac{1}{\alpha_2}X_v'''(+0)\right]\sin\alpha_2 x\Bigg\} \end{aligned} \tag{11}$$

Since $\alpha_1^2 + \frac{N}{EI} = \alpha_2^2$ and $\alpha_2^2 - \frac{N}{EI} = \alpha_1^2$, (*11*) may be written as (*7-7*) with symbolic notations given in (*7-10*).

7.10. Derive the transport matrix for the shape function (*11*) of Problem 7.9.

Using (*7-11*) in (*11*) of Problem 7.9 gives

$$\begin{aligned} X_v(x) = \lambda\Bigg[&(\alpha_2^2\cosh\alpha_1 x + \alpha_1^2\cos\alpha_2 x)v_L + \left(\frac{\beta'}{\alpha_1}\sinh\alpha_1 x + \frac{\beta}{\alpha_2}\sin\alpha_2 x\right)\theta_L + (\cosh\alpha_1 x - \cos\alpha_2 x)\frac{M_L}{EI} \\ &- \left(\frac{1}{\alpha_1}\sinh\alpha_1 x - \frac{1}{\alpha_2}\sin\alpha_2 x\right)\frac{V_L}{EI}\Bigg] \end{aligned} \tag{1}$$

where $\lambda = \frac{1}{\alpha_1^2 + \alpha_2^2} \qquad \beta = \alpha_1^2 - \frac{N}{EI} \qquad \beta' = \alpha_2^2 + \frac{N}{EI}$

Thus $X_v'(x)$, $X_v''(x)$, and $X_v'''(x)$ may be easily obtained from (*1*). For the equation of shear, we use (*7-11*) as

$$\begin{aligned} -\frac{V}{EI} &= X_v'''(x) - \frac{N}{EI}X_v'(x) \\ &= \lambda\Big[(\alpha_2^2\alpha_1\beta\sinh\alpha_1 x + \alpha_1^2\alpha_2\beta'\sin\alpha_2 x)v_L + (\beta\beta'\cosh\alpha_1 x - \beta\beta'\cos\alpha_2 x)\theta_L \\ &\quad + (\alpha_1\beta\sinh\alpha_1 x - \alpha_2\beta'\sin\alpha_2 x)\frac{M_L}{EI} - (\beta\beta'\cosh\alpha_1 x + \beta'\cos\alpha_2 x)\frac{V_L}{EI}\Big] \end{aligned} \tag{2}$$

The boundary conditions at the right end are

$$v_R = X_v(l) \qquad \theta_R = X_v'(l) \qquad \frac{M_R}{EI} = X_v''(l) \qquad -\frac{V_R}{EI} = X_v'''(l) - \frac{N}{EI}X_v'(l)$$

Thus the transport matrix is

$$\begin{bmatrix} v_R \\ \theta_R \\ \dfrac{M_R}{EI} \\ \dfrac{-V_R}{EI} \end{bmatrix} = \lambda \begin{bmatrix} \alpha_2^2 \text{cs1} + \alpha_1^2 \text{cs2} & \dfrac{\beta' \text{sn1}}{\alpha_1} + \dfrac{\beta \text{sn2}}{\alpha_2} & \text{cs1} - \text{cs2} & \dfrac{\text{sn1}}{\alpha_1} - \dfrac{\text{sn2}}{\alpha_2} \\ \alpha_1\alpha_2(\alpha_2 \text{sn1} - \alpha_1 \text{sn2}) & \beta' \text{cs1} + \beta \text{cs2} & \alpha_1 \text{sn1} + \alpha_2 \text{sn2} & \text{cs1} - \text{cs2} \\ \alpha_1^2\alpha_2^2(\text{cs1} - \text{cs2}) & \beta'\alpha_1 \text{sn1} - \beta\alpha_2 \text{sn2} & \alpha_1^2 \text{cs1} + \alpha_2^2 \text{cs2} & \alpha_1 \text{sn1} + \alpha_2 \text{sn2} \\ \alpha_1\alpha_2(\alpha_2\beta \text{sn1} + \alpha_1\beta' \text{sn2}) & \beta\beta'(\text{cs1} - \text{cs2}) & \alpha_1\beta \text{sn1} - \alpha_2\beta' \text{sn2} & \beta \text{cs1} + \beta' \text{cs2} \end{bmatrix} \begin{bmatrix} v_L \\ \theta_L \\ \dfrac{M_L}{EI} \\ \dfrac{-V_L}{EI} \end{bmatrix} \tag{3}$$

where $\text{sn1} = \sinh \alpha_1 l$, $\text{sn2} = \sin \alpha_2 l$, $\text{cs1} = \cosh \alpha_1 l$, $\text{cs2} = \cos \alpha_2 l$. For any mode m, we use $\alpha_{1,m}$ and $\alpha_{2,m}$ in (3).

7.11. Using the transport matrix (3) of Problem 7.10, find the angular frequencies of the beam shown in Fig. P-7.11. Assume the member properties to be EI, l, and μ.

Applying the boundary conditions $v_L = v_R = M_L = M_R = 0$ in (3) of Problem 7.10 gives

$$\begin{bmatrix} 0 \\ 0 \end{bmatrix} = \lambda \begin{bmatrix} \dfrac{\beta' \text{sn1}}{\alpha_1} + \dfrac{\beta \text{sn2}}{\alpha_2} & \dfrac{\text{sn1}}{\alpha_1} - \dfrac{\text{sn2}}{\alpha_2} \\ \beta'\alpha_1 \text{sn1} - \beta\alpha_2 \text{sn2} & \alpha_1 \text{sn1} + \alpha_2 \text{sn2} \end{bmatrix} \begin{bmatrix} \theta_L \\ \dfrac{-V_L}{EI} \end{bmatrix} \tag{1}$$

The frequency equation becomes

$$\lambda \text{ sn1 sn2} [(\beta + \beta')(\alpha_1^2 + \alpha_2^2)]/\alpha_1\alpha_2 = 0$$

Since $\lambda \neq 0$, $\alpha_1\alpha_2 \neq 0$, $\text{sn1} \neq 0$, $(\beta + \beta')(\alpha_1^2 + \alpha_2^2) \neq 0$, we have

$$\text{sn2} = 0 \qquad \text{or} \qquad \alpha_2 l = n\pi \tag{2}$$

From (7-8),

$$\frac{\mu\omega_m^2}{EI} = \alpha_2^4 - \alpha_2^2 \frac{N}{EI} \qquad \text{or} \qquad \omega_m^2 = \frac{EI}{\mu}\left(\frac{m\pi}{l}\right)^4 \left(1 - \frac{N}{N_{cr}}\right) \tag{3}$$

where $N_{cr} = (m\pi)^2 EI/l^2$ is defined as the Euler buckling load. If N reaches the value of N_{cr}, then $\omega_m^2 = 0$, no vibration exists, and static instability will occur. If N is tensile, the terms in the last parentheses of (3) will add and the frequency will increase.

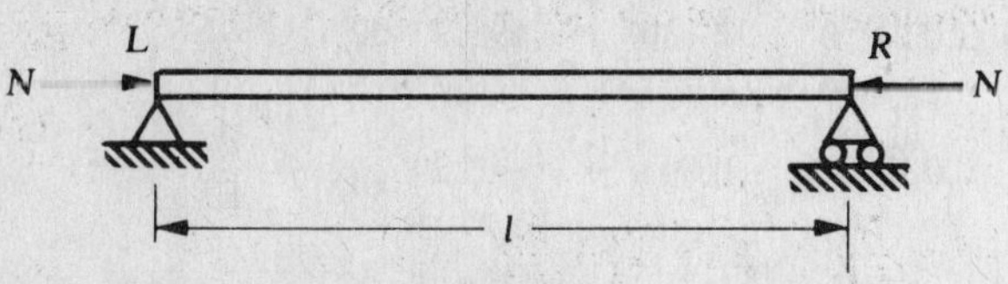

Fig. P-7.11

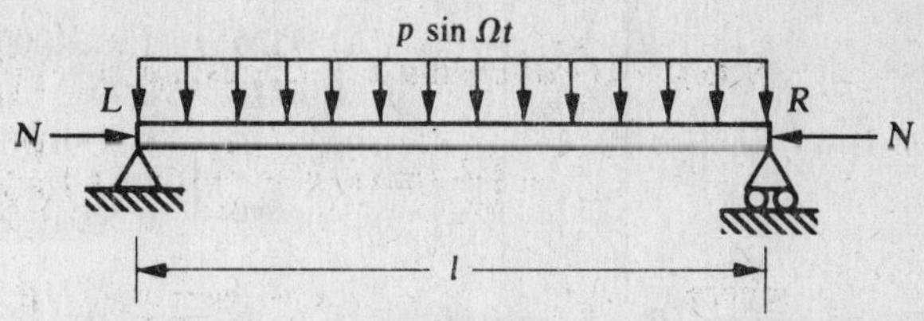

Fig. P-7.12

7.12. Using the transport matrix find the responses of the steady state vibration of the beam shown in Fig. P-7.12. Given: $l = 40$ ft, $A = 36$ in^2, $I = 243$ in^4, $\gamma = 490$ lb/ft^3, $p = 40$ lb/ft, $\Omega = (\pi^2/2l^2)\sqrt{EI/\mu}$, and $N = \pi^2 EI/4l^2$.

For the purpose of finding the load function $\Phi_v(x)$, let us study the following differential equation for the general harmonic force $p(x, t) = p(x) \sin \Omega t$:

$$X_v^{\text{iv}}(x) + \frac{N}{EI} X_v''(x) - \frac{\mu\omega^2}{EI} x_v(x) = \frac{p(x)}{EI} \tag{1}$$

of which the homogeneous solution and its associated transport matrix were shown in Problems 7.9 and 7.10 respectively.

The particular solution of (*1*) may be obtained as

$$\mathscr{L}\{x_v(s)\} = \frac{1}{\alpha_1^2+\alpha_2^2}\left(\frac{1}{s^2-\alpha_1^2} - \frac{1}{s^2+\alpha_2^2}\right)\frac{1}{EI}\mathscr{L}\{p(s)\} \tag{2}$$

The inverse of (*2*) is

$$\Phi_v(x) = \frac{1}{EI(\alpha_1^2+\alpha_2^2)}\int_0^x \left[\frac{1}{\alpha_1}\sinh\alpha_1(x-\tau) - \frac{1}{\alpha_2}\sin\alpha_2(x-\tau)\right]p(\tau)\,d\tau \tag{3}$$

For $p(x) = p$,

$$\Phi_v(x) = \frac{p}{EI(\alpha_1^2+\alpha_2^2)}\left[\frac{1}{\alpha_1^2}(\cosh\alpha_1 x - 1) + \frac{1}{\alpha_2^2}(\cos\alpha_2 x - 1)\right] \tag{4}$$

$$\Phi_v'(x) = \frac{p}{EI(\alpha_1^2+\alpha_2^2)}\left(\frac{1}{\alpha_1}\sinh\alpha_1 x - \frac{1}{\alpha_2}\sin\alpha_2 x\right) \tag{5}$$

$$\Phi_v''(x) = \frac{p}{EI(\alpha_1^2+\alpha_2^2)}(\cosh\alpha_1 x - \cos\alpha_2 x) \tag{6}$$

$$\Phi_v'''(x) = \frac{p}{EI(\alpha_1^2+\alpha_2^2)}(\alpha_1\sinh\alpha_1 x + \alpha_2\sin\alpha_2 x) \tag{7}$$

Knowing (*4*), (*5*), (*6*), and (*7*) above and (*3*) of Problem 7.10, we may write

$$\begin{bmatrix} 1 \\ \hline v_R \\ \theta_R \\ \dfrac{M_R}{EI} \\ \dfrac{-V_R}{EI} \end{bmatrix} = \left[\begin{array}{c|cccc} 1 & 0 & 0 & 0 & 0 \\ \hline \Phi_v(l) & a_{11} & a_{12} & a_{13} & a_{14} \\ \Phi_v'(l) & a_{21} & a_{22} & a_{23} & a_{24} \\ \Phi_v''(l) & a_{31} & a_{32} & a_{33} & a_{34} \\ \Phi_v'''(l) - \dfrac{l\Phi_v'(l)}{EI} & a_{41} & a_{42} & a_{43} & a_{44} \end{array}\right] \begin{bmatrix} 1 \\ \hline v_L \\ \theta_L \\ \dfrac{M_L}{EI} \\ \dfrac{-V_L}{EI} \end{bmatrix} \tag{8}$$

where a_{ij} represents the element at ith row and jth column in (*3*) of Problem 7.10. Using the given member properties we calculate $\Omega = 11.25$ rad/s, $N = 78.068$ k, $\alpha_1 = 0.04906$, $\alpha_2 = 0.06284$, $\lambda = 157.3597$, $\beta = 0.0008639$, $\beta' = 0.005491$, cs1 = 3.6284, cs2 = −0.8092, sn1 = 3.4879, and sn2 = 0.5875. Thus (*8*) becomes

$$\begin{bmatrix} 1 \\ \hline 0 \\ \theta_R \\ 0 \\ \dfrac{-V_R}{EI} \end{bmatrix} = \left[\begin{array}{c|cccc} 1 & 0 & 0 & 0 & 0 \\ \hline 0.0788500 & 1.9481000 & 62.700000 & 698.29900 & 9716.14300 \\ 0.0076810 & 0.0923400 & 3.024900 & 32.73080 & 698.29900 \\ 0.0005520 & 0.0066360 & 0.142830 & 0.87146 & 32.73080 \\ 0.0000259 & 0.0001686 & 0.002104 & -0.00863 & -0.20614 \end{array}\right] \begin{bmatrix} 1 \\ \hline 0 \\ \theta_L \\ 0 \\ \dfrac{-V_L}{EI} \end{bmatrix} \tag{9}$$

From (*9*) $\theta_L = 0.004185$ rad, $\theta_R = -0.004185$ rad, $V_L = 1.78$ k, and $V_R = -1.78$ k.

TIMOSHENKO'S BEAMS

7.13. Derive the differential equation of transverse vibration (*7-14*) of a finite prismatic bar assuming shear and bending deformations. Then find the general integrals of the differential equation by using Laplace transforms.

Let the total deflection v be composed of bending deflection v_1 and shear deflection v_2. Then the shear and moment may be expressed as

$$V = A^*G\frac{\partial v_2}{\partial x} \qquad M = EI\frac{\partial^2 v_1}{\partial x^2} \tag{1, 2}$$

where $A^* = kA$, k is shape factor, and $G(\partial v_2/\partial x)$ represents the average shear stress. As discussed previously, when we consider transverse vibration, the equilibrium equations of vertical forces and moments are respectively

$$\frac{\partial V}{\partial x} = \mu\frac{\partial^2 v}{\partial t^2} \qquad \frac{\partial M}{\partial x} = -V \tag{3, 4}$$

Thus substituting (1) into (3) and combining (1), (2), and (4) yields (5, 6) respectively

$$\frac{\partial^2 v_2}{\partial x^2} = \frac{\mu}{A^*G}\frac{\partial^2 v}{\partial t^2} \qquad EI\frac{\partial^3 v_1}{\partial x^3} + A^*G\frac{\partial v_2}{\partial x} = 0 \tag{5, 6}$$

Using $v = v_1 + v_2$, the fourth derivative of (6) gives

$$EI\frac{\partial^4 v}{\partial x^4} - EI\frac{\partial^4 v_2}{\partial x^4} + A^*G\frac{\partial^2 v_2}{\partial x^2} = 0 \tag{7}$$

Finally, substituting (5) into (7) yields (7-14) as

$$EI\frac{\partial^4 v}{\partial x^4} - \frac{EI\mu}{A^*G}\frac{\partial^4 v}{\partial x^2 \partial t^2} + \mu\frac{\partial^2 v}{\partial t^2} = 0 \tag{8}$$

Let $v(x, t) = X_v(x)Y_v(t)$ be substituted into (8) which may then be written as

$$\frac{-X_v^{iv}(x)}{\dfrac{\mu}{A^*G}X_v''(x) - \dfrac{\mu}{EI}X_v(x)} = -\frac{Y_v''(t)}{Y_v(t)} = \omega^2 \tag{9}$$

from which

$$X_v^{iv}(x) + \frac{\mu}{A^*G}\omega^2 X_v''(x) - \frac{\mu\omega^2}{EI}X_v(x) = 0 \tag{10}$$

$$Y_v''(t) + \omega^2 Y_v(t) = 0 \tag{11}$$

Note that (10) is similar to (5) of Problem 7.9. Thus the solution of (10) is formally identical to (7-7) with the shape parameters shown in (7-15). The solution of (11) is formally identical to (9) of Problem 7.1.

7.14. **Derive the differential equation of transverse vibration (7-16) of a finite straight prismatic bar assuming bending deformation and rotatory inertia. Find the solution of the differential equation.**

For the transverse vibration of bending deformation only, each cross section of a vibrating member rotates through an angle $\partial v_1/\partial x$. Thus the horizontal deformations above and below the neutral axis may be expressed as $z(\partial v_1/\partial x)$, where z is the fiber coordinate measured from the neutral axis. For an element $(dx)(dA)$, the magnitude of the inertia force is

$$\frac{\gamma}{g}(dA)(dx)(z)\frac{\partial^3 v_1}{\partial x \partial t^2} \tag{1}$$

where $z(\partial^3 v_1/\partial x \partial t^2)$ is the acceleration of the horizontal fibers. The resultant effect of the inertia forces on a cross section will cause a time-dependent couple as

$$dM = \frac{\gamma}{g}\int_A (dA)(dx)(z^2)\frac{\partial^3 v_1}{\partial x \partial t^2} = \frac{\gamma}{g}\frac{\partial^3 v_1}{\partial x \partial t^2}\left(\int_A z^2\,dA\right)dx \tag{2}$$

Since $\int_A z^2\,dA = I$, (2) becomes

$$\frac{\partial M}{\partial x} = \left(\frac{\gamma I}{g}\right)\frac{\partial^3 v_1}{\partial x \partial t^2} \tag{3}$$

which is the shear due to rotatory inertia. To include the regular shear, we write

$$\frac{\partial M}{\partial x} = -V + \left(\frac{\gamma I}{g}\right)\frac{\partial^3 v_1}{\partial x \partial t^2} \tag{4}$$

With $\dfrac{\partial^2 v_1}{\partial x^2} = \dfrac{M}{EI}$, $\dfrac{\partial V}{\partial x} = \mu \dfrac{\partial^2 v_1}{\partial t^2}$, $\dfrac{\gamma I}{g} = \mu r^2$, and $r^2 = \dfrac{I}{A}$, (4) may be rewritten as

$$EI\frac{\partial^4 v_1}{\partial x^4} - \mu r^2 \frac{\partial^4 v_1}{\partial x^2 \partial t^2} + \mu \frac{\partial^2 v_1}{\partial t^2} = 0 \tag{5}$$

which is identical to (*7-16*) and formally identical to (*7-14*).

Substituting $v_1(x,t) = X_v(x)Y_v(t)$ into (*5*), we have

$$\frac{-X_v^{\text{iv}}(x)}{\dfrac{\mu r^2}{EI}X_v''(x) - \dfrac{\mu}{EI}X_v(x)} = \frac{-Y_v''(t)}{Y_v(t)} = \omega^2 \tag{6}$$

Thus
$$X_v^{\text{iv}}(x) + \frac{\mu r^2 \omega^2}{EI}X_v''(x) - \frac{\mu\omega^2}{EI}X_v(x) = 0 \qquad Y_v''(t) + \omega^2 Y_v(t) = 0 \tag{7, 8}$$

Comparing (*7*) with (*5*) of Problem 7.9 or (*10*) of Problem 7.13 clearly reveals that the solution of (*7*) is formally identical to (*7-7*). However, the shape parameters should be

$$\alpha_{1,m} = \sqrt{-\frac{\mu r^2\omega_m^2}{2EI} + \sqrt{\frac{\mu\omega_m^2}{EI} + \left(\frac{\mu r^2\omega_m^2}{2EI}\right)^2}} \qquad \alpha_{2,m} = \sqrt{\frac{\mu r^2\omega_m^2}{2EI} + \sqrt{\frac{\mu\omega_m^2}{EI} + \left(\frac{\mu r^2\omega_m^2}{2EI}\right)^2}} \tag{9, 10}$$

7.15. **Derive the differential equation of transverse vibration (*7-17*) of a finite straight prismatic bar assuming rotatory inertia, shear, and bending deformations. Find the solution of the differential equation.**

From Problems 7.13 and 7.14 we have

$$V = A^*G\left(\frac{\partial v}{\partial x} - \frac{\partial v_1}{\partial x}\right) \qquad \frac{\partial M}{\partial x} = -V + \mu r^2 \frac{\partial^3 v_1}{\partial x \partial t^2} \tag{1, 2}$$

where (*1*) is the shear equation and (*2*) is obtained from the equilibrium equation of moments. Using $M = EI(\partial^2 v_1/\partial x^2)$ and (*1*) in (*2*) gives

$$EI\frac{\partial^3 v_1}{\partial x^3} + A^*G\left(\frac{\partial v}{\partial x} - \frac{\partial v_1}{\partial x}\right) - \mu r^2 \frac{\partial^3 v_1}{\partial x \partial t^2} = 0 \tag{3}$$

Substituting the equilibrium equation of shear $\dfrac{\partial V}{\partial x} = \mu \dfrac{\partial^2 v}{\partial x^2}$ into (*1*) results in

$$\mu\frac{\partial^2 v}{\partial t^2} - A^*G\left(\frac{\partial^2 v}{\partial x^2} - \frac{\partial^2 v_1}{\partial x^2}\right) = 0 \qquad \text{or} \qquad \frac{\partial^4 v_1}{\partial x^4} = -\frac{\mu}{A^*G}\frac{\partial^4 v}{\partial x^2 \partial t^2} + \frac{\partial^4 v}{\partial x^4} \tag{4, 5}$$

The substitution of (*4, 5*) into (*3*) leads to the final differential equation (*7-17*) as

$$\frac{\partial^4 v}{\partial x^4} - \left(\frac{\mu}{A^*G} + \frac{\mu r^2}{EI}\right)\frac{\partial^4 v}{\partial x^2 \partial t^2} + \frac{\mu}{EI}\frac{\partial^2 v}{\partial t^2} + \frac{\mu^2 r^2}{A^*GEI}\frac{\partial^4 v}{\partial t^4} = 0 \tag{6}$$

Let $v(x,t) = X_v(x)Y_v(t)$; $Y(t)$ is sinusoidal and may be expressed in terms of $e^{i\omega t}$. Thus (*6*) becomes

$$X_v^{\text{iv}}(x) + \omega^2\left(\frac{\mu}{A^*G} + \frac{\mu r^2}{EI}\right)X_v''(x) - \frac{\mu\omega^2}{EI}X_v(x) + \frac{\mu^2 r^2 \omega^4}{A^*GEI}X_v(x) = 0 \tag{7}$$

$$\beta_1 = \frac{\mu\omega^2}{A^*G} \qquad \beta_2 = \frac{\mu r^2\omega^2}{EI} \qquad \beta_3 = \frac{\mu\omega^2}{EI} \tag{8}$$

Substituting into (*7*), we have

$$X_v^{\text{iv}}(x) + (\beta_1 + \beta_2)X_v''(x) - (\beta_3 - \beta_1\beta_2)X_v(x) = 0 \tag{9}$$

which is similar to (*5*) of Problem 7.9. Consequently, (*7-7*) is the solution of (*9*), having the integration constants formally identical to (*7-10*) with the following shape parameters:

$$\alpha_{1,2,m} = \sqrt{\pm\frac{\beta_{1,m} + \beta_{2,m}}{2} + \sqrt{\left(\frac{\beta_{1,m} - \beta_{2,m}}{2}\right)^2 + \beta_{3,m}}} \tag{10}$$

7.16. Derive the transport matrix based on (*7-17*).

The solution of (*9*) of the preceding problem is

$$X_v(x) = \frac{1}{\alpha_1^2 + \alpha_2^2}\Big\{[\alpha_1^2 \cosh \alpha_1 x + \alpha_2^2 \cos \alpha_2 x + (\beta_1 + \beta_2)(\cosh \alpha_1 x - \cos \alpha_2 x)]X_v(+0)$$
$$+ \left[\alpha_1 \sinh \alpha_1 x + \alpha_2 \sin \alpha_2 x + (\beta_1 + \beta_2)\left(\frac{1}{\alpha_1}\sinh \alpha_1 x - \frac{1}{\alpha_2}\sin \alpha_2 x\right]X_v'(+0)\right.$$
$$+ (\cosh \alpha_1 x - \cos \alpha_2 x)X_v''(+0) + \left(\frac{1}{\alpha_1}\sinh \alpha_1 x - \frac{1}{\alpha_2}\sin \alpha_2 x\right)X_v'''(+0)\Big\} \qquad (1)$$

From (*7-19*),

$$\frac{M(x)}{EI} = X_v''(x) + \beta_1 X_v(x) \qquad -\frac{V(x)}{EI} = \frac{1}{1-\beta_1 r^2}[X_v'''(x) + (\beta_1+\beta_2)X_v'(x)] \qquad (2, 3)$$

$$\psi(x) = X_v'(x) - \frac{EI}{A^*G}\frac{V(x)}{EI} \qquad (4)$$

Substituting $x = l$ into (*1*), (*2*), (*3*), and (*4*) yields the following transport matrix:

$$\begin{bmatrix} v_R \\ \psi_R \\ \dfrac{M_R}{EI} \\ \dfrac{-V_R}{EI} \end{bmatrix} = \begin{bmatrix} a_{11} & a_{12} & a_{13} & a_{14} \\ a_{21} & a_{22} & a_{23} & a_{24} \\ a_{31} & a_{32} & a_{33} & a_{34} \\ a_{41} & a_{42} & a_{43} & a_{44} \end{bmatrix} \begin{bmatrix} v_L \\ \psi_L \\ \dfrac{M_L}{EI} \\ \dfrac{-V_L}{EI} \end{bmatrix} \qquad (5)$$

where $a_{11} = \phi_1 \cosh \alpha_1 l + \phi_2 \cos \alpha_2 l$

$a_{12} = \alpha_1 \sinh \alpha_1 l + \alpha_2 \sin \alpha_2 l$

$a_{13} = \cosh \alpha_1 l - \cos \alpha_2 l$

$$a_{14} = \frac{\sinh \alpha_1 l}{\alpha_1} - \frac{\sin \alpha_2 l}{\alpha_2} - \frac{EI}{A^*G}\left(\frac{\phi_1}{\alpha_1}\sinh \alpha_1 l + \frac{\phi_2}{\alpha_2}\sin \alpha_2 l\right)$$

$a_{21} = \phi_1\tau_1 \sinh \alpha_1 l - \phi_2\tau_2 \sin \alpha_2 l$

$a_{22} = \alpha_1\tau_1 \cosh \alpha_1 l + \alpha_2\tau_2 \cos \alpha_2 l$

$a_{23} = \tau_1 \sinh \alpha_1 l + \tau_2 \sin \alpha_2 l$

$$a_{24} = \tau_3 \cosh \alpha_1 l + \tau_4 \cos \alpha_2 l - (\phi_1\tau_3 \cosh \alpha_1 l - \phi_2\tau_4 \cos \alpha_2 l)\frac{EI}{A^*G}$$

$a_{31} = (\alpha_1^2\phi_1 + \beta_1^2 + \beta_1\beta_2)\cosh \alpha_1 l - (\alpha_2^2\phi_2 + \beta_1^2 + \beta_1\beta_2)\cos \alpha_2 l$

$a_{32} = \alpha_1\phi_2 \sinh \alpha_1 l - \alpha_2\phi_1 \sin \alpha_2 l$

$a_{33} = \phi_2 \cosh \alpha_1 l + \phi_1 \cos \alpha_2 l$

$$a_{34} = \frac{\phi_2}{\alpha_1}\sinh \alpha_1 l + \frac{\phi_1}{\alpha_2}\sin \alpha_2 l - \left[\left(\alpha_1\alpha_2^2 + \frac{\beta_1\beta_2}{\alpha_2}\right)\sinh \alpha_1 l - \left(\alpha_1^2\alpha_2 + \frac{\beta_1\beta_2}{\alpha_2}\right)\sin \alpha_2 l\right]\frac{EI}{A^*G}$$

$a_{41} = c_2\alpha_1\alpha_2^2\phi_1 \sinh \alpha_1 l + c_2\alpha_2\alpha_1^2\phi_2 \sin \alpha_2 l$

$a_{42} = c_2\alpha_1^2\alpha_2^2 \cosh \alpha_1 l - c_2\alpha_2^2\alpha_1^2 \cosh \alpha_2 l$

$a_{43} = c_2\alpha_1\alpha_2^2 \sinh \alpha_1 l - c_2\alpha_2\alpha_1^2 \sin \alpha_2 l$

$$a_{44} = c_2(\alpha_2^2 \cosh \alpha_1 l + \alpha_1^2 \cos \alpha_2 l) - (\alpha_2^2\phi_1 \cosh \alpha_1 l - \phi_2\alpha_1^2 \cos \alpha_2 l)\frac{EIc_2}{A^*G}$$

in which $\phi_1 = \alpha_1^2 + \beta_2 \qquad \tau_1 = \alpha_1 + c_1\alpha_1\alpha_2^2 \qquad c_1 = EI/(1-\beta_1 r^2)GA^*$

$\phi_2 = \alpha_2^2 - \beta_2 \qquad \tau_2 = \alpha_2 - c_1\alpha_2\alpha_1^2 \qquad c_2 = 1/(1-\beta_1 r^2)$

$\tau_3 = 1 + c_1\alpha_2^2$

$\tau_4 = c_1\alpha_1^2 - 1$

Four typical end conditions of this matrix equation are defined in the closing statement of Section 7.4 as Eq. (7-19). The fixed end and fixed axis conditions are also shown graphically in Fig. P-7.16.

(a) Fixed End (b) Fixed Axis

Fig. P-7.16

7.17. Find the first three angular frequencies of a simple beam for (a) bending deformations only, (b) bending deformations and rotatory inertia, (c) bending and shear deformations, and (d) bending and shear deformations and rotatory inertia. Let the member properties be $A = 36\text{ in}^2$, $l = 7\text{ ft}$, $I = 243\text{ in}^4$, $\gamma = 490\text{ lbs/ft}^3$, $E = 30\,000\text{ ksi}$, $G = 11\,500\text{ ksi}$, and $k = 2/3$.

Applying the boundary conditions $v_R = v_L = M_R = M_L = 0$ to (5) of Problem 7.16 gives

$$\begin{vmatrix} a_{12} & a_{14} \\ a_{32} & a_{34} \end{vmatrix} = 0 \tag{1}$$

from which the frequency equation is

$$(\alpha_1 \sinh \alpha_1 l + \alpha_2 \sin \alpha_2 l)\left\{\left[\left(\alpha_1\alpha_2^2 + \frac{\beta_1^2 + \beta_1 + \beta_2}{\alpha_1}\right)\sinh \alpha_1 l - \left(\alpha_1^2\alpha_2 + \frac{\beta_1^2 + \beta_1 + \beta_2}{\alpha_2}\right)\sin \alpha_2 l\right]\left(-\frac{EI}{A^*G}\right) + \frac{\phi_2}{\alpha_1}\sinh \alpha_1 l + \frac{\phi_1}{\alpha_2}\sin \alpha_2 l\right\} = 0 \tag{2}$$

(a) For bending deformations only,

$\beta_1 = \beta_2 = 0, \quad \beta_3 \neq 0$

Then (2) becomes

$$\sin (\sqrt[4]{\beta_3})^l = 0 \qquad \text{or} \qquad \sqrt[4]{\beta_3} = m\pi/l \tag{3}$$

Thus
$$\omega = (m\pi/l)^2\sqrt{EI/\mu} \tag{4}$$

(b) For bending deformations and rotatory inertia,

$\beta_1 = 0, \quad \beta_2 \neq 0, \quad \beta_3 \neq 0$

Then (2) becomes

$$\sin \alpha_2 l = 0 \qquad \text{or} \qquad \alpha_2 l = m\pi \tag{5}$$

Thus
$$\frac{\beta_2}{2} + \sqrt{\left(\frac{\beta_2}{2}\right)^2 + \beta_3} = \left(\frac{m\pi}{l}\right)^2 \qquad \text{or} \qquad \omega^2 = \left(\frac{m\pi}{l}\right)^4 \bigg/ \left[\frac{\mu}{EI} + \frac{\mu r^2}{EI}\left(\frac{m\pi}{l}\right)^2\right] \tag{6}$$

(c) For bending and shear deformations,

$\beta_2 = 0, \quad \beta_1 \neq 0, \quad \beta_3 \neq 0$

From (5),

$$\frac{\beta_1}{2} + \sqrt{\left(\frac{\beta_1}{2}\right)^2 + \beta_3} = \left(\frac{m\pi}{l}\right)^2 \qquad \text{or} \qquad \omega^2 = \left(\frac{m\pi}{l}\right)^4 \bigg/ \left[\frac{\mu}{EI} + \frac{\mu}{GA^*}\left(\frac{m\pi}{l}\right)^2\right] \tag{7}$$

(d) For bending and shear deformations and rotatory inertia,

$\beta_1 \neq 0, \quad \beta_2 \neq 0, \quad \beta_3 \neq 0$

Then (5) yields

$$\frac{\beta_1 + \beta_2}{2} + \sqrt{\left(\frac{\beta_1 - \beta_2}{2}\right)^2 + \beta_3} = \left(\frac{m\pi}{l}\right)^2$$

or
$$\left\{\frac{\mu r^2}{EIA^*G}\right\}\omega^4 - \left[\frac{\mu}{EI} + \left(\frac{\mu}{GA^*} + \frac{\mu r^2}{EI}\right)\frac{m^2\pi^2}{l^2}\right]\omega^2 + \left(\frac{m\pi}{l}\right)^4 = 0 \tag{8}$$

The frequencies obtained from (4), (6), (7), and (8) are summarized below.

mth mode	Angular frequencies, ω, rad/s			
	Case (a)	Case (b)	Case (c)	Case (d)
1st	734.58	731.10	721.57	717.937
2nd	2938.95	2884.34	2743.21	2704.74
3rd	6611.21	6347.00	5728.70	5594.51

Supplementary Problems

RESISTING MEDIUM

7.18. Using the transport matrix, find the first three fundamental longitudinal frequencies of the beam shown in Fig. P-7.18. Given: $l = 20$ ft, $A = 36\text{ in}^2$, $\gamma = 490\text{ lbs/ft}^3$, $E = 30\,000$ ksi, and $k_u = AE\pi^2/8l^2$.

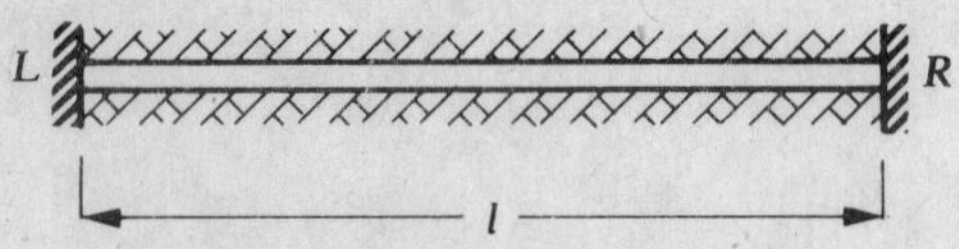

Fig. P-7.18

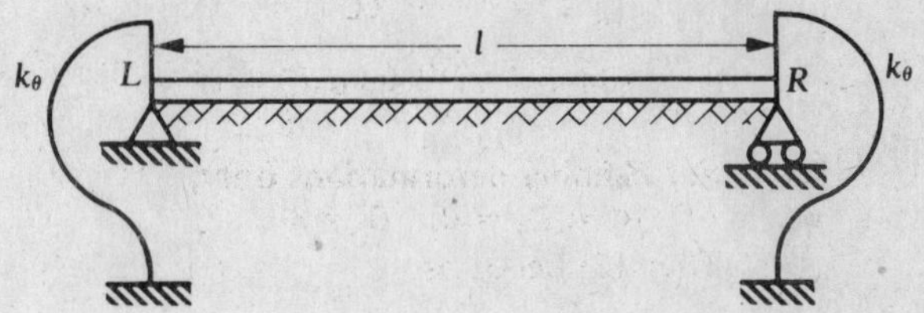

Fig. P-7.19

7.19. Determine the transverse frequencies of the beam shown in Fig. P-7.19 by using $k_\theta = 5EI/l$, $l = 40$ ft, $I = 243\text{ in}^4$, $A = 36\text{ in}^2$, $\gamma = 490\text{ lbs/ft}^3$, $k_v = 250.273\,EI/l^4$, and $E = 30\,000$ ksi.

7.20. Using the transport matrix, find the responses at both ends of the beam shown in Fig. P-7.20 with the member properties given in Problem 7.19, $p = 40$ lbs/ft and $\Omega = 36.53$ rad/s.

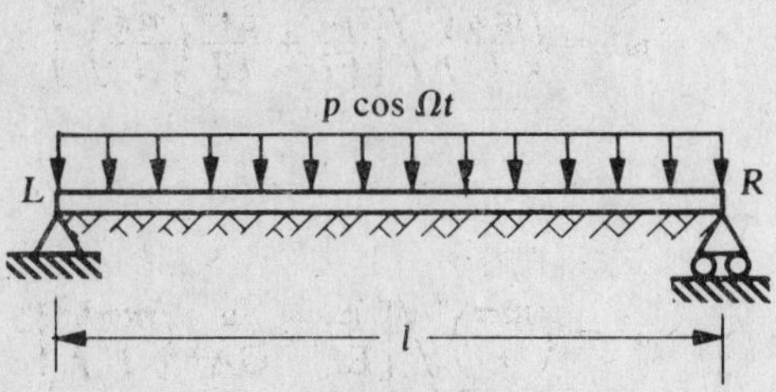

Fig. P-7.20

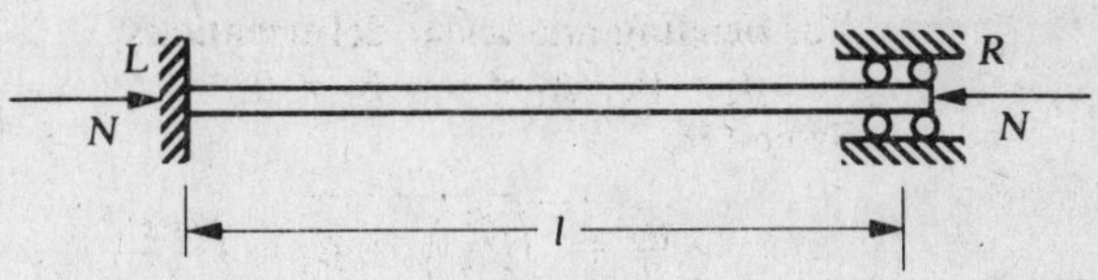

Fig. P-7.21

BEAM-COLUMNS

7.21. Calculate the first three transverse frequencies of the beam-column shown in Fig. P-7.21. Given: $l = 40$ ft, $A = 36\text{ in}^2$, $I = 243\text{ in}^4$, $\gamma = 490\text{ lbs/ft}^3$, and $N = \pi^2 EI/5l^2$.

7.22. Find the first three transverse frequencies of the member shown in Fig. P-7.22. Given: $l = 40$ ft, $A = 36\text{ in}^2$, $I = 243\text{ in}^4$, $\gamma = 490\text{ lbs/ft}^3$, $k_v = 150EI/l^4$, and $N = \pi^2 EI/4l^2$.

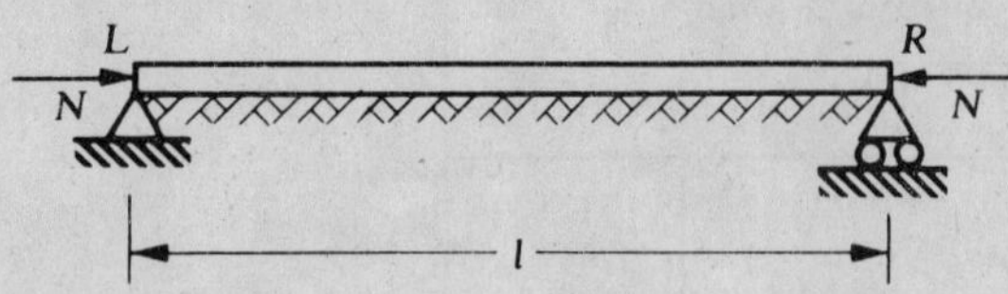

Fig. P-7.22

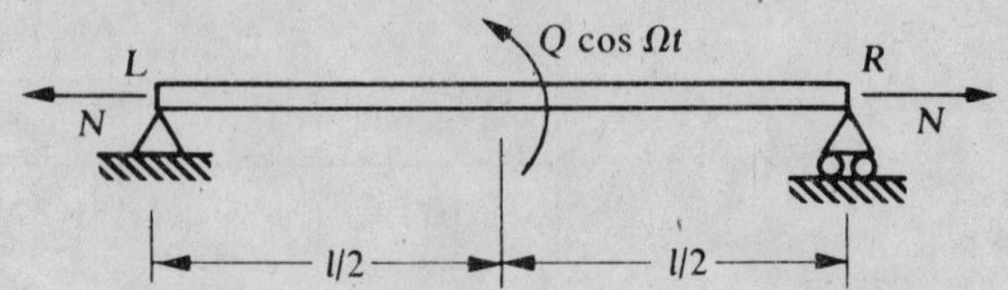

Fig. P-7.23

7.23. Determine the end rotations and reactions of the beam shown in Fig. P-7.23. Given: $Q = 40$ ft-k, $\Omega = 40$ rad/s, $N = \pi^2 EI/4l^2$, $l = 40$ ft, $A = 36\text{ in}^2$, $I = 243\text{ in}^4$, and $\gamma = 490\text{ lbs/ft}^3$.

TIMOSHENKO'S BEAMS

7.24. Find the first three fundamental frequencies of a fixed-end beam by the Timoshenko theory and then compare the solution with that obtained by the Bernoulli–Euler theory. Given: $l = 7$ ft, $A = 36\text{ in}^2$, $I = 243\text{ in}^4$, $\gamma = 490\text{ lbs/ft}^3$, $E = 30\,000$ ksi, and $G = 11\,500$ ksi.

7.25. Calculate the first three fundamental frequencies of a free-end beam by the Timoshenko theory for the member properties given in Problem 7.24. Compare the solution with that obtained by considering bending deformation only.

Chapter 8

Dynamic Transport Method

8.1 TRANSPORT MATRIX

Dimensionless Matrices

The *dimensionless dynamic transport matrix* connecting the dimensionless state vectors $\bar{H}_R^l$ and $\bar{H}_L^l$ is obtained by matrix composition of the component transport matrices (*6-12*) and (*6-31*) or (*6-13*) and (*6-32*) as shown in Table 8.1.

In *symbolic form,*

$$\bar{H}^l_{R,m} = \bar{U}^l_{RL,m}\bar{H}^l_{L,m} \qquad \bar{H}^l_{L,m} = \bar{U}^l_{LR,m}\bar{H}^l_{R,m} \tag{8-1, 2}$$

where $\bar{H}^l_{R,m}$ and $\bar{H}^l_{L,m}$ are the dimensionless state vectors of the right and left ends respectively and $\bar{U}^l_{RL,m}$ and $\bar{U}^l_{LR,m}$ are the dimensionless dynamic transport matrices, all in the mth mode and in the l-system.

These *transport matrices,* $\bar{U}^l_{RL,m}$ and $\bar{U}^l_{LR,m}$, are the *distributed mass equivalents* of their lumped mass counterparts $\bar{L}_{RL,m}$ and $\bar{L}_{LR,m}$, given in (*5-1, 2*). They are characteristic for the given bar and are independent of their end conditions.

Their product is

$$\bar{U}^l_{RL,m}\bar{U}^l_{LR,m} = I \tag{8-3}$$

which shows their *inverse relationship.*

Since the inverse of the matrix is obtained by reversing the sign of l in each dynamic parameter of the respective matrix, the inverse is said to be equal to the *reverse.* This statement is true for all transport matrices.

Dimensioning and Scaling

The *absolute state vectors* H_R^l and H_L^l are related to their dimensionless counterparts as

$$H^l_{R,m} = \kappa_{H,m}\bar{H}^l_{R,m} \qquad H^l_{L,m} = \kappa_{H,m}\bar{H}^l_{L,m} \tag{8-4, 5}$$

where

$$\kappa_{H,m} = \left[\begin{array}{ccc|ccc} EA\bar{\lambda}_{u,m}/l & & & & & \\ & -EI\lambda^3_{v,m}/l^3 & & & 0 & \\ & & EI\bar{\lambda}^2_{v,m}/l^2 & & & \\ \hline & & & 1 & & \\ & 0 & & & 1 & \\ & & & & & \bar{\lambda}_{v,m}/l \end{array}\right] = \left[\begin{array}{c|c} \kappa_{\sigma,m} & 0 \\ \hline 0 & \kappa_{\Delta,m} \end{array}\right] \tag{8-6}$$

is called the *dimensioning matrix.*

The *dimensionless state vectors* $\bar{H}^l_R$ and $\bar{H}^l_L$ are related to their absolute counterparts as

$$\bar{H}^l_{R,m} = \lambda_{H,m} H^l_{R,m} \qquad \bar{H}^l_{L,m} = \lambda_{H,m} H^l_{L,m} \tag{8-7, 8}$$

where

$$\lambda_{H,m} = \left[\begin{array}{ccc|ccc} l/EA\bar{\lambda}_{u,m} & & & & & \\ & -l^3/EI\bar{\lambda}^3_{v,m} & & & 0 & \\ & & l^2/EI\bar{\lambda}^2_{v,m} & & & \\ \hline & & & 1 & & \\ & 0 & & & 1 & \\ & & & & & l/\bar{\lambda}_{v,m} \end{array}\right] = \left[\begin{array}{c|c} \lambda_{\sigma,m} & 0 \\ \hline 0 & \lambda_{\Delta,m} \end{array}\right] \tag{8-9}$$

is called the *scaling matrix.*

The comparison of (*8-6*) and (*8-9*) reveals that

$$\kappa_{H,m}\lambda_{H,m} = I \tag{8-10}$$

which facilitates the subsequent transformations (Problem 8.1).

Table 8-1 Dimensionless Dynamic Transport Matrix

(*a*) Standard Matrix Form, $\bar{U}_{RL}$

$$\begin{bmatrix} \bar{N}_R \\ \bar{V}_R \\ \bar{M}_R \\ \hline \bar{u}_R \\ \bar{v}_R \\ \bar{\theta}_R \end{bmatrix} = \left[\begin{array}{ccc|ccc} a_u & 0 & 0 & -d_u & 0 & 0 \\ 0 & a_v & d_v & 0 & b_v & c_v \\ 0 & b_v & a_v & 0 & c_v & d_v \\ \hline d_u & 0 & 0 & a_u & 0 & 0 \\ 0 & d_v & c_v & 0 & a_v & b_v \\ 0 & c_v & b_v & 0 & d_v & a_v \end{array}\right] \begin{bmatrix} \bar{N}_L \\ \bar{V}_L \\ \bar{M}_L \\ \hline \bar{u}_L \\ \bar{v}_L \\ \bar{\theta}_L \end{bmatrix}$$

(*b*) Standard Matrix Form, $\bar{U}_{LR}$

$$\begin{bmatrix} \bar{N}_L \\ \bar{V}_L \\ \bar{M}_L \\ \hline \bar{u}_L \\ \bar{v}_L \\ \bar{\theta}_L \end{bmatrix} = \left[\begin{array}{ccc|ccc} a_u & 0 & 0 & d_u & 0 & 0 \\ 0 & a_v & -d_v & 0 & -b_v & c_v \\ 0 & -b_v & a_v & 0 & c_v & -d_v \\ \hline -d_u & 0 & 0 & a_u & 0 & 0 \\ 0 & -d_v & c_v & 0 & a_v & -b_v \\ 0 & c_v & -b_v & 0 & -d_v & a_v \end{array}\right] \begin{bmatrix} \bar{N}_R \\ \bar{V}_R \\ \bar{M}_R \\ \hline \bar{u}_R \\ \bar{v}_R \\ \bar{\theta}_R \end{bmatrix}$$

(*c*) Submatrix Form, $\bar{U}_{RL}$

$$\underbrace{\begin{bmatrix} \bar{\sigma}_R \\ \hline \bar{\Delta}_R \end{bmatrix}}_{\bar{H}_R} = \underbrace{\left[\begin{array}{c|c} \bar{U}_{\sigma\sigma,RL} & \bar{U}_{\sigma\Delta,RL} \\ \hline \bar{U}_{\Delta\sigma,RL} & \bar{U}_{\Delta\Delta,RL} \end{array}\right]}_{\bar{U}_{RL}} \underbrace{\begin{bmatrix} \bar{\sigma}_L \\ \hline \bar{\Delta}_L \end{bmatrix}}_{\bar{H}_L}$$

(*d*) Submatrix Form, $\bar{U}_{LR}$

$$\underbrace{\begin{bmatrix} \bar{\sigma}_L \\ \hline \bar{\Delta}_L \end{bmatrix}}_{\bar{H}_L} = \underbrace{\left[\begin{array}{c|c} \bar{U}_{\sigma\sigma,LR} & \bar{U}_{\sigma\Delta,LR} \\ \hline \bar{U}_{\Delta\sigma,LR} & \bar{U}_{\Delta\Delta,LR} \end{array}\right]}_{\bar{U}_{LR}} \underbrace{\begin{bmatrix} \bar{\sigma}_R \\ \hline \bar{\Delta}_R \end{bmatrix}}_{\bar{H}_R}$$

Equivalents:

$$\begin{array}{llll} a_u = \cos\bar{\lambda}_u & a_v = \tfrac{1}{2}(\cosh\bar{\lambda}_v + \cos\bar{\lambda}_v) & c_v = \tfrac{1}{2}(\cosh\bar{\lambda}_v - \cos\bar{\lambda}_v) & \bar{\lambda}_u = l\lambda_u = l\sqrt{\mu\omega^2/EA} \\ d_u = \sin\bar{\lambda}_u & b_v = \tfrac{1}{2}(\sinh\bar{\lambda}_v + \sin\bar{\lambda}_v) & d_v = \tfrac{1}{2}(\sinh\bar{\lambda}_v - \sin\bar{\lambda}_v) & \bar{\lambda}_v = l\lambda_v = l\sqrt[4]{\mu\omega^2/EI} \end{array}$$

Note: All terms in these matrices correspond to the mth mode and all vectorial quantities are in the l-system. For the origin of a_u, d_u, a_v, b_v, c_v, d_v and $\bar{\lambda}_u$, $\bar{\lambda}_v$ refer to (*6-6*), (*6-12, 13*), and (*6-14*) and (*6-25*), (*6-31, 32*), and (*6-34*).

Absolute Matrices

The *absolute transport matrix equations* (*8-1, 2*) in terms of (*8-6*) and (*8-9*) are

$$H^{l}_{R,m} = \underbrace{\kappa_{H,m}\bar{U}^{l}_{RL,m}\lambda_{H,m}}_{U^{l}_{RL,m}}H^{l}_{L,m} \qquad H^{l}_{L,m} = \underbrace{\kappa_{H,m}\bar{U}^{l}_{LR,m}\lambda_{H,m}}_{U^{l}_{LR,m}}H^{l}_{R,m} \qquad (8\text{-}11, 12)$$

These matrices have all the characteristics of $\bar{U}^{l}_{RL}$ and $\bar{U}^{l}_{LR}$ and consequently

$$U^{l}_{RL,m}U^{l}_{LR,m} = I \qquad (8\text{-}13)$$

Angular Transformations

Since the state vectors of the respective ends are not always in the bar system the angular transformation matrices must be employed in (*8-1, 2*) and (*8-11, 12*).

Thus (*8-1, 2*) become

$$\bar{H}^{R}_{R,m} = \underbrace{\lambda_{H,m}\pi^{Rl}\kappa_{H,m}\bar{U}^{l}_{RL,m}\lambda_{H,m}\pi^{lL}\kappa_{H,m}}_{\bar{U}^{RL}_{RL,m}}\bar{H}^{L}_{L,m} \qquad (8\text{-}14)$$

$$\bar{H}^{L}_{L,m} = \underbrace{\lambda_{H,m}\pi^{Ll}\kappa_{H,m}\bar{U}^{l}_{LR,m}\lambda_{H,m}\pi^{lR}\kappa_{H,m}}_{\bar{U}^{LR}_{LR,m}}\bar{H}^{R}_{R,m} \qquad (8\text{-}15)$$

and (*8-11, 12*) become

$$H^{R}_{R,m} = \underbrace{\pi^{Rl}U^{l}_{RL,m}\pi^{lL}}_{U^{RL}_{RL,m}}H^{L}_{L,m} \qquad H^{L}_{L,m} = \underbrace{\pi^{Ll}U^{l}_{LR,m}\pi^{lR}}_{U^{LR}_{LR,m}}H^{R}_{R,m} \qquad (8\text{-}16, 17)$$

where

$$\pi^{Rl} = \left[\begin{array}{c|c} \pi^{Rl} & 0 \\ \hline 0 & \pi^{Rl} \end{array}\right]_{[6\times 6]} \qquad \pi^{Ll} = \left[\begin{array}{c|c} \pi^{Ll} & \\ \hline & \pi^{Ll} \end{array}\right]_{[6\times 6]} \qquad (8\text{-}18, 19)$$

These angular transformation matrices are compositions of submatrices π^{Rl} and π^{Ll} and π^{lR} and π^{lL} defined by (*1-3, 4*). Their coefficients are constants based on the angle of rotation (Problem 8.2).

The new transport matrices $\bar{U}^{RL}_{RL,m}$, $\bar{U}^{LR}_{LR,m}$, $U^{RL}_{RL,m}$, and $U^{LR}_{LR,m}$ are called general transport matrices, the application of which is shown in Sec. 8.4.

8.2 SERIES EXPANSION

Concept

The closed form of the transport matrix coefficients may always be expressed by a series, which can be a *coefficient series* or a *matrix series.*

Coefficient Series

The *dynamic parameters* can be expanded into the respective MacLaurin's series about $\bar{\lambda}_u = 0$ and $\bar{\lambda}_v = 0$ respectively as

$$\begin{aligned}
a_u &= \cos\bar{\lambda}_u = \bar{\gamma}_{u,0} - \bar{\gamma}_{u,2} + \bar{\gamma}_{u,4} - \cdots \\
d_u &= \sin\bar{\lambda}_u = \bar{\gamma}_{u,1} - \bar{\gamma}_{u,3} + \bar{\gamma}_{u,5} - \cdots \\
a_v &= \tfrac{1}{2}(\cosh\bar{\lambda}_v + \cos\bar{\lambda}_v) = \bar{\gamma}_{v,0} + \bar{\gamma}_{v,4} + \bar{\gamma}_{v,8} + \cdots \\
b_v &= \tfrac{1}{2}(\sinh\bar{\lambda}_v + \sin\bar{\lambda}_v) = \bar{\gamma}_{v,1} + \bar{\gamma}_{v,5} + \bar{\gamma}_{v,9} + \cdots \\
c_v &= \tfrac{1}{2}(\cosh\bar{\lambda}_v - \cos\bar{\lambda}_v) = \bar{\gamma}_{v,2} + \bar{\gamma}_{v,6} + \bar{\gamma}_{v,10} + \cdots \\
d_v &= \tfrac{1}{2}(\sinh\bar{\lambda}_v - \sin\bar{\lambda}_v) = \bar{\gamma}_{v,3} + \bar{\gamma}_{v,7} + \bar{\gamma}_{v,11} + \cdots
\end{aligned} \qquad (8\text{-}20)$$

where $$\bar{\gamma}_{u,n} = \frac{(\bar{\lambda}_u)^n}{n!} \qquad \bar{\gamma}_{v,n} = \frac{(\bar{\lambda}_v)^n}{n!}, \quad n = 0, 1, 2, \ldots$$

are dimensionless factors in terms of $\bar{\lambda}_u$ and $\bar{\lambda}_v$, all in the respective mode, the subscript of which is omitted (Problem 8.5).

Matrix Series

The alternate approach is based on the *homogeneous linear differential matrix equation* stating the stereo- and elasto-dynamic relations of the bar element as

$$\underbrace{\frac{d}{dx}\begin{bmatrix} N \\ -V \\ M \\ \hline u \\ v \\ \theta \end{bmatrix}}_{\frac{dH}{dx}} = \underbrace{\left[\begin{array}{ccc|ccc} 0 & 0 & 0 & -\mu\omega^2 & 0 & 0 \\ 0 & 0 & 0 & 0 & \mu\omega^2 & 0 \\ 0 & 1 & 0 & 0 & 0 & 0 \\ \hline 1/EA & 0 & 0 & 0 & 0 & 0 \\ 0 & 0 & 0 & 0 & 0 & 1 \\ 0 & 0 & 1/EI & 0 & 0 & 0 \end{array}\right]}_{D} \underbrace{\begin{bmatrix} N \\ -V \\ M \\ \hline u \\ v \\ \theta \end{bmatrix}}_{H} \tag{8-21}$$

The solution of this equation is

$$H = e^{Dx}H_0 \tag{8-22}$$

where H and H_0 are the state vectors at $x > 0$ and $x = 0$ respectively and

$$e^{Dx} = U = I + Dx + (Dx)^2/2! + (Dx)^3/3! + \cdots \tag{8-23}$$

must be the absolute transport matrix given as the *matrix power series* in $(Dx)^n/n!$, $n = 0, 1, 2, \ldots$

For $x = l$,

$$\{H_R\} = [\underbrace{I + Dl + (Dl)^2/2! + (Dl)^3/3! + \cdots}_{U_{RL}}]\{H_L\} \tag{8-24a}$$

and inversely, for $x = -l$,

$$\{H_L\} = [\underbrace{I - Dl + (Dl)^2/2! - (Dl)^3/3! + \cdots}_{U_{LR}}]\{H_R\} \tag{8-25a}$$

which can be scaled if desired (Problem 8.10) or expressed in nested parenthetic form (Ref. 8.11, p. 102) as

$$\{H_R\} = [\underbrace{I + S(I + \tfrac{1}{2}S(I + \tfrac{1}{3}S(I + \cdots)))}_{U_{RL}}]\{H_L\} \tag{8-25b}$$

$$\{H_L\} = [I - S(I - \tfrac{1}{2}S(I - \tfrac{1}{3}S(I - \cdots)))]\{H_R\} \tag{8-25b}$$

where $S = Dl$ is a matrix constant.

The application of *(8-23)* in connection with the Cayley-Hamilton theorem leads to the finite form of transport matrix construction. The real practical value of *(8-23)* is the latitude of application. As required by the definition of the problem and the imposed assumptions, the differential transport matrix *(8-21)* may accommodate readily the secondary effects (shear deformation, rotatory inertia, etc.) and will allow the elimination of the algebraic labor involved in the closed form construction of the transport matrix.

8.3 LOAD FUNCTIONS

Load Effect Matrix

The application of transport matrices (*8-1, 2*) is restricted only to the analysis of free vibrations. In cases of *forced vibration*, the respective transport matrix must be expanded to accommoderate the load effect vector.

$$\underbrace{\begin{bmatrix} 1 \\ \hline \bar{\sigma}_R \\ \bar{\Delta}_R \end{bmatrix}}_{\hat{\bar{H}}_R} = \underbrace{\left[\begin{array}{c|cc} 1 & 0 & 0 \\ \hline \bar{\Phi}_{\sigma,RL} & \bar{U}_{\sigma\sigma,RL} & \bar{U}_{\sigma\Delta,RL} \\ \bar{\Phi}_{\Delta,RL} & \bar{U}_{\Delta\sigma,RL} & \bar{U}_{\Delta\Delta,RL} \end{array}\right]}_{\hat{\bar{U}}_{RL}} \underbrace{\begin{bmatrix} 1 \\ \hline \bar{\sigma}_L \\ \bar{\Delta}_L \end{bmatrix}}_{\hat{\bar{H}}_L} \tag{8-26}$$

$$\underbrace{\begin{bmatrix} 1 \\ \hline \bar{\sigma}_L \\ \bar{\Delta}_L \end{bmatrix}}_{\hat{\bar{H}}_L} = \underbrace{\left[\begin{array}{c|cc} 1 & 0 & 0 \\ \hline \bar{\Phi}_{\sigma,LR} & \bar{U}_{\sigma\sigma,LR} & \bar{U}_{\sigma\Delta,LR} \\ \bar{\Phi}_{\Delta,LR} & \bar{U}_{\Delta\sigma,LR} & \bar{U}_{\Delta\Delta,LR} \end{array}\right]}_{\hat{\bar{U}}_{LR}} \underbrace{\begin{bmatrix} 1 \\ \hline \bar{\sigma}_R \\ \bar{\Delta}_R \end{bmatrix}}_{\hat{\bar{H}}_R} \tag{8-27}$$

where $\bar{\Phi}_{\sigma,RL}$ and $\bar{\Phi}_{\sigma,LR}$ are the dynamic stress vectors due to loads [3 × 1], and $\bar{\Phi}_{\Delta,RL}$ and $\bar{\Phi}_{\Delta,LR}$ are the dynamic displacements due to loads [3 × 1].

The load effect functions forming the components of these load effect vectors can be obtained directly from the particular integral of the differential equation of motion or by a matrix series method similar to the procedure of Section 8.2.

Matrix Series

This alternate approach, as in (*8-21*), is based on the *nonhomogeneous linear differential matrix equation* stating the stereo- and elasto-dynamic relations of the bar element as

$$\underbrace{\frac{d}{dx}\begin{bmatrix} N \\ -V \\ M \\ \hline u \\ v \\ \theta \end{bmatrix}}_{\frac{dH}{dx}} = \underbrace{\left[\begin{array}{ccc|ccc} 0 & 0 & 0 & -\mu\omega^2 & 0 & 0 \\ 0 & 0 & 0 & 0 & \mu\omega^2 & 0 \\ 0 & 1 & 0 & 0 & 0 & -\rho\omega^2 \\ \hline 1/EA & 0 & 0 & 0 & 0 & 0 \\ 0 & 0 & 0 & 0 & 0 & 1 \\ 0 & 0 & 1/EI & 0 & 0 & 0 \end{array}\right]}_{D} \underbrace{\begin{bmatrix} N \\ -V \\ M \\ \hline u \\ v \\ \theta \end{bmatrix}}_{H} + \underbrace{\begin{bmatrix} P_u(x) \\ P_v(x) \\ P_\theta(x) \\ \hline 0 \\ 0 \\ 0 \end{bmatrix}}_{W(x)} \tag{8-28}$$

The solution of this equation is

$$H = e^{Dx}H_0 + e^{Dx}\int_0^x e^{-D\tau}W(\tau)\,d\tau \tag{8-29}$$

where again

$$e^{Dx} = U(x) \qquad \text{and} \qquad e^{-Dx} = U(-x) \tag{8-30}$$

The *general integral* of (*8-28*) in terms of the above is then, for $x = l$,

$$H_R = U_{RL}H_L + \underbrace{U_{RL}\int_0^l U(-\tau)W(\tau)\,d\tau}_{\Phi_{RL}} \tag{8-31}$$

where Φ_{RL} is the absolute load effect vector and U_{RL} and U_{LR} are the absolute matrices, the matrix series of which are given in (*8-24*) and (*8-25*) respectively.

The *particular integral* Φ_{RL} in (*8-31*) is useful not only in connection with the matrix series expansion but also in cases where U_{RL} and U_{LR} are known and only the load effect vector is required (Problems 8.8 and 8.9).

The *inverse relation* is even simpler since

$$H_L = U_{LR}H_R - \underbrace{\int_0^l U(-\tau)W(\tau)\,d\tau}_{\Phi_{LR}} \tag{8-32}$$

is obtained from (*8-31*) by premultiplying both sides by U_{LR}.

8.4 TRANSPORT CHAINS

Direct Transport

As in static analysis, once the dynamic transport is available for each segment of the structure, the transport of the state vector from one end to another end of a topological chain is accomplished by matrix multiplication.

If the structure is continuous between two supports (Fig. 8-1) and *no internal conditions* (such as intermediate supports, hinges, etc.) occur within, the end state vectors relationship can be stated as

$$H_R^R = \underbrace{U_{Rk}^{Rk}U_{kj}^{kj}U_{ji}^{ji}U_{iL}^{iL}}_{U_{RL}^{RL}}H_L^L \tag{8-33}$$

where U_{RL}^{RL} is the direct transport matrix *RL*, whose superscripts identify the systems of the end state vectors, and the matrix product is called the *direct transport chain* (Problem 8.3).

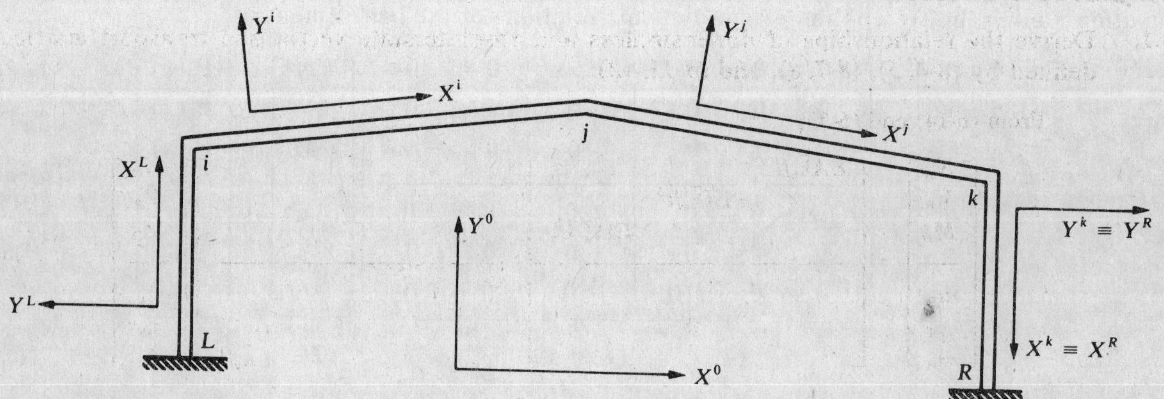

Fig. 8-1. Direct Transport Chain

Indirect Transport

When the internal conditions are imposed on the transport by the intermediate reactions, connections, or changes in mass, the joint effect matrices introduced in Sec. 5.2 must be employed. Their inclusion changes the structure of the transport flow which then is called an *indirect transport*. For the beam in Fig. 8-2,

$$H_R^0 = J_R^0U_{Rl}^0J_l^0U_{lk}^0J_k^0U_{kj}^0J_j^0U_{ji}^0J_i^0U_{iL}^0J_L^0H_L^0 \tag{8-34}$$

where all vectors remain in the 0-system (Problem 8.4).

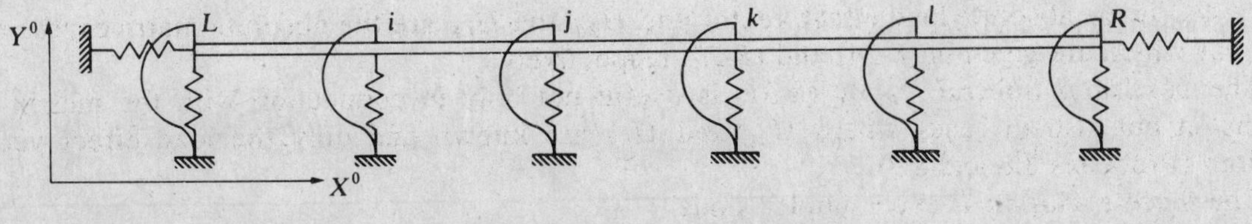

Fig. 8-2. Indirect Transport Chain

Free and Forced Motion

The *direct and indirect transport chains* (*8-28*) and (*8-29*) are applicable in the analysis of free and forced vibrations of structures with predominantly linear topology such as beams and single span and multispan one-story frames. The extension of the transport method to more complicated systems is possible but rarely employed.

In *cases of free vibrations,* the segmental transport matrices are of the U^{RL}_{RL} or $\bar{U}^{RL}_{RL}$ type, [6 × 6], and lead to the frequency determinant described in Sec. 5.2.

In *cases of forced harmonic vibrations,* the segmental transport matrices are of the $\hat{U}^{RL}_{RL}$ or $\hat{\bar{U}}^{RL}_{RL}$ type, [7 × 7], and include the load effect vector.

Solved Problems

TRANSPORT MATRIX

8.1. Derive the relationships of dimensionless and absolute state vector and transport matrices defined by (*8-4, 5*), (*8-7, 8*), and (*8-11, 12*).

From (*6-14*) and (*6-33*)

$$\begin{bmatrix} N_R \\ V_R \\ M_R \\ \hline u_R \\ v_R \\ \theta_R \end{bmatrix} = \left[\begin{array}{ccc|ccc} EA\bar{\lambda}_u/l & & & & & \\ & -EI\bar{\lambda}_v^3/l^3 & & & 0 & \\ & & EI\bar{\lambda}_v^2/l^2 & & & \\ \hline & & & 1 & & \\ & 0 & & & 1 & \\ & & & & & \bar{\lambda}_v/l \end{array}\right] \begin{bmatrix} \bar{N}_R \\ \bar{V}_R \\ \bar{M}_R \\ \hline \bar{u}_R \\ \bar{v}_R \\ \bar{\theta}_R \end{bmatrix} \tag{1}$$

which is identical to (*8-4*). Equation (*8-5*) may be similarly derived.

For (*8-7*) or (*8-8*), we may also use (*6-14*) and (*6-33*) as

$$\begin{bmatrix} \bar{N}_R \\ \bar{V}_R \\ \bar{M}_R \\ \hline \bar{u}_R \\ \bar{v}_R \\ \bar{\theta}_R \end{bmatrix} = \left[\begin{array}{ccc|ccc} l/EA\bar{\lambda}_u & & & & & \\ & -l^3/EI\bar{\lambda}_v^3 & & & 0 & \\ & & l^2/EI\bar{\lambda}_v^2 & & & \\ \hline & & & 1 & & \\ & 0 & & & 1 & \\ & & & & & l/\bar{\lambda}_v \end{array}\right] \begin{bmatrix} N_R \\ V_R \\ M_R \\ \hline u_R \\ v_R \\ \theta_R \end{bmatrix} \tag{2}$$

Equation (*8-11*) may be obtained by substituting (*8-1*) and (*8-8*) into (*8-4*) as follows:

$$H^l_{R,m} = \kappa_{H,m}\bar{H}^l_{R,m} = \kappa_{H,m}\bar{U}^l_{RL,m}\bar{H}^l_{L,m} = \kappa_{H,m}\bar{U}^l_{RL,m}\lambda_{H,m}H^l_{L,m} \tag{3}$$

Manipulating $\kappa_{H,m}\bar{U}_{RL,m}\lambda_{H,m}$ by using (*8-6*) and (*8-9*) and Table 8-1(*a*) gives

$$U^l_{RL,m} = \left[\begin{array}{ccc|ccc} a_u & 0 & 0 & -EAd_u\bar{\lambda}_{u,m}/l & 0 & 0 \\ 0 & a_v & -d_v\bar{\lambda}_{v,m}/l & 0 & -b_vEI\bar{\lambda}^3_{v,m}/l^3 & -c_vEI\bar{\lambda}^2_{v,m}/l^2 \\ 0 & -lb_v/\bar{\lambda}_{v,m} & a_v & 0 & c_vEI\bar{\lambda}^2_{v,m}/l^2 & d_vEI\bar{\lambda}_{v,m}/l \\ \hline ld_u/EA\bar{\lambda}_{u,m} & 0 & 0 & a_u & 0 & 0 \\ 0 & -d_vl^3/EI\bar{\lambda}^3_{v,m} & c_vl^2/EI\bar{\lambda}^2_{v,m} & 0 & a_v & b_vl/\bar{\lambda}_{v,m} \\ 0 & -c_vl^2/EI\bar{\lambda}^2_{v,m} & b_vl/EI\bar{\lambda}_{v,m} & 0 & d_v\bar{\lambda}_{v,m}/l & a_v \end{array}\right] \tag{4}$$

8.2. Derive the transformation relationships (*8-14, 15*) and (*8-16, 17*).

The absolute state vector transport matrix equations in the l system (*8-11, 12*) are

$$H^l_{R,m} = U^l_{RL,m}H^l_{L,m} \qquad H^l_{L,m} = U^l_{LR,m}H^l_{R,m} \tag{1, 2}$$

where $H^l_{R,m}$ and $H^l_{L,m}$ may be written in their respective systems as

$$H^l_{R,m} = \pi^{lR}H^R_{R,m} \qquad H^l_{L,m} = \pi^{lL}H^L_{L,m} \tag{3, 4}$$

where π^{lR} and π^{lL} are the inverses (transposes) of the orthogonal angular transformation matrices (*8-18, 19*).

Then (*1, 2*) in terms of (*3, 4*) become

$$\pi^{lR}H^R_{R,m} = U^l_{RL,m}\pi^{lL}H^L_{L,m} \qquad \pi^{lL}H^L_{L,m} = U^l_{L,m} = U^l_{LR,m}\pi^{lR}H^R_{R,m} \tag{5, 6}$$

which, after premulitplying by π^{Rl} and π^{Ll} respectively, yield (*8-16, 17*).

$$H^R_{R,m} = \pi^{Rl}U^l_{RL,m}\pi^{lL}H^L_{L,m} \qquad H^L_{L,m} = \pi^{Ll}U^l_{LR,m}\pi^{lR}H^R_{R,m} \tag{7, 8}$$

where by (*8-4, 5*)

$$H^R_{R,m} = \kappa_{H,m}\bar{H}^R_{R,m} \qquad H^L_{L,m} = \kappa_{H,m}\bar{H}^L_{L,m} \tag{9, 10}$$

and by (*8-11, 12*)

$$U^l_{RL,m} = \kappa_{H,m}\bar{U}^l_{RL,m}\lambda_{H,m} \qquad U^l_{LR} = \kappa_{H,m}\bar{U}^l_{LR}\lambda_{H,m} \tag{11, 12}$$

In terms of (*9, 10*) and (*11, 12*), (*7, 8*) become

$$\kappa_{H,m}\bar{H}^R_{R,m} = \pi^{Rl}\kappa_{H,m}\bar{U}^l_{RL,m}\lambda_{H,m}\pi^{lL}\kappa_{H,m}\bar{H}^L_{L,m} \tag{13}$$

$$\kappa_{H,m}\bar{H}^L_{L,m} = \pi^{Ll}\kappa_{H,m}\bar{U}^l_{LR,m}\lambda_{H,m}\pi^{lR}\kappa_{H,m}\bar{H}^R_{R,m} \tag{14}$$

Since by (*8-10*) $\lambda_{H,m} = \kappa^{-1}_{H,m}$, (*13*) and (*14*) reduce to (*8-14, 15*), i.e.,

$$\bar{H}^R_{R,m} = \lambda_{H,m}\pi^{Rl}\kappa_{H,m}\bar{U}^l_{RI,m}\lambda_{H,m}\pi^{lL}\kappa_{H,m}\bar{H}^L_{L,m} \tag{15}$$

$$\bar{H}^L_{L,m} = \lambda_{H,m}\pi^{Ll}\kappa_{H,m}U^l_{LR,m}\pi^{lR}\kappa_{H,m}\bar{H}^R_{R,m} \tag{16}$$

TRANSPORT CHAIN-FREE VIBRATION

8.3. Show symbolically the direct transport chains of free vibration for the systems of Fig. P-8.3.

(*a*) Using the absolute matrix we write

$$H^0_3 = U^0_{32}U^0_{21}H^0_1 \tag{1}$$

where U^0_{21} and U^0_{32} are typically shown in (*4*) of Problem 8.1 and correspond to members a and b respectively. For the dimensionless matrix we write

$$\bar{H}^0_3 = \bar{U}^0_{32}\bar{U}^0_{21}\bar{H}^0_1 \tag{2}$$

The boundary conditions are

$$\bar{H}^0_1 = [N_1 \quad V_1 \quad M_1 \quad 0 \quad 0 \quad 0]^T \qquad H^0_3 = [0 \quad 0 \quad 0 \quad u_3 \quad v_3 \quad \theta_3]^T \tag{3}$$

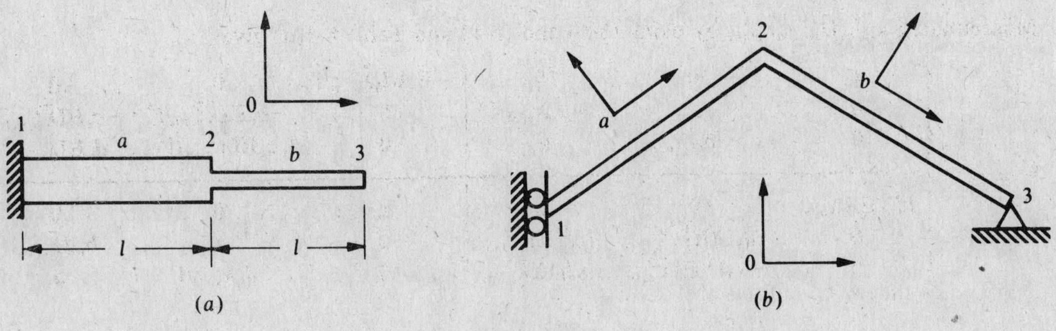

Fig. P-8.3

(b) The direct transport chain for Fig. P-8.3b may be expressed in dimensionless matrix form as

$$\bar{H}_3^0 = \lambda_{32}\pi^{0b}\kappa_{32}\bar{U}_{32}^{b}\lambda_{21}\pi^{ba}\kappa_{21}\bar{U}_{21}^{a}\lambda_{21}\pi^{a0}\kappa_{21}\bar{H}_1^0 \tag{4}$$

where

$$\pi^{0b} = \begin{bmatrix} \cos\phi^{0b} & -\sin\phi^{0b} & 0 & 0 & 0 & 0 \\ \sin\phi^{0b} & \cos\phi^{0b} & 0 & 0 & 0 & 0 \\ 0 & 0 & 1 & 0 & 0 & 0 \\ 0 & 0 & 0 & \cos\phi^{0b} & -\sin\phi^{0b} & 0 \\ 0 & 0 & 0 & \sin\phi^{0b} & \cos\phi^{0b} & 0 \\ 0 & 0 & 0 & 0 & 0 & 1 \end{bmatrix} \tag{5}$$

The matrices π^{ba} and π^{a0} are similar to (5) with proper substitution of ϕ^{ba} and ϕ^{a0}. The boundary conditions are

$$H_3^0 = [0 \quad V_3 \quad 0 \quad u_3 \quad 0 \quad \theta_3]^T \qquad H_1^0 = [N_1 \quad 0 \quad M_1 \quad 0 \quad v_1 \quad 0]^T \tag{6}$$

8.4. Show symbolically the indirect transport chains of free vibration for the systems of Fig. P-8.4a and b.

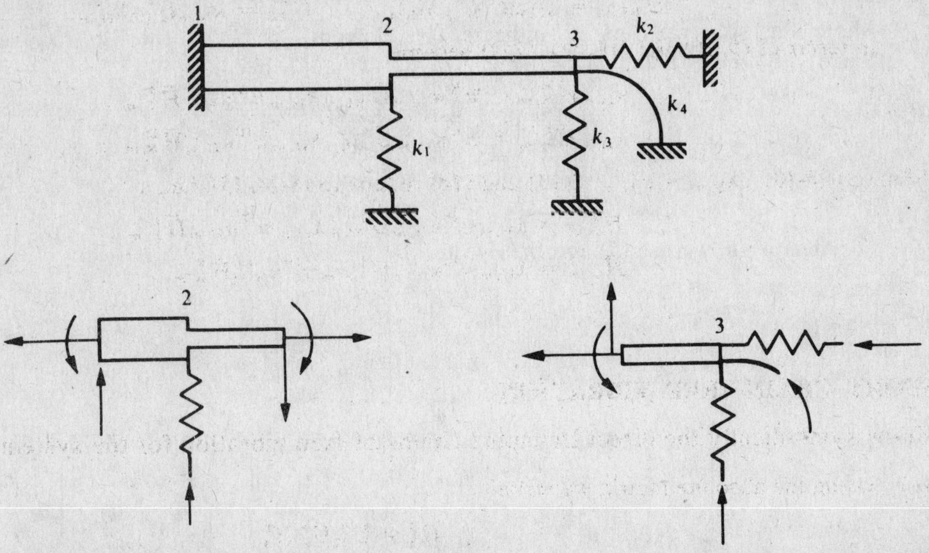

(a)

Fig. P-8.4

(a) For Fig. P-8.4a, the indirect transport chain may be expressed as

$$H_3^0 = J_3^0 U_{32}^0 J_2^0 U_{21}^0 H_1^0 \tag{1}$$

where U_{32}^0 and U_{21}^0 are formally identical to (4) of Problem 8.1; J_2^0 and J_3^0 may be established from the free-body diagrams:

$$J_2 = \begin{bmatrix} 1 & 0 & 0 & 0 & 0 & 0 \\ 0 & 1 & 0 & 0 & k_1 & 0 \\ 0 & 0 & 1 & 0 & 0 & 0 \\ 0 & 0 & 0 & 1 & 0 & 0 \\ 0 & 0 & 0 & 0 & 1 & 0 \\ 0 & 0 & 0 & 0 & 0 & 1 \end{bmatrix} \qquad J_3 = \begin{bmatrix} 1 & 0 & 0 & k_2 & 0 & 0 \\ 0 & 1 & 0 & 0 & k_3 & 0 \\ 0 & 0 & 1 & 0 & 0 & k_4 \\ 0 & 0 & 0 & 1 & 0 & 0 \\ 0 & 0 & 0 & 0 & 1 & 0 \\ 0 & 0 & 0 & 0 & 0 & 1 \end{bmatrix} \tag{2}$$

The boundary conditions are

$$H_1^0 = [N_1 \quad V_1 \quad M_1 \quad 0 \quad 0 \quad 0]^T \qquad H_3^0 = [0 \quad 0 \quad 0 \quad u_3 \quad v_3 \quad \theta_3]^T$$

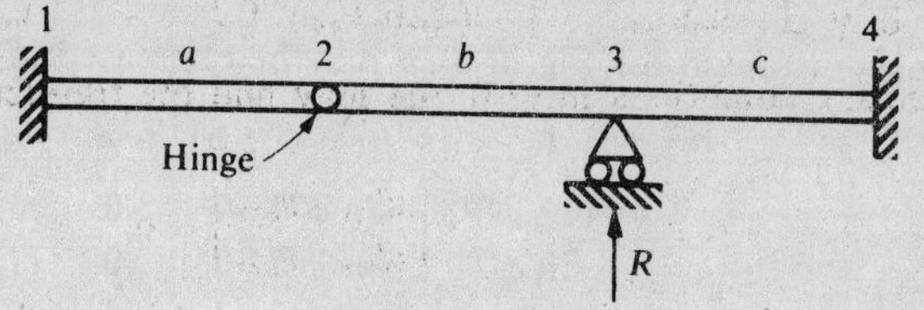

Fig. P-8.4 (*b*)

(*b*) For Fig. P-8.4*b*, let us assume that the transport matrices for the members between 1 and 2, 2 and 3, and 3 and 4 respectively, are

$$H_{21} = U_a H_{12} \qquad H_{32} = U_b H_{23} \qquad H_4 = U_c H_{34} \tag{3}$$

where U, for simplicity, has dimension $[4 \times 4]$ for bending deformation only as shown in (4) of Problem 8.1 by neglecting the elements a_u and d_u. Because of discontinuity, the matrices relating H_{21} to H_{23} and H_{32} to H_{34} are missing. Since $v_{12} = \theta_{12} = 0$, we write

$$\begin{bmatrix} V_{21} \\ M_{21} \\ v_{21} \\ \theta_{21} \end{bmatrix} = \begin{bmatrix} u_{a11} & u_{a12} \\ u_{a21} & u_{a22} \\ u_{a31} & u_{a32} \\ u_{a41} & u_{a42} \end{bmatrix} \begin{bmatrix} V_{12} \\ M_{12} \end{bmatrix} \tag{4}$$

where the subscripts of u_a indicate the element of U_a at the ith row and jth column. Since $M_{21} = 0$, then $M_1 = -u_{a21}V_1/u_{a22}$. Thus (2) becomes

$$\begin{bmatrix} V_{21} \\ M_{21} \\ v_{21} \\ \theta_{21} \end{bmatrix} = \begin{bmatrix} u_{a11} - u_{a12}u_{a21}/u_{a22} \\ 0 \\ u_{a31} - u_{a32}u_{a21}/u_{a22} \\ u_{a41} - u_{a42}u_{a21}/u_{a22} \end{bmatrix} V[U_a']V_1 \tag{5}$$

For member b, we use (5) and then write

$$\begin{bmatrix} V_{23} \\ M_{23} \\ v_{23} \\ \theta_{23} \end{bmatrix} = \begin{bmatrix} u'_{a11} & 0 \\ u'_{a21} & 0 \\ u'_{a31} & 0 \\ u'_{a41} & 1 \end{bmatrix} \begin{bmatrix} V_1 \\ \theta_2 \end{bmatrix} \tag{6}$$

and

$$\begin{bmatrix} V_{32} \\ M_{32} \\ v_{32} \\ \theta_{32} \end{bmatrix} = \begin{bmatrix} u_{b11} & \cdots & u_{b14} \\ u_{b21} & \cdots & u_{b24} \\ u_{b31} & \cdots & u_{b34} \\ u_{b41} & \cdots & u_{b44} \end{bmatrix} \begin{bmatrix} u'_{a11} & 0 \\ u'_{a21} & 0 \\ u'_{a31} & 0 \\ u'_{a41} & 1 \end{bmatrix} \begin{bmatrix} V_1 \\ \theta_2 \end{bmatrix} = \begin{bmatrix} u'_{b11} & u'_{b12} \\ u'_{b21} & u'_{b22} \\ u'_{b31} & u'_{b32} \\ u'_{b41} & u'_{b42} \end{bmatrix} \begin{bmatrix} V_1 \\ \theta_2 \end{bmatrix} \tag{7}$$

Since $v_{32} = 0$, then $\theta_2 = -u'_{b31}v_1/u'_{b32}$. Consequently (7) becomes

$$\begin{bmatrix} V_{32} \\ M_{32} \\ v_{32} \\ \theta_{32} \end{bmatrix} = \begin{bmatrix} u'_{b11} - u'_{b12}u'_{b31}/u'_{b32} \\ u'_{b21} - u'_{b22}u'_{b31}/u'_{b32} \\ 0 \\ u'_{b41} - u'_{b42}u'_{b31}/u'_{b32} \end{bmatrix} V_{12} = [U_b'']V_{12} \tag{8}$$

At point 3, the shear is not continuous and may be expressed as $V_{34} = V_{32} + R$. Thus we use (8) for the following:

$$\begin{bmatrix} V_{34} \\ M_{34} \\ v_{34} \\ \theta_{34} \end{bmatrix} = \begin{bmatrix} u''_{b11} & 1 \\ u''_{b21} & 0 \\ u''_{b31} & 0 \\ u''_{b41} & 0 \end{bmatrix} \begin{bmatrix} V_{12} \\ R \end{bmatrix} \tag{9}$$

Finally, the state vector at point 4 may be

$$\begin{bmatrix} V_{43} \\ M_{43} \\ 0 \\ 0 \end{bmatrix} = \begin{bmatrix} u_{c11} & \cdots & u_{c14} \\ u_{c21} & \cdots & u_{c24} \\ u_{c31} & \cdots & u_{c34} \\ u_{c41} & \cdots & u_{c44} \end{bmatrix} \begin{bmatrix} u''_{b11} & 1 \\ u''_{b21} & 0 \\ u''_{b31} & 0 \\ u''_{b41} & 0 \end{bmatrix} \begin{bmatrix} V_{12} \\ R \end{bmatrix} = \begin{bmatrix} u'_{c11} & u'_{c12} \\ u'_{c21} & u'_{c22} \\ u'_{c31} & u'_{c32} \\ u'_{c41} & u'_{c42} \end{bmatrix} \begin{bmatrix} V_{12} \\ R \end{bmatrix} \tag{10}$$

Using the boundary conditions in (10) one may find the frequency equation from the following determinant:

$$\begin{vmatrix} u'_{c31} & u'_{c32} \\ u'_{c41} & u'_{c42} \end{vmatrix} = 0 \tag{11}$$

This example shows that the transport matrix method is suitable for continuous beams without rigid intermediate conditions and releases.

Table P-8.5 Sensitivity of Series

$\bar{\lambda}$	a_u	d_u	a_v	b_v	c_v	d_v
0.5	0.877 58	0.479 43	1.002 60	0.500 26	0.125 02	0.020 84
1.0	0.540 30	0.841 47	1.041 69	1.008 34	0.501 39	0.166 87
2.0	−0.416 15	0.909 30	1.673 02	2.268 08	2.089 17	1.358 78
3.0	−0.989 99	0.141 12	4.538 83	5.079 50	5.528 83	4.938 38
4.0	−0.653 64	−0.756 80	13.327 29	13.266 56	13.980 94	14.023 36
5.0	0.283 66	−0.958 92	37.246 80	36.622 14	37.054 82	38.539 98
10.0	−0.839 07	−0.544 02	5 506.196 89	5 506.344 42	5 507.035 96	5 506.888 44
20.0	0.408 08	0.912 94	121 291 299.056 48	121 291 299.308 91	121 291 298.648 40	121 291 298.395 97

8.5. Show the sensitivity of the series (8-20).

The sensitivity of the series (8-20) is shown in Table P-8.5 based on the following approximation:

For $0 \le \bar{\lambda} \le 1$, $N = 3$ for all functions in (8-20)

where N is the number of terms of the series required to obtain a good approximation.

For $1 < \bar{\lambda} \le 4$, $N = 10$ for all functions in (8-20)

For $4 < \bar{\lambda} \le \infty$ (trigonometric functions),

$$\cos \bar{\lambda} = \cos(2n\pi + \bar{\tau}) = \cos \bar{\tau} \qquad \text{and} \qquad \sin \bar{\lambda} = \sin(2n\pi + \bar{\tau}) = \sin \bar{\tau}$$

For $4 < \bar{\lambda} \le 12$ (hyperbolic functions),

$$\cosh \bar{\lambda} = \cosh(2n\pi + \bar{\tau}) \quad \text{and} \quad \sinh \bar{\lambda} = \sinh(2n\pi + \bar{\tau}).$$

For instance,

$$a_v = \frac{\cosh \bar{\lambda} + \cos \bar{\lambda}}{2} = \frac{\cosh(2n\pi + \bar{\tau}) + \cos \bar{\tau}}{2} = \frac{\cosh 2n\pi \cosh \bar{\tau} + \sinh 2n\pi \sinh \bar{\tau} + \cos \bar{\tau}}{2}$$

in which

$$\cosh 2n\pi = \tfrac{1}{2}[(e^{2\pi})^n + (e^{2\pi})^{-n}] \qquad \text{and} \qquad \sinh 2n\pi = \tfrac{1}{2}[(e^{2\pi})^n - (e^{2\pi})^{-n}]$$

where $e^{2\pi}$ may be found in all mathematical tables and n is an integer.

For $12 < \bar{\lambda} \le \infty$ (hyperbolic functions), $\cosh \bar{\lambda} = e^{\bar{\lambda}}/2$ and $\sinh \bar{\lambda} = e^{\bar{\lambda}}/2$. For instance,

$$a_v = \frac{\cosh \bar{\lambda} + \cos \bar{\lambda}}{2} = \frac{e^{\bar{\lambda}}/2 + \cos \bar{\lambda}}{2} = \frac{e^{2n\pi}e^{\bar{\tau}} + 2\cos \bar{\tau}}{4}$$

8.6. Discuss the numerical problem involved in the evaluation of the frequency determinant of a transport chain.

From the preceding problem, one may observe that when the argument $\bar{\lambda}_v$ increases, $\sinh \bar{\lambda}_v$ and $\cosh \bar{\lambda}_v$ increase continuously; however, $\sin \bar{\lambda}_v$ and $\cos \bar{\lambda}_v$ cannot exceed unity. For instance, when $\bar{\lambda}_v = 10$, $\cosh \bar{\lambda}_v = \sinh \bar{\lambda}_v = 11\,013$, $\cos \bar{\lambda}_v = -0.839\,03$, and $\sin \bar{\lambda}_v = -0.544\,077$. Therefore the arguments in a_v, b_v, c_v, and d_v must be carried out to at least five significant digits for the influence of the sine and cosine terms to be felt at all. Since higher arguments usually correspond to higher modes, the transport matrix method may fail to determine the eigenvalues beyond the third or fourth mode. In order to overcome the numerical difficulties, we may divide the member into a number of segments N. For the present case, say $N = 8$, then $\bar{\lambda}_v/N = 1.25$, for which $\sinh \bar{\lambda}_v = 1.6019$, $\cosh \bar{\lambda}_v = 1.8884$, $\sin \bar{\lambda}_v = 0.948\,98$, and $\cos \bar{\lambda}_v = 0.315\,32$. Thus the hyperbolic functions and trigonometric functions will have the same order of magnitude.

8.7. Find the angular frequencies of a beam with left end fixed and right end hinged of which the member properties are $A = 232.26\ \text{cm}^2$, $I = 1.01\ \text{dm}^4$, $l = 12.19\ \text{m}$, $\gamma = 76.973\ \text{kN/m}^3$, and $E = 206.84\ \text{GN/m}^2$.

Since the longitudinal and transverse vibrations are not coupled for straight members, the coefficients a_u and d_u may not be necessarily included in a transport matrix. Thus the transport matrix for the present case is

$$\begin{bmatrix} \bar{V}_R \\ \bar{M}_R \\ \bar{v}_R \\ \bar{\theta}_R \end{bmatrix} = \begin{bmatrix} a_v & d_v & b_v & c_v \\ b_v & a_v & c_v & d_v \\ d_v & c_v & a_v & b_v \\ c_v & b_v & d_v & a_v \end{bmatrix} \begin{bmatrix} \bar{V}_L \\ \bar{M}_L \\ \bar{v}_L \\ \bar{\theta}_L \end{bmatrix} \tag{1}$$

Using the boundary conditions $\bar{M}_R = \bar{v}_R = \bar{v}_L = \bar{\theta}_L = 0$ yields

$$\begin{vmatrix} b_v & a_v \\ d_v & c_v \end{vmatrix} = 0 \qquad \text{or} \qquad \tanh \bar{\lambda}_v = \tan \bar{\lambda}_v \tag{2}$$

By trials (2) gives

$$\bar{\lambda}_{v1} = 3.9266, \qquad \bar{\lambda}_{v2} = 7.0686, \qquad \bar{\lambda}_{v3} = 10.2102, \qquad \bar{\lambda}_{v4} = 13.3518, \qquad \ldots$$

Since $\omega = \bar{\lambda}_v^2/\sqrt{\mu l^4/EI}$ and $\sqrt{EI/\mu l^4} = 2.279\,59$, we have the following frequencies in radians per second:

$$\omega_1 = 35.417, \qquad \omega_2 = 113.900, \qquad \omega_3 = 237.643, \qquad \omega_4 = 406.384, \qquad \ldots$$

TRANSPORT CHAIN-FORCED VIBRATION

8.8. Find the reactions at both ends of the beam shown in Fig. P-8.8. Let $p_v = 40 \sin \Omega t$ k/ft, $\Omega = (\pi/l)^2\sqrt{EI/\mu}$, $l = 40$ ft, $I = 243\ \text{in}^4$, and $\mu = 3.806$ lb-sec/ft^2.

From (8-31),

$$H_R = U_{RL}H_L + U_{RL}\int_0^l U(-\tau)W(\tau)\,d\tau \tag{1}$$

in which U_{RL} is typically shown in (4) of Problem 8.1, $W(\tau) = [0 \quad -40. \quad 0 \quad 0 \quad 0 \quad 0]^T$, and $U(-\tau)$ is similar to U_{RL}. The parameters $\lambda_u l$ in a_u and d_u and $\lambda_v l$ in a_v, b_v, c_v, and d_v are to be replaced by $\lambda_u(-\tau)$ and $\lambda_v(-\tau)$ respectively for the following integration:

$$\int_0^l U(-\tau)W(\tau)\,d\tau = \int_0^l \begin{bmatrix} 0 \\ -40a_v \\ 40b_v/\lambda_v \\ 0 \\ 40d_v/EI\lambda_v^3 \\ 40c_v/EI\lambda_v^2 \end{bmatrix} dt \tag{2}$$

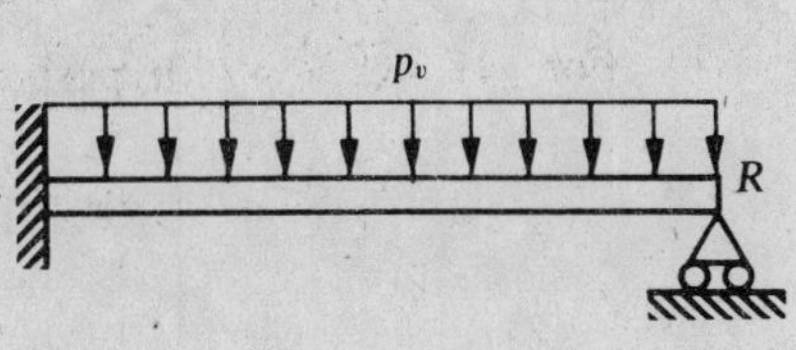

Fig. P-8.8

For the given forcing frequencies Ω of steady state vibration, we have $\omega = \Omega$ and $\bar{\lambda}_v = \pi$. Consequently $\sinh \bar{\lambda}_v = 11.5488$, $\cosh \bar{\lambda}_v = 11.592$, $\sin \bar{\lambda}_v = 0$, and $\cos \bar{\lambda}_v = -1$. Since the longitudinal and the transverse vibrations are not coupled for the straight beam, the longitudinal frequency may be different from Ω. Let us assume the longitudinal frequency to be Ω; then $\bar{\lambda}_u = 0.001\,335$, $\sin \bar{\lambda}_u = 0.0014$, and $\cos \bar{\lambda}_u = 1.0$. Thus

$$U_{RL} = \begin{bmatrix} 1 & 0 & 0 & 50.4630 & 0 & 0 \\ 0 & 5.2958 & -0.4535 & 0 & -141.6190 & -1966.0644 \\ 0 & -73.5180 & 5.2958 & 0 & 1966.0644 & 22958.3260 \\ 0.4 \times 10^{-6} & 0 & 0 & 1. & 0 & 0 \\ 0 & -0.2354 & 0.0202 & 0 & 5.2958 & 73.5179 \\ 0 & -0.0202 & 0.0615 & 0 & 0.4535 & 5.2958 \end{bmatrix} \tag{3a}$$

$$\underbrace{U_{RL}\int_0^l U(-\tau)W(\tau)\,d\tau}_{\Phi_{RL}} = \begin{bmatrix} 0 \\ -2.9486 \\ 40.8940 \\ 0 \\ 0.0895 \\ 0.0095 \end{bmatrix} \tag{3b}$$

Substituting (3) into (1) and then using the boundary conditions:

$$H_R = [0 \quad V_R \quad 0 \quad u_R \quad 0 \quad \theta_R]^T \qquad \text{and} \qquad H_L = [N_L \quad V_L \quad M_L \quad 0 \quad 0 \quad 0]^T$$

we have

$$\begin{aligned} -73.518\,V_L + 5.2958\,M_L + 40.894 &= 0 \\ -0.2354\,V_L + 0.020\,15\,M_L + 0.089\,55 &= 0 \\ 5.2958\,V_L - 0.453\,49\,M_L - 2.9486 &= V_R \end{aligned} \tag{4}$$

from which $V_L = 1.488$ k, $M_L = 12.9359$ ft-k, and $V_R = 0.9349$ k.

8.9. Find the moment and shear at the left end of the beam shown in Fig. P-8.9 where $p_{va} = 40 \sin \Omega t$ k/ft, $p_{vb} = 20 \sin \Omega t$ k/ft, $\Omega = (\pi/l)^2\sqrt{EI_a/\mu_a}$, $l = 40$ ft, $I_a = 243$ in^4, $\mu_a = 3.806$ lb-sec^2/ft, $A_a = 36$ in^2, $I_b = I_a/2$, $\mu_b = \mu_a/2$, and $A_b = A_a/2$.

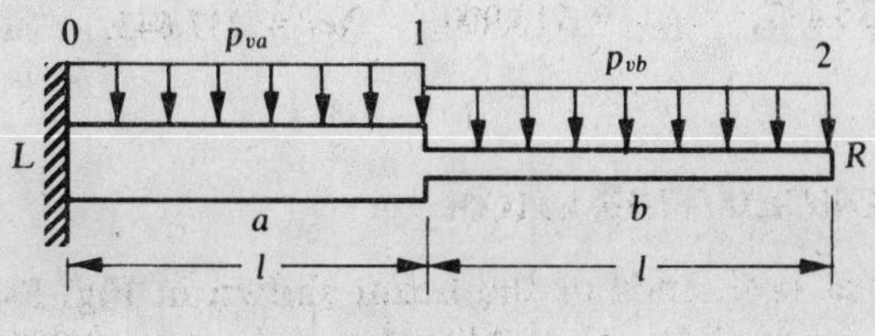

Fig. P-8.9

From (8-31), we write

$$H_2 = U_{21}(H_1 + \Phi_{21}) = U_{21}[U_{10}(H_0 + \Phi_{10}) + \Phi_{21}] = U_{21}U_{10}H_0 + U_{21}U_{10}\Phi_{10} + U_{21}\Phi_{21} \tag{1}$$

where U_{21}, U_{10}, Φ_{10}, and Φ_{21} are similar to U_{RL} and Φ_{RL} used in Problem 8.8. For the given member properties and $\omega = \Omega$, we have $\bar{\lambda}_{va} = l\sqrt[4]{\mu_a\omega^2/EI_a} = \pi$, $\bar{\lambda}_{vb} = \pi$, $\bar{\lambda}_{ua} = l\sqrt{\mu_a\omega^2/EA_a} = 0.001\,335$, and $\bar{\lambda}_{ub} = 0.001\,335$. Consequently,

$$U_{21} = \begin{bmatrix} 1 & 0 & 0 & 25.2365 & 0 & 0 \\ 0 & 5.2958 & -0.4535 & 0 & -70.8095 & -983.0322 \\ 0 & -73.5180 & 5.2958 & 0 & 983.0322 & 11479.1630 \\ 0.8\times10^{-6} & 0 & 0 & 1 & 0 & 0 \\ 0 & -0.4708 & 0.0403 & 0 & 5.2958 & 73.5179 \\ 0 & -0.0430 & 0.0029 & 0 & 0.4535 & 5.2958 \end{bmatrix}$$

$$\Phi_{21} = \begin{bmatrix} 0 \\ -1.4705 \\ -20.4130 \\ 0 \\ -0.0892 \\ 0.0094 \end{bmatrix} \qquad \Phi_{10} = \begin{bmatrix} 0 \\ -2.941 \\ 40.826 \\ 0 \\ -0.089 \\ 0.009 \end{bmatrix} \tag{2}$$

U_{10} is exactly the same as shown in (*3*) of the preceding problem.

$$\begin{bmatrix} 0 \\ 0 \\ 0 \\ u_R \\ v_R \\ \theta_R \end{bmatrix} = \begin{bmatrix} 1 & 0 & 0 & 75.7095 & 0 & 0 \\ 0 & 98.573 & -7.716 & 0 & -2478.260 & -31455.6 \\ 0 & -1241.383 & 97.843 & 0 & 31235.170 & 399185.7 \\ 1.2\times10^{-6} & 0 & 0 & 1 & 0 & 0 \\ 0 & -8.185 & 0.640 & 0 & 207.312 & 2629.7 \\ 0 & -0.6402 & 0.050 & 0 & 16.215 & 207.2 \end{bmatrix} \begin{bmatrix} N_L \\ V_L \\ M_L \\ 0 \\ 0 \\ 0 \end{bmatrix} + \begin{bmatrix} 0 \\ -51.592 \\ 651.013 \\ 0 \\ 4.297 \\ 0.338 \end{bmatrix} \tag{3}$$

from which

$$\begin{aligned} 98.573\,V_L - 7.716\,M_L - 51.592 &= 0 \\ -124.383\,V_L + 97.843\,M_L + 651.013 &= 0 \\ -8.185\,V_L + 0.6403\,M_L + 4.297 &= v_R \end{aligned}$$

Thus $V_L = 0.3727$ k, $M_L = -1.926$ ft-k, and $v_R = 0.0146$ ft.

SERIES EXPANSION

8.10. Using (*8-24*) find the angular frequencies of a simply supported beam with $\mu = 55.544$ N/m, $l = 12.192$ m, $I = 1.01$ dm^4, and $E = 206.84$ GN/m^2.

Using (*8-28*) and neglecting axial deformation and rotatory inertia, we write

$$Dl = \begin{bmatrix} 0 & 0 & \mu\omega^2 l & 0 \\ l & 0 & 0 & 0 \\ 0 & 0 & 0 & l \\ 0 & l/EI & 0 & 0 \end{bmatrix} \qquad (Dl)^2/2! = \frac{1}{2}\begin{bmatrix} 0 & 0 & 0 & \mu\omega^2 l^2 \\ 0 & 0 & \mu\omega^2 l^2 & 0 \\ 0 & l^2/EI & 0 & 0 \\ l^2/EI & 0 & 0 & 0 \end{bmatrix}$$

$$(Dl)^3/3! = \frac{1}{6}\begin{bmatrix} 0 & \mu\omega^2 l^3/EI & 0 & 0 \\ 0 & 0 & 0 & \mu\omega^2 l^3 \\ l^3/EI & 0 & 0 & 0 \\ 0 & 0 & \mu\omega^2 l^3/EI & 0 \end{bmatrix}$$

$$(Dl)^4/4! = \frac{1}{24}\begin{bmatrix} \mu\omega^2 l^4/EI & 0 & 0 & 0 \\ 0 & \mu\omega^2 l^4/EI & 0 & 0 \\ 0 & 0 & \mu\omega^2 l^4/EI & 0 \\ 0 & 0 & 0 & \mu\omega^2 l^4/EI \end{bmatrix} = \frac{\bar{\lambda}_v^4}{4!}[I]$$

which leads to the conclusion that

$$\frac{(Dl)^{4n+1}}{(4n+1)!} = \frac{(Dl)\bar{\lambda}_v^{4n}}{(4n+1)!} \qquad \frac{(Dl)^{4n+2}}{(4n+2)!} = \frac{(Dl)^2\bar{\lambda}_v^{4n}}{(4n+2)!} \qquad \frac{(Dl)^{4n+3}}{(4n+3)!} = \frac{(Dl)^3\bar{\lambda}_v^{4n}}{(4n+3)!} \qquad \frac{(Dl)^{4n+4}}{(4n+4)!} = \frac{[I]\bar{\lambda}_v^{4n}}{(4n+4)!}$$

where $\bar{\lambda}_v = l\sqrt[4]{\mu\omega^2/EI}$ and $n = 0, 1, 2, \ldots$. If the first four terms are used, (8-24) gives

$$\begin{bmatrix} V_R \\ 0 \\ 0 \\ \theta_R \end{bmatrix} = \begin{bmatrix} 1 & \mu\omega^2 l^3/6EI & \mu\omega^2 l & \mu\omega^2 l^2/2 \\ l & 1 & \mu\omega^2 l^2/2 & \mu\omega^2 l^3/6 \\ l^3/6EI & l^2/2EI & 1 & l \\ l^2/2EI & l/EI & \mu\omega^2 l^3/6EI & 1 \end{bmatrix} \begin{bmatrix} V_L \\ 0 \\ 0 \\ \theta_L \end{bmatrix} \quad \text{and} \quad \begin{vmatrix} l & \bar{\lambda}^4 EI/6l \\ l^3/6EI & l \end{vmatrix} = 0 \tag{1}$$

from which

$$\omega = G\sqrt{EI/\mu l^4} \qquad \text{or} \qquad \omega = 3.798 \text{ rad/s} \tag{2}$$

For the first eight terms the determinant of (1) becomes

$$\begin{vmatrix} l\left(\frac{1}{1!} + \frac{\bar{\lambda}_v^4}{5!}\right) & \frac{\lambda_v EI}{l}\left(\frac{1}{3!} + \frac{\bar{\lambda}_v^4}{7!}\right) \\ \frac{l^3}{EI}\left(\frac{1}{3!} + \frac{\bar{\lambda}_v^4}{7!}\right) & l\left(\frac{1}{1!} + \frac{\bar{\lambda}_v^4}{5!}\right) \end{vmatrix} = 0 \tag{3}$$

and yields

$$\omega^6 - 84(EI/\mu l^4)\omega^4 + 282\,240(EI/\mu l^4)^2\omega^2 - 254\,016\,00(EI/\mu l^4)^3 = 0 \tag{4}$$

Since $EI/\mu l^4 = 0.4007$, the solution of (4) is $\omega_1 = 6.247$ rad/s, ...

For any higher sum of $4n$ terms the determinant (1) can be expressed in the series form as

$$\begin{vmatrix} l\left(\frac{1}{1!} + \frac{\bar{\lambda}_v^4}{5!} + \frac{\bar{\lambda}_v^8}{9!} + \cdots\right) & \frac{\bar{\lambda}_v^4 EI}{l}\left(\frac{1}{3!} + \frac{\bar{\lambda}_v^4}{7!} + \frac{\bar{\lambda}_v^8}{11!} + \cdots\right) \\ \frac{l^3}{EI}\left(\frac{1}{3!} + \frac{\bar{\lambda}_v^4}{7!} + \frac{\bar{\lambda}_v^8}{11!} + \cdots\right) & l\left(\frac{1}{1!} + \frac{\bar{\lambda}_v^4}{5!} + \frac{\bar{\lambda}_v^8}{9!} + \cdots\right) \end{vmatrix} = 0 \tag{5}$$

where each series converges to the respective dynamic parameters of (8-20), i.e.,

$$\frac{1}{1!} + \frac{\bar{\lambda}_v^4}{5!} + \frac{\bar{\lambda}_v^8}{9!} + \cdots = 2b_v \qquad \text{and} \qquad \frac{1}{3!} + \frac{\bar{\lambda}_v^4}{7!} + \frac{\bar{\lambda}_v^8}{11!} + \cdots = \frac{2d_v}{\bar{\lambda}_v^3} \tag{6}$$

Consequently the exact form of the frequency equation is

$$(\bar{\lambda}_v b_v)^2 - (d_v)^2 = 0 \tag{7}$$

from which $\omega_1 = 6.247$ rad/s, ...

The sensitivity of series (8-20) was discussed in Problem 8.5.

8.11. Using (8-23) and the Cayley-Hamilton theorem derive the transport matrix in finite form.

Let (8-21) be expressed for bending deformation only as

$$D = \begin{bmatrix} 0 & 0 & -\mu\omega^2 & 0 \\ -1 & 0 & 0 & 0 \\ 0 & 0 & 0 & 1 \\ 0 & 1/EI & 0 & 0 \end{bmatrix} \tag{1}$$

For $U_{RL} = e^{Dl}$, Equation (8-24), the characteristic equation for Dl, is

$$[\gamma I - Dl] = \begin{bmatrix} \gamma & 0 & \mu\omega^2 l & 0 \\ l & \gamma & 0 & 0 \\ 0 & 0 & \gamma & -l \\ 0 & -l/EI & 0 & \gamma \end{bmatrix} = \gamma^4 - \frac{\mu\omega^2 l^4}{EI} = 0 \tag{2}$$

Let $\mu\omega^2 l^4/EI = \bar{\lambda}^4$ as shown in (6-25) and (6-33); then $\gamma_{1,2} = \pm\bar{\lambda}$ and $\gamma_{3,4} = \pm j\bar{\lambda}$, where $j = \sqrt{-1}$. Since Dl is a fourth order matrix, the series of e^{Dl} may be expressed as

$$e^{Dl} = a_0 I + a_1 Dl + a_2(Dl)^2 + a_3(Dl)^3 \tag{3}$$

Applying the Cayley-Hamilton theorem,

$$\begin{aligned} e^{\bar\lambda} &= a_0 + a_1\bar\lambda + a_2\bar\lambda^2 + a_3\bar\lambda^3 \\ e^{-\bar\lambda} &= a_0 - a_1\bar\lambda + a_2\bar\lambda^2 - a_3\bar\lambda^3 \\ e^{j\bar\lambda} &= a_0 + a_1 j\bar\lambda - a_2\bar\lambda^2 - a_3 j\bar\lambda^3 \\ e^{-j\bar\lambda} &= a_0 - a_1 j\bar\lambda - a_2\bar\lambda^2 + a_3 j\bar\lambda^3 \end{aligned}$$

from which

$$\begin{aligned} a_0 &= (\cosh\bar\lambda + \cos\bar\lambda)/2 \qquad & a_1 &= (\sinh\bar\lambda + \sin\bar\lambda)/2\bar\lambda \\ a_2 &= (\cosh\bar\lambda - \cos\bar\lambda)/2\bar\lambda^2 \qquad & a_3 &= (\sinh\bar\lambda - \sin\bar\lambda)/2\bar\lambda^3 \end{aligned} \tag{4}$$

Substitution of (3) gives

$$e^{Dl} = \begin{bmatrix} a_0 & 0 & 0 & 0 \\ 0 & a_0 & 0 & 0 \\ 0 & 0 & a_0 & 0 \\ 0 & 0 & 0 & a_0 \end{bmatrix} + a_1 \begin{bmatrix} 0 & 0 & -\mu\omega^2 & 0 \\ -l & 0 & 0 & 0 \\ 0 & 0 & 0 & l \\ 0 & l/EI & 0 & 0 \end{bmatrix} + a_2 \begin{bmatrix} 0 & 0 & 0 & -\mu\omega^2 l^2 \\ 0 & 0 & \mu\omega^2 l^2 & 0 \\ 0 & l^2/EI & 0 & 0 \\ -l^2/EI & 0 & 0 & 0 \end{bmatrix}$$

$$+ a_3 \begin{bmatrix} 0 & -\mu\omega^2 l^3/EI & 0 & 0 \\ 0 & 0 & 0 & \mu\omega^2 l^3 \\ -l^3/EI & 0 & 0 & 0 \\ 0 & 0 & \mu\omega^2 l^3/EI & 0 \end{bmatrix} \tag{5}$$

and finally

$$e^{Dl} = \begin{bmatrix} a_0 & -\bar\lambda^4 a_3/l & -a_1 EI\bar\lambda^4/l^3 & -a_2 EI\bar\lambda^4/l^2 \\ -a_1 l & a_0 & a_2 EI\bar\lambda^4/l^2 & a_3 EI\bar\lambda^4/l \\ -a_3 l^3/EI & l^2 a_2/EI & a_0 & l a_1 \\ -a_2 l^2/EI & l a_1/EI & a_3\bar\lambda^4/l & a_0 \end{bmatrix} \tag{6}$$

Substituting (5) into (6) and then using the notations in Table 8-1, we write

$$\begin{bmatrix} V_R \\ M_R \\ v_R \\ \theta_R \end{bmatrix} = \begin{bmatrix} a_v & -d_v\bar\lambda/l & -b_v EI\bar\lambda^3/l^3 & -c_v EI\bar\lambda^2/l^2 \\ -l b_v/\bar\lambda & a_v & c_v EI\bar\lambda^2/l^2 & EI d_v\bar\lambda/l \\ -d_v l^3/EI\bar\lambda^3 & c_v l^2/EI\bar\lambda^2 & a_v & b_v l/\bar\lambda \\ -c_v l^2/EI\bar\lambda^2 & b_v l/EI\bar\lambda & d_v\bar\lambda/l & a_v \end{bmatrix} \begin{bmatrix} V_L \\ M_L \\ v_L \\ \theta_L \end{bmatrix} \tag{7}$$

which is identical to the part in (4), Problem 8-1, representing the transport matrix U_{RL} for bending deformation only.

Supplementary Problems

8.12. Show symbolically the transport chain of free vibration for the structures shown in Fig. P-8.12. Given: a, b, h, A_a, A_b, I_a, I_b, k_u, k_v, k_θ, μ_a, and μ_b.

8.13. Show symbolically the frequency determinants for the structures of Fig. P-8.12.

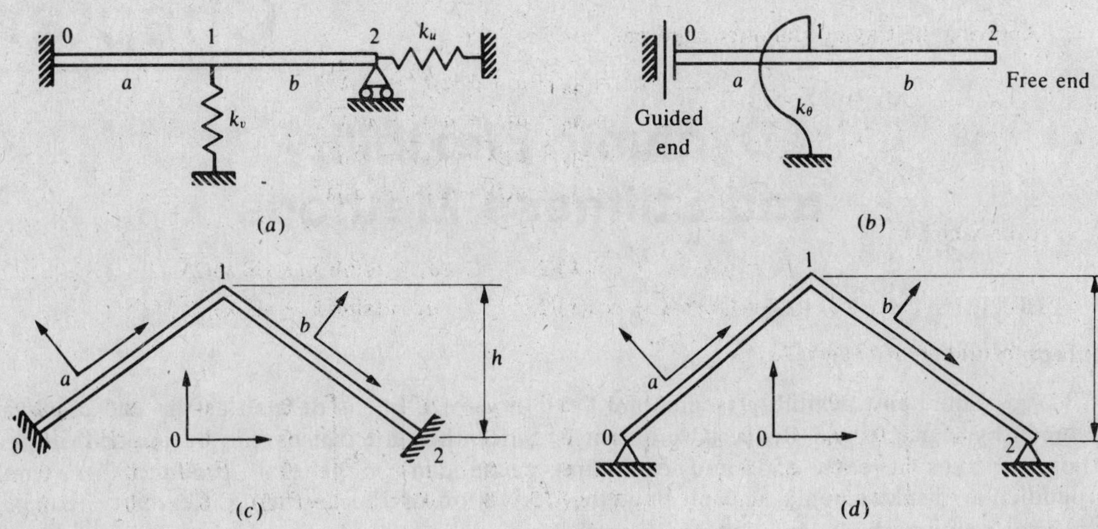

Fig. P-8.12

8.14. Find the first three angular frequencies of the rigid frame shown in Fig. P-8.14. Let $L_{01} = 15$ ft, $L_{12} = 20$ ft, $L_{02} = 25$ ft, $\mu_{01} = 1.8418 \times 10^{-4}$ k-sec^2/ft^2, $\mu_{12} = 2.368 \times 10^{-4}$ k-sec^2/ft^2, $\mu_{02} = 2.892 \times 10^{-4}$ k-sec^2/ft^2, $EI_{01} = 18\,988$ k-ft^2, $EI_{12} = 40\,039$ k-ft^2, $EI_{02} = 72\,802$ k-ft^2, $EA_{01} = 52\,500$ k, $EA_{12} = 67\,500$ k, and $EA_{02} = 82\,500$ k.

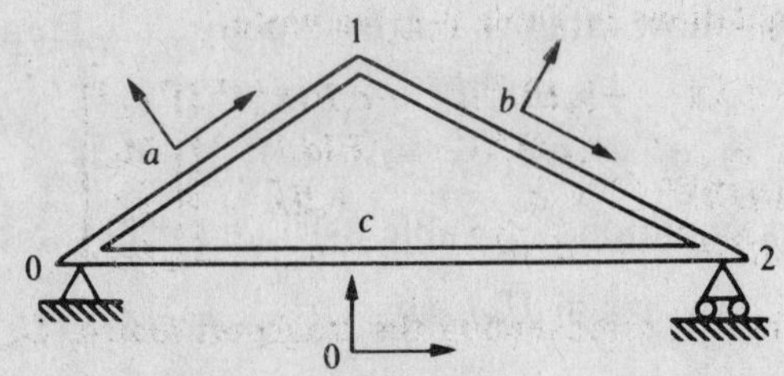

Fig. P-8.14

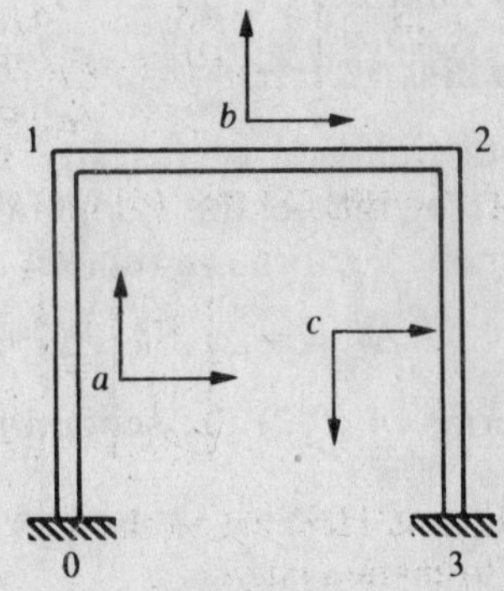

Fig. P-8.15

8.15. Determine the first three angular frequencies of the portal frame shown in Fig. P-8.15. Let $L_{01} = L_{23} = L_{12} = 9.1$ in. $EI_{01} = EI_{12} = EI_{23} = 2060$ k/in^2; $\mu_{01} = \mu_{12} = \mu_{23} = 6.88 \times 10^{-5}$ lb-s^2/in^2.

8.16. Show graphically the first three modes of the elastic curve of the structures of Fig. P-8.14 and P-8.15.

Chapter 9

Dynamic Flexibility and Stiffness Methods

9.1 FLEXIBILITY MATRIX

Concept

The dynamic end flexibility (segmental flexibility) of a bar is defined as the end displacement produced by a unit cause (force or moment). Since the unit causes can be applied along three orthogonal axes at each end, and each unit cause can (in general) produce three mutually perpendicular displacements at each end, the derivation of the segmental flexibility matrix for a planar bar requires the calculation of *36 coefficients.*

If the derivation is carried out in the member system (principal flexibilities) and the reciprocal relationships are observed, the total number of distinct coefficients in a straight bar of constant section reduces to *eight.*

Derivation

These coefficients can be obtained from the respective equations of motion or directly from the transport matrices of Table 8-1. The latter approach is introduced below.

Dimensionless Matrix

The comparison of the stress and displacement vectors of the transport method (Table 9-1a) with those of the flexibility method (Table 9-1b) shows that the stress vectors are identical and the displacement vectors are related as

$$\Delta_L = -\Delta_{LR} \qquad \Delta_R = \Delta_{RL} \qquad \text{or} \qquad \bar{\Delta}_L = -\bar{\Delta}_{LR} \qquad \bar{\Delta}_R = \bar{\Delta}_{RL} \tag{9-1, 2}$$

In terms of (*9-2*) the second row submatrix equations of Table 8-1c and d become

$$\bar{\sigma}_L = \bar{U}_{\sigma\sigma,LR}\bar{\sigma}_R + \bar{U}_{\sigma\Delta,LR}\bar{\Delta}_{RL} \qquad \bar{\sigma}_R = \bar{U}_{\sigma\sigma,RL}\bar{\sigma}_L - \bar{U}_{\sigma\Delta,RL}\bar{\Delta}_{LR} \tag{9-3}$$

and yield

$$\bar{\Delta}_{RL} = \underbrace{-\bar{U}_{\sigma\Delta,LR}^{)-1}\bar{U}_{\sigma\sigma,LR}}_{\bar{f}_{RR}}\bar{\sigma}_R + \underbrace{\bar{U}_{\sigma\Delta,LR}^{)-1}}_{\bar{f}_{RL}}\bar{\sigma}_L \qquad \bar{\Delta}_{LR} = \underbrace{-\bar{U}_{\sigma\Delta,RL}^{)-1}}_{\bar{f}_{LR}}\bar{\sigma}_R + \underbrace{\bar{U}_{\sigma\Delta,RL}^{)-1}\bar{U}_{\sigma\sigma,RL}}_{\bar{f}_{LL}}\bar{\sigma}_L \tag{9-4}$$

where $\bar{f}_{RR}$ and $\bar{f}_{LL}$ are the dimensionless near end (direct) flexibility submatrices, [3 × 3], at R and L respectively, and $\bar{f}_{RL}$ and $\bar{f}_{LR}$ are the dimensionless far end (indirect) flexibility submatrices, [3 × 3], at R and L respectively.

These four submatrices define symbolically the *dimensionless dynamic flexibility matrix*, [6 × 6], of the given bar in the l-system, corresponding to the mth mode. The coefficients of these submatrices are functions of $\bar{\lambda}_u$ or $\bar{\lambda}_v$; their analytical expressions are given in Table 9-2a and their derivation is shown in Problem 9.1.

Absolute Matrix

The dimensioning of the flexibility matrix of Table 9-2a is accomplished by matrices κ_Δ and λ_σ introduced in (*8-6*) and (*8-9*) respectively. Thus,

$$\Delta_{RL} = \underbrace{\kappa_\Delta\bar{f}_{RR}\lambda_\sigma}_{f_{RR}}\sigma_R + \underbrace{\kappa_\Delta\bar{f}_{RL}\lambda_\sigma}_{f_{RL}}\sigma_L \qquad \Delta_{LR} = \underbrace{\kappa_\Delta\bar{f}_{LR}\lambda_\sigma}_{f_{LR}}\sigma_R + \underbrace{\kappa_\Delta\bar{f}_{LL}\lambda_\sigma}_{f_{LL}}\sigma_L \tag{9-5}$$

where f_{RR} and f_{LL} are the absolute near end (direct) flexibility submatrices, [3 × 3], at R and L respectively, and f_{RL} and f_{LR} are the absolute far end (indirect) flexibility submatrices, [3 × 3], at R and L respectively, again all in the member system and corresponding to the mth mode. Their analytical expressions are given in Table 9-3d and their derivation is shown in Problem 9.2.

Table 9-1 Stress and Displacement Sign Conventions

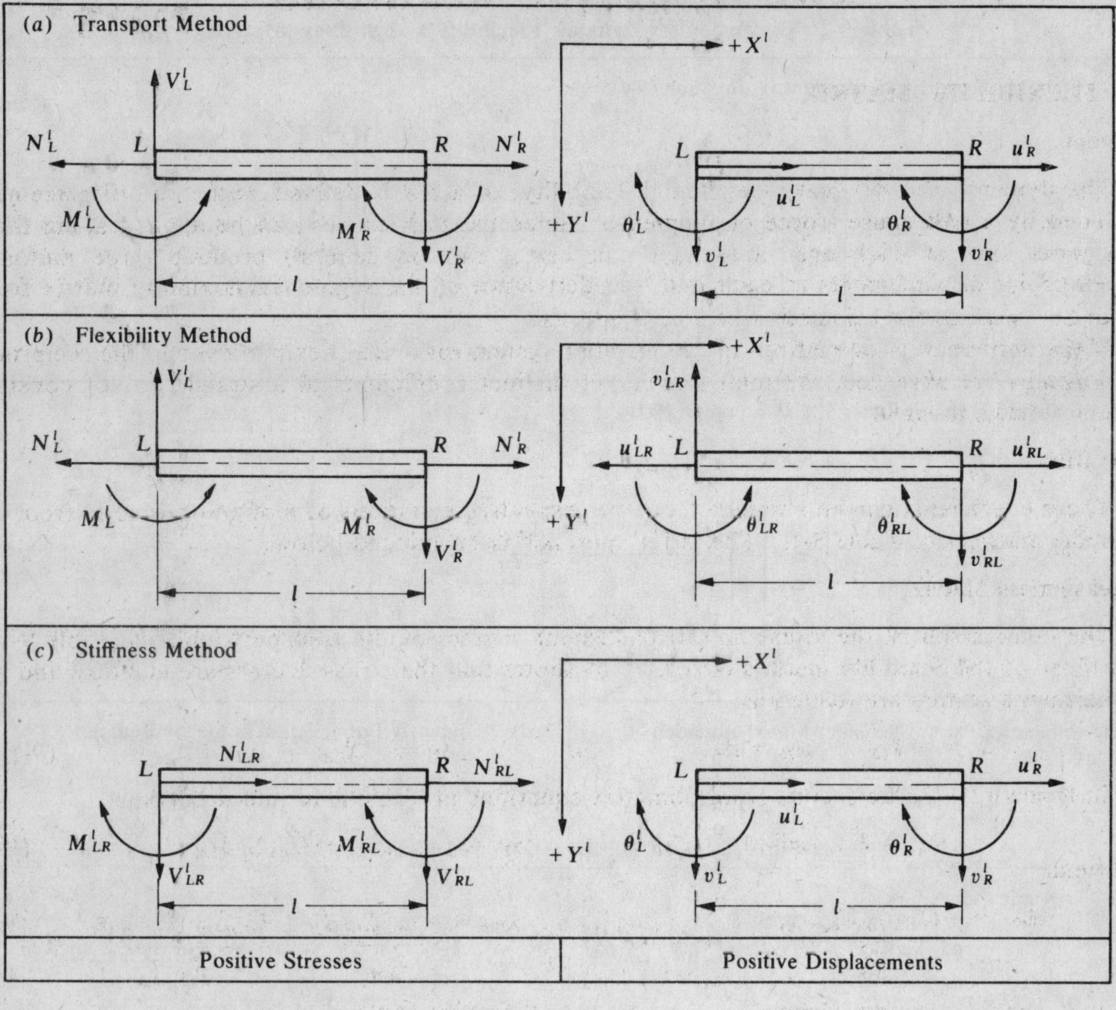

Angular Transformations

Since the end displacement vectors and the end stress vectors are not always in the bar system, the angular transformation matrices must be employed in (9-4) and (9-5).

The angular transformation of (9-4) is

$$\begin{aligned}
\bar{\Delta}^R_{RL} &= \underbrace{\lambda_\Delta \pi^{Rl} \kappa_\Delta \bar{f}^l_{RR} \lambda_\sigma \pi^{lR} \kappa_\sigma}_{\bar{f}^{RR}_{RR}} \bar{\sigma}^R_R + \underbrace{\lambda_\Delta \pi^{Rl} \kappa_\Delta \bar{f}^l_{RL} \lambda_\sigma \pi^{lL} \kappa_\sigma}_{\bar{f}^{RL}_{RL}} \bar{\sigma}^L_L \\
\bar{\Delta}^L_{LR} &= \underbrace{\lambda_\Delta \pi^{Ll} \kappa_\Delta \bar{f}^l_{LR} \lambda_\sigma \pi^{lR} \kappa_\sigma}_{\bar{f}^{LR}_{LR}} \bar{\sigma}^R_R + \underbrace{\lambda_\Delta \pi^{Ll} \kappa_\Delta \bar{f}^l_{LL} \lambda_\sigma \pi^{lL} \kappa_\sigma}_{\bar{f}^{LL}_{LL}} \bar{\sigma}^L_L
\end{aligned} \tag{9-6}$$

where the superscripts R, L, and l identify the right end system, left end system, and member system respectively. The π-matrices are those used in (*8-18, 19*) and which were first introduced in (*1-3, 4*).

The derivation of (*9-6*) and the angular transformation of (*9-5*) are shown in Problem 9.3.

Table 9-2 Dimensionless Dynamic Flexibility and Stiffness Matrix

(*a*) Standard Form, Flexibility Matrix Equation (*9-4*)*

$$\begin{bmatrix} \bar{u}_{RL} \\ \bar{v}_{RL} \\ \bar{\theta}_{RL} \\ \hline \bar{u}_{LR} \\ \bar{v}_{LR} \\ \bar{\theta}_{LR} \end{bmatrix} = \left[\begin{array}{ccc|ccc} -\bar{\psi}_1 & & & \bar{\psi}_5 & & \\ & \bar{\psi}_2 & -\bar{\psi}_3 & & \bar{\psi}_6 & \bar{\psi}_7 \\ & \bar{\psi}_3 & -\bar{\psi}_4 & & \bar{\psi}_7 & \bar{\psi}_8 \\ \hline \bar{\psi}_5 & & & -\bar{\psi}_1 & & \\ & \bar{\psi}_6 & -\bar{\psi}_7 & & \bar{\psi}_2 & \bar{\psi}_3 \\ & -\bar{\psi}_7 & \bar{\psi}_8 & & -\bar{\psi}_3 & -\bar{\psi}_4 \end{array}\right] \begin{bmatrix} \bar{N}_R \\ \bar{V}_R \\ \bar{M}_R \\ \hline \bar{N}_L \\ \bar{V}_L \\ \bar{M}_L \end{bmatrix}$$

(*b*) Standard Form, Stiffness Matrix Equation (*9-10*)*

$$\begin{bmatrix} \bar{N}_{RL} \\ \bar{V}_{RL} \\ \bar{M}_{RL} \\ \hline \bar{N}_{LR} \\ \bar{V}_{LR} \\ \bar{M}_{LR} \end{bmatrix} = \left[\begin{array}{ccc|ccc} \bar{\psi}_1 & & & -\bar{\psi}_5 & & \\ & -\bar{\psi}_4 & \bar{\psi}_3 & & \bar{\psi}_8 & \bar{\psi}_7 \\ & -\bar{\psi}_3 & \bar{\psi}_2 & & \bar{\psi}_7 & \bar{\psi}_6 \\ \hline -\bar{\psi}_5 & & & \bar{\psi}_1 & & \\ & \bar{\psi}_8 & -\bar{\psi}_7 & & -\bar{\psi}_4 & \bar{\psi}_3 \\ & -\bar{\psi}_7 & \bar{\psi}_6 & & \bar{\psi}_3 & \bar{\psi}_2 \end{array}\right] \begin{bmatrix} \bar{u}_R \\ \bar{v}_R \\ \bar{\theta}_R \\ \hline \bar{u}_L \\ \bar{v}_L \\ \bar{\theta}_L \end{bmatrix}$$

(*c*) Submatrix Form, Flexibility Matrix Equation

$$\begin{bmatrix} \bar{\Delta}_{RL} \\ \bar{\Delta}_{LR} \end{bmatrix} = \begin{bmatrix} \bar{f}_{RR} & \bar{f}_{RL} \\ \bar{f}_{LR} & \bar{f}_{LL} \end{bmatrix} \begin{bmatrix} \bar{\sigma}_R \\ \bar{\sigma}_L \end{bmatrix}$$

(*d*) Submatrix Form, Stiffness Matrix Equation

$$\begin{bmatrix} \bar{\sigma}_{RL} \\ \bar{\sigma}_{LR} \end{bmatrix} = \begin{bmatrix} \bar{k}_{RR} & \bar{k}_{RL} \\ \bar{k}_{LR} & \bar{k}_{LL} \end{bmatrix} \begin{bmatrix} \bar{\Delta}_R \\ \bar{\Delta}_L \end{bmatrix}$$

(*e*) Dynamic Parameters

$$\bar{\psi}_1 = \frac{a_u}{d_u} = \frac{\cos \bar{\lambda}_u}{\sin \bar{\lambda}_u} \qquad \bar{\psi}_2 = \frac{b_v c_v - a_v d_v}{c_v^2 - b_v d_v} = \frac{\cosh \bar{\lambda}_v \sin \bar{\lambda}_v - \sinh \bar{\lambda}_v \cos \bar{\lambda}_v}{1 - \cosh \bar{\lambda}_v \cos \bar{\lambda}_v}$$

$$\bar{\psi}_3 = \frac{a_v c_v - d_v^2}{c_v^2 - b_v d_v} = \frac{\sinh \bar{\lambda}_v \sin \bar{\lambda}_v}{1 - \cosh \bar{\lambda}_v \cos \bar{\lambda}_v} \qquad \bar{\psi}_4 = \frac{a_v b_v - c_v d_v}{c_v^2 - b_v d_v} = \frac{\cosh \bar{\lambda}_v \sin \bar{\lambda}_v + \sinh \bar{\lambda}_v \cos \bar{\lambda}_v}{1 - \cosh \bar{\lambda}_v \cos \bar{\lambda}_v}$$

$$\bar{\psi}_5 = \frac{1}{d_u} = \frac{1}{\sin \bar{\lambda}_u} \qquad \bar{\psi}_6 = \frac{d_v}{c_v^2 - b_v d_v} = \frac{\sinh \bar{\lambda}_v - \sin \bar{\lambda}_v}{1 - \cosh \bar{\lambda}_v \cos \bar{\lambda}_v}$$

$$\bar{\psi}_7 = \frac{c_v}{c_v^2 - b_v d_v} = \frac{\cosh \bar{\lambda}_v - \cos \bar{\lambda}_v}{1 - \cosh \bar{\lambda}_u \cos \bar{\lambda}_v} \qquad \bar{\psi}_8 = \frac{b_v}{c_v^2 - b_v d_v} = \frac{\sinh \bar{\lambda}_v + \sin \bar{\lambda}_v}{1 - \cosh \bar{\lambda}_v \cos \bar{\lambda}_v}$$

(*f*) Shape Parameters

$$\bar{\lambda}_u = l\lambda_u = l\sqrt{\frac{\mu\omega^2}{EA}} \qquad \bar{\lambda}_v = l\lambda_v = l\sqrt[4]{\frac{\mu\omega^2}{EI}}$$

* All terms in these matrices correspond to the mth mode and all quantities are in the l-system. For the origin of a_u, d_u, a_v, b_v, c_v, and d_v and $\bar{\lambda}_u$ and $\bar{\lambda}_v$ refer to (*6-6*), (*6-12, 13*), and (*6-14*) and (*6-25*), (*6-31, 32*) and (*6-34*).

9.2 FLEXIBILITY METHOD

Free Vibration

Once the dynamic flexibility matrix is available for a single straight segment, the analysis of multisegment open bars or of multisegment closed rings is accomplished by *chain overlapping* of single member matrices and is analogous to the static flexibility method (Problem 9.4).

Table 9-3 Absolute Dynamic Flexibility and Stiffness Matrix

(*a*) Equivalents

$$r^2 = I/A, \quad \bar{\lambda}_u = l\sqrt{\mu\omega^2/EA}, \quad \bar{\lambda}_v = l\sqrt[4]{\mu\omega^2/EI}, \quad \bar{\eta} = 1 - \cosh\bar{\lambda}_v \cos\bar{\lambda}_v$$

(*b*) Dimensionless Flexibility Coefficients

$$\bar{f}_1 = \frac{\bar{\psi}_1}{\bar{\lambda}_u} = \frac{\cos\bar{\lambda}_u}{\bar{\lambda}_u \sin\bar{\lambda}_u}$$

$$\bar{f}_2 = \frac{\bar{\psi}_2}{\bar{\lambda}_v^3} = \frac{\cosh\bar{\lambda}_v \sin\bar{\lambda}_v - \sinh\bar{\lambda}_v \cos\bar{\lambda}_v}{\bar{\eta}\bar{\lambda}_v^3}$$

$$\bar{f}_3 = \frac{\bar{\psi}_3}{\bar{\lambda}_v^2} = \frac{\sinh\bar{\lambda}_v \sin\bar{\lambda}_v}{\bar{\eta}\bar{\lambda}_v^3}$$

$$\bar{f}_4 = \frac{\bar{\psi}_4}{\bar{\lambda}_v} = \frac{\cosh\bar{\lambda}_v \sin\bar{\lambda}_v + \sinh\bar{\lambda}_v \cos\bar{\lambda}_v}{\bar{\eta}\bar{\lambda}_v^2}$$

$$\bar{f}_5 = \frac{\bar{\psi}_5}{\bar{\lambda}_u} = \frac{1}{\bar{\lambda}_u \sin\bar{\lambda}_u}$$

$$\bar{f}_6 = \frac{\bar{\psi}_6}{\bar{\lambda}_v^3} = \frac{\sinh\bar{\lambda}_v - \sin\bar{\lambda}_v}{\bar{\eta}\bar{\lambda}_v^3}$$

$$\bar{f}_7 = \frac{\bar{\psi}_7}{\bar{\lambda}_v^2} = \frac{\cosh\bar{\lambda}_v - \cos\bar{\lambda}_v}{\bar{\eta}\bar{\lambda}_v^2}$$

$$\bar{f}_8 = \frac{\bar{\psi}_8}{\bar{\lambda}_v} = \frac{\sinh\bar{\lambda}_v + \sin\bar{\lambda}_v}{\bar{\eta}\bar{\lambda}_v}$$

(*c*) Dimensionless Stiffness Coefficients

$$\bar{k}_1 = \bar{\psi}_1\bar{\lambda}_u = \frac{\bar{\lambda}_u \cos\bar{\lambda}_u}{\sin\bar{\lambda}_u}$$

$$\bar{k}_2 = \bar{\psi}_2\bar{\lambda}_v = \frac{\bar{\lambda}_v(\cosh\bar{\lambda}_v \sin\bar{\lambda}_v - \sinh\bar{\lambda}_v \cos\bar{\lambda}_v)}{\bar{\eta}}$$

$$\bar{k}_3 = \bar{\psi}_3\bar{\lambda}_v^2 = \frac{\bar{\lambda}_v^2(\sinh\bar{\lambda}_v \sin\bar{\lambda}_v)}{\bar{\eta}}$$

$$\bar{k}_4 = \bar{\psi}_4\bar{\lambda}_v^3 = \frac{\bar{\lambda}_v^3(\cosh\bar{\lambda}_v \sin\bar{\lambda}_v + \sinh\bar{\lambda}_v \cos\bar{\lambda}_v)}{\bar{\eta}}$$

$$\bar{k}_5 = \bar{\psi}_5\bar{\lambda}_u = \frac{\bar{\lambda}_u}{\sin\bar{\lambda}_u}$$

$$\bar{k}_6 = \bar{\psi}_6\bar{\lambda}_v = \frac{\bar{\lambda}_v(\sinh\bar{\lambda}_v - \sin\bar{\lambda}_v)}{\bar{\eta}}$$

$$\bar{k}_7 = \bar{\psi}_7\bar{\lambda}_v^2 = \frac{\bar{\lambda}_v^2(\cosh\bar{\lambda}_v - \cos\bar{\lambda}_v)}{\bar{\eta}}$$

$$\bar{k}_8 = \bar{\psi}_8\bar{\lambda}_v^3 = \frac{\bar{\lambda}_v^3(\sinh\bar{\lambda}_v + \sin\bar{\lambda}_v)}{\bar{\eta}}$$

(*d*) Absolute Flexibility Matrix Equation (*9-5*)*

$$\begin{bmatrix} u_{RL} \\ v_{RL} \\ \theta_{RL} \\ \hline u_{LR} \\ v_{LR} \\ \theta_{LR} \end{bmatrix} = \frac{l}{EI} \left[\begin{array}{ccc|ccc} -r^2\bar{f}_1 & 0 & 0 & r^2\bar{f}_5 & 0 & 0 \\ 0 & -l^2\bar{f}_2 & -l\bar{f}_3 & 0 & -l^2\bar{f}_6 & l\bar{f}_7 \\ 0 & -l\bar{f}_3 & -\bar{f}_4 & 0 & -l\bar{f}_7 & \bar{f}_8 \\ \hline r^2\bar{f}_5 & 0 & 0 & -r^2\bar{f}_1 & 0 & 0 \\ 0 & -l^2\bar{f}_6 & -l\bar{f}_7 & 0 & -l^2\bar{f}_2 & l\bar{f}_3 \\ 0 & l\bar{f}_7 & \bar{f}_8 & 0 & l\bar{f}_3 & -\bar{f}_4 \end{array}\right] \begin{bmatrix} N_R \\ V_R \\ M_R \\ \hline N_L \\ V_L \\ M_L \end{bmatrix} + \begin{bmatrix} u_{R0} \\ v_{R0} \\ \theta_{R0} \\ \hline u_{L0} \\ v_{L0} \\ \theta_{L0} \end{bmatrix}$$

(*e*) Absolute Stiffness Matrix Equation (*9-11*)*

$$\begin{bmatrix} N_{RL} \\ V_{RL} \\ M_{RL} \\ \hline N_{LR} \\ V_{LR} \\ M_{LR} \end{bmatrix} = \frac{EI}{l} \left[\begin{array}{ccc|ccc} \bar{k}_1/r^2 & 0 & 0 & -\bar{k}_5/r^2 & 0 & 0 \\ 0 & \bar{k}_4/l^2 & -\bar{k}_3/l & 0 & -\bar{k}_8/l^2 & -\bar{k}_7/l \\ 0 & -\bar{k}_3/l & \bar{k}_2 & 0 & \bar{k}_7/l & \bar{k}_6 \\ \hline -\bar{k}_5/r^2 & 0 & 0 & \bar{k}_1/r^2 & 0 & 0 \\ 0 & -\bar{k}_8/l^2 & \bar{k}_7/l & 0 & \bar{k}_4/l^2 & \bar{k}_3/l \\ 0 & -\bar{k}_7/l & \bar{k}_6 & 0 & \bar{k}_3/l & \bar{k}_2 \end{array}\right] \begin{bmatrix} u_R \\ v_R \\ \theta_R \\ \hline u_L \\ v_L \\ \theta_L \end{bmatrix} + \begin{bmatrix} N_{R0} \\ V_{R0} \\ M_{R0} \\ \hline N_{L0} \\ V_{L0} \\ M_{L0} \end{bmatrix}$$

* For particular forms of load functions $u_{R0}, v_{R0}, \ldots$, and $N_{R0}, V_{R0}, \ldots$, refer to Tables P-9.20 and P-9.26 respectively.

The mechanics of construction of system flexibility matrices based on the conditions of joint deformation compatibility follows the pattern of static analysis and leads eventually to the frequency determinant equation.

Forced Vibration

In addition to the treatment of joint loads, the distributed mass bars are able to account for *dynamic loads* acting between the joints.

For dynamic loads, the second submatrix equations in (*8-26, 27*) with $\bar{\Delta}_L = -\bar{\Delta}_{LR}$ and $\bar{\Delta}_R = \bar{\Delta}_{RL}$ yield

$$\begin{aligned}
\bar{\Delta}_{RL} &= \underbrace{-\bar{U}^{)-1}_{\sigma\Delta, LR}\bar{U}_{\sigma\sigma, LR}}_{\bar{f}_{RR}}\bar{\sigma}_R + \underbrace{\bar{U}^{)-1}_{\sigma\Delta, LR}}_{\bar{f}_{RL}}\bar{\sigma}_L - \underbrace{\bar{U}^{)-1}_{\sigma\Delta, LR}\bar{\Phi}_{\sigma, LR}}_{\bar{\Delta}_{R0}} \\
\bar{\Delta}_{LR} &= \underbrace{-\bar{U}^{)-1}_{\sigma\Delta, RL}}_{\bar{f}_{LR}}\bar{\sigma}_R + \underbrace{\bar{U}^{)-1}_{\sigma\Delta, RL}\bar{U}_{\sigma\Delta, RL}}_{\bar{f}_{LL}}\bar{\sigma}_L + \underbrace{\bar{U}^{)-1}_{\sigma\Delta, RL}\bar{\Phi}_{\sigma, RL}}_{\bar{\Delta}_{L0}}
\end{aligned} \tag{9-7}$$

where the last matrix term in each equation is the *dimensionless load term* (Problems 9.5 and 9.6). A special application of (*9-7*) is the *dynamic three-moment equation* derived in Problem 9.7 and applied in Problem 9.8.

9.3 STIFFNESS MATRIX

Concept

The dynamic end stiffness (segmental stiffness) of a bar is defined as the end stress produced by a unit displacement (deflection or slope). Since the unit displacement can be applied along three orthogonal axes at each end, and each unit displacement can in general produce three mutually perpendicular stresses at each end, the derivation of a stiffness matrix for a planar bar requires the calculation of *36 coefficients.*

If the derivation is carried out in the member system (principal stiffnesses) and the reciprocal relationships are observed, the total number of distinct coefficients in a straight bar of constant section reduces to *eight.*

Derivation

These coefficients can be obtained (as in Sec. 9.1) from the respective equations of motion or directly from the transport matrices of Table 8-1. The latter approach is introduced below.

Dimensionless Matrix

The comparison of the stress and displacement vectors of the transport method (Table 9-1*a*) with those of the stiffness method (Table 9-1*c*) shows that the displacement vectors are identical and the stress vectors are related as

$$\sigma_L = -\sigma_{LR}, \qquad \sigma_R = \sigma_{RL} \qquad \text{or} \qquad \bar{\sigma}_L = -\bar{\sigma}_{LR}, \qquad \bar{\sigma}_R = \bar{\sigma}_{RL} \tag{9-8, 9}$$

In terms of (*9-9*) the third row submatrix equations of Table 8-1*c* and *d* become

$$\bar{\Delta}_L = \bar{U}_{\Delta\sigma, LR}\bar{\sigma}_{RL} + \bar{U}_{\Delta\Delta, LR}\bar{\Delta}_R \qquad \bar{\Delta}_R = -\bar{U}_{\Delta\sigma, RL}\bar{\sigma}_{LR} + \bar{U}_{\Delta\Delta, RL}\bar{\Delta}_L \tag{9-10}$$

and yield

$$\bar{\sigma}_{RL} = \underbrace{-\bar{U}^{)-1}_{\Delta\sigma, LR}\bar{U}_{\Delta\Delta, LR}}_{\bar{k}_{RR}}\bar{\Delta}_R + \underbrace{\bar{U}^{)-1}_{\Delta\sigma, LR}}_{\bar{k}_{RL}}\bar{\Delta}_L \qquad \bar{\sigma}_{LR} = \underbrace{-\bar{U}^{)-1}_{\Delta\sigma, RL}}_{\bar{k}_{LR}}\bar{\Delta}_R + \underbrace{\bar{U}^{)-1}_{\Delta\sigma, RL}\bar{U}_{\Delta\Delta, RL}}_{\bar{k}_{LL}}\bar{\Delta}_L \tag{9-11}$$

where $\bar{k}_{RR}$ and $\bar{k}_{LL}$ are the dimensionless near end (direct) stiffness submatrices, [3 × 3], at R and L respectively, and $\bar{k}_{RL}$ and $\bar{k}_{LR}$ are the dimensionless far end (indirect) stiffness submatrices, [3 × 3], at R and L respectively.

These four submatrices define symbolically the *dimensionless dynamic stiffness matrix*, $[6 \times 6]$, of the given bar in the l-system, corresponding to the mth mode. The coefficients of these submatrices are again functions of $\bar{\lambda}_u$ or $\bar{\lambda}_v$; their analytical expressions are given in Table 9-2*b* and their derivation is shown in Problem 9.9.

Absolute Matrix

The dimensioning of the stiffness matrix of Table 9-2*b* is accomplished by matrices κ_σ and λ_Δ introduced in (*8-6*) and (*8-9*) respectively. Thus,

$$\sigma_{RL} = \underbrace{\kappa_\sigma \bar{k}_{RR} \lambda_\Delta}_{k_{RR}} \Delta_R + \underbrace{\kappa_\sigma \bar{k}_{RL} \lambda_\Delta}_{k_{RL}} \Delta_L \qquad \sigma_{LR} = \underbrace{\kappa_\sigma \bar{k}_{LR} \lambda_\Delta}_{k_{LR}} \Delta_R + \underbrace{\kappa_\sigma \bar{k}_{LL} \lambda_\Delta}_{k_{LL}} \Delta_L \tag{9-12}$$

where k_{RR} and k_{LL} are the absolute near end (direct) stiffness submatrices, $[3 \times 3]$, at R and L respectively, and k_{RL} and k_{LR} are the absolute far end (indirect) stiffness submatrices, $[3 \times 3]$ at R and L respectively; again all are in the member system and correspond to the mth mode. Their analytical expressions are given in Table 9-3*e* and their derivation is shown in Problem 9.10.

Angular Transformations

The angular transformation of (*9-11*) is analogous to (*9-6*), i.e.,

$$\begin{aligned} \bar{\sigma}^R_{RL} &= \underbrace{\lambda_\sigma \pi^{Rl} \kappa_\sigma \bar{k}^l_{RR} \lambda_\Delta \pi^{lR} \kappa_\Delta}_{\bar{k}^{RR}_{RR}} \bar{\Delta}^R_R + \underbrace{\lambda_\sigma \pi^{Rl} \kappa_\sigma \bar{k}^l_{RL} \lambda_\Delta \pi^{lL} \kappa_\Delta}_{\bar{k}^{RL}_{RL}} \bar{\Delta}^l_L \\ \bar{\sigma}^L_{LR} &= \underbrace{\lambda_\sigma \pi^{Ll} \kappa_\sigma \bar{k}^l_{LR} \lambda_\Delta \pi^{lR} \kappa_\Delta}_{\bar{k}^{LR}_{LR}} \bar{\Delta}^R_R + \underbrace{\lambda_\sigma \pi^{Ll} \kappa_\sigma \bar{k}^l_{LL} \lambda_\Delta \pi^{lL} \kappa_\Delta}_{\bar{k}^{LL}_{LL}} \bar{\Delta}^L_L \end{aligned} \tag{9-13}$$

where the superscripts R, L, and l and the π-matrices have the same meaning as in (*9-6*).

The derivation of (*9-13*) and the angular transformation of (*9-12*) are shown in Problem 9.11.

9.4 STIFFNESS METHOD

Free Vibration

Once the dynamic stiffness matrix is available for a single straight segment, the analysis of complex frames is accomplished by *overlapping* the single member matrices and is analogous to the static stiffness method (Problem 9.12).

The mechanics of construction of system stiffness matrices based on the conditions of joint equilibrium follows the pattern of static analysis and leads eventually to the frequency determinant equation.

Forced Vibration

In addition to the treatment of joint loads, the distributed mass bars are able to account for dynamic loads acting between the joints.

For dynamic loads, the third submatrix equations in (*8-26, 27*) with $\bar{\sigma}_L = -\bar{\sigma}_{LR}$ and $\bar{\sigma}_R = \bar{\sigma}_{RL}$ yield

$$\begin{aligned} \bar{\sigma}_{RL} &= \underbrace{-\bar{U}^{-1}_{\Delta\sigma,LR} \bar{U}_{\Delta\Delta,LR}}_{\bar{k}_{RR}} \bar{\Delta}_R + \underbrace{\bar{U}^{-1}_{\Delta\sigma,LR}}_{\bar{k}_{RL}} \bar{\Delta}_L - \underbrace{\bar{U}^{-1}_{\Delta\sigma,LR} \bar{\Phi}_{\Delta,LR}}_{\bar{\sigma}_{R0}} \\ \bar{\sigma}_{LR} &= \underbrace{-\bar{U}^{-1}_{\Delta\sigma,RL}}_{\bar{k}_{LR}} \bar{\Delta}_R + \underbrace{\bar{U}^{-1}_{\Delta\sigma,RL} \bar{U}_{\Delta\Delta,RL}}_{\bar{k}_{LL}} \bar{\Delta}_L + \underbrace{\bar{U}^{-1}_{\Delta\sigma,RL} \bar{\Phi}_{\Delta,RL}}_{\bar{\sigma}_{L0}} \end{aligned} \tag{9-14}$$

where the last term in each equation is the dimensionless load term (Problems 9.13 and 9.14). Again a special application of (*9-14*) is the *dynamic three-slope equation* derived in Problem 9.15 and applied in Problem 9.16.

9.5 ANALYSIS OF FUNCTIONS

Dimensionless Matrices

The dimensionless flexibility and stiffness matrices of Table 9-2*a* and *b* are functions of eight dimensionless parameters $\bar{\psi}_1, \bar{\psi}_2, \ldots, \bar{\psi}_8$, which in turn are functions of six dimensionless parameters a_u, d_u, a_v, b_v, c_v, d_v forming the coefficients of the dimensionless transport matrix of Table 8-1.

Absolute Matrices

The absolute flexibility and stiffness matrices of Table 9-3*d* and *e* are again functions of $\bar{\psi}_1, \bar{\psi}_2, \ldots, \bar{\psi}_8$ divided or multiplied by the power of the respective dimensionless shape parameter $\bar{\lambda}_u$ or $\bar{\lambda}_v$. For convenience they are designated as the *dimensionless flexibility coefficients* $\bar{f}_1, \bar{f}_2, \ldots, \bar{f}_8$ and the *dimensionless stiffness coefficients* $\bar{k}_1, \bar{k}_2, \ldots, \bar{k}_8$. As is apparent from Table 9-3*b* and *c*,

$$\frac{\bar{k}_i}{\bar{f}_i} = \bar{\lambda}_u^2 \qquad \frac{\bar{k}_j}{\bar{f}_j} = \bar{\lambda}_v^4 \tag{9-15}$$

where $i = 1, 5$ and $j = 2, 3, 4, 6, 7, 8$.

Matrix Structures

It may be easily verified that these matrices are unsymmetrical in the dimensionless form (Table 9-2) but are symmetrical in the absolute form (Table 9-3), and the position of the negative quantities in Table 9-3*d* and *e* is identical to the position of their static counterparts.

The coefficients $\bar{f}_1, \bar{f}_2, \ldots, \bar{f}_8$ (Table 9-3*b*) can be expanded into the respective McLaurin's series about $\bar{\lambda}_u = 0$ and $\bar{\lambda}_v = 0$ as

$$\bar{f}_i = \frac{(n_{i,0} + n_{i,2}\bar{\lambda}_u^2 + n_{i,4}\bar{\lambda}_u^4 + \cdots)}{\bar{\lambda}_u^2}, \quad i = 1, 5$$
$$\bar{f}_j = \frac{(n_{j,0} + n_{j,2}\bar{\lambda}_v^4 + n_{j,4}\bar{\lambda}_v^8 + \cdots)}{\bar{\lambda}_v^4}, \quad j = 2, 3, 4, 6, 7, 8 \tag{9-16}$$

where all odd derivatives of $\bar{f}_i(0)$ and $\bar{f}_j(0)$ equal zero (Problem 9.18).

Similarly, the coefficients $\bar{k}_1, \bar{k}_2, \ldots, \bar{k}_8$ (Table 9-3*c*) can be expanded as

$$\bar{k}_i = n_{i,0} + n_{i,2}\bar{\lambda}_u^2 + n_{i,4}\bar{\lambda}_u^4 + \cdots, \quad i = 1, 5$$
$$\bar{k}_j = n_{j,0} + n_{j,2}\bar{\lambda}_v^4 + n_{j,4}\bar{\lambda}_v^8 + \cdots, \quad j = 2, 3, 4, 6, 7, 8 \tag{9-17}$$

where again all odd derivatives of $\bar{k}_i(0)$ and $\bar{k}_j(0)$ equal zero (Problem 9.17).

Solved Problems

FLEXIBILITY MATRIX

9.1. Derive the dimensionless dynamic flexibility matrix (*9-4*) of the prismatic bar shown in Table 9-1*b*.

From Table 8-1*c* and *d* of the dimensionless dynamic transport matrix,

$$\bar{\sigma}_R = \bar{U}_{\sigma\sigma,RL}\bar{\sigma}_L + \bar{U}_{\sigma\Delta,RL}\bar{\Delta}_L \qquad \bar{\sigma}_L = \bar{U}_{\sigma\sigma,LR}\bar{\sigma}_R + \bar{U}_{\sigma\Delta,LR}\bar{\Delta}_R \tag{1, 2}$$

Changing the sign convention according to Table 9-1*a* and *b* as $\bar{\Delta}_L = -\bar{\Delta}_{LR}$ and $\bar{\Delta}_R = \bar{\Delta}_{RL}$, (*1, 2*) become

$$\bar{\sigma}_R = \bar{U}_{\sigma\sigma,RL}\bar{\sigma}_L - \bar{U}_{\sigma\Delta,RL}\bar{\Delta}_{LR} \qquad \bar{\sigma}_L = \bar{U}_{\sigma\sigma,LR}\bar{\sigma}_R + \bar{U}_{\sigma\Delta,LR}\bar{\Delta}_{RL} \tag{3, 4}$$

Since the flexibility matrix expresses the displacement vectors in terms of stress vectors, (*3, 4*) should be arranged as

$$\bar{\Delta}_{LR} = \underbrace{-\bar{U}^{)-1}_{\sigma\Delta,RL}}_{\bar{f}_{LR}}\bar{\sigma}_R + \underbrace{\bar{U}^{)-1}_{\sigma\Delta,RL}\bar{U}_{\sigma\sigma,RL}}_{\bar{f}_{LL}}\bar{\sigma}_L \qquad \bar{\Delta}_{RL} = \underbrace{-\bar{U}^{)-1}_{\sigma\Delta,LR}\bar{U}_{\sigma\sigma,LR}}_{\bar{f}_{RR}}\bar{\sigma}_R + \underbrace{\bar{U}^{)-1}_{\sigma\Delta,LR}}_{\bar{f}_{RL}}\bar{\sigma}_L \tag{5, 6}$$

The inverse matrix terms in (5, 6) are

$$\bar{U}^{)-1}_{\sigma\Delta,RL} = \begin{bmatrix} -\dfrac{b_v d_v - c_v^2}{d_u} & 0 & 0 \\ 0 & d_v & -c_v \\ 0 & -c_v & b_v \end{bmatrix} \frac{1}{b_v d_v - c_v^2} \qquad \bar{U}^{)-1}_{\sigma\Delta,LR} = \begin{bmatrix} \dfrac{b_v d_v - c_v^2}{d_u} & 0 & 0 \\ 0 & -d_v & -c_v \\ 0 & -c_v & -b_v \end{bmatrix} \frac{1}{b_v d_v - c_v^2} \tag{7, 8}$$

for which the notational equivalents were given in Table 8-1*e*. Manipulating (5, 6) and using the dynamic parameters (Table 9-2*e*) yield the desired flexibility coefficients as

$$\bar{f}_{LR} = \begin{bmatrix} \bar{\psi}_5 & 0 & 0 \\ 0 & \bar{\psi}_6 & -\bar{\psi}_7 \\ 0 & -\bar{\psi}_7 & \bar{\psi}_8 \end{bmatrix} \qquad \bar{f}_{LL} = \begin{bmatrix} -\bar{\psi}_1 & 0 & 0 \\ 0 & \bar{\psi}_2 & \bar{\psi}_3 \\ 0 & -\bar{\psi}_3 & -\bar{\psi}_4 \end{bmatrix} \tag{9, 10}$$

$$\bar{f}_{RR} = \begin{bmatrix} -\bar{\psi}_1 & 0 & 0 \\ 0 & \bar{\psi}_2 & -\bar{\psi}_3 \\ 0 & \bar{\psi}_3 & -\bar{\psi}_4 \end{bmatrix} \qquad \bar{f}_{RL} = \begin{bmatrix} \bar{\psi}_5 & 0 & 0 \\ 0 & \bar{\psi}_6 & \bar{\psi}_7 \\ 0 & \bar{\psi}_7 & \bar{\psi}_8 \end{bmatrix} \tag{11, 12}$$

9.2. Derive the dimensional flexibility matrix (9-5) of a prismatic bar in the *l*-system.

In terms of the scaling matrix (8-9), the stress and displacement vectors in (9-4) may be expressed as

$$\bar{\sigma}_R = \lambda_\sigma \sigma_R \qquad \bar{\sigma}_L = \lambda_\sigma \sigma_L \tag{1, 2}$$

Thus (9-4) becomes

$$\bar{\Delta}_{RL} = \bar{f}_{RR}\lambda_\sigma\sigma_R + \bar{f}_{RL}\lambda_\sigma\sigma_L \qquad \bar{\Delta}_{LR} = \bar{f}_{LR}\lambda_\sigma\sigma_R + \bar{f}_{LL}\lambda_\sigma\sigma_L \tag{3, 4}$$

Using the dimensioning matrix (8-6) for $\Delta_{RL} = \kappa_\Delta \bar{\Delta}_{RL}$ and $\Delta_{LR} = \kappa_\Delta \bar{\Delta}_{LR}$ yields the following flexibility matrix:

$$\Delta_{RL} = \underbrace{\kappa_\Delta \bar{f}_{RR}\lambda_\sigma}_{f_{RR}}\sigma_R + \underbrace{\kappa_\Delta \bar{f}_{RL}\lambda_\sigma}_{f_{RL}}\sigma_L \qquad \Delta_{LR} = \underbrace{\kappa_\Delta \bar{f}_{LR}\lambda_\sigma}_{f_{LR}}\sigma_R + \underbrace{\kappa_\Delta \bar{f}_{LL}\lambda_\sigma}_{f_{LL}}\sigma_L \tag{5, 6}$$

where

$$f_{RR} = \begin{bmatrix} -\bar{\psi}_1/\lambda_u AE & 0 & 0 \\ 0 & -\bar{\psi}_2/\lambda_v^3 EI & -\bar{\psi}_3/\lambda_v^2 EI \\ 0 & \bar{\psi}_3/\lambda_v^2 EI & -\bar{\psi}_4/\lambda_v EI \end{bmatrix} \qquad f_{RL} = f_{LR}^{)T} = \begin{bmatrix} \bar{\psi}_5/\lambda_u AE & 0 & 0 \\ 0 & -\bar{\psi}_6/\lambda_v^3 EI & \bar{\psi}_7/\lambda_v^2 EI \\ 0 & -\bar{\psi}_7/\lambda_v^2 EI & \bar{\psi}_8/\lambda_v EI \end{bmatrix} \tag{7, 8}$$

and

$$f_{LL} = \begin{bmatrix} -\bar{\psi}_1/\lambda_u AE & 0 & 0 \\ 0 & -\bar{\psi}_2/\lambda_v^3 EI & \bar{\psi}_3/\lambda_v^2 EI \\ 0 & \bar{\psi}_3/\lambda_v^2 EI & -\bar{\psi}_4/\lambda_v EI \end{bmatrix} \tag{9}$$

9.3. Derive the angular transformation of the flexibility matrix (9-6).

According to (9-5),

$$\Delta^l_{RL} = f^l_{RR}\sigma^l_R + f^l_{RL}\sigma^l_L \tag{1}$$

where by (8-6) and (8-9)

$$\Delta^l_{RL} = \kappa_\Delta \bar{\Delta}^l_{RL} \qquad \sigma^l_R = \kappa_\sigma \bar{\sigma}^l_R \qquad \sigma^l_L = \kappa_\sigma \bar{\sigma}^l_L \tag{2}$$

Then

$$\Delta^l_{RL} = \pi^{lR}\Delta^R_{RL} = \pi^{lR}\kappa_\Delta \bar{\Delta}^R_{RL} \qquad \sigma^l_R = \pi^{lR}\sigma^R_R = \pi^{lR}\kappa_\sigma \bar{\sigma}^R_R \qquad \sigma^l_L = \pi^{lL}\sigma^L_L = \pi^{lL}\kappa_\sigma \bar{\sigma}^L_L \tag{3}$$

In terms of (3) the initial flexibility equation (1) becomes

$$\pi^{Rl}\kappa_\Delta \bar{\Delta}^l_{RL} = f^l_{RR}\pi^{lR}\kappa_\sigma \bar{\sigma}^R_R + f^l_{RL}\pi^{lL}\kappa_\sigma \bar{\sigma}^L_L \tag{4}$$

where by (9-4)

$$f^l_{RR} = \kappa_\Delta \bar{f}^l_{RR}\lambda_\sigma \qquad f^l_{RL} = \kappa_\Delta \bar{f}^l_{RL}\lambda_\sigma \tag{5}$$

On premultiplying (4) by $\lambda_\Delta \pi^{lR}$ the desired relation is

$$\bar{\Delta}^R_{RL} = \underbrace{\lambda_\Delta \pi^{Rl}\kappa_\Delta}_{\bar{\pi}^{Rl}_\Delta} \bar{f}^l_{RR} \underbrace{\lambda_\sigma \pi^{lR}\kappa_\sigma}_{\bar{\pi}^{lR}_\sigma} \bar{\sigma}^R_R + \underbrace{\lambda_\Delta \pi^{Rl}\kappa_\Delta}_{\bar{\pi}^{Rl}_\Delta} \bar{f}^l_{RL} \underbrace{\lambda_\sigma \pi^{lL}\kappa_\sigma}_{\bar{\pi}^{lL}_\sigma} \bar{\sigma}^L_L \tag{6}$$

where $\bar{\pi}^{Rl}_\Delta$, $\bar{\pi}^{lR}_\sigma$, and $\bar{\pi}^{lL}_\sigma$ are modified angular transformation matrices. The second equation in (9-6) can be derived from (6) by cyclic substitution.

The transformation of (9-5) is accomplished by means of relations (3) as

$$\pi^{lR}\Delta^R_{RL} = f^l_{RR}\pi^{lR}\sigma^R_R + f^l_{RL}\pi^{lL}\sigma^L_L \tag{7}$$

which, on premultiplying by π^{Rl}, becomes

$$\Delta^R_{RL} = \pi^{Rl} f^l_{RR}\pi^{lR}\sigma^R_R + \pi^{Rl} f^l_{RL}\pi^{lL}\sigma^L_L \tag{8}$$

which is formally similar to (6). The transformation of the second equation in (9-5) is also similar.

FLEXIBILITY METHOD

9.4. Formulate the system flexibility matrix for the frame of Fig. P-9.4 and then find the fundamental frequency of free vibration. Given for all members:

$l = 19$ in $\qquad \mu = 13.76 \times 10^{-5}$ lb-s^2/in^2 $\qquad A = 0.1875$ in^2

$I = 13.732 \times 10^{-5}$ in^4 $\qquad E = 30 \times 10^6$ psi

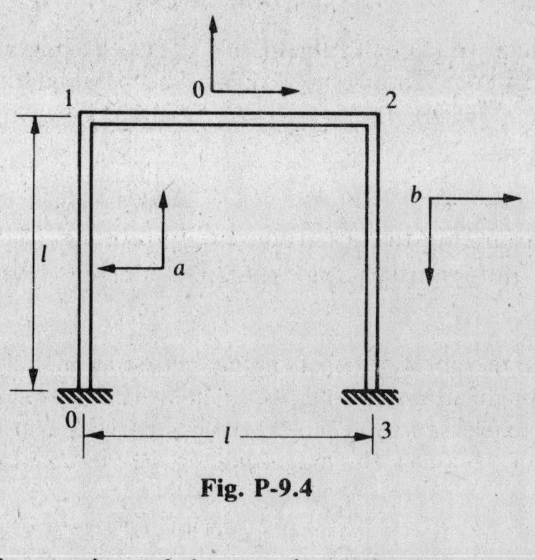

Fig. P-9.4

The dimensional flexibility matrices of the constituent members of Fig. P-9.4 may be formulated according to (8) of the preceding problem as

$$\begin{bmatrix} \Delta^0_{10} \\ \Delta^0_{01} \end{bmatrix} = [\pi^{0a}] \begin{bmatrix} f^a_{11} & f^a_{10} \\ f^a_{01} & f^a_{00} \end{bmatrix} [\pi^{a0}] \begin{bmatrix} \sigma^0_{10} \\ \sigma^0_{01} \end{bmatrix} = \begin{bmatrix} A^0_1 & B^0_1 \\ C^0_1 & D^0_1 \end{bmatrix} \begin{bmatrix} \sigma^0_{10} \\ \sigma^0_{01} \end{bmatrix} \tag{1}$$

$$\begin{bmatrix} \Delta^0_{21} \\ \Delta^0_{12} \end{bmatrix} = \begin{bmatrix} f^0_{22} & f^0_{21} \\ f^0_{12} & f^0_{11} \end{bmatrix} \begin{bmatrix} \sigma^0_{21} \\ \sigma^0_{12} \end{bmatrix} = \begin{bmatrix} A^0_2 & B^0_2 \\ C^0_2 & D^0_2 \end{bmatrix} \begin{bmatrix} \sigma^0_{21} \\ \sigma^0_{12} \end{bmatrix} \tag{2}$$

$$\begin{bmatrix} \Delta^0_{32} \\ \Delta^0_{23} \end{bmatrix} = [\pi^{0b}] \begin{bmatrix} f^b_{33} & f^b_{32} \\ f^b_{23} & f^b_{22} \end{bmatrix} [\pi^{b0}] \begin{bmatrix} \sigma^0_{32} \\ \sigma^0_{23} \end{bmatrix} = \begin{bmatrix} A^0_3 & B^0_3 \\ C^0_3 & D^0_3 \end{bmatrix} \begin{bmatrix} \sigma^0_{32} \\ \sigma^0_{23} \end{bmatrix} \tag{3}$$

The continuity at joints 0, 1, 2, and 3 requires that

$$\{\Delta^0_0\} = \{\Delta^0_{01}\} = 0 \qquad \{\Delta^0_1\} = \{\Delta^0_{10} + \Delta^0_{12}\} = 0 \qquad \{\Delta^0_2\} = \{\Delta^0_{21} + \Delta^0_{23}\} = 0 \qquad \{\Delta^0_3\} = \{\Delta^0_{32}\} = 0 \tag{4}$$

Substituting (*1*), (*2*), and (*3*) into (*4*) yields the following system flexibility matrix:

$$\underbrace{\begin{bmatrix} \Delta_0^0 \\ \Delta_1^0 \\ \Delta_2^0 \\ \Delta_3^0 \end{bmatrix}}_{\{\Delta\}} = \underbrace{\begin{bmatrix} 0 \\ 0 \\ 0 \\ 0 \end{bmatrix}}_{\{0\}} = \underbrace{\begin{bmatrix} D_1^0 & C_1^0 & 0 & 0 \\ B_1^0 & A_1^0 + D_2^0 & C_2^0 & 0 \\ 0 & B_2^0 & A_2^0 + D_3^0 & C_3^0 \\ 0 & 0 & B_3^0 & A_3^0 \end{bmatrix}}_{[\Phi]} \underbrace{\begin{bmatrix} \sigma_0^0 \\ \sigma_1^0 \\ \sigma_2^0 \\ \sigma_3^0 \end{bmatrix}}_{\{\sigma\}} \tag{5}$$

which shows that $[\Phi]$ is formed by matrix chains, laid out diagonally like a deck of cards, with overlapping corners. The dynamic parameters in $[\Phi]$ are

$$A_1^0 = \begin{bmatrix} -\psi_2/\lambda_v^3 EI & 0 & -\bar{\psi}_3/\lambda_v^2 EI \\ 0 & -\bar{\psi}_1/\lambda_u EA & 0 \\ -\bar{\psi}_3/\lambda_v^2 EI & 0 & -\bar{\psi}_4/\lambda_v EI \end{bmatrix} \qquad B_1^0 = C_1^{0)T} = \begin{bmatrix} -\bar{\psi}_6/\lambda_v^3 EI & 0 & \bar{\psi}_7/\lambda_v^2 EI \\ 0 & \bar{\psi}_5/\lambda_u EA & 0 \\ -\bar{\psi}_7/\lambda_v^2 EI & 0 & \bar{\psi}_8/\lambda_v EI \end{bmatrix}$$

$$D_1^0 = \begin{bmatrix} -\bar{\psi}_2/\lambda_v^3 EI & 0 & \bar{\psi}_3/\lambda_v^2 EI \\ 0 & -\bar{\psi}_1/\lambda_u AE & 0 \\ \bar{\psi}_3/\lambda_v^2 EI & 0 & -\bar{\psi}_4/\lambda_v EI \end{bmatrix} \qquad B_2^0 = C_2^{0)T} = \begin{bmatrix} \bar{\psi}_5/\lambda_u AE & 0 & 0 \\ 0 & -\bar{\psi}_6/\lambda_v^3 EI & \bar{\psi}_7/\lambda_v^2 EI \\ 0 & -\bar{\psi}_7/\lambda_v^2 EI & \bar{\psi}_8/\lambda_v EI \end{bmatrix}$$

$$A_2^0 = \begin{bmatrix} -\bar{\psi}_1/\lambda_u AE & 0 & 0 \\ 0 & -\bar{\psi}_2/\lambda_v^3 EI & -\bar{\psi}_3/\lambda_v^2 EI \\ 0 & -\bar{\psi}_3/\lambda_v^2 EI & -\bar{\psi}_4/\lambda_v EI \end{bmatrix} \qquad D_2^0 = \begin{bmatrix} -\bar{\psi}_1/\lambda_u AE & 0 & 0 \\ 0 & -\bar{\psi}_2/\lambda_v^3 EI & \bar{\psi}_3/\lambda_v^2 EI \\ 0 & \bar{\psi}_3/\lambda_v^2 EI & -\bar{\psi}_4/\lambda_v EI \end{bmatrix}$$

$$A_3^0 = D_1^0 \qquad B_3^0 = B_1^{0)T} \qquad C_3^0 = B_1^0 \qquad \text{and} \qquad D_3^0 = A_1^0$$

The solution of (*5*) can exist only if

$$\text{Det.}\,[\Phi] = 0 \tag{6}$$

in which the only variable is the angular frequency ω that is a function of λ_v and λ_u. Thus, varying ω by increasing $\Delta\omega$ (or λ_v by $\Delta\lambda_v$, $\lambda_u = \lambda_v^2\sqrt{I/A}$), one may calculate the determinants (*6*) corresponding to various values of $\omega + \Delta\omega$ (or $\lambda_v + \Delta\lambda_v$). For the given problem, the typical values of $\bar{\lambda}_v$ and determinant are shown below.

Det.	$\bar{\lambda}_v$	Det.	$\bar{\lambda}_v$
2.841×10^{-19}	2.5	-4.735×10^{-27}	4.5
2.387×10^{-20}	2.7	3.092×10^{-25}	4.7
2.203×10^{-21}	2.9	6.666×10^{-27}	4.9
2.121×10^{-22}	3.1	6.753×10^{-28}	5.1
1.880×10^{-23}	3.3	1.213×10^{-28}	5.3
0.992×10^{-25}	3.5	2.673×10^{-29}	5.5
-3.733×10^{-25}	3.7	6.522×10^{-30}	5.7
-1.974×10^{-25}	3.9	1.674×10^{-30}	5.9
-7.590×10^{-26}	4.1	4.333×10^{-31}	6.1
-2.659×10^{-26}	4.3		

(7)

It is obvious that $\text{Det.}\,(\Phi) = 0$ is between $\bar{\lambda}_v = 3.5$ and $\bar{\lambda}_v = 3.7$. By linear interpolation, $\bar{\lambda}_v = 3.542$ is the solution. Thus using the given member properties gives

$$\omega = \bar{\lambda}_v^2/l^2 \sqrt{\frac{EI}{\mu}} = 190.2 \text{ rad/s}$$

9.5. Derive the load functions (*9-7*) for a prismatic bar subjected to dynamic loads acting between two joints.

From the second submatrix equations in (*8-26, 27*),

$$\bar{\sigma}_R = \bar{\Phi}_{\sigma,RL} + \bar{U}_{\sigma\sigma,RL}\bar{\sigma}_L + \bar{U}_{\sigma\Delta,RL}\bar{\Delta}_L \qquad \bar{\sigma}_L = \bar{\Phi}_{\sigma,LR} + \bar{U}_{\sigma\sigma,LR}\bar{\sigma}_R + \bar{U}_{\sigma\Delta,LR}\bar{\Delta}_R \tag{1, 2}$$

in which the displacement vectors should be modified in accordance with the flexibility method (Table 9-1b) as $\bar{\Delta}_L = -\bar{\Delta}_{LR}$ and $\bar{\Delta}_R = -\bar{\Delta}_{RL}$. Consequently (*1, 2*) become

$$\bar{\Delta}_{RL} = \underbrace{-\bar{U}^{)-1}_{\sigma\Delta,LR}\bar{U}_{\sigma\sigma,LR}}_{\bar{f}_{RR}}\bar{\sigma}_R + \underbrace{\bar{U}^{)-1}_{\sigma\Delta,LR}}_{\bar{f}_{RL}}\bar{\sigma}_L \underbrace{- \bar{U}^{)-1}_{\sigma\Delta,LR}\bar{\Phi}_{\sigma,LR}}_{\bar{\Delta}_{RO}} \tag{3}$$

$$\bar{\Delta}_{LR} = \underbrace{-\bar{U}^{)-1}_{\sigma\Delta,RL}}_{\bar{f}_{LR}}\bar{\sigma}_R + \underbrace{\bar{U}^{)-1}_{\sigma\Delta,RL}\bar{U}_{\sigma\sigma,RL}}_{\bar{f}_{LL}}\bar{\sigma}_L + \underbrace{\bar{U}^{)-1}_{\sigma\Delta,RL}\bar{\Phi}_{\sigma,RL}}_{\bar{\Delta}_{LO}} \tag{4}$$

The last matrix terms of (*3*) and (*4*) are load functions. The dimensionless load effect vectors $\bar{\Phi}_{\sigma,RL}$ were derived in Problems 6.10 and 6.22 as

$$\bar{\Phi}_{\sigma,RL} = [\bar{\Phi}'_{Ru}/\lambda_u \quad \bar{\Phi}'''_{Rv}/\lambda_v^3 \quad \bar{\Phi}''_{Rv}/\lambda_v^2]^T \tag{5}$$

and $\bar{\Phi}_{\sigma,LR}$ may be obtained from the inverse of the transport matrix (*8-26, 27*), $\bar{\Phi}_{LR} = -\bar{U}^{)-1}_{RL}\bar{\Phi}_{RL} = -\bar{U}_{LR}\bar{\Phi}_{RL}$. Thus

$$\bar{\Phi}_{\sigma,LR} = -\bar{U}_{\sigma\sigma,LR}\bar{\Phi}_{\sigma,RL} - \bar{U}_{\sigma\Delta,LR}\bar{\Phi}_{\Delta,RL} \tag{6}$$

Substituting (*5, 6*) and (*7, 8*) of Problem 9.1 into (*3*) and (*4*) yields the desired load functions of Table 9-3d:

$$\bar{\Delta}_{RO} = \bar{U}^{)-1}_{\sigma\Delta,LR}[\bar{U}_{\sigma\sigma,LR} \mid \bar{U}_{\sigma\Delta,LR}][\bar{\Phi}_{\sigma,RL} \mid \bar{\Phi}_{\Delta,RL}]^T = \begin{bmatrix} \bar{\psi}_1\bar{\Phi}'_{Ru} + \bar{\Phi}_{Ru} \\ -\bar{\psi}_2\bar{\Phi}'''_{Rv} + \bar{\psi}_3\bar{\Phi}''_{Rv} + \bar{\Phi}_{Rv} \\ -\bar{\psi}_3\bar{\Phi}'''_{Rv} + \bar{\psi}_4\bar{\Phi}''_{Rv} + \bar{\Phi}'_{Rv} \end{bmatrix} = \begin{bmatrix} \bar{u}_{RO} \\ \bar{v}_{RO} \\ \bar{\theta}_{RO} \end{bmatrix} \tag{7}$$

$$\bar{\Delta}_{LO} = \bar{U}^{)-1}_{\sigma\Delta,RL}\bar{\Phi}_{\sigma,RL} = \begin{bmatrix} -\bar{\psi}_5\bar{\Phi}'_{Ru} \\ -\bar{\psi}_6\bar{\Phi}'''_{Rv} + \bar{\psi}_7\bar{\Phi}''_{Rv} \\ \bar{\psi}_7\bar{\Phi}'''_{Rv} - \bar{\psi}_8\bar{\Phi}''_{Rv} \end{bmatrix} = \begin{bmatrix} \bar{u}_{LO} \\ \bar{v}_{LO} \\ \bar{\theta}_{LO} \end{bmatrix} \tag{8}$$

Several particular cases of (*7*) and (*8*) are given in dimensional form in Table P-9.20.

9.6. **Find the internal forces and moments at the ends of the constituent members of the frame shown in Fig. P-9.6. Given:**

$$p_v(t) = 40 \sin \Omega t \text{ lb/in}$$
$$l = 19 \text{ in}$$
$$I = 13.732 \times 10^{-5} \text{ in}^4$$
$$\mu = 13.76 \times 10^{-5} \text{ lb-s}^2/\text{in}^2$$
$$E = 30 \times 10^6 \text{ psi}$$
$$A = 0.1875 \text{ in}^2$$
$$\Omega = 49.108 \text{ rad/s}$$

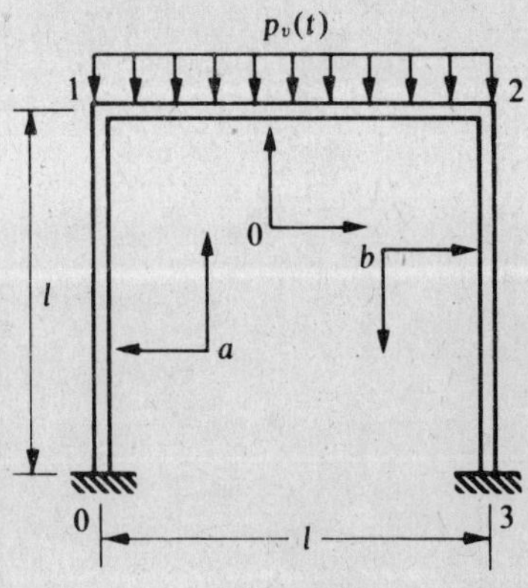

Fig. P-9.6

Since the structural configuration of Fig. P-9.6 is the same as Fig. P-9.4, the joint force-moment matrix (*5*) of Problem 9.4 can be directly used here with inclusion of the effect of external loads $\{\Delta_0\}$ as

$$\{\Delta\} = [\Phi]\{\sigma\} + \{\Delta_0\} \tag{1}$$

in which $\{\Delta_0\} = [0 \quad \Sigma\Delta_{10} \quad \Sigma\Delta_{20} \quad 0]^T$ of dimensional form. Thus by (*7*) and (*8*) of Problem 9.5, the effect of external loads for member 12 may be written as

$$\Delta_{10,2} = \kappa_\Delta\bar{\Delta}_{10,2} = \begin{bmatrix} 0 \\ \bar{\psi}_7\bar{\Phi}''_{2,v} - \bar{\psi}_6\bar{\Phi}'''_{2,v} \\ (\bar{\psi}_7\bar{\Phi}'''_{2,v} - \bar{\psi}_8\bar{\Phi}''_{2,v})\lambda_v \end{bmatrix} \qquad \Delta_{20,1} = \kappa_\Delta\bar{\Delta}_{20,1} = \begin{bmatrix} 0 \\ -\bar{\psi}_2\bar{\Phi}'''_{2,v} + \bar{\Phi}_{2,v} + \bar{\psi}_3\bar{\Phi}''_{2,v} \\ (-\bar{\psi}_3\bar{\Phi}'''_{2,v} + \bar{\psi}_4\bar{\Phi}''_{2,v} + \bar{\Phi}'_{2,v})\lambda_v \end{bmatrix} \tag{2, 3}$$

in which the load effect vectors due to the uniform load were given in Problem 6.22 as

$$\bar{\Phi}_{2,v} = -\frac{p_v[1 - a(\lambda_v l)]}{EI\lambda_v^4} \qquad \bar{\Phi}'_{2,v} = \frac{p_v d(\lambda_v l)}{EI\lambda_v^4} \qquad \bar{\Phi}''_{2,v} = \frac{p_v c(\lambda_v l)}{EI\lambda_v^4} \qquad \bar{\Phi}'''_{2,v} = \frac{p_v b(\lambda_v l)}{EI\lambda_v^4} \tag{4}$$

Considering the compatibility condition at joints gives $\{\Delta\} = 0$. Thus (*1*) becomes

$$\{\sigma\} = -[\Phi]^{-1}\{\Delta_0\} \tag{5}$$

For the given structure, we have

$$\begin{array}{lll} \bar{\psi}_1 = 216.69 & \bar{\psi}_5 = 216.69 & \lambda_v l = 1.8 \\ \bar{\psi}_2 = 2.166 & \bar{\psi}_6 = 1.1541 & \lambda_u l = 0.4615 \times 10^{-2} \\ \bar{\psi}_3 = 1.68 & \bar{\psi}_7 = 1.9551 & \\ \bar{\psi}_4 = 1.3828 & \bar{\psi}_8 = 1.7099 & \end{array}$$

and Φ is the system flexibility matrix constructed in (*5*) of Problem 9.4.

In terms of the inverse of this matrix and load functions given in (*2, 3*), the end forces and moments at the fixed end 0 and at the joint 1 are

$$\sigma_0^0 = \begin{bmatrix} 6.85 \\ -4.52 \\ 67.87 \end{bmatrix} \qquad \sigma_1^0 = \begin{bmatrix} 3.47 \\ -4.52 \\ -42.89 \end{bmatrix}$$

Their counterparts at 3 and 2 must be symmetrical. All vectors are in the 0-system, all forces are in kips, and all moments in kip-inches.

9.7. Derive the three-moment equation for dynamic analysis of the continuous beam of Fig. P-9.7.

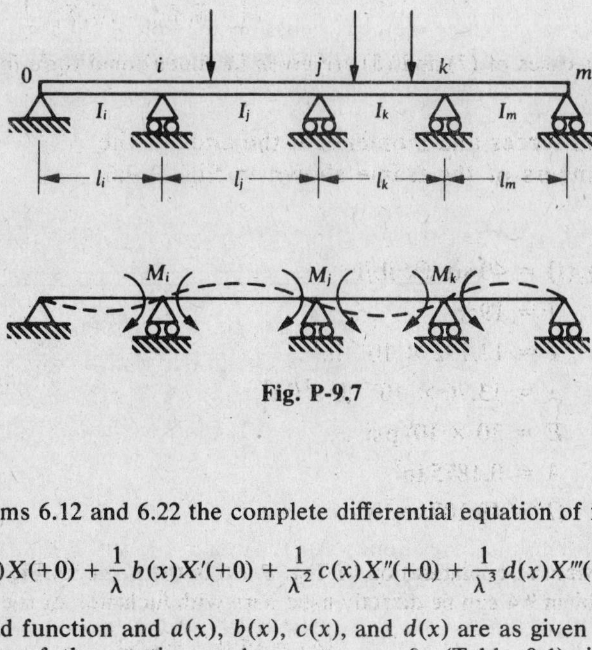

Fig. P-9.7

As shown in Problems 6.12 and 6.22 the complete differential equation of forced vibration is

$$X(x) = a(x)X(+0) + \frac{1}{\lambda}b(x)X'(+0) + \frac{1}{\lambda^2}c(x)X''(+0) + \frac{1}{\lambda^3}d(x)X'''(+0) + \Phi(x) \tag{1}$$

in which $\Phi(x)$ is the load function and $a(x)$, $b(x)$, $c(x)$, and $d(x)$ are as given in (*6-24*). Substituting the boundary conditions of the continuous beam at $x = 0$ (Table 9-1), i.e., $X(+0) = -v_L = 0$ and $X''(+0) = M_L/EI$, into (*1*) yields

$$X(x) = \frac{1}{\lambda}b(x)X'(+0) + \frac{1}{\lambda^2}c(x)\frac{M_L}{EI} + \frac{1}{\lambda^3}d(x)X'''(+0) + \Phi(x) \tag{2}$$

After imposing the boundary conditions at $x = l$, i.e., $X(+l) = v_R = 0$ and $X''(+l) = M_R/EI$, on (*2*), we find

$$\begin{bmatrix} X'(+0) \\ X'''(+0) \end{bmatrix} = \frac{\lambda^2}{\sinh \lambda l \sin \lambda l} \begin{bmatrix} -\frac{1}{\lambda}b(l)\left[\frac{b(l)}{\lambda^2}\frac{M_L}{EI} + \Phi(l)\right] + \frac{1}{\lambda^3}d(l)\left[a(l)\frac{M_L}{EI} + \Phi''(l)\right] \\ \lambda d(l)\left[\frac{b(l)}{\lambda^2}\frac{M_L}{EI} + \Phi(l)\right] - \frac{1}{\lambda}b(l)\left[a(l)\frac{M_L}{EI} + \Phi''(l)\right] \end{bmatrix} \tag{3}$$

Substituting (3) into (2) yields

$$X(x) = \frac{1}{2EI\lambda^2}\left[\frac{\sinh\lambda(l-x)}{\sinh\lambda l} - \frac{\sin\lambda(l-x)}{\sin\lambda l}\right]M_L + \frac{1}{2EI\lambda^2}\left[\frac{\sinh\lambda x}{\sinh\lambda l} - \frac{\sin\lambda x}{\sin\lambda l}\right]M_R + \frac{1}{\lambda}b(x)q + \frac{1}{\lambda^3}d(x)r + \Phi(x) \tag{4}$$

in which

$$q = \frac{\lambda}{\sinh\lambda l\sin\lambda l}\left[\frac{1}{\lambda^2}d(l)\Phi''(l) - b(l)\Phi(l)\right] \qquad r = \frac{\lambda^2}{\sinh\lambda l\sin\lambda l}\left[\lambda d(l)\Phi(l) - \frac{1}{\lambda}b(l)\Phi''(l)\right] \tag{5}$$

Introducing $X'(+0) = -\theta_L$ and $X'(+l) = -\theta_R$ into (4) yields

$$\theta_L = F_{LL}M_L + F_{LR}M_R + \tau_{LO} \qquad \theta_R = -F_{RR}M_R - F_{RL}M_L - \tau_{RO} \tag{6}$$

where in terms of $\lambda l = \bar{\lambda}$, $a(l) = a$ and $d(l) = d$,

$$F_{LL} = F_{RR} = \frac{\coth\bar{\lambda} - \cot\bar{\lambda}}{2EI\lambda} \qquad F_{LR} = F_{RL} = \frac{\csc\bar{\lambda} - \operatorname{csch}\bar{\lambda}}{2EI\lambda}$$

are the angular flexibilities and in terms of q and r given in (5), $a(l) = a$ and $c(l) = c$,

$$\tau_{LO} = -q + \Phi'(0) \qquad \tau_{RO} = aq + \frac{cr}{\lambda^2} + \Phi'(l)$$

are the angular load functions given in particular forms in Table P-9.21.

The condition of continuity at a support j requires that $\theta_{j\,(\text{left})} = \theta_{j\,(\text{right})}$ and lead to

$$F_{ji}M_i + \Sigma F_j M_j + F_{jk}M_k + \Sigma\tau_{jo} = 0 \tag{7}$$

known as the dynamic three-moment equation in which $\Sigma F_j = F_{jj,j} + F_{jj,k}$ is the joint angular flexibility and $\Sigma\tau_{jo} + \tau_{jo,j} + \tau_{jo,k}$ is the joint angular load function. Their physical interpretations are analogous to those of the static analysis.

For free vibration, $\Sigma\tau_{jo} = 0$.

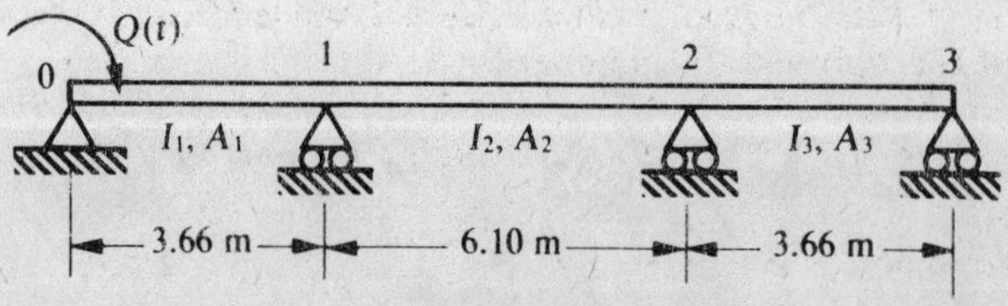

Fig. P-9.8

9.8. Find (a) the fundamental frequency for the beam of Fig. P-9.8 for free vibration $(Q(t) = 0)$ and (b) the internal moments due to the applied moment $Q(t)$. Given:

$I_1 = I_3 = 10.11\ \text{dm}^4$ $\qquad E = 199.95\ \text{GN/m}^2$ $\qquad Q(t) = 67.79\cos\omega t$ kN-m

$I_2 = 20.23\ \text{dm}^4$ $\qquad \gamma = 76.973\ \text{kN/m}^3$ $\qquad \omega = 1057.5$ rad/s

$A_1 = A_2 = A_3 = 1161.3\ \text{cm}^2$

(a) Applying the three-moment equation at joints 1 and 2 of Fig. P-9.8 gives

$$\begin{bmatrix} F_{11,1} + F_{11,2} & F_{12} \\ F_{21} & F_{22,2} + F_{22,3} \end{bmatrix}\begin{bmatrix} M_1 \\ M_2 \end{bmatrix} = 0 \tag{1}$$

in which $\mu_1 = \mu_2 = \mu_3 = \mu$ and

$$\lambda l_2 = l_2\sqrt[4]{\frac{\mu\omega^2}{EI_2}} \qquad \lambda l_1 = l_1\sqrt[4]{\frac{\mu\omega^2}{EI_1}} = \frac{l_1}{l_2}\sqrt[4]{\frac{I_2}{I_1}}(\lambda l_2) = \lambda l_3 \tag{2}$$

Let $\lambda l_2 = 3.5$; then $\lambda l_1 = \lambda l_3 = (3.66/6.10)\sqrt[4]{20.23/10.11}(3.5) = 2.5$

$$\sinh \lambda l_1 = 6.0525 \qquad \sin \lambda l_1 = 0.5985 \qquad \sinh \lambda l_2 = 16.543 \qquad \sin \lambda l_2 = -0.3522$$
$$\cosh \lambda l_1 = 6.13225 \qquad \cos \lambda l_1 = -0.8011 \qquad \cosh \lambda l_2 = 16.573 \qquad \cos \lambda l_2 = -0.93645$$

Consequently (*1*) becomes

$$\frac{l_1}{2EI_1}\begin{bmatrix} 0.9408 - 0.3945 & -0.60409 \\ -0.60409 & 0.9408 - 0.3945 \end{bmatrix}\begin{bmatrix} M_1 \\ M_2 \end{bmatrix} = 0 \tag{3}$$

The determinant of the coefficients in (*3*) is Det. $= -0.0665$.
Now let $\lambda l_2 = 3.6$; then $\lambda l_1 = \lambda l_3 = 2.568$.

$$\sinh \lambda l_1 = 6.48165 \qquad \sin \lambda l_1 = 0.5426 \qquad \sinh \lambda l_2 = 18.2853 \qquad \sin \lambda l_2 = -0.4446$$
$$\cosh \lambda l_1 = 6.55835 \qquad \cos \lambda l_1 = -0.840 \qquad \cosh \lambda l_2 = 18.3126 \qquad \cos \lambda l_2 = -0.8968$$

Thus (*1*) may be rewritten as

$$\frac{l_1}{2EI_1}\begin{bmatrix} 0.99686 - 0.23506 & -0.5333 \\ -0.5333 & 0.99686 - 0.23506 \end{bmatrix}\begin{bmatrix} M_1 \\ M_2 \end{bmatrix} = 0 \tag{4}$$

where the determinant is Det. $= 0.2965$.
By linear interpolation, the approximate solution is

$$\lambda l_2 = 3.5002 \qquad \omega = (\lambda l_2)^2\sqrt{\frac{EI_2}{\mu_2 l_2^4}} = 999.57 \text{ rad/s}$$

(*b*) Using (*9*) of the preceding problem at joints 1 and 2 of Fig. P-9.8 we write

$$\begin{bmatrix} F_{11,1} + F_{11,2} & F_{12} \\ F_{21} & F_{22,2} + F_{22,3} \end{bmatrix}\begin{bmatrix} M_1 \\ M_2 \end{bmatrix} + \begin{bmatrix} \tau_{10,1} \\ 0 \end{bmatrix} = 0 \tag{5}$$

in which $\omega = 1057.5$ rad/s. Thus $\lambda l_2 = l_2\sqrt[4]{\mu_2\omega^2/EI_2} = 3.6$ and $\lambda l_1 = \lambda l_3 = 2.568$. The load effect at the right end of member 0—1 is

$$\tau_{10,1} = a(l_1)q + \frac{1}{\lambda^2}c(l_1)r + \Phi'(l_1) \tag{6}$$

From Table P-6.35,

$$\Phi(l_1) = -\frac{Qc(l_1)}{EI_1\lambda^2} \qquad \Phi'(l_1) = -\frac{Qb(l_1)}{EI_1\lambda} \qquad \Phi''(l_1) = -\frac{Qa(l_1)}{EI_1\lambda^2} \tag{7}$$

Thus

$$q = \frac{\lambda^2}{\sinh \lambda l_1 \sin \lambda l_1}\left[\frac{1}{\lambda}b(l_1)\frac{Qc(l_1)}{EI_1\lambda^2} - \frac{1}{\lambda^3}d(l_1)\frac{Qa(l_1)}{EI_1}\right] = \frac{Q}{2EI_1\lambda}(\coth \lambda l_1 - \cot \lambda l_1) \tag{8}$$

$$r = \frac{-\lambda^2}{\sinh \lambda l_1 \sin \lambda l_1}\left[\lambda d(l_1)\frac{Qc(l_1)}{EI_1\lambda^2} - \frac{1}{\lambda}b(l_1)\frac{Qa(l_1)}{EI_1}\right] = \frac{\lambda Q}{2EI_1}(\coth \lambda l_1 + \cot \lambda l_1) \tag{9}$$

Substituting (*8, 9*) into (*6*) yields

$$\tau_{10,1} = \frac{Q}{2EI_1\lambda}(\cosh \lambda l_1 \coth \lambda l_1 - \cos \lambda l_1 \cot \lambda l_1) - \Phi'(l_1) = \frac{Q}{2EI_1\lambda}(\text{csch } \lambda l_1 - \csc \lambda l_1) \tag{10}$$

Since the shape parameter $\bar{\lambda}$ is the same in (*4*) and (*5*), (*5*) may be written as

$$\frac{l_1}{2EI_1}\begin{bmatrix} 0.99686 - 0.23506 & -0.5333 \\ -0.5333 & 0.99686 - 0.23506 \end{bmatrix}\begin{bmatrix} M_1 \\ M_2 \end{bmatrix} + \frac{l}{2EI_1}\begin{bmatrix} \frac{-67.79}{2.968}(-1.6886) \\ 0 \end{bmatrix} = 0 \tag{11}$$

from which $M_1 = -137.5$ kN-m and $M_2 = -962$ kN-m.

STIFFNESS MATRIX

9.9. Derive the dimensionless stiffness matrix (*9-11*) of the prismatic bar shown in Table 9-1*c*.

From the dimensionless dynamic transport matrix in Table 8-1*c* and *d*, we write

$$\bar{\Delta}_R = \bar{U}_{\Delta\sigma,RL}\bar{\sigma}_L + \bar{U}_{\Delta\Delta,RL}\bar{\Delta}_L \qquad \bar{\Delta}_L = \bar{U}_{\Delta\sigma,LR}\bar{\sigma}_R + \bar{U}_{\Delta\Delta,LR}\bar{\Delta}_R \tag{1, 2}$$

In terms of the sign convention $\bar{\sigma}_L = -\bar{\sigma}_{LR}$ and $\bar{\sigma}_R = \bar{\sigma}_{RL}$ (Table 9-1*a* and *c*), equations (*1, 2*) become

$$\bar{\Delta}_R = -\bar{U}_{\Delta\sigma,RL}\bar{\sigma}_{LR} + \bar{U}_{\Delta\Delta,RL}\bar{\Delta}_L \qquad \bar{\Delta}_L = \bar{U}_{\Delta\sigma,LR}\bar{\sigma}_{RL} + \bar{U}_{\Delta\Delta,LR}\bar{\Delta}_R \tag{3, 4}$$

By the definition of stiffness, arrange (*3, 4*) as

$$\bar{\sigma}_{RL} = \underbrace{-\bar{U}^{)-1}_{\Delta\sigma,LR}\bar{U}_{\Delta\Delta,LR}}_{\bar{k}_{RR}}\bar{\Delta}_R + \underbrace{\bar{U}^{)-1}_{\Delta\sigma,LR}}_{\bar{k}_{RL}}\bar{\Delta}_L \qquad \bar{\sigma}_{LR} = \underbrace{-\bar{U}^{)-1}_{\Delta\sigma,RL}}_{\bar{k}_{LR}}\bar{\Delta}_R + \underbrace{\bar{U}^{)-1}_{\Delta\sigma,RL}\bar{U}_{\Delta\Delta,RL}}_{\bar{k}_{LL}}\bar{\Delta}_L \tag{5, 6}$$

in which the inverse terms are

$$\bar{U}^{)-1}_{\Delta\sigma,LR} = \begin{bmatrix} -\dfrac{b_v d_v - c_v^2}{d_u} & 0 & 0 \\ 0 & -b_v & -c_v \\ 0 & -c_v & -d_v \end{bmatrix} \frac{1}{b_v d_v - c_v^2} \qquad \bar{U}^{)-1}_{\Delta\sigma,RL} = \begin{bmatrix} \dfrac{b_v d_v - c_v^2}{d_u} & 0 & 0 \\ 0 & b_v & -c_v \\ 0 & -c_v & d_v \end{bmatrix} \frac{1}{b_v d_v - c_v^2} \tag{7, 8}$$

The notational equivalents in (*7, 8*) may be found in Table 8-1*e*. Substituting (*7, 8*) into (*5, 6*) gives the desired stiffness coefficients as

$$\bar{k}_{RR} = \begin{bmatrix} \bar{\psi}_1 & 0 & 0 \\ 0 & -\bar{\psi}_4 & \bar{\psi}_3 \\ 0 & -\bar{\psi}_3 & \bar{\psi}_2 \end{bmatrix} \qquad \bar{k}_{RL} = \begin{bmatrix} -\bar{\psi}_5 & 0 & 0 \\ 0 & \bar{\psi}_8 & \bar{\psi}_7 \\ 0 & \bar{\psi}_7 & \bar{\psi}_6 \end{bmatrix} \tag{9, 10}$$

$$\bar{k}_{LR} = \begin{bmatrix} -\bar{\psi}_5 & 0 & 0 \\ 0 & \bar{\psi}_8 & -\bar{\psi}_7 \\ 0 & -\bar{\psi}_7 & \bar{\psi}_6 \end{bmatrix} \qquad \bar{k}_{LL} = \begin{bmatrix} \bar{\psi}_1 & 0 & 0 \\ 0 & -\bar{\psi}_4 & \bar{\psi}_3 \\ 0 & \bar{\psi}_3 & \bar{\psi}_2 \end{bmatrix} \tag{11, 12}$$

where the dynamic parameters $\bar{\psi}_1, \bar{\psi}_2, \ldots, \bar{\psi}_8$ are those given in Table 9-2.

9.10. Derive the dimensional stiffness matrix (*9-12*) of a prismatic bar in *l*-system.

Let the displacement vectors in (*9-11*) be expressed according to (*8-9*) as

$$\bar{\Delta}_R = \lambda_\Delta \Delta_R \qquad \bar{\Delta}_L = \lambda_\Delta \Delta_L \tag{1, 2}$$

Then (*9-11*) becomes

$$\bar{\sigma}_{RL} = \bar{k}_{RR}\lambda_\Delta\Delta_R + \bar{k}_{RL}\lambda_\Delta\Delta_L \qquad \bar{\sigma}_{LR} = \bar{k}_{LR}\lambda_\Delta\Delta_R + \bar{k}_{RL}\lambda_\Delta\Delta_L \tag{3, 4}$$

Multiplying (*3, 4*) by the dimensioning matrix κ_σ, (*8-6*) yields the desired stiffness matrix as

$$\sigma_{RL} = \underbrace{\kappa_\sigma\bar{k}_{RR}\lambda_\Delta}_{k_{RR}}\Delta_R + \underbrace{\kappa_\sigma\bar{k}_{RL}\lambda_\Delta}_{k_{RL}}\Delta_L \qquad \sigma_{LR} = \underbrace{\kappa_\sigma\bar{k}_{LR}\lambda_\Delta}_{k_{LR}}\Delta_R + \underbrace{\kappa_\sigma\bar{k}_{RL}\lambda_\Delta}_{k_{LL}}\Delta_L \tag{5}$$

where in terms of the dynamic parameters of Table 9-2*c*

$$k_{RR} = \begin{bmatrix} \bar{\psi}_1 EA\lambda_u & 0 & 0 \\ 0 & \bar{\psi}_4\lambda_v^3 EI & -\bar{\psi}_3\lambda_v^2 EI \\ 0 & -\bar{\psi}_3\lambda_v^2 EI & \bar{\psi}_2\lambda_v EI \end{bmatrix} \qquad k_{LL} = \begin{bmatrix} \bar{\psi}_1 EA\lambda_u & 0 & 0 \\ 0 & \bar{\psi}_4\lambda_v^3 EI & \bar{\psi}_3\lambda_v^2 EI \\ 0 & \bar{\psi}_3\lambda_v^2 EI & \bar{\psi}_2\lambda_v EI \end{bmatrix}$$

$$k_{RL} = k_{LR}^{)T} = \begin{bmatrix} -\bar{\psi}_5\lambda_u EA & 0 & 0 \\ 0 & -\bar{\psi}_8\lambda_v^3 EI & -\bar{\psi}_7\lambda_v^2 EI \\ 0 & \bar{\psi}_7\lambda_v^2 EI & \bar{\psi}_6\lambda_v EI \end{bmatrix}$$

9.11. Derive the angular transformation of stiffness matrix (*9-11*) and (*9-12*) in *l*-system.

According to (*9-12*),

$$\sigma^{l}_{RL} = k^{l}_{RR}\Delta^{l}_{R} + k^{l}_{RL}\Delta^{l}_{L} \tag{1}$$

where by (*8-6*) and (*8-9*)

$$\sigma^{l}_{RL} = \kappa_\sigma \bar{\sigma}^{l}_{RL} \qquad \Delta^{l}_{R} = \kappa_\Delta \bar{\Delta}^{l}_{R} \qquad \bar{\Delta}^{l}_{L} = \kappa_\Delta \bar{\Delta}^{l}_{L} \tag{2}$$

Then

$$\sigma^{l}_{RL} = \pi^{lR}\sigma^{R}_{RL} = \pi^{lR}\kappa_\sigma\bar{\sigma}^{R}_{RL} \qquad \Delta^{l}_{R} = \pi^{lR}\Delta^{R}_{R} = \pi^{lR}\kappa_\Delta\bar{\Delta}^{R}_{R} \qquad \Delta^{l}_{L} = \pi^{lL}\Delta^{L}_{L} = \pi^{lL}\kappa_\Delta\bar{\Delta}^{L}_{L} \tag{3}$$

Expressing (*1*) in terms of (*3*) gives

$$\pi^{lR}\kappa_\sigma\bar{\sigma}^{R}_{RL} = k^{l}_{RR}\pi^{lR}\kappa_\Delta\bar{\Delta}^{R}_{R} + k^{l}_{RL}\pi^{lR}\kappa_\Delta\bar{\Delta}^{L}_{L} \tag{4}$$

where by (*9-12*)

$$k^{l}_{RR} = \kappa_\sigma \bar{k}^{l}_{RR}\lambda_\Delta \qquad k^{l}_{RL} = \kappa_\sigma \bar{k}^{l}_{RL}\lambda_\Delta \tag{5}$$

On premultiplying (*4*) by $\lambda_\sigma\pi^{Rl}$, the desired relation is

$$\bar{\sigma}^{R}_{RL} = \underbrace{\lambda_\sigma\pi^{Rl}\kappa_\sigma}_{\bar{\pi}^{Rl}_{\sigma}}\bar{k}^{l}_{RR}\underbrace{\lambda_\Delta\pi^{lR}\kappa_\Delta}_{\bar{\pi}^{lR}_{\Delta}}\bar{\Delta}^{R}_{R} + \underbrace{\lambda_\sigma\pi^{Rl}\kappa_\sigma}_{\bar{\pi}^{Rl}_{\sigma}}\bar{k}^{l}_{RL}\underbrace{\lambda_\Delta\pi^{lR}\kappa_\Delta}_{\bar{\pi}^{lL}_{\Delta}}\bar{\Delta}^{L}_{L} \tag{6}$$

which confirms the first equation of (*9-13*). The second equation of (*9-13*) can be derived from (6) by cyclic substitution.

The transformation of (*9-12*) is accomplished by means of relations (*3*) as

$$\pi^{lR}\sigma^{R}_{RL} - k^{l}_{RR}\pi^{lR}\Delta^{R}_{R} + k^{l}_{RL}\pi^{lL}\Delta^{L}_{L} \tag{7}$$

and on premultiplying by π^{Rl} becomes

$$\sigma^{R}_{RL} = \pi^{Rl}k^{l}_{RR}\pi^{lR}\Delta^{R}_{R} + \pi^{Rl}k^{l}_{RL}\pi^{lL}\Delta^{L}_{L} \tag{8}$$

which is formally similar to (*6*). The transformation of the second equation in (*9-12*) is similar.

STIFFNESS METHOD

9.12. Formulate the system stiffness matrix for the frame of Fig. P-9.12 and then find the fundamental angular frequency of free vibration. Given for all members:

$$l = 19\text{ in}$$
$$I = 13.732 \times 10^{-5}\text{ in}^4$$
$$\mu = 13.76 \times 10^{-5}\text{ lb-s}^2/\text{in}^2$$
$$E = 30 \times 10^6\text{ psi}$$
$$A = 0.1875\text{ in}^2$$

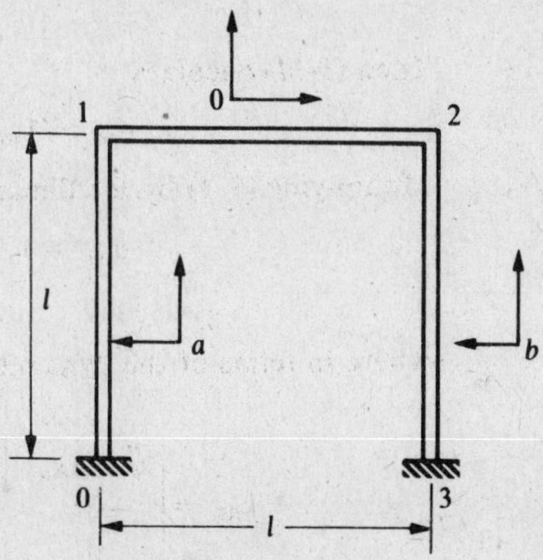

Fig. P-9.12

According to (*8*) of the preceding problem, the stiffness matrix of the constituent members of Fig. P-9.12 may be written in 0-system as

$$\begin{bmatrix}\sigma^{0}_{10}\\ \sigma^{0}_{01}\end{bmatrix} = [\pi^{0a}]\begin{bmatrix}k^{a}_{11} & k^{a}_{10}\\ k^{a}_{01} & k^{a}_{00}\end{bmatrix}[\pi^{a0}]\begin{bmatrix}\Delta^{0}_{10}\\ \Delta^{0}_{01}\end{bmatrix} = \begin{bmatrix}E^{0}_{1} & F^{0}_{1}\\ G^{0}_{1} & H^{0}_{1}\end{bmatrix}\begin{bmatrix}\Delta^{0}_{10}\\ \Delta^{0}_{01}\end{bmatrix} \tag{1}$$

$$\begin{bmatrix}\sigma^{0}_{21}\\ \sigma^{0}_{12}\end{bmatrix} = \begin{bmatrix}k^{0}_{22} & k^{0}_{21}\\ k^{0}_{12} & k^{0}_{11}\end{bmatrix}\begin{bmatrix}\Delta^{0}_{21}\\ \Delta^{0}_{12}\end{bmatrix} = \begin{bmatrix}E^{0}_{2} & F^{0}_{2}\\ G^{0}_{2} & H^{0}_{2}\end{bmatrix}\begin{bmatrix}\Delta^{0}_{21}\\ \Delta^{0}_{12}\end{bmatrix} \tag{2}$$

$$\begin{bmatrix}\sigma^{0}_{32}\\ \sigma^{0}_{23}\end{bmatrix} = [\pi^{0b}]\begin{bmatrix}k^{b}_{33} & k^{b}_{32}\\ k^{b}_{23} & k^{b}_{22}\end{bmatrix}[\pi^{b0}]\begin{bmatrix}\Delta^{0}_{32}\\ \Delta^{0}_{23}\end{bmatrix} = \begin{bmatrix}E^{0}_{3} & F^{0}_{3}\\ G^{0}_{3} & H^{0}_{3}\end{bmatrix}\begin{bmatrix}\Delta^{0}_{32}\\ \Delta^{0}_{23}\end{bmatrix} \tag{3}$$

where the angular transformation matrices are based on (*1-3, 4*) and the stiffness coefficients are

$$E_1^0 = \begin{bmatrix} \bar\psi_4\lambda_v^3EI & 0 & -\bar\psi_3\lambda_v^2EI \\ 0 & \bar\psi_1\lambda_uAE & 0 \\ -\bar\psi_3\lambda_v^2EI & 0 & \bar\psi_2\lambda_vEI \end{bmatrix} \qquad E_2^0 = \begin{bmatrix} \bar\psi_1\lambda_uAE & 0 & 0 \\ 0 & \bar\psi_4\lambda_v^3EI & -\bar\psi_3\lambda_v^2EI \\ 0 & -\bar\psi_3\lambda_v^2EI & \bar\psi_2\lambda_v^2EI \end{bmatrix}$$

$$F_1^0 = G_1^{0)T} = \begin{bmatrix} -\bar\psi_8\lambda_v^3EI & 0 & -\bar\psi_7\lambda_v^2EI \\ 0 & -\bar\psi_5\lambda_uEA & 0 \\ \bar\psi_7\lambda_v^2EI & 0 & \bar\psi_6\lambda_vEI \end{bmatrix} \qquad F_2^0 = G_2^{0)T} = \begin{bmatrix} -\bar\psi_5\lambda_uEA & 0 & 0 \\ 0 & -\bar\psi_8\lambda_v^3EI & \bar\psi_7\lambda_v^2EI \\ 0 & \bar\psi_7\lambda_v^2EI & \bar\psi_6\lambda_vEI \end{bmatrix}$$

$$H_1^0 = \begin{bmatrix} \bar\psi_4\lambda_v^3EI & 0 & \bar\psi_3\lambda_v^2EI \\ 0 & \bar\psi_1\lambda_uAE & 0 \\ \bar\psi_3\lambda_v^2EI & 0 & -\bar\psi_2\lambda_vEI \end{bmatrix} \qquad H_2^0 = \begin{bmatrix} \bar\psi_1\lambda_uAE & 0 & 0 \\ 0 & \bar\psi_4\lambda_v^3EI & \bar\psi_3\lambda_v^2EI \\ 0 & \bar\psi_3\lambda_v^2EI & \bar\psi_2\lambda_vEI \end{bmatrix}$$

$$E_3^0 = E_1^0 \qquad F_3^0 = F_1^0 \qquad G_3^0 = G_1^0 \qquad \text{and} \qquad H_3^0 = H_1^0$$

Using the equilibrium conditions at joints 1 and 2 gives

$$\sigma_1^0 = \sigma_{10}^0 + \sigma_{12}^0 = 0 \qquad \sigma_2^0 = \sigma_{21}^0 + \sigma_{23}^0 = 0 \tag{4}$$

In terms of (*1*), (*2*), and (*3*), equation (*4*) becomes

$$\underbrace{\begin{bmatrix} \sigma_1^0 \\ \sigma_2^0 \end{bmatrix}}_{\{\sigma\}} = \underbrace{\begin{bmatrix} 0 \\ 0 \end{bmatrix}}_{\{0\}} = \underbrace{\begin{bmatrix} E_1^0 + H_2^0 & G_2^0 \\ F_2^0 & E_2^0 + H_3^0 \end{bmatrix}}_{[\kappa]} \underbrace{\begin{bmatrix} \Delta_1^0 \\ \Delta_2^0 \end{bmatrix}}_{\{\Delta\}} \tag{5}$$

The solution of (*5*) can exist only if

$$\text{Det.}\,(\kappa) = 0 \tag{6}$$

Since the angular frequency ω is the only variable in (*7*), one may vary ω by increasing $\Delta\omega$ or λ_v by $\Delta\lambda_v$ (λ_u can be expressed in terms of λ_v as $\lambda_u = \lambda_v^2\sqrt{I/A}$) for the corresponding determinants (*7*). For the given problem, the typical solutions of $\bar\lambda_v$ vs. Det. (κ) are shown in Table P-9.12 which indicates that Det. $(\kappa) = 0$ is between $\bar\lambda_v = 3.5$ and 3.7. Using linear interpolation gives $\bar\lambda_v = 3.542$.

Thus

$$\omega = \frac{\bar\lambda_v^2\sqrt{EI/\mu}}{l^2} = 190.2 \text{ rad/s}$$

which must confirm as of Problem 9.4.

Table P-9.12

Det.	$\bar\lambda_v$	Det.	$\bar\lambda_v$	Det.	$\bar\lambda_v$
-1.650×10^{26}	2.5	1.027×10^{27}	3.9	-5.496×10^{28}	5.1
-2.214×10^{26}	2.7	2.391×10^{27}	4.1	-3.942×10^{28}	5.3
-2.675×10^{26}	2.9	4.652×10^{27}	4.3	-3.297×10^{28}	5.5
-2.843×10^{26}	3.1	4.257×10^{27}	4.5	-2.910×10^{28}	5.7
-2.392×10^{26}	3.3	-1.330×10^{30}	4.7	-2.585×10^{28}	5.9
-7.780×10^{25}	3.5	-1.285×10^{29}	4.9	-2.222×10^{28}	6.1
2.920×10^{26}	3.7				

The same result may be obtained by the overlapping of the segmental stiffness matrices (*1*), (*2*), and (*3*) as

$$\begin{bmatrix} \sigma_0^0 \\ \sigma_1^0 \\ \sigma_2^0 \\ \sigma_3^0 \end{bmatrix} = \begin{bmatrix} \sigma_0^0 \\ 0 \\ 0 \\ \sigma_3^0 \end{bmatrix} = \begin{bmatrix} H_1^0 & G_1^0 & 0 & 0 \\ F_1^0 & E_1^0 + H_2^0 & G_2^0 & 0 \\ 0 & F_2^0 & E_2^0 + H_3^0 & G_3^0 \\ 0 & 0 & F_3^0 & E_3^0 \end{bmatrix} \begin{bmatrix} \Delta_0^0 \\ \Delta_1^0 \\ \Delta_2^0 \\ \Delta_3^0 \end{bmatrix} \tag{7}$$

in which $\Delta_0^0 = \Delta_3^0 = 0$, (*6*) may be written as (*5*).

9.13. Derive the load functions (*9-14*) of a prismatic bar subjected to dynamic loads acting between two joints.

From the third submatrix equation in (*8-26, 27*),

$$\bar{\Delta}_R = \bar{\Phi}_{\Delta,RL} + \bar{U}_{\Delta\sigma,RL}\bar{\sigma}_L + U_{\Delta\Delta,RL}\bar{\Delta}_L \qquad \bar{\Delta}_L = \bar{\Phi}_{\Delta,LR} + \bar{U}_{\Delta\sigma,LR}\bar{\sigma}_R + \bar{U}_{\Delta\Delta,LR}\bar{\Delta}_R \tag{1, 2}$$

in which the stress vectors corresponding to those of the stiffness matrix (Table 9-1*c*) should be modified as $\bar{\sigma}_L = -\bar{\sigma}_{LR}$ and $\bar{\sigma}_R = \bar{\sigma}_{RL}$. Then (*1, 2*) becomes

$$\bar{\sigma}_{RL} = \underbrace{-\bar{U})^{-1}_{\Delta\sigma,LR}\bar{U}_{\Delta\Delta,LR}}_{\bar{k}_{RR}}\bar{\Delta}_R + \underbrace{\bar{U})^{-1}_{\Delta\sigma,LR}}_{\bar{k}_{RL}}\bar{\Delta}_L - \underbrace{\bar{U})^{-1}_{\Delta\sigma,LR}\bar{\Phi}_{\Delta,LR}}_{\bar{\sigma}_{R0}} \tag{3}$$

$$\bar{\sigma}_{LR} = \underbrace{-\bar{U})^{-1}_{\Delta\sigma,RL}}_{\bar{k}_{LR}}\bar{\Delta}_R + \underbrace{\bar{U})^{-1}_{\Delta\sigma,RL}\bar{U}_{\Delta\Delta,RL}}_{\bar{k}_{LL}}\bar{\Delta}_L + \underbrace{\bar{U})^{-1}_{\Delta\sigma,RL}\bar{\Phi}_{\Delta,RL}}_{\bar{\sigma}_{L0}} \tag{4}$$

where $\bar{\sigma}_{R0}$ and $\bar{\sigma}_{L0}$ are load functions. The dimensionless load effect vectors $\bar{\Phi}_{\Delta,RL}$ may be found from Problems 6.10 and 6.22 as

$$\bar{\Phi}_{\Delta,RL} = [\Phi'_{Ru}/\lambda_u \quad \Phi_{Rv} \quad \Phi'_{Rv}/\lambda v]^T \tag{5}$$

and $\bar{\Phi}_{\Delta,LR}$ may be obtained from the inverse of the transport matrix (*8-26, 27*) as $\bar{\Phi}_{LR} = -\bar{U})^{-1}_{RL}\bar{\Phi}_{RL} = -\bar{U}_{LR}\bar{\Phi}_{RL}$. Thus

$$\bar{\Phi}_{\Delta,LR} = -\bar{U}_{\Delta\sigma,LR}\bar{\Phi}_{\sigma,RL} - \bar{U}_{\Delta\Delta,LR}\bar{\Phi}_{\Delta,RL} \tag{6}$$

Using $\bar{U})^{-1}_{\Delta\sigma,LR}$ and $\bar{U})^{-1}_{\Delta\sigma,RL}$ (*7, 8* of Problem 9.9) in (*3*) and (*4*) gives the following load functions of Table 9-3*e*:

$$\bar{\sigma}_{R0} = \begin{bmatrix} \bar{\Phi}'_{Ru} - \bar{\psi}_1\bar{\bar{\Phi}}_{Ru} \\ \bar{\Phi}'''_{Rv} + \bar{\psi}_4\bar{\Phi}_{Rv} - \bar{\psi}_3\bar{\Phi}'_{Rv} \\ \bar{\Phi}''_{Rv} + \bar{\psi}_3\bar{\Phi}_{Rv} - \bar{\psi}_2\bar{\Phi}'_{Rv} \end{bmatrix} = \begin{bmatrix} \bar{N}_{R0} \\ \bar{V}_{R0} \\ \bar{M}_{R0} \end{bmatrix} \qquad \bar{\sigma}_{L0} = \begin{bmatrix} \psi_5\bar{\Phi}_{Ru} \\ -\bar{\psi}_8\bar{\Phi}_{Rv} + \bar{\psi}_7\bar{\Phi}'_{Rv} \\ \bar{\psi}_7\bar{\Phi}_{Rv} - \bar{\psi}_6\bar{\Phi}'_{Rv} \end{bmatrix} = \begin{bmatrix} \bar{N}_{L0} \\ \bar{V}_{L0} \\ \bar{M}_{L0} \end{bmatrix} \tag{7, 8}$$

Several particular cases of (*7*) and (*8*) are given in dimensional form in Table P-9.26.

9.14. Find the internal moments of the continuous beam shown in Fig. P-9.14. Given:

$l = 3.66$ m $\qquad \mu = 2533.82$ kN-s^2/m^2 $\qquad p = 17.513$ kN/m

$I = 1.01$ dm^4 $\qquad p_v(t) = p \cos \Omega t$ $\qquad \Omega = 3.085$ rad/s

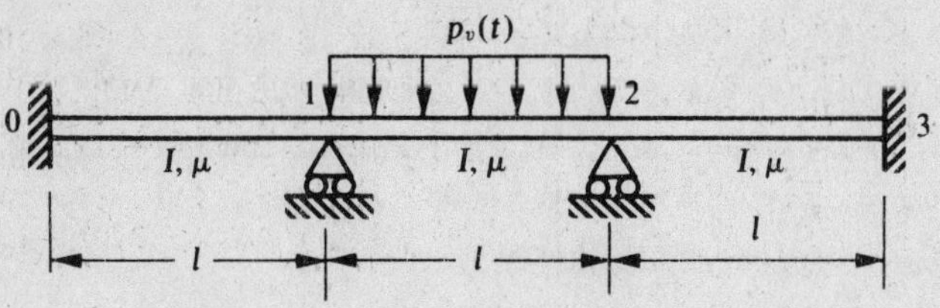

Fig. P-9.14

From (*3*) and (*4*) of the preceding problem, the dimensional segmental stiffness matrix may be written as

$$\begin{bmatrix} \sigma_{RL} \\ \sigma_{LR} \end{bmatrix} = \begin{bmatrix} k_{RR} & k_{RL} \\ k_{LR} & k_{LL} \end{bmatrix}\begin{bmatrix} \Delta_R \\ \Delta_L \end{bmatrix} + \begin{bmatrix} \sigma_{R0} \\ \sigma_{L0} \end{bmatrix} \tag{1}$$

Since the displacement vectors of a continuous beam have joint rotations only, (*1*) may be modified in accordance with (*5*) and (*6*) of Problem 9.10 as

$$\begin{bmatrix} M_{RL} \\ M_{LR} \end{bmatrix} = \begin{bmatrix} \bar{\psi}_2\lambda_v EI & \bar{\psi}_6\lambda_v EI \\ \bar{\psi}_6\lambda_v EI & \bar{\psi}_2\lambda_v EI \end{bmatrix}\begin{bmatrix} \theta_{RL} \\ \theta_{LR} \end{bmatrix} + \begin{bmatrix} (\bar{\Phi}''_{Rv} + \bar{\psi}_3\bar{\Phi}_{Rv} - \bar{\psi}_2\Phi'_{Rv})EI\lambda_v^2 \\ (\bar{\psi}_7\bar{\Phi}_{Rv} - \bar{\psi}_6\bar{\Phi}'_{Rv})EI\lambda_v^2 \end{bmatrix} \tag{2}$$

or in symbolic form

$$\begin{bmatrix} M_{RL} \\ M_{LR} \end{bmatrix} = \begin{bmatrix} k_{RR} & k_{RL} \\ k_{LR} & k_{LL} \end{bmatrix} \begin{bmatrix} \theta_{RL} \\ \theta_{LR} \end{bmatrix} + \begin{bmatrix} M_{RO} \\ M_{LO} \end{bmatrix} \tag{3}$$

where k_{RR} and k_{LL} are the near end moment stiffnesses at R and L respectively, k_{RL} and k_{LR} are the far end stiffnesses at R and L respectively, and M_{RO} and M_{LO} are the fixed-end moments due to loads at R and L respectively, given for particular cases in Table P-9.26.

Applying the equilibrium conditions at joints 1 and 2 of Fig. P-9.14 gives

$$\Sigma M_1 = 0 \quad \text{and} \quad \Sigma M_2 = 0 \quad \text{or} \quad M_{10} + M_{12} = 0 \quad \text{and} \quad M_{21} + M_{23} = 0 \tag{4}$$

In terms of (3), (4) becomes

$$\begin{bmatrix} \Sigma k_1 & k_{12} \\ k_{21} & \Sigma k_2 \end{bmatrix} \begin{bmatrix} \theta_1 \\ \theta_2 \end{bmatrix} + \begin{bmatrix} \Sigma M_{10} \\ \Sigma M_{20} \end{bmatrix} = 0 \tag{5}$$

where $\Sigma k_1 = k_{11,1} + k_{11,2}$ and $\Sigma k_2 = k_{22,2} + k_{22,3}$ are the joint moment stiffnesses at 1 and 2 respectively, and ΣM_{10} and ΣM_{20} are the sums of fixed-end moments due to loads at 1 and 2 respectively.

Since the members 01, 12, and 23 are identical, $k_{11,1} = k_{11,2} = k_{22,2} = k_{22,3} = \bar{\psi}_2 \lambda EI$ and $k_{01} = k_{10} = k_{12} = k_{21} = k_{23} = k_{32} = \bar{\psi}_6 \lambda EI$.

The fixed-end moments in the loadless spans 01 and 23 are zero and for the span 12 are computed by (2) in terms of

$$\bar{\Phi}_{R,v} = -p(1-a)/EI\lambda_v^4 \qquad \bar{\Phi}'_{R,v} = pd/EI\lambda_v^4 \qquad \bar{\Phi}''_{R,v} = pc/EI\lambda_v^4 \tag{6}$$

where $a = a_v(l)$, $c = c_v(l)$, and $d = d_v(l)$.

Knowing $\Omega = 3.085$ rad/s, we have

$$\begin{array}{lll} \lambda_v l = 3.079 & \sin \lambda_v l = 0.062\,56 & \bar{\psi}_3 = 0.057\,16 \\ \sinh \lambda_v l = 10.8453 & \cos \lambda_v l = -0.998 & \bar{\psi}_6 = 0.9084 \\ \cosh \lambda_v l = 10.891 & \bar{\psi}_2 = -0.967 & \bar{\psi}_7 = 1.0016 \end{array}$$

Consequently (5) becomes

$$10^5 \begin{bmatrix} 0.34141 & 0.15998 \\ 0.15998 & 0.34141 \end{bmatrix} \begin{bmatrix} \theta_1 \\ \theta_2 \end{bmatrix} + \begin{bmatrix} -23.342 \\ 23.342 \end{bmatrix} = 0 \tag{7}$$

from which $\theta_1 = 1.287 \times 10^{-3}$ rad and $\theta_2 = -1.287 \times 10^{-3}$ rad.

By (2), the internal moments are

$$\begin{aligned} M_{01} &= k_{01}\theta_1 = 20.586 \text{ kN-m} \\ M_{10} &= k_{11,1}\theta_1 = 21.964 \text{ kN-m} \\ M_{12} &= k_{11,2}\theta_2 + k_{12}\theta_2 + M_{10,2} = 21.964 - 20.586 - 23.342 = -21.964 \text{ kN-m} \\ M_{21} &= k_{22,1}\theta_2 + k_{21}\theta_1 + M_{20,2} = -21.964 + 20.586 + 23.342 = -21.964 \text{ kN-m} \\ M_{23} &= k_{22,3}\theta_2 = -21.964 \text{ kN-m} \\ M_{32} &= k_{32}\theta_2 = -20.586 \text{ kN-m} \end{aligned}$$

9.15. Derive the three-slope equation for dynamic analysis of a continuous prismatic beam.

For the continuous beam having rigid supports, the typical moment equilibrium equations at joint j are

$$\Sigma M_j = 0 \quad \text{or} \quad M_{ji} + M_{jk} = 0 \tag{1}$$

and in terms of (3) and (5) of the preceding problem they become

$$k_{ji}\theta_i + \Sigma k_j \theta_j + k_{jk}\theta_k + \Sigma M_{jo} = 0 \tag{2}$$

Equation (2) is known as the *three-slope equation* and it is a very useful tool of continuous beam analysis.

9.16. Solve Problem P-9.14 by the three-slope equation derived in Problem 9.15.

Using (2) of Problem 9.15, we have at 1

$$\Sigma k_1\theta_1 + k_{12}\theta_2 + \Sigma M_{10} = 0 \tag{1}$$

where $i = 0$, $j = 1$, $k = 2$, and $M_{10,1} = 0$ but $M_{10,2} \neq 0$.
Similarly at 2,

$$k_{21}\theta_2 + \Sigma k_2\theta_2 + \Sigma M_{20} = 0 \tag{2}$$

where $i = 1$, $j = 2$, $k = 3$, and $M_{20,2} \neq 0$ but $M_{20,3} = 0$.
In matrix form (1) and (2) give

$$\begin{bmatrix} \Sigma k_1 & k_{12} \\ k_{21} & \Sigma k_2 \end{bmatrix}\begin{bmatrix} \theta_1 \\ \theta_2 \end{bmatrix} + \begin{bmatrix} \Sigma M_{10} \\ \Sigma M_{20} \end{bmatrix} = 0 \tag{3}$$

9.17. Formulate the system stiffness matrix in symbolic form for the gable frame of Fig. P-9.17. Assume all members to have the same A, l, I, and μ; $P(t) = P\cos\Omega t$.

The typical transformation segmental stiffness matrix is

$$\sigma^R_{RL} = \pi^{Rl}k^l_{RR}\pi^{lR}\Delta^R_R + \pi^{Rl}k^l_{RL}\pi^{lL}\Delta^L_L + \pi^{Rl}\kappa_\sigma\bar{\sigma}^l_{R0} \tag{1}$$

$$\sigma^L_{LR} = \pi^{Ll}k^l_{LR}\pi^{lR}\Delta^R_R + \pi^{Ll}k^l_{LL}\pi^{lL}\Delta^L_L + \pi^{Ll}\kappa_\sigma\bar{\sigma}^l_{L0} \tag{2}$$

For a typical member m with left end i and right end j, (1) and (2) may be expressed as

$$\begin{bmatrix} \sigma^0_{ji} \\ \sigma^0_{ij} \end{bmatrix} = [\pi^{0m}]\begin{bmatrix} k^m_{jj} & k^m_{ji} \\ k^m_{ij} & k^m_{ii} \end{bmatrix}[\pi^{m0}]\begin{bmatrix} \Delta^0_{ji} \\ \Delta^0_{ij} \end{bmatrix} + [\pi^{0m}][\kappa_0]\begin{bmatrix} \bar{\sigma}^m_{j0,m} \\ \sigma^m_{i0,m} \end{bmatrix} = \begin{bmatrix} E^0_m & F^0_m \\ G^0_m & H^0_m \end{bmatrix}\begin{bmatrix} \Delta^0_{ji} \\ \Delta^0_{ij} \end{bmatrix} + \begin{bmatrix} \sigma^0_{j0,m} \\ \sigma^0_{i0,m} \end{bmatrix} \tag{3}$$

where π^{0m} and π^{m0} were typically given in (1-3, 4).

Using $\bar{\sigma}_{R0}$ and $\bar{\sigma}_{L0}$ (7, 8 of Problem 9.13) and κ_σ (8-6) we may find the fixed-end stress vectors of member a as

$$[\pi^{0a}][\kappa_\sigma]\begin{bmatrix} \bar{\sigma}^a_{10,a} \\ \bar{\sigma}^a_{01,a} \end{bmatrix} = \begin{bmatrix} -EI\lambda^3_v(\bar{\Phi}'''_{Rv} + \bar{\psi}_4\bar{\Phi}_{Rv} - \bar{\psi}_3\bar{\Phi}'_{Rv}) \\ 0 \\ EI\lambda^2_v(\bar{\Phi}''_{Rv} + \bar{\psi}_3\bar{\Phi}_{Rv} - \bar{\psi}_2\bar{\Phi}'_{Rv}) \\ -EI\lambda^3_v(-\bar{\psi}_8\bar{\Phi}_{Rv} + \bar{\psi}_7\bar{\Phi}'_{Rv}) \\ 0 \\ EI\lambda^2_v(\bar{\psi}_7\bar{\Phi}_{Rv} - \bar{\psi}_6\bar{\Phi}'_{Rv}) \end{bmatrix} = \begin{bmatrix} V^0_{10,a} \\ 0 \\ M^0_{10,a} \\ V^0_{00,a} \\ 0 \\ M^0_{00,a} \end{bmatrix} \tag{4}$$

The fixed-end stress vectors of other members are zero.

The dimensionless load effects may be obtained from Problem 6.22 as

$$\bar{\Phi}_{Rv} = d(\lambda_v l/2)P/EI\lambda^3_v$$

$$\bar{\Phi}'_{Rv} = c(\lambda_v l/2)P/EI\lambda^3_v$$

$$\bar{\Phi}''_{Rv} = b(\lambda_v l/2)P/EI\lambda^3_v$$

$$\bar{\Phi}'''_{Rv} = a(\lambda_v l/2)P/EI\lambda^3_v$$

Introducing the equilibrium equations at joints 1, 2, and 3, we write

$$\{\sigma^0_{10}\} + \{\sigma^0_{12}\} = 0$$
$$\{\sigma^0_{21}\} + \{\sigma^0_{23}\} = 0 \tag{5}$$
$$\{\sigma^0_{32}\} + \{\sigma^0_{34}\} = 0$$

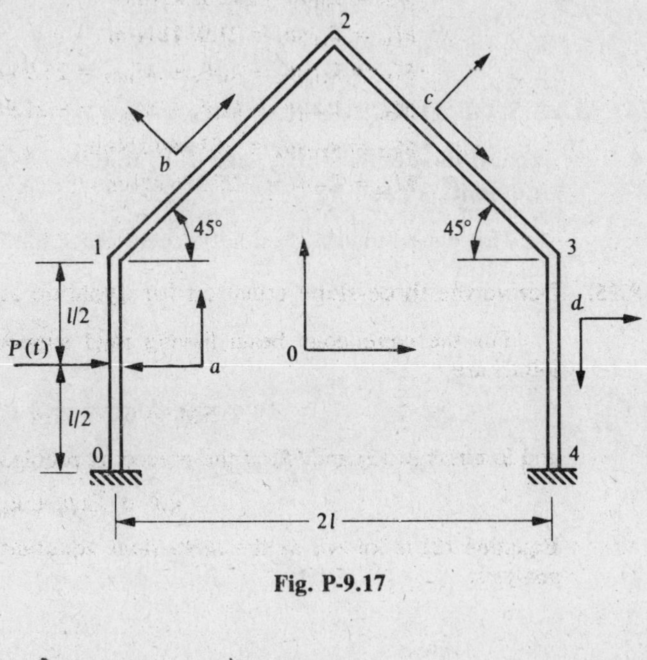

Fig. P-9.17

In terms of (3) and (4), equation (5) becomes

$$\left[\begin{array}{c|c|c} E_a^0 + H_b^0 & G_b^0 & 0 \\ \hline F_b^0 & E_b^0 + H_c^0 & G_c^0 \\ \hline 0 & F_c^0 & E_c^0 + H_d^0 \end{array}\right]\left[\begin{array}{c} \Delta_1^0 \\ \hline \Delta_2^0 \\ \hline \Delta_3^0 \end{array}\right] + \left[\begin{array}{c} V_{10,a}^0 \\ 0 \\ M_{10,a}^0 \\ \hline 0 \\ 0 \\ \hline 0 \end{array}\right] = 0 \tag{6}$$

in which

$$E_a^0 = \begin{bmatrix} \bar{\psi}_4\lambda_v^3 EI & 0 & -\bar{\psi}_3\lambda_v^2 EI \\ 0 & \bar{\psi}_1\lambda_u AE & 0 \\ -\bar{\psi}_3\lambda_v^2 EI & 0 & \bar{\psi}_2\lambda_v EI \end{bmatrix}$$

$$F_b^0 = G_b^{0)T} = \frac{1}{2}\begin{bmatrix} -\bar{\psi}_5 AE\lambda_u - \bar{\psi}_8\lambda_v^3 EI & \bar{\psi}_5 AE\lambda_u - \bar{\psi}_8\lambda_v^3 EI & -\sqrt{2}\bar{\psi}_7\lambda_v^2 EI \\ \bar{\psi}_5 AE\lambda_u - \bar{\psi}_8\lambda_v^3 EI & -\bar{\psi}_5 AE\lambda_u - \bar{\psi}_8\lambda_v^3 EI & -\sqrt{2}\bar{\psi}_7\lambda_v^2 EI \\ \sqrt{2}\bar{\psi}_7\lambda_v^2 EI & \sqrt{2}\bar{\psi}_7\lambda_v^2 EI & 2\bar{\psi}_6\lambda_v EI \end{bmatrix}$$

$$E_b^0 = \frac{1}{2}\begin{bmatrix} \bar{\psi}_1 AE\lambda_u + \bar{\psi}_4\lambda_v^3 EI & -\bar{\psi}_1 AE\lambda_u + \bar{\psi}_4\lambda_v^3 EI & -\sqrt{2}\bar{\psi}_3\lambda_v^2 EI \\ -\bar{\psi}_1 AE\lambda_u + \bar{\psi}_4\lambda_v^3 EI & \bar{\psi}_1 AE\lambda_u + \bar{\psi}_4\lambda_v^3 EI & -\sqrt{2}\bar{\psi}_3\lambda_v^2 EI \\ -\sqrt{2}\bar{\psi}_3\lambda_v^2 EI & -\sqrt{2}\bar{\psi}_3\lambda_v^2 EI & 2\bar{\psi}_2\lambda_v EI \end{bmatrix}$$

$$H_b^0 = \frac{1}{2}\begin{bmatrix} \bar{\psi}_1 AE\lambda_u + \bar{\psi}_4\lambda_v^3 EI & -\bar{\psi}_1 AE\lambda_u + \bar{\psi}_4\lambda_v^3 EI & \sqrt{2}\bar{\psi}_3\lambda_v^2 EI \\ -\bar{\psi}_1 AE\lambda_u + \bar{\psi}_4\lambda_v^3 EI & \bar{\psi}_1 AE\lambda_u + \bar{\psi}_4\lambda_v^3 EI & \sqrt{2}\bar{\psi}_3\lambda_v^2 EI \\ \sqrt{2}\bar{\psi}_3\lambda_v^2 EI & \sqrt{2}\bar{\psi}_3\lambda_v^2 EI & 2\bar{\psi}_2\lambda_v^2 EI \end{bmatrix}$$

$$E_c^0 = \frac{1}{2}\begin{bmatrix} \bar{\psi}_1 AE\lambda_u + \bar{\psi}_4\lambda_v^3 EI & \bar{\psi}_1 AE\lambda_u - \bar{\psi}_4\lambda_v^3 EI & \sqrt{2}\bar{\psi}_3\lambda_v^2 EI \\ \bar{\psi}_1 AE\lambda_u - \bar{\psi}_4\lambda_v^3 EI & \bar{\psi}_1 AE\lambda_u + \bar{\psi}_4\lambda_v^3 EI & -\sqrt{2}\bar{\psi}_3\lambda_v^2 EI \\ \sqrt{2}\bar{\psi}_3\lambda_v^2 EI & -\sqrt{2}\bar{\psi}_3\lambda_v^2 EI & 2\bar{\psi}_2\lambda_v EI \end{bmatrix}$$

$$F_c^0 = G_c^{0)T} = \frac{1}{2}\begin{bmatrix} -\bar{\psi}_5 AE\lambda_u - \bar{\psi}_8\lambda_v^3 EI & -\bar{\psi}_5 AE\lambda_u + \bar{\psi}_8\lambda_v^3 EI & \sqrt{2}\bar{\psi}_7\lambda_v^2 EI \\ -\bar{\psi}_5 AE\lambda_u + \bar{\psi}_8\lambda_v^3 EI & -\bar{\psi}_5 AE\lambda_u - \bar{\psi}_8\lambda_v^3 EI & -\sqrt{2}\bar{\psi}_7\lambda_v^2 EI \\ -\sqrt{2}\bar{\psi}_7\lambda_v^2 EI & \sqrt{2}\bar{\psi}_7\lambda_v^2 EI & 2\bar{\psi}_6\lambda_v EI \end{bmatrix}$$

$$H_c^0 = \frac{1}{2}\begin{bmatrix} \bar{\psi}_1 AE\lambda_u + \bar{\psi}_4\lambda_v^3 EI & \bar{\psi}_1 AE\lambda_u - \bar{\psi}_4\lambda_v^3 EI & -\sqrt{2}\bar{\psi}_3\lambda_v^2 EI \\ \bar{\psi}_1 AE\lambda_u - \bar{\psi}_4\lambda_v^3 EI & \bar{\psi}_1 AE\lambda_u + \bar{\psi}_4\lambda_v^3 EI & \sqrt{2}\bar{\psi}_3\lambda_v^2 EI \\ -\sqrt{2}\bar{\psi}_3\lambda_v^2 EI & \sqrt{2}\bar{\psi}_3\lambda_v^2 EI & 2\bar{\psi}_2\lambda_v EI \end{bmatrix}$$

$$H_a^0 = E_a^0$$

ANALYSIS OF FUNCTIONS

9.18. Expand the coefficients $\bar{f}_i$ and $\bar{f}_j$ (Table 9-3*b*) into the respective McLaurin's series for small arguments $\bar{\lambda}_u$ and $\bar{\lambda}_v$.

The dimensionless flexibility coefficients of Table 9-3*b* can be expressed into the following McLaurin's series:

$$\bar{f}_1 = \frac{\bar{\psi}_1}{\bar{\lambda}_u} = \left(1 - \frac{\bar{\lambda}_u^2}{3} - \frac{\bar{\lambda}_u^4}{45} - \frac{2\bar{\lambda}_u^6}{945} - \cdots\right)\Big/\bar{\lambda}_u^2 \tag{1a}$$

$$\bar{f}_2 = \frac{\bar{\psi}_2}{\bar{\lambda}_v^3} = 4\left(\frac{1}{3!} - \frac{2^2\bar{\lambda}_v^4}{7!} + \frac{2^4\bar{\lambda}_v^8}{11!} - \frac{2^6\bar{\lambda}_v^{12}}{15!} + \cdots\right)\Big/\bar{\eta} \tag{2a}$$

$$\bar{f}_3 = \frac{\bar{\psi}_3}{\bar{\lambda}_v^2} = 2\left(\frac{1}{2!} - \frac{2^2\bar{\lambda}_v^4}{6!} + \frac{2^4\bar{\lambda}_v^8}{10!} - \frac{2^6\bar{\lambda}_v^{12}}{14!} + \cdots\right)\Big/\bar{\eta} \tag{3a}$$

$$\bar{f}_4 = \frac{\bar{\psi}_4}{\bar{\lambda}_v} = 2\left(\frac{1}{1!} - \frac{2^2\bar{\lambda}_v^4}{5!} + \frac{2^4\bar{\lambda}_v^8}{9!} - \frac{2^6\bar{\lambda}_v^{12}}{13!} + \cdots\right)\Big/\bar{\eta} \tag{4a}$$

$$\bar{f}_5 = \frac{\bar{\psi}_5}{\bar{\lambda}_u} = \left(1 + \frac{\bar{\lambda}_u^2}{6} + \frac{7\bar{\lambda}_u^4}{360} + \frac{81\bar{\lambda}_u^6}{15120} + \cdots\right) \Big/ \bar{\lambda}_u^2 \qquad (5a)$$

$$\bar{f}_6 = \frac{\bar{\psi}_6}{\bar{\lambda}_v^3} = 2\left(\frac{1}{3!} + \frac{\bar{\lambda}_v^4}{7!} + \frac{\bar{\lambda}_v^8}{11!} + \frac{\bar{\lambda}_v^{12}}{15!} + \cdots\right) \Big/ \bar{\eta} \qquad (6a)$$

$$\bar{f}_7 = \frac{\bar{\psi}_7}{\bar{\lambda}_v^2} = 2\left(\frac{1}{2!} + \frac{\bar{\lambda}_v^4}{6!} + \frac{\bar{\lambda}_v^8}{10!} + \frac{\bar{\lambda}_v^{12}}{14!} + \cdots\right) \Big/ \bar{\eta} \qquad (7a)$$

$$\bar{f}_8 = \frac{\bar{\psi}_8}{\bar{\lambda}_v} = 2\left(\frac{1}{1!} + \frac{\bar{\lambda}_v^4}{5!} + \frac{\bar{\lambda}_v^8}{9!} + \frac{\bar{\lambda}_v^{12}}{13!} + \cdots\right) \Big/ \bar{\eta} \qquad (8a)$$

where the series in $\bar{\lambda}_u$ are simple expansions and the series in $\bar{\lambda}_v$ are fractions in which the numerator is a McLaurin's series of the numerator of the respective $\bar{\psi}$ and the denominator is the McLaurin's series of the denominator $\bar{\eta}$ which is the same series for $\bar{\psi}_2, \bar{\psi}_3, \bar{\psi}_4, \bar{\psi}_6, \bar{\psi}_7, \bar{\psi}_8$, i.e.,

$$\bar{\eta} = 4\bar{\lambda}_v^4\left(\frac{1}{4!} - \frac{2^2\bar{\lambda}_v^4}{8!} + \frac{2^4\bar{\lambda}_v^8}{12!} - \frac{2^6\bar{\lambda}_v^{12}}{16!} + \cdots\right) \qquad (9)$$

The general form of series in $\bar{\lambda}_u$ is readily available (Ref. 9.8, p. 106) and in symbolic form,

$$\bar{f}_j = \left[C_{j0} + \sum_{r=1}^{\infty} C_{jr}\left(\frac{\bar{\lambda}_u^2}{10}\right)^r\right]\frac{1}{\bar{\lambda}_u^2} \qquad j = 1, 5 \qquad (1b, 5b)$$

The general form of series in $\bar{\lambda}_v$ is derived by series division (Ref. 9.8, p. 105) and in symbolic form,

$$\bar{f}_j = \left[C_{j0} + \sum_{r=1}^{\infty} C_{jr}\left(\frac{\bar{\lambda}_v^4}{100}\right)^r\right]\frac{1}{\bar{\lambda}_v^4} \qquad j = 2, 3, 4, 6, 7, 8 \qquad (2b, 3b, 4b, 6b, 7b, 8b)$$

The first five coefficients of each series are tabulated below.

$\bar{f}_j$	C_{j0}	C_{j1}	C_{j2}	C_{j3}	C_{j4}
$\bar{f}_1$	1	−3.33333 33333	−2.22222 22222	−2.11640 21164	−2.11640 21164
$\bar{f}_2$	4	−0.95238 08956	−0.16262 39100	−0.03196 60103	−0.00637 31298
$\bar{f}_3$	6	−5.23809 49260	−0.76616 48851	−0.14879 71986	−0.02962 38589
$\bar{f}_4$	12	−37.14285 49301	−3.64873 09264	−0.69343 58794	−0.13771 91694
$\bar{f}_5$	1	1.66666 66667	1.94444 44444	2.05026 45502	2.09986 77250
$\bar{f}_6$	2	0.71428 56717	0.15704 07767	0.03239 86171	0.00650 94690
$\bar{f}_7$	6	3.09523 79110	0.72192 99732	0.14565 26563	0.02959 38163
$\bar{f}_8$	12	12.85713 91100	3.29571 07760	0.68442 98355	0.13748 30250

The series in $\bar{\lambda}_u$ apply for $0 \le \bar{\lambda}_u \le \pi$ and the sries in $\bar{\lambda}_v$ apply for $0 \le \bar{\lambda}_v \le 4.73$.

For $0 \le \bar{\lambda}_u < 0.8$ and $0 \le \bar{\lambda}_v < 2.0$, the first two terms of ($1b$ to $8b$) give good approximations of $\bar{f}$ and allow the formulation of the system flexibility matrix in terms of ω^2 which is a great simplification (Ref. 9.3, pp. 117–125).

9.19. Expand the coefficients of $\bar{k}_i$ and $\bar{k}_j$ (Table 9-3c) into the respective McLaurin's series in $\bar{\lambda}_u$ and $\bar{\lambda}_v$.

From Table 9-3a and b, we find

$$\frac{\bar{k}_1}{\bar{f}_1} = \frac{\bar{k}_5}{\bar{f}_5} = \bar{\lambda}_u^2 \qquad \frac{\bar{k}_2}{\bar{f}_2} = \frac{\bar{k}_3}{\bar{f}_3} = \frac{\bar{k}_4}{\bar{f}_4} = \frac{\bar{k}_6}{\bar{f}_6} = \frac{\bar{k}_7}{\bar{f}_7} = \frac{\bar{k}_8}{\bar{f}_8} = \bar{\lambda}_v^4 \qquad (1, 2)$$

Thus the series obtained in (*1*)–(*8*) of Problem 9.18 may be directly used here by multiplying $\bar{\lambda}_u^2$ in $\bar{f}_1$ and $\bar{f}_5$ and $\bar{\lambda}_v^4$, in $\bar{f}_2$, $\bar{f}_3$, $\bar{f}_4$, $\bar{f}_6$, $\bar{f}_7$, and $\bar{f}_8$. Again, for $0 < \bar{\lambda}_u < 0.8$ and $0 < \bar{\lambda}_v < 2.0$, the first two terms of the series will give good approximation of $\bar{k}$ and allow the formulation of the system stiffness matrix in terms of ω^2.

It is important to observe that the first numerical term in the series of (*1b*) through (*8b*) in Problem 9.18 is the numerical coefficient of the respective static stiffness and the sum of the remaining terms is the dynamic correction.

For example, from Table 9-3*c* in terms of (*2b*) of Problem 9.18,

$$\frac{EI\bar{k}_2}{l} = \frac{EI}{l}\left[4 - 0.952\,381\left(\frac{\bar{\lambda}_v^4}{100}\right) - \cdots\right]$$

where $4EI/l$ is the static stiffness due to $\theta_R = 1$ rad where u_R, u_L, v_R, v_L, and θ_L are zero and $[-0.952\,381(\bar{\lambda}_v^4/100) - \cdots]$ is the dynamic correction under the same condition.

Supplementary Problems

FLEXIBILITY METHOD

9.20. Find the dynamic load functions Δ_{RO} and Δ_{LO} of the members shown in Table P-9.20 for flexibility matrix analysis.

9.21. Find the dynamic load functions τ_{RO} and τ_{LO} [Eq. (*6*) of Problem 9.7] of the members given in Table P-9.21 for the method of three-moment equation.

9.22. Formulate the symbolic equations for dynamic analysis of the structures shown in Fig. P-9.22.

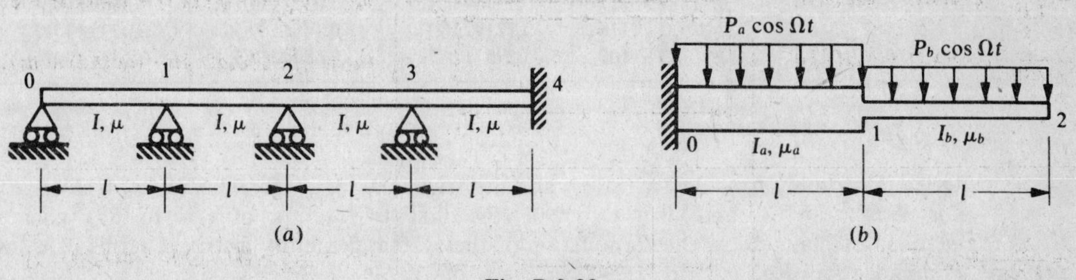

Fig. P-9.22

9.23. Using the three-moment equation, find the natural frequencies of the continuous beam of Fig. P-9.22*a*. Given: $E = 29\,000$ ksi, $I = 243$ in^4, $\mu = 0.3675$ k-s^2/in^2, and $l = 12$ ft.

9.24. Rework Problem 8.13 using the flexibility method.

9.25. Rework Problem 3.1 using the flexibility method and including the column mass.

Table P-9.20 Flexibility Method, Load Functions, Δ_{L0} and Δ_{R0}

Ω = Forcing Frequency $\quad \lambda_u = \sqrt{\mu\Omega^2/EA} \quad \lambda_v = \sqrt[4]{\mu\Omega^2/EI}$

Δ_{L0}	Case	Δ_{R0}
$u_{L0} = \frac{P_u}{AE\lambda_u}\bar{\psi}_5 d(\lambda_u l_2)$ $v_{L0} = \frac{P_v}{EI\lambda_v^3}[\bar{\psi}_7 b(\lambda_v l_2) - \bar{\psi}_6 a(\lambda_v l_2)]$ $\theta_{L0} = \frac{P_v}{EI\lambda_v^2}[\bar{\psi}_7 a(\lambda_v l_2) - \bar{\psi}_8 b(\lambda_v l_2)]$	1. 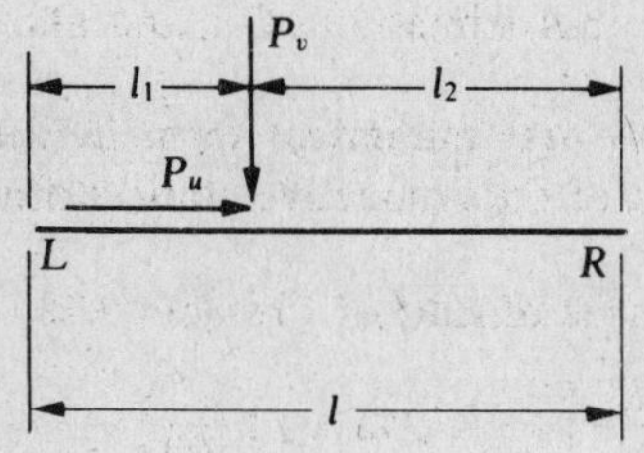 	$u_{R0} = -\frac{P_u}{AE\lambda_u}[d(\lambda_u l_2) + \bar{\psi}_1 a(\lambda_u l_2)]$ $v_{R0} = \frac{P_v}{EI\lambda_v^3}[-\bar{\psi}_2 a(\lambda_v l_2) + \bar{\psi}_3 b(\lambda_v l_2) + d(\lambda_v l_2)]$ $\theta_{R0} = \frac{P_v}{EI\lambda_v^2}[\bar{\psi}_4 b(\lambda_v l_2) - \bar{\psi}_3 a(\lambda_v l_2) + c(\lambda_v l_2)]$
$u_{L0} = 0$ $v_{L0} = -\frac{Q}{EI\lambda_v^2}[\bar{\psi}_7 a(\lambda_v l_2) - d(\lambda_v l_2)\bar{\psi}_6]$ $\theta_{L0} = \frac{Q}{EI\lambda_v}[\bar{\psi}_8 a(\lambda_v l_2) - \bar{\psi}_7 d(\lambda_v l_2)]$	2. 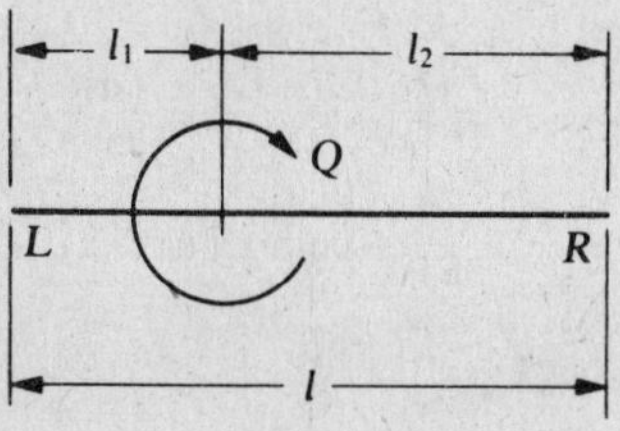 	$u_{R0} = 0$ $v_{R0} = -\frac{Q}{EI\lambda_v^2}[-\bar{\psi}_2 d(\lambda_v l_2) + \bar{\psi}_3 a(\lambda_v l_2) + c(\lambda_v l_2)]$ $\theta_{R0} = \frac{Q}{EI\lambda_v}[\bar{\psi}_3 d(\lambda_v l_2) - \bar{\psi}_4 a(\lambda_v l_2) - b(\lambda_v l_2)]$
$u_{L0} = \frac{p_u}{EA\lambda_u^2} d(\lambda_u l)\bar{\psi}_5$ $v_{L0} = \frac{p_v}{EI\lambda_v^4}[\bar{\psi}_7 c(\lambda_v l) - \bar{\psi}_6 b(\lambda_v l)]$ $\theta_{L0} = \frac{p_v}{EI\lambda_v^3}[\bar{\psi}_7 b(\lambda_v l) - \bar{\psi}_8 c(\lambda_v l)]$	3. 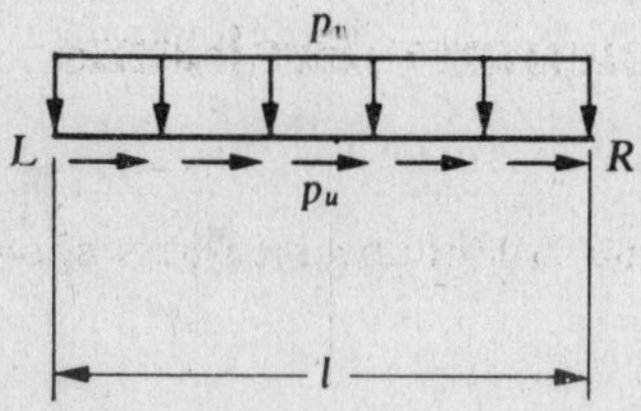 	$u_{R0} = -\frac{p_u}{AE\lambda_u^2}[\bar{\psi}_1 d(\lambda_u l) - a(\lambda_u l) + 1]$ $v_{R0} = \frac{p_v}{EI\lambda_v^4}[-\bar{\psi}_2 b(\lambda_v l) + \bar{\psi}_3 c(\lambda_v l) + a(\lambda_v l) - 1]$ $\theta_{R0} = \frac{p_v}{EI\lambda_v^2}[\bar{\psi}_4 c(\lambda_v l) - \bar{\psi}_3 b(\lambda_v l) + d(\lambda_v l)]$
$u_{L0} = \frac{p_u}{EA\lambda_u^2} d(\lambda_u l)\bar{\psi}_5$ $v_{L0} = \frac{p_v}{EIl\lambda_v^5}[\bar{\psi}_7 d(\lambda_v l) - \bar{\psi}_6 c(\lambda_v l)]$ $\theta_{L0} = \frac{p_v}{EIl\lambda_v^4}[\bar{\psi}_7 c(\lambda_v l) - \bar{\psi}_8 d(\lambda_v l)]$	4. 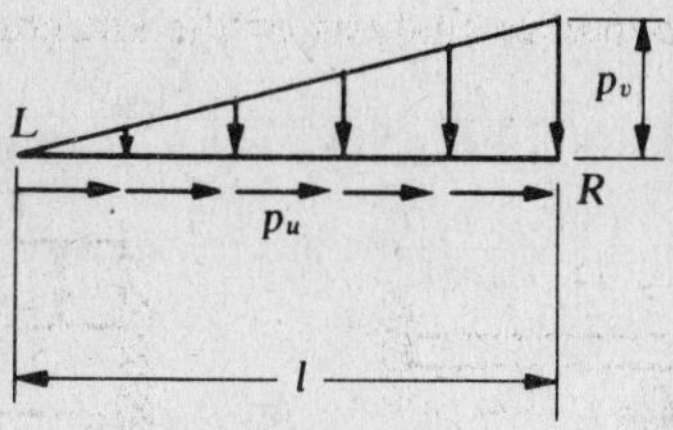 	$u_{R0} = \frac{-p_u}{EA\lambda_u^2}[\bar{\psi}_1 d(\lambda_u l) - a(\lambda_u l) + 1]$ $v_{R0} = \frac{p_v}{EIl\lambda_v^5}[-\bar{\psi}_2 c(\lambda_v l) + \bar{\psi}_3 d(\lambda_v l) + b(\lambda_v l) - \lambda_v l]$ $\theta_{R0} = \frac{p_v}{EIl\lambda_v^4}[\bar{\psi}_4 d(\lambda_v l) - \psi_3 c(\lambda_v l) + a(\lambda_v l) - 1]$
$u_{L0} = \frac{p_v}{EA\lambda_u^2} d(\lambda_u l_2)\bar{\psi}_5$ $v_{L0} = \frac{p_v}{EI\lambda_v^4}[\bar{\psi}_7 c(\lambda_v l_2) - \bar{\psi}_6 b(\lambda_v l_2)]$ $\theta_{L0} = \frac{p_v}{EI\lambda_v^3}[\bar{\psi}_7 b(\lambda_v l_2) - \bar{\psi}_8 c(\lambda_v l_2)]$	5. 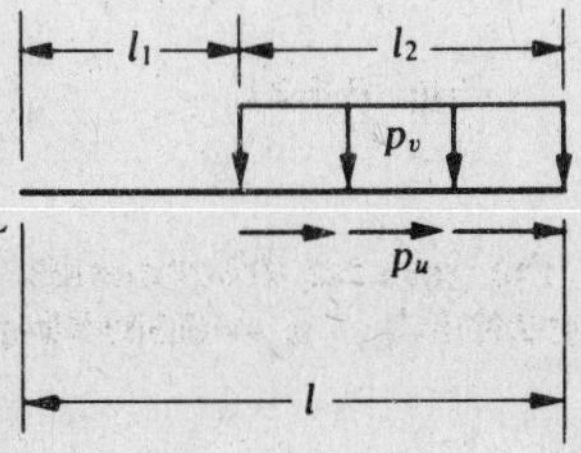 	$u_{R0} = \frac{-p_u}{EA\lambda_u^2}[d(\lambda_u l_2)\bar{\psi}_1 + a(\lambda_u l_2) - 1]$ $v_{R0} = \frac{p_v}{EI\lambda_v^4}[-\psi_2 b(\lambda_v l_2) + \bar{\psi}_3 c(\lambda_v l_2) + a(\lambda_v l_2) - 1]$ $\theta_{R0} = \frac{p_v}{EI\lambda_v^3}[-\bar{\psi}_3 b(\lambda_v l_2) + \bar{\psi}_4 c(\lambda_v l_2) + d(\lambda_v l_2)]$

Note: The dynamic parameters a_u and d_u and a_v, b_v, c_v, and d_v are given analytically in (6-5) and (6-24) respectively. The dynamic parameters $\bar{\psi}_1, \bar{\psi}_2, \ldots, \bar{\psi}_8$ are given analytically in Table 9-2c.

Table P-9.21 Three Moment Method, Load Functions, τ_{L0} and τ_{R0}

Ω = Forcing Frequency $\quad \lambda_v = \sqrt[4]{\mu\Omega^2/EI}$

1. 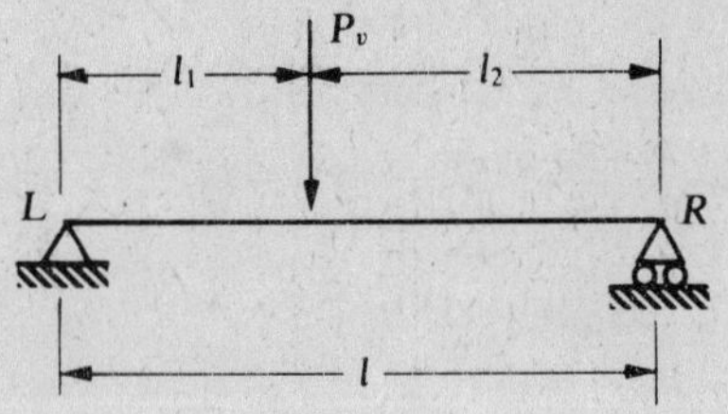	$\tau_{L0} = -\dfrac{P_v}{2EI\lambda_v^2}\left[\dfrac{\sin\lambda_v l_2}{\sin\lambda_v l} - \dfrac{\sinh\lambda_v l_2}{\sinh\lambda_v l}\right]$ $\tau_{R0} = \dfrac{P_v}{2EI\lambda_v^2}[\sin\lambda_v l_2\cot\lambda_v l - \sinh\lambda_v l_2\coth\lambda_v l - 2c(\lambda_v l)]$
2. 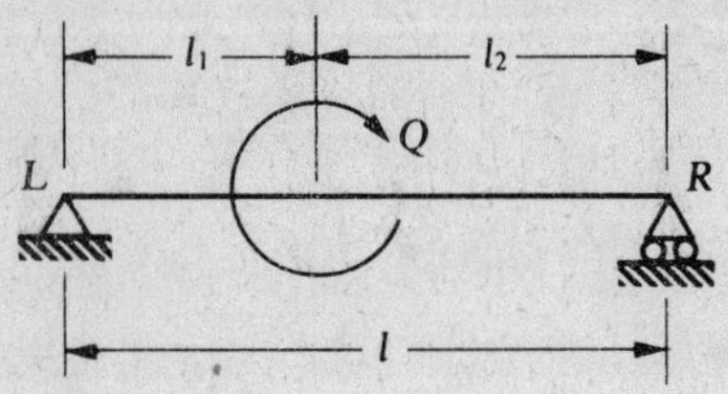	$\tau_{L0} = \dfrac{Q}{2EI\lambda_v}\left[\dfrac{\cos\lambda_v l_2}{\sin\lambda_v l} - \dfrac{\cosh\lambda_v l_2}{\sinh\lambda_v l}\right]$ $\tau_{R0} = \dfrac{Q}{2EI\lambda_v}[\coth\lambda_v l\cosh\lambda_v l - \cot\lambda_v l\cos\lambda_v l_2 - 2b(\lambda_v l_2)]$
3. 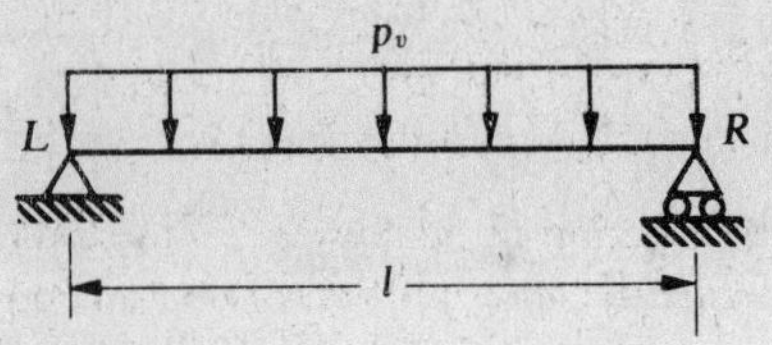	$\tau_{L0} = \dfrac{p_v}{2EI\lambda_v^3}\left[\coth\lambda_v l + \cot\lambda_v l - \dfrac{2b(\lambda_v l)}{\sinh\lambda_v l\sin\lambda_v l}\right]$ $\tau_{R0} = \dfrac{p_v}{2EI\lambda_v^3}\left[\coth\lambda_v l(1-\cosh\lambda_v l) + \cot\lambda_v l(1-\cos\lambda_v l) + 2d(\lambda_v l)\right]$
4. 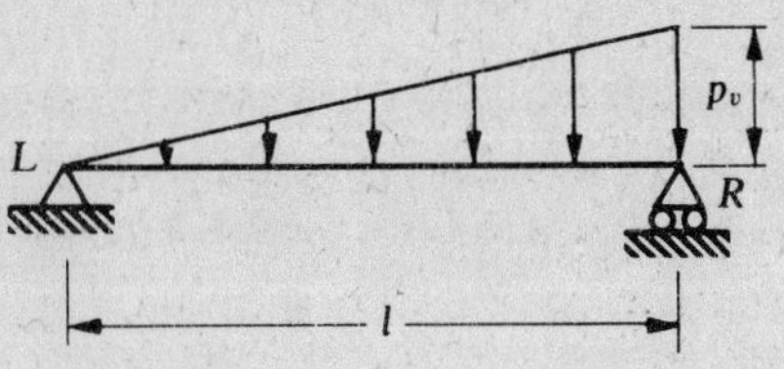	$\tau_{L0} = \dfrac{p_v}{EI\lambda_v^3}\left[\lambda_v l - \dfrac{b(\lambda_v l)}{\sinh\lambda_v l\sin\lambda_v l}\right]$ $\tau_{R0} = \dfrac{p_v}{2EI\lambda_v^3}[\coth\lambda_v l + \cot\lambda_v l]$
5. 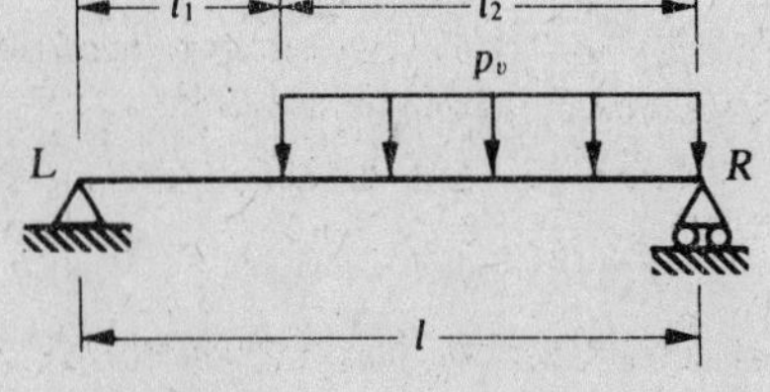	$\tau_{L0} = \dfrac{p_v}{2EI\lambda_v^3}\left[\dfrac{\cos\lambda_v l_2}{\sin\lambda_v l} + \dfrac{\cosh\lambda_v l_2}{\sinh\lambda_v l} - \dfrac{2b(\lambda_v l)}{\sinh\lambda_v l\sin\lambda_v l}\right]$ $\tau_{R0} = \dfrac{p_v}{2EI\lambda_v^3}[\coth\lambda_v l(1-\cosh\lambda_v l_2) + \cot\lambda_v l(1-\cos\lambda_v l_2) + 2d(\lambda_v l)]$

Note: The shape functions b, c, and d are given in (6-24).

Table P-9.26 Stiffness Method, Load Functions, σ_{L0} and σ_{R0}

Ω = Forcing Frequency $\lambda_u = \sqrt{\mu\Omega^2/EA}$ $\lambda_v = \sqrt[4]{\mu\Omega^2/EI}$		
$N_{L0} = -P_u\bar{\psi}_5 d(\lambda_u l_2)$ $V_{L0} = P_v[\bar{\psi}_8 d(\lambda_v l_2) - \bar{\psi}_7 c(\lambda_v l_2)]$ $M_{L0} = \frac{P_v}{\lambda_v}[\bar{\psi}_7 d(\lambda_v l_2) - \bar{\psi}_6 c(\lambda_v l_2)]$	1. 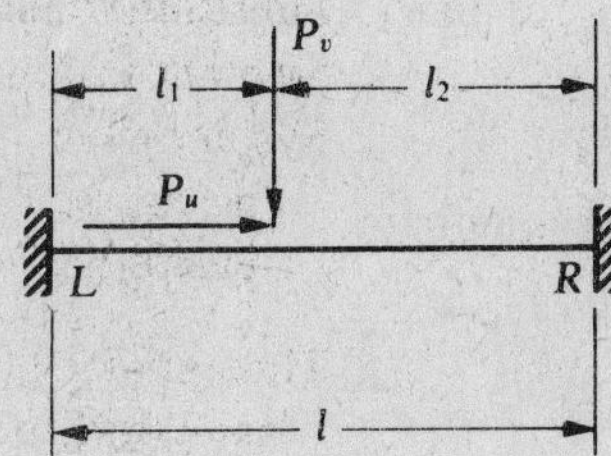 	$N_{R0} = P_u[\bar{\psi}_1 d(\lambda_u l_2) - a(\lambda_u l_2)]$ $V_{R0} = P_v[\bar{\psi}_3 c(\lambda_v l_2) - \bar{\psi}_4 d(\lambda_v l_2) - a(\lambda_v l_2)]$ $M_{R0} = \frac{P_v}{\lambda_v}[b(\lambda_v l_2) + \bar{\psi}_3 d(\lambda_v l_2) - \bar{\psi}_2 c(\lambda_v l_2)]$
$N_{L0} = 0$ $V_{L0} = Q\lambda_v[\bar{\psi}_7 b(\lambda_v l_2) - \bar{\psi}_8 c(\lambda_v l)]$ $M_{L0} = Q[\bar{\psi}_6 b(\lambda_v l_2) - \bar{\psi}_7 c(\lambda_v l_2)]$	2. 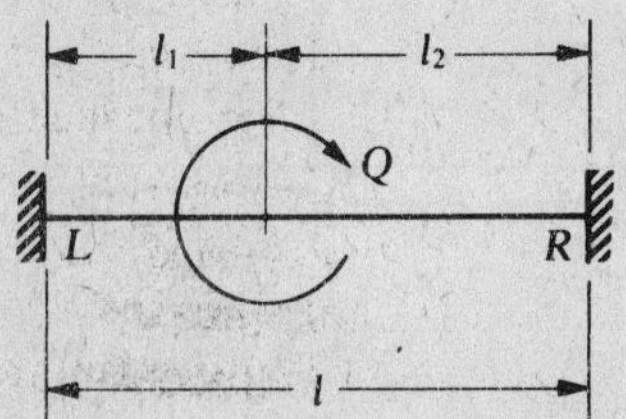 	$N_{R0} = 0$ $V_{R0} = Q\lambda_v[d(\lambda_v l_2) + \bar{\psi}_4 c(\lambda_v l_2) - \bar{\psi}_3 b(\lambda_v l_2)]$ $M_{R0} = -Q[a(\lambda_v l_2) + \bar{\psi}_3 c(\lambda_v l_2) - \bar{\psi}_2 b(\lambda_v l_2)]$
$N_{L0} = \frac{p_u}{\lambda_u}[a(\lambda_u l) - 1]\bar{\psi}_5$ $V_{L0} = \frac{p_v}{\lambda_v}\{\bar{\psi}_8[a(\lambda_v l) - 1] - \bar{\psi}_7 d(\lambda_v l)\}$ $M_{L0} = \frac{p_v}{\lambda_v^2}\{\bar{\psi}_7[a(\lambda_v l) - 1] - \bar{\psi}_6 d(\lambda_v l)\}$	3. 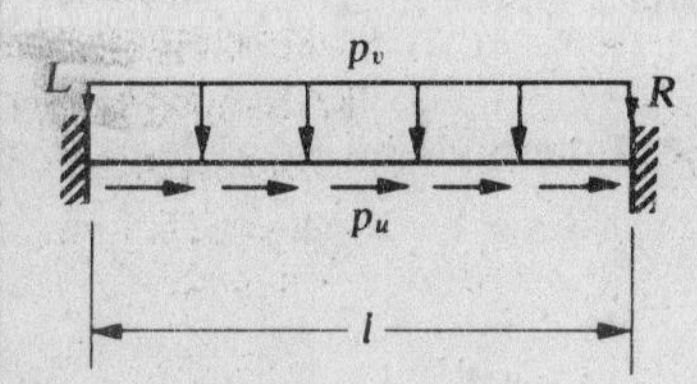 	$N_{R0} = \frac{p_u}{\lambda_u}\{\bar{\psi}_1[1 - a(\lambda_u l)] - d(\lambda_u l)\}$ $V_{R0} = \frac{p_v}{\lambda_v}\{\bar{\psi}_4[1 - a(\lambda_v l)] - b(\lambda_v l) + \bar{\psi}_3 d(\lambda_v l)\}$ $M_{R0} = \frac{p_v}{\lambda_v^2}\{c(\lambda_v l) + \bar{\psi}_3[a(\lambda_v l) - 1] + \bar{\psi}_2 d(\lambda_v l)\}$
$N_{L0} = \frac{p_u}{\lambda_u}[a(\lambda_u l) - 1]\bar{\psi}_5$ $V_{L0} = \frac{p_v}{l\lambda_v^2}\{\bar{\psi}_8[b(\lambda_v l) - \lambda_v l] - \bar{\psi}_7[a(\lambda_v l) - 1]\}$ $M_{L0} = \frac{p_v}{l\lambda_v^3}\{\bar{\psi}_7[b(\lambda_v l) - \lambda_v l] - \bar{\psi}_6[a(\lambda_v l) - 1]\}$	4. 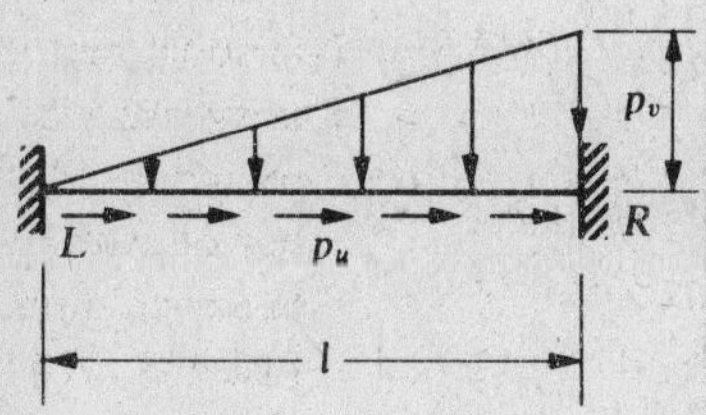 	$N_{R0} = \frac{p_u}{\lambda_u}\{\bar{\psi}_1[1 - a(\lambda_u l)] - d(\lambda_u l)\}$ $V_{R0} = \frac{p_v}{l\lambda_v^2}\{\bar{\psi}_3[a(\lambda_v l) - 1] - \bar{\psi}_4[b(\lambda_v l) - \lambda_v l] - c(\lambda_v l)\}$ $M_{R0} = \frac{p_v}{l\lambda_v^3}\{d(\lambda_v l) + \bar{\psi}_3[b(\lambda_v l) - \lambda_v l] - \bar{\psi}_2[a(\lambda_v l) - 1]\}$
$N_{L0} = \frac{p_u}{\lambda_u}\bar{\psi}_5[a(\lambda_u l_2) - 1]$ $V_{L0} = \frac{p_v}{\lambda_v}\{\bar{\psi}_8[a(\lambda_v l_2) - 1] - \bar{\psi}_7 d(\lambda_v l_2)\}$ $M_{L0} = \frac{p_v}{\lambda_v^2}\{[a(\lambda_v l_2) - 1]\bar{\psi}_7 - \bar{\psi}_6 d(\lambda_v l_2)\}$	5. 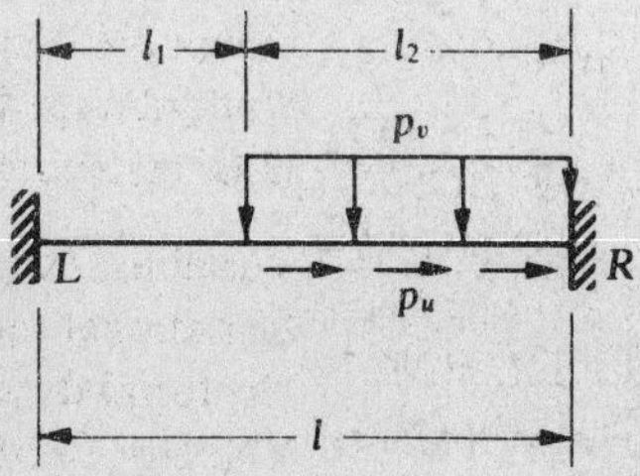 	$N_{R0} = \frac{p_u}{\lambda_u}\{\bar{\psi}_1[1 - a(\lambda_u l_2)] - d(\lambda_u l_2)\}$ $V_{R0} = \frac{p_v}{\lambda_v}\{\bar{\psi}_4[1 - a(\lambda_v l_2)] - \bar{\psi}_3 d(\lambda_v l_2) - b(\lambda_v l_2)\}$ $M_{R0} = \frac{p_v}{\lambda_v^2}\{c(\lambda_v l_2) + \bar{\psi}_3[a(\lambda_v l_2) - 1] - \bar{\psi}_2 d(\lambda_v l_2)\}$

Note: The dynamic parameters a_u and d_u and a_v, b_v, c_v, and d_v are given analytically in (6-5) and (6-24) respectively. The dynamic parameters $\bar{\psi}_1, \bar{\psi}_2, \ldots, \bar{\psi}_8$ are given analytically in Table 9-2*c*.

STIFFNESS METHOD

9.26. Derive the load functions σ_{R0} and σ_{L0} of the prismatic bars shown in Table P-9.26 for the stiffness method.

9.27. Rework Problem 9.22 using the stiffness method.

9.28. Rework Problem 9.23 using the three-slope equation.

9.29. Find the moments at L and R of Fig. P-9.29. Given: $l = 12$ ft, $I_1 = 486$ in^4, $I_2 = I_1/2$, $E = 29\,000$ ksi, $\mu_1 = 0.735$ k-s^2/in^2, $\mu_2 = \mu_1/2$, $P(t) = P \cos \Omega t$, $P = 40$ k, and $\Omega = \pi/2$ rad/s.

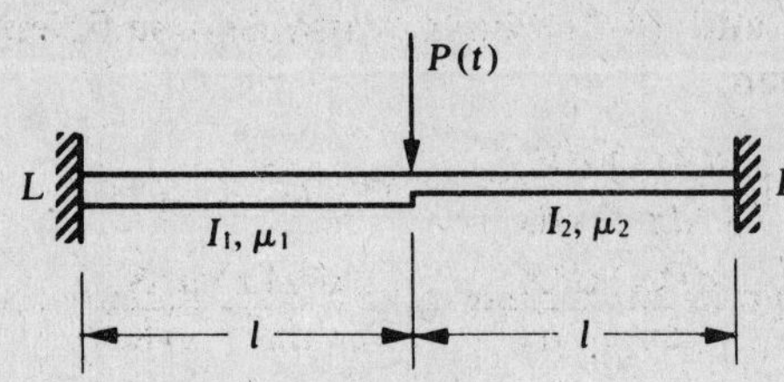

Fig. P-9.29

9.30. Rework Problem 8.14 using the stiffness method.

9.31. Rework Problem 3.1 using the stiffness method and including the column mass.

Chapter 10

Variable Cross Section

10.1 EFFECT OF VARIATION

Differential Equations

When the cross section of the bars in Chapter 6 varies along the entire length or along one or more segments of the bar (Fig. 10-1), the area $A(x)$ of the cross section, its moment of inertia $I(x)$, and the intensity of mass $\mu(x)$ become functions of the position coordinate x and the differential equations of motion (*6-2*) and (*6-21*) have variable coefficients. The respective *eigenfunction differential equations* are then

$$EA(x)\frac{d^2X_{u,m}}{dx^2} + E\frac{dA(x)}{dx}\frac{dX_{u,m}}{dx} + \mu(x)\omega_m^2 X_{u,m} = 0 \tag{10-1}$$

$$EI(x)\frac{d^4X_{v,m}}{dx^4} + 2E\frac{dI(x)}{dx}\frac{d^3X_{v,m}}{dx^3} + E\frac{d^2I(x)}{dx^2}\frac{d^2X_{v,m}}{dx^2} - \mu(x)\omega_m^2 X_{v,m} = 0 \tag{10-2}$$

where the subscript m designating the mth node is omitted in the subsequent formulations (Problems 10.1 and 10.2).

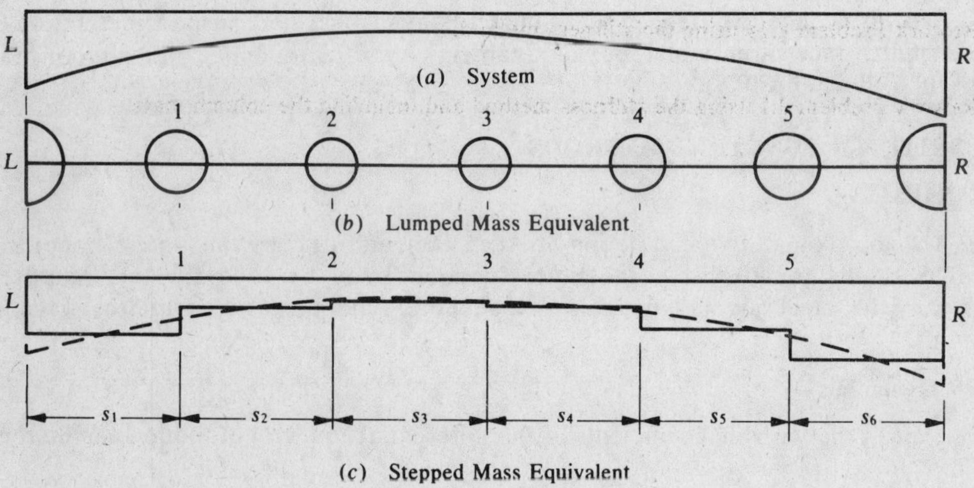

Fig. 10-1. Equivalent Systems

It is not possible to find closed form general integrals of such differential equations and *substitute models* or *substitute functions* must be introduced for their solution.

Methods of Solution

Among many methods available for the solution of such cases, three classes of methods are the most commonly used:

(1) The lumped mass methods
(2) The stepped mass methods
(3) The substitute function methods

A brief summary of the theoretical background of these methods and their applications is given in this chapter.

10.2 LUMPED MASS METHODS

Mathematical Model

For analysis, the real bar is divided in an arbitrarily selected number of segments (Fig. 10-1*b*) and the distributed mass is lumped at the segment boundaries (joints, nodes). This model is essentially the lumped mass system of Chapter 5. Since variable mass is involved, the spacing of nodes and/or the magnitudes of lumped mass must vary to closely approximate the variation of $A(x)$, $I(x)$, and $\mu(x)$.

Methods of Analysis

Once the lumped mass model is available, the transport method, the flexibility method, and the stiffness method of Chapter 5 are directly applicable without modification (Problems 10.3, 10.4, and 10.5).

10.3 STEPPED MASS METHODS

Mathematical Model

The real bar is replaced by an equivalent stepped bar (Fig. 10-1*c*), the dynamic parameters of which are computed on the basis of average values of A, I, and μ. For shallow longitudinal profiles good results are obtained even if only two or three segments are used.

Methods of Analysis

As in Sec. 10.2, once the stepped mass model is available, the transport method of Chapter 8 and the flexibility and stiffness methods of Chapter 9 are directly applicable without modification (Problems 10.6, 10.7, and 10.8).

10.4 SUBSTITUTE FUNCTION METHODS

Mathematical Model

Instead of approximating the variation of $A(x)$, $I(x)$, and $\mu(x)$ by the finite element (substitute) models of Sec. 10.2 or 10.3, the substitute functions may be used for this purpose. These functions may be algebraic polynomials, exponential functions, trigonometric series, or their combinations.

Algebraic Polynomials

In some cases the variable coefficients of the differential equation of motion can be represented (Ref. 10.3, p. 16) as

$$A(x) = A_L(a + bx/s)^{n-4} \qquad I(x) = I_L(a + bx/s)^n \qquad \mu(x) = \mu_L(a + bx/s)^{n-4} \tag{10-3}$$

where n is a real number,

$$a = \frac{1 - (c/s)\sqrt[n]{I_R/I_L}}{1 - c/s} \qquad b = \frac{\sqrt[n]{I_R/I_L} - 1}{1 - c/s} \tag{10-4}$$

and A_L, I_L, I_R, and μ_L are respectively the cross-sectional area, the moments of inertia, and the intensity of mass at the respective end of the bar, and c/s is a geometrical constant taken from Fig. 10-2.

Exponential Functions

In other cases the variable coefficients can be represented (Ref. 10.3, p. 18) as

$$A(x) = A_L\alpha e^{\beta x/s} \qquad I(x) = I_L\alpha e^{\beta x/s} \qquad \mu(x) = \mu_L\alpha e^{\beta x/s} \tag{10-5}$$

where

$$\alpha = \left(\frac{I_L}{I_R}\right)^{(c/s)(1-c/s)} \qquad \beta = \frac{\ln I_R - \ln I_L}{1 - c/s} \tag{10-6}$$

and the remaining symbols are those of (*10-3*) (Problem 10.10).

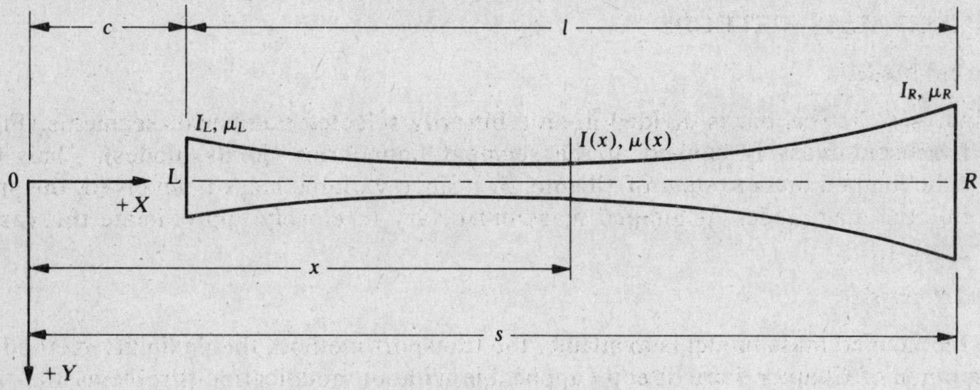

Fig. 10-2. Bar of Variable Mass

Fourier Series

For simply supported bars, the trigonometric series representation is based on

$$X_u(x) = \sum_{m=1}^{\infty} X_{u,m} \sin a_m x \qquad X_v(x) = \sum_{m=1}^{\infty} X_{v,m} \sin a_m x \tag{10-7, 8}$$

$$A(x) = \sum_{m=0}^{\infty} A_m \cos a_m x \qquad I(x) = \sum_{m=0}^{\infty} I_m \cos a_m x \qquad \mu(x) = \sum_{m=0}^{\infty} \mu_m \cos a_m x \tag{10-9}$$

where

$$a_m = m\pi/l \qquad A_m = \frac{2\int_0^l A(x)\cos a_m x\, dx}{\int_0^l A(x)\, dx} \qquad I_m = \frac{2\int_0^l I(x)\cos a_m x\, dx}{\int_0^l I(x)\, dx} \qquad \mu_m = \frac{2\int_0^l \mu(x)\cos a_m x\, dx}{\int_0^l \mu(x)\, dx} \tag{10-10}$$

and $X_{u,m}$ and $X_{v,m}$ are the unknown constants.

The substitution of the respective functions given above in the corresponding differential equation of motion (*10-1*) or (*10-2*) leads to a trigonometric identity in which the coefficients of identical trigonometric functions must be equal.

If the respective series are restricted to r terms, r algebraic simultaneous equations yield r values of $X_{u,m}$ or $X_{v,m}$, which when inserted in (*10-7*) or (*10-8*) respectively allow the construction of the frequency determinant.

Ritz Method

Finally, the eigenfunctions of the respective motion can also be represented as

$$X_u(x) = X_{u,1}f_1(x) + X_{u,2}f_2(x) + \cdots \tag{10-11}$$

$$X_v(x) = X_{v,1}g_1(x) + X_{v,2}g_2(x) + \cdots \tag{10-12}$$

where $f_1(x), f_2(x), \ldots, g_1(x), g_2(x), \ldots$ are selected functions satisfying the boundaries of the problem and $X_{u,1}, X_{u,2}, \ldots, X_{v,1}, X_{v,2}$ are unknown constants.

These constants are then computed from the system of simultaneous equations derived respectively from the conditions

$$\frac{\partial}{\partial X_{u,m}} \int \left[EA(x)\left[\frac{dX_u(x)}{dx}\right]^2 + \mu(x)\omega^2 X_u^2(x) \right] dx = 0 \tag{10-13}$$

$$\frac{\partial}{\partial X_{v,m}} \int \left[EI(x)\left[\frac{d^2X_v(x)}{dx^2}\right]^2 - \mu(x)\omega^2 X_v^2(x) \right] dx = 0 \tag{10-14}$$

where $m = 1, 2, 3, \ldots, r$.

As in the Fourier series solution, r values of $X_{u,m}$ or $X_{v,m}$ inserted in (*10-11*) or (*10-12*) respectively allow the construction of the frequency determinant (Problem 10.12).

Solved Problems

DIFFERENTIAL EQUATIONS

10.1. Derive the differential equation (*10-1*) of longitudinal vibration of a bar with variable cross section $A(x)$ and mass per unit length $\mu(x)$.

Referring to Fig. 6-1, the unit elongation e and tensile force N at any cross section may be expressed as

$$e = \frac{\partial u}{\partial x} \qquad N = A(x)E\frac{\partial u}{\partial x} \tag{1, 2}$$

and for the cross section at the incremental distance dx, the tensile force is then

$$N + \frac{\partial N}{\partial x}dx = A(x)E\frac{\partial u}{\partial x} + \frac{\partial}{\partial x}\left[A(x)E\frac{\partial u}{\partial x}\right]dx \tag{3}$$

The equilibrium condition of the incremental element yields

$$E\frac{\partial A(x)}{\partial x}\frac{\partial u}{\partial x} + EA(x)\frac{\partial^2 u}{\partial x^2} - \mu(x)\frac{\partial^2 u}{\partial t^2} = 0 \tag{4}$$

As shown in Problem 6.1, we let $u(x,t) = X_u(x)Y_u(t)$ and then use the separation of variables to express (*4*) as

$$\left[EA(x)\frac{d^2X_u(x)}{dx^2} + E\frac{dA(x)}{dx}\frac{dX_u(x)}{dx}\right]\frac{1}{X_u(x)\mu(x)} = \frac{1}{Y_u(t)}\frac{d^2Y_u(t)}{dt^2} = -\omega^2 \tag{5}$$

from which the desired result (*10-1*) is

$$EA(x)\frac{d^2X_u(x)}{dx^2} + E\frac{dA(x)}{dx}\frac{dX_u(x)}{dx} + \mu(x)\omega^2X_u(x) = 0 \tag{6}$$

10.2. Derive the differential equation (*10-2*) of transverse vibration of a member with variable cross section $A(x)$, moment of inertia $I(x)$, and mass per unit length $\mu(x)$.

As shown in Problem 6.12, by using the sign convention of Fig. 6-2, we may write the equations of moment and shear as

$$EI(x)\frac{\partial^2 v}{\partial x^2} = M \qquad -\frac{\partial^2 M}{\partial x^2} = \mu(x)\frac{\partial^2 v}{\partial t^2} \tag{1, 2}$$

Substituting (*2*) into (*1*) gives

$$\frac{\partial^2}{\partial x^2}\left[EI(x)\frac{\partial^2 v}{\partial x^2}\right] + \mu(x)\frac{\partial^2 v}{\partial t^2} = 0 \tag{3}$$

With $v(x,t) = X_v(x)Y_v(t)$, (*3*) becomes

$$\frac{d^2}{dx^2}\left[EI(x)\frac{d^2X_v(x)}{dx^2}Y_v(t)\right] + \mu(x)X_v(x)\frac{d^2Y_v(t)}{dt^2} = 0 \tag{4}$$

Operating derivation on (*4*) gives

$$\left[EI(x)\frac{d^4X_v(x)}{dx^4} + 2E\frac{dI(x)}{dx}\frac{d^3X_v(x)}{dx^3} + E\frac{d^2I(x)}{dx^2}\frac{d^2X_v(x)}{dx^2}\right]\frac{1}{\mu(x)X_v(x)} = -\frac{d^2Y_v(t)}{dt^2}\frac{1}{Y_v(t)} \tag{5}$$

Since the terms on the left side and the term on the right side of (*5*) depend on x and t respectively, they can be equal only if they are equal to a positive constant (say ω^2; zero and negative constant are not the solutions of vibration).

Thus from (*5*) the desired result (*10-2*) is

$$EI(x)\frac{d^4X_v(x)}{dx^4} + 2E\frac{dI(x)}{dx}\frac{d^3X_v(x)}{dx^3} + E\frac{d^2I(x)}{dx^2}\frac{d^2X_v(x)}{dx^2} - \mu(x)\omega^2X_v(x) = 0 \tag{6}$$

LUMPED MASS METHODS

10.3. Using the stiffness matrix find the fundamental frequency of the nonuniform beam with constant thickness b shown in Fig. P-10.3a by lumping the masses at one-third point (Fig. P-10.3b). Let $l = 9.15$ m, $h = 38.1$ cm, $b = 38.1$ cm, $\gamma = 76.973\ \text{kN/m}^3$, and $E = 199.95\ \text{GN/m}^2$. Consider transverse motion only.

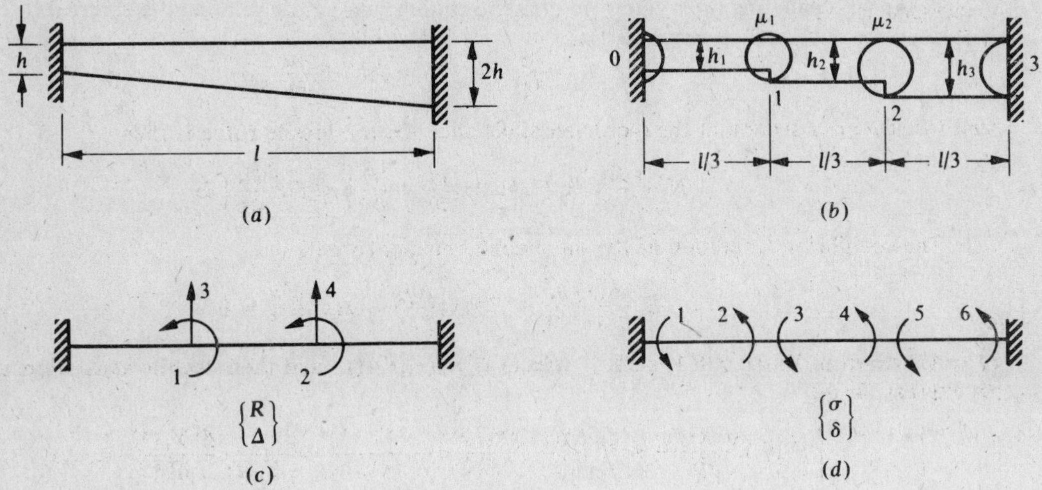

Fig. P-10.3

Let the nonuniform beam be idealized as shown in Fig. P-10.3b of which the segments 01, 12 and 23 are massless and uniform with the following member properties:

$$h_1 = (38.1 + 50.8)/2 = 44.45\ \text{cm} \qquad h_2 = (50.8 + 63.5)/2 = 57.15\ \text{cm}$$
$$h_3 = (63.5 + 76.2)/2 = 69.85\ \text{cm} \tag{1}$$

$$I_1 = bh_1^3/12 = 27.884\ \text{dm}^4 \qquad I_2 = bh_2^3/12 = 59.264\ \text{dm}^4 \qquad I_3 = bh_3^3/12 = 108.204\ \text{dm}^4 \tag{2}$$

The masses lumped at the nodes are

$$\mu_1 = \frac{\gamma}{g}(A_1 + A_2)\frac{l}{3} = 4.619\ \text{kN-s}^2/\text{m} \qquad \mu_2 = \frac{\gamma}{g}(A_2 + A_3)\frac{l}{3} = 5.779\ \text{kN-s}^2/\text{m} \tag{3}$$

Now sketch the generalized coordinates and internal forces as shown in Fig. P-10.3c and d of which the equilibrium conditions, $\Sigma M = 0$ and $\Sigma F_y = 0$, at the nodes give

$$\underbrace{\begin{bmatrix} R_1 \\ R_2 \\ R_3 \\ R_4 \end{bmatrix}}_{\{R\}} = \underbrace{\begin{bmatrix} 0 & 1 & 1 & 0 & 0 & 0 \\ 0 & 0 & 0 & 1 & 1 & 0 \\ -1/3.05 & -1/3.05 & 1/3.05 & 1/3.05 & 0 & 0 \\ 0 & 0 & -1/3.05 & -1/3.05 & 1/3.05 & 1/3.05 \end{bmatrix}}_{[j]} \underbrace{[\sigma_1 \quad \sigma_2 \quad \sigma_3 \quad \sigma_4 \quad \sigma_5 \quad \sigma_6]^T}_{\{\sigma\}} \tag{4}$$

The stiffness coefficients of the idealized segments are

$$\underbrace{\begin{bmatrix} \sigma_1 \\ \sigma_2 \\ \sigma_3 \\ \sigma_4 \\ \sigma_5 \\ \sigma_6 \end{bmatrix}}_{\{\sigma\}} = \frac{E}{3.05} \underbrace{\begin{bmatrix} 4I_1 & 2I_1 & 0 & 0 & 0 & 0 \\ 2I_1 & 4I_1 & 0 & 0 & 0 & 0 \\ 0 & 0 & 4I_2 & 2I_2 & 0 & 0 \\ 0 & 0 & 2I_2 & 4I_2 & 0 & 0 \\ 0 & 0 & 0 & 0 & 4I_3 & 2I_3 \\ 0 & 0 & 0 & 0 & 2I_3 & 4I_3 \end{bmatrix}}_{[k]} \underbrace{\begin{bmatrix} \delta_1 \\ \delta_2 \\ \delta_3 \\ \delta_4 \\ \delta_5 \\ \delta_6 \end{bmatrix}}_{\{\delta\}} \tag{5}$$

where $4EI_1/3.05 = 731.20$ MN-m, $4EI_2/3.05 = 1554.08$ MN-m, and $4EI_3/3.05 = 2837.43$ MN-m.

Thus the system stiffness matrix becomes

$$[\Omega] = [j][k][j]^T = \begin{bmatrix} 2285.28 & 777.04 & 404.70 & -764.30 \\ & 4391.51 & 764.30 & 631.16 \\ \text{Sym.} & & 736.99 & -501.18 \\ & & & 1416.24 \end{bmatrix} \times 10^6 \tag{6}$$

Without considering the rotatory inertia, the equilibrium matrix relating the external forces $\{R\}$ and displacements $\{\Delta\}$ may be expressed as

$$\underbrace{\begin{bmatrix} 0 \\ 0 \\ \hline \mu_1\omega^2 X_3 \\ \mu_2\omega^2 X_4 \end{bmatrix}}_{\{R\}} = \underbrace{\left[\begin{array}{c|c} \Omega_{11} & \Omega_{12} \\ \hline \Omega_{21} & \Omega_{22} \end{array}\right]}_{[\Omega]} \underbrace{\begin{bmatrix} X_1 \\ X_2 \\ \hline X_3 \\ X_4 \end{bmatrix}}_{\{\Delta\}} \tag{7}$$

from which

$$\begin{bmatrix} \mu_1\omega^2 X_3 \\ \mu_2\omega^2 X_4 \end{bmatrix} = [\Omega_{22} - \Omega_{21}\Omega_{11}^{-1}\Omega_{12}]\begin{bmatrix} X_3 \\ X_4 \end{bmatrix} = [\Omega_R]\begin{bmatrix} X_3 \\ X_4 \end{bmatrix} \tag{8}$$

in which Ω_{11}, Ω_{22}, Ω_{21}, and Ω_{12} may be found in (6). Thus Ω_{11}^{-1} and Ω_R are

$$\Omega_{11}^{-1} = \begin{bmatrix} 4.656 & -0.824 \\ -0.824 & 2.423 \end{bmatrix} \times 10^{-10} \qquad \Omega_R = \begin{bmatrix} 570.16 & -501.18 \\ -501.18 & 968.26 \end{bmatrix} \times 10^6 \tag{9, 10}$$

From (8),

$$\left([\Omega_R] - \omega^2\begin{bmatrix} \mu_1 & 0 \\ 0 & \mu_2 \end{bmatrix}\right)\begin{bmatrix} X_3 \\ X_4 \end{bmatrix} = 0 \tag{11}$$

The determinant of the coefficients of (11) yields the following frequency equation:

$$\omega^4 - 290\,986\omega^2 + 1\,127\,184.94 \times 10^4 = 0 \tag{12}$$

from which

$$\omega_1 = 214 \text{ rad/s} \qquad \omega_2 = 495.3 \text{ rad/s}$$

10.4. Rework Problem 10.3 using the flexibility matrix.

Let the indeterminate beam (Fig. P-10.4*a*) be changed to a stable determinate beam with external inertia forces $\{R_g\}$ and redundant forces $\{R_x\}$ as shown in Fig. P-10.4*b*.

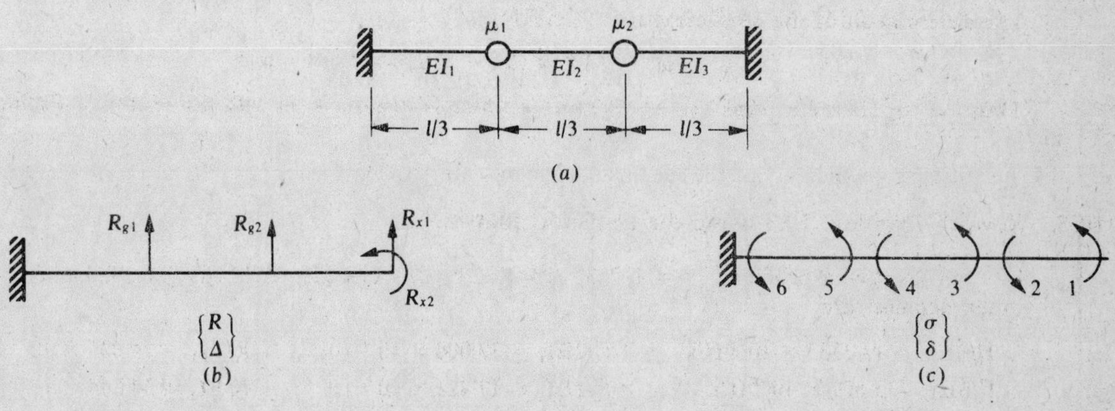

Fig. P-10.4

$$\{R_g\} = \omega^2\begin{bmatrix}\mu_1 & 0\\ 0 & \mu_2\end{bmatrix}\begin{bmatrix}\Delta_{R1}\\ \Delta_{R2}\end{bmatrix} = \omega^2[\mu]\{\Delta_R\} \qquad \{R_x\} = \begin{bmatrix}R_{x1}\\ R_{x2}\end{bmatrix} \tag{1, 2}$$

The relationship between the internal forces $\{\sigma\}$ (Fig. P-10.4c) and the external forces $\begin{bmatrix}R_g\\ R_x\end{bmatrix}$ may be established by equilibrium condition as

$$\underbrace{\begin{bmatrix}\sigma_1\\ \sigma_2\\ \sigma_3\\ \sigma_4\\ \sigma_5\\ \sigma_6\end{bmatrix}}_{\{\sigma\}} = \underbrace{\left[\begin{array}{cc|cc} 0 & 0 & 0 & 1\\ 0 & 0 & -3.05 & -1\\ 0 & 0 & 3.05 & 1\\ 0 & -3.05 & -6.10 & -1\\ 0 & 3.05 & 6.10 & 1\\ -3.05 & -6.10 & -9.15 & -1\end{array}\right]}_{[q_R \,\vdots\, q_x]} \underbrace{\left[\begin{array}{c} R_{g1}\\ R_{g2}\\ \hline R_{x1}\\ R_{x2}\end{array}\right]}_{\begin{Bmatrix}R_g\\ R_x\end{Bmatrix}} \tag{3}$$

For the given idealized system, the unit stress flexibility chain is

$$\underbrace{\begin{bmatrix}\delta_1\\ \delta_2\\ \delta_3\\ \delta_4\\ \delta_5\\ \delta_6\end{bmatrix}}_{\{\delta\}} = \frac{3.05}{E}\underbrace{\begin{bmatrix} 1/3I_3 & -1/6I_3 & 0 & 0 & 0 & 0\\ -1/6I_3 & 1/3I_3 & 0 & 0 & 0 & 0\\ 0 & 0 & 1/3I_2 & -1/6I_2 & 0 & 0\\ 0 & 0 & -1/6I_2 & 1/3I_2 & 0 & 0\\ 0 & 0 & 0 & 0 & 1/3I_1 & -1/6I_1\\ 0 & 0 & 0 & 0 & -1/6I_1 & 1/3I_1\end{bmatrix}}_{[f]}\underbrace{\begin{bmatrix}\sigma_1\\ \sigma_2\\ \sigma_3\\ \sigma_4\\ \sigma_5\\ \sigma_6\end{bmatrix}}_{\{\sigma\}} \tag{4}$$

where $EI_1 = 557.54\ \text{MN-m}^2$, $EI_2 = 1184.98\ \text{MN-m}^2$, and $EI_3 = 2163.54\ \text{MN-m}^2$.

Performing $[q_R \quad q_x]^T[f][q_R \quad q_x]$ yields

$$\begin{bmatrix}\Delta_{R1}\\ \Delta_{R2}\\ \Delta_{x1}\\ \Delta_{x2}\end{bmatrix} = \begin{bmatrix} 16.9630 & 42.4075 & 67.8520 & 8.3425\\ & 126.7220 & 215.0280 & 28.9526\\ \text{Sym.} & & 382.5360 & 55.6376\\ & & & 9.4541\end{bmatrix} \times 10^{-9}\begin{bmatrix}R_{g1}\\ R_{g2}\\ R_{x1}\\ R_{x2}\end{bmatrix} \quad \text{or} \quad \begin{bmatrix}\Delta_R\\ \Delta_x\end{bmatrix} = \begin{bmatrix}\Phi_{RR} & \Phi_{Rx}\\ \Phi_{xR} & \Phi_{xx}\end{bmatrix}\begin{bmatrix}R_g\\ R_x\end{bmatrix} \tag{5}$$

Since $\{\Delta_x\} = 0$, (5) may be rewritten as

$$\{\Delta_R\} = [\Phi_{RR} - \Phi_{Rx}\Phi_{xx}^{-1}\Phi_{xR}]\{R_g\} = [\Phi_R]\{R_g\} \tag{6}$$

where
$$\Phi_{xx}^{-1} = \begin{bmatrix}18.1467 & -106.794\\ -106.794 & 734.262\end{bmatrix} \times 10^6 \qquad \Phi_R = \begin{bmatrix}3.2174 & 1.665\\ 1.665 & 1.893\end{bmatrix} \times 10^{-9} \tag{7, 8}$$

Substituting (*1*) into (*6*) yields

$$(\omega^2[\Phi_R][\mu] - [I])\{\Delta_R\} = 0, \qquad [I] = \text{unit matrix} \tag{9}$$

The determinant of the coefficients of (*9*) yields

$$\omega^4 - 291\,284.3\,\omega^2 + 11\,288\,974\,000 = 0 \tag{10}$$

Thus $\omega_1 = 214$ rad/s and $\omega_2 = 496$ rad/s, which is in agreement with the results of Problem 10.3.

10.5. Rework Problem 10.3 using the transport matrix.

For the given problem, $i = 1, 2, 3$, $j = 1, 2$, $\mu_1 = 0.3165\ \text{k-s}^2/\text{ft}$, $\mu_2 = 0.396\ \text{k-s}^2/\text{ft}$, and the other elements are

$l_1^3/6EI_1 = 12.3335 \times 10^{-5}$ ft/k $\qquad l_1^2/2EI_1 = 37.000 \times 10^{-6}$ 1/k $\qquad l_1/EI_1 = 74.005 \times 10^{-7}$ 1/k-ft

$l_2^3/6EI_2 = 5.807 \times 10^{-5}$ ft/k $\qquad l_2^2/2EI_2 = 17.422 \times 10^{-6}$ 1/k $\qquad l_2/EI_2 = 34.844 \times 10^{-7}$ 1/k-ft

$l_3^3/6EI_3 = 3.183 \times 10^{-5}$ ft/k $\qquad l_3^2/2EI_2 = 9.549 \times 10^{-6}$ 1/k $\qquad l_3/EI_3 = 19.098 \times 10^{-7}$ 1/k-ft

By (5-6), the transport chain for Fig. P-10.4*a* is

$$\hat{H}_0 = \hat{T}_{01}\hat{I}_1\hat{T}_{12}\hat{I}_2\hat{T}_{23}\hat{H}_3 = \hat{L}_{03}\hat{H}_3 \tag{1}$$

in which the typical $\hat{T}_{Li}$ and $\hat{I}_{ik}$ are

$$\underbrace{\begin{bmatrix} V_L \\ M_L \\ v_L \\ \theta_L \end{bmatrix}}_{\hat{H}_L} = \underbrace{\begin{bmatrix} 1 & 0 & 0 & 0 \\ l_i & 1 & 0 & 0 \\ l_i^3/6EI_i & l_i^2/2EI_i & 1 & -l_i \\ -l_i^2/2EI_i & -l_i/EI_i & 0 & 1 \end{bmatrix}}_{\hat{T}_{Li}} \underbrace{\begin{bmatrix} V_i \\ M_i \\ v_i \\ \theta_i \end{bmatrix}}_{\hat{H}_i} \qquad \underbrace{\begin{bmatrix} V_i \\ M_i \\ v_i \\ \theta_i \end{bmatrix}}_{\hat{H}_i} = \underbrace{\begin{bmatrix} 1 & 0 & \mu_j\omega^2 & 0 \\ 0 & 1 & 0 & 0 \\ 0 & 0 & 1 & 0 \\ 0 & 0 & 0 & 0 \end{bmatrix}}_{\hat{I}_{ik}} \underbrace{\begin{bmatrix} V_k \\ M_k \\ v_k \\ \theta_k \end{bmatrix}}_{\hat{H}_k} \tag{2, 3}$$

Considering the boundary conditions,

$$\hat{H}_0 = [V_0 \quad M_0 \quad 0 \quad 0]^T \qquad \hat{H}_3 = [V_3 \quad M_3 \quad 0 \quad 0]^T \tag{4, 5}$$

and manipulating (*1*) yield two homogeneous equations of which the determinant gives $\omega_1 = 214$ rad/s and $\omega_2 = 496$ rad/s, which is again in agreement with the results of Problems 10.3 and 10.4.

STEPPED MASS METHODS

10.6. Using the stiffness method, find the fundamental frequency of the nonuniform beam of Fig. P-10.6*a* by idealizing the beam into three stepped segments as shown in Fig. P-10.6*b*. Let $l = 30$ ft, $h = 15$ in, $b = 15$ in, $\gamma = 490$ lb/ft³, and $E = 29 \times 10^6$ psi. The thickness of the segments is a constant b.

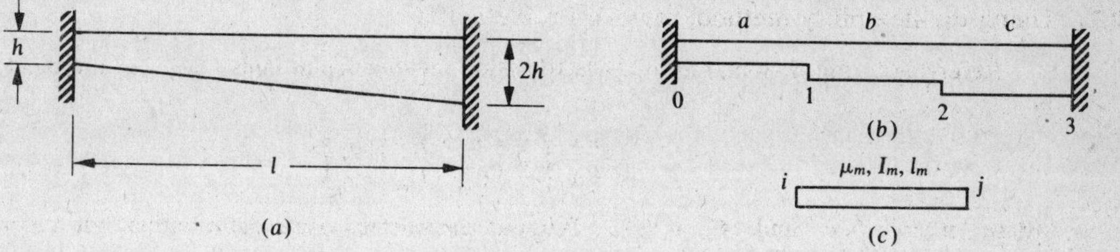

Fig. P-10.6

For a typical segment in Fig. P-10.6*c* without considering axial deformation, the relationship between the stress vector and displacement vector may be expressed as

$$\begin{bmatrix} \sigma_{ji} \\ \sigma_{ij} \end{bmatrix} = \begin{bmatrix} E_m & F_m \\ G_m & H_m \end{bmatrix} \begin{bmatrix} \Delta_{ji} \\ \Delta_{ij} \end{bmatrix} \tag{1}$$

where $m = a, b, c$; $G = F_m^T$; and

$$E_m = \begin{bmatrix} (\bar{\psi}_4\lambda^3EI)_m & (-\bar{\psi}_3\lambda^2EI)_m \\ (-\bar{\psi}_3\lambda^2EI)_m & (-\bar{\psi}_2\lambda EI)_m \end{bmatrix} \qquad F_m = \begin{bmatrix} (-\bar{\psi}_8\lambda^3EI)_m & (-\bar{\psi}_7\lambda^2EI)_m \\ (\bar{\psi}_7\lambda^2EI)_m & (\bar{\psi}_3\lambda EI)_m \end{bmatrix}$$

$$H_m = \begin{bmatrix} (\bar{\psi}_4\lambda^3EI)_m & (\bar{\psi}_3\lambda^2EI)_m \\ (\bar{\psi}_3\lambda^2EI)_m & (-\bar{\psi}_2\lambda EI)_m \end{bmatrix} \tag{2}$$

in which the coefficients may be found in Table 9-2*b*, *e*, and *f*.

Considering the equilibrium conditions at nodes 1 and 2 (Fig. P-10.6*b*) gives

$$\{\Sigma\sigma_1\} = \{\sigma_{10} + \sigma_{12}\} = 0 \qquad \{\Sigma\sigma_2\} = \{\sigma_{21} + \sigma_{23}\} = 0 \tag{3}$$

Substituting (*1*) and (*2*) into (*3*) yields

$$\begin{bmatrix} \Sigma V_1 \\ \Sigma M_1 \\ \hline \Sigma V_2 \\ \Sigma M_2 \end{bmatrix} = \begin{bmatrix} 0 \\ 0 \\ \hline 0 \\ 0 \end{bmatrix} = \left[\begin{array}{c|c} E_a + H_b & G_b \\ \hline F_b & E_b + H_c \end{array}\right] \begin{bmatrix} v_1 \\ \theta_1 \\ \hline v_2 \\ \theta_2 \end{bmatrix} \qquad \text{or} \qquad \begin{bmatrix} \Sigma\sigma_1 \\ \Sigma\sigma_2 \end{bmatrix} = \{0\} = [\kappa] \begin{bmatrix} \Delta_1 \\ \Delta_2 \end{bmatrix} \tag{4}$$

Consequently, the solution of (*4*) can exist only if

$$\text{Det.}\,[\kappa] = 0 \tag{5}$$

As discussed in Chapters 8 and 9, the frequency may be obtained by changing $\bar{\lambda}$ for evaluating the zero determinant of (*5*). Since $\mu_m = bh_m\gamma/g$ and $I_m = bh_m^3/12$, we may write $\bar{\lambda}_m = l_m\sqrt[4]{12\omega^2\gamma/(Egh_m^2)}$. For illustration purposes, try $\omega = 217$ rad/s. Then $\bar{\lambda}_a = 1.766$, $\bar{\lambda}_b = 1.555$, $\bar{\lambda}_c = 1.409$, and

$\sinh\bar{\lambda}_a = 2.838$ $\sin\bar{\lambda}_a = 0.981$ $\sinh\bar{\lambda}_b = 2.262$ $\sin\bar{\lambda}_b = 0.999$ $\sinh\bar{\lambda}_c = 1.924$ $\sin\bar{\lambda}_c = 0.987$

$\cosh\bar{\lambda}_a = 3.009$ $\cos\bar{\lambda}_a = -0.194$ $\cosh\bar{\lambda}_b = 2.473$ $\cos\bar{\lambda}_b = 0.016$ $\cosh\bar{\lambda}_c = 2.168$ $\cos\bar{\lambda}_c = 0.161$

Substituting the above numerical values into the appropriate coefficients shown in (*2*) and (*4*), we have

$$[\kappa] = \begin{bmatrix} 3.284 & 897. & -3.056 & 1772. \\ & 19900. & -1772.000 & 7040. \\ \text{Sym.} & & 6.920 & 1390. \\ & & & 38500. \end{bmatrix} \tag{6}$$

from which $\text{Det.}\,[\kappa] = 2460.851 \times 10^{10}$. Similarly, by trying $\omega = 211$ rad/s, $\text{Det.}\,[\kappa] = -1230 \times 10^{10}$. It is obvious that the zero determinant is between $\omega = 217$ and $\omega = 211$. By linear interpolation, the desired result is $\omega = 213$ rad/s.

10.7. Using the flexibility method, rework Problem 10.6.

Referring to Fig. P-10.6, the force-deformation relationship at ends i and j of a typical segment m is

$$\begin{bmatrix} \Delta_{ji} \\ \Delta_{ij} \end{bmatrix} = \begin{bmatrix} A_m & B_m \\ C_m & D_m \end{bmatrix} \begin{bmatrix} \sigma_{ji} \\ \sigma_{ij} \end{bmatrix} \tag{1}$$

where $m = a, b, c$ and $C_m = B_m^T$. Without considering axial deformation, the elements in (*1*) may be obtained from Table 9-2*b*, *e*, and *f* as

$$A_m = \begin{bmatrix} \left(\dfrac{\bar{\psi}_2}{EI\lambda^3}\right)_m & \left(\dfrac{-\bar{\psi}_3}{EI\lambda^2}\right)_m \\ \left(\dfrac{-\bar{\psi}_3}{EI\lambda^2}\right)_m & \left(\dfrac{\bar{\psi}_4}{EI\lambda}\right)_m \end{bmatrix} \quad B_m = \begin{bmatrix} \left(\dfrac{-\bar{\psi}_6}{EI\lambda^3}\right)_m & \left(\dfrac{\bar{\psi}_7}{EI\lambda^2}\right)_m \\ \left(\dfrac{-\bar{\psi}_7}{EI\lambda^2}\right)_m & \left(\dfrac{\bar{\psi}_8}{EI\lambda}\right)_m \end{bmatrix} \quad D_m = \begin{bmatrix} \left(\dfrac{\bar{\psi}_2}{EI\lambda^3}\right)_m & \left(\dfrac{\bar{\psi}_3}{EI\lambda^2}\right)_m \\ \left(\dfrac{\bar{\psi}_3}{EI\lambda^2}\right)_m & \left(\dfrac{-\bar{\psi}_4}{EI\lambda}\right)_m \end{bmatrix} \tag{2}$$

Considering the continuity at nodes 0, 1, 2, and 3 (Fig. P-10.6*b*) requires

$$\begin{aligned} \{\Sigma\Delta_0\} &= \{\Delta_{01}\} = 0 & \{\Sigma\Delta_2\} &= \{\Delta_{21} + \Delta_{23}\} = 0 \\ \{\Sigma\Delta_1\} &= \{\Delta_{10} + \Delta_{12}\} = 0 & \{\Sigma\Delta_3\} &= \{\Delta_{32}\} = 0 \end{aligned} \tag{3}$$

Substituting (*1*) and (*2*) into (*3*) gives

$$\underbrace{\begin{bmatrix} \Sigma\Delta_0 \\ \Sigma\Delta_1 \\ \Sigma\Delta_2 \\ \Sigma\Delta_3 \end{bmatrix}}_{\{\Sigma\Delta\}} = \begin{bmatrix} 0 \\ 0 \\ 0 \\ 0 \end{bmatrix} = \underbrace{\begin{bmatrix} D_a & C_a & 0 & 0 \\ B_a & A_a + D_b & C_b & 0 \\ 0 & B_b & A_b + D_c & C_c \\ 0 & 0 & B_c & A_c \end{bmatrix}}_{[\Phi]} \underbrace{\begin{bmatrix} \sigma_0 \\ \sigma_1 \\ \sigma_2 \\ \sigma_3 \end{bmatrix}}_{\{\sigma\}} \tag{4}$$

whose solution can exist only if

$$\text{Det.}\,[\Phi] = 0 \tag{5}$$

By varying ω or λ in $[\Phi]$, the zero determinants of (*5*) may be obtained. Similar to Problem 10.6, the positive and negative determinants correspond to $\omega = 217$ rad/s and $\omega = 211$ rad/s respectively. Thus the desired result is $\omega = 213$ rad/s.

10.8. Rework Problem 10.6 using the transport method.

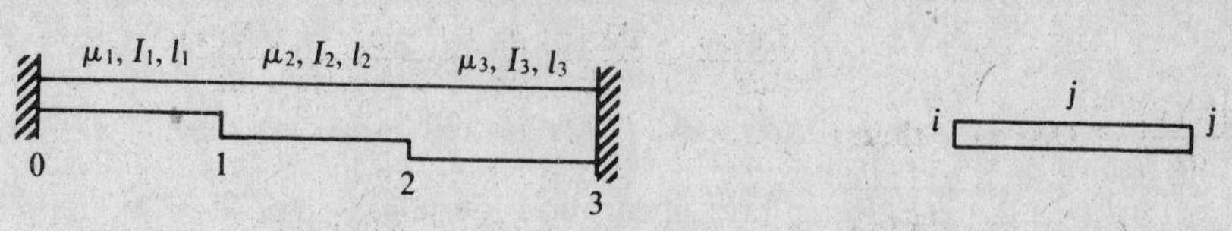

(a) (b)

Fig. P-10.8

For the idealized beam of Fig. P-10.8a, the transport chain is

$$H_3 = [U_{32}^3 \quad U_{21}^2 \quad U_{10}^1 \quad H_0] \tag{1}$$

in which the notations in U_{ji}^j correspond to the accompanied typical segment. Since $v_0 = \theta_0 = v_3 = \theta_3 = 0$, H_3 and H_0 may be written as

$$H_3 = [V_3 \quad M_3 \quad 0 \quad 0]^T \qquad H_0 = [V_0 \quad M_0 \quad 0 \quad 0]^T \tag{2}$$

From Problem 8.1, for a typical segment j with both ends denoted by ij (Fig. P-10.8b), the transport matrix for bending deformation and transverse inertia force only is

$$U_{ji}^j = \begin{bmatrix} a_j & (-d\lambda)_j & (-bEI\lambda^3)_j & (-cEI\lambda^2)_j \\ (-b/\lambda)_j & a_j & (cEI\lambda^2)_j & (dEI\lambda)_j \\ (-d/EI\lambda^3)_j & (a/EI\lambda^2)_j & a_j & (b/\lambda)_j \\ (-c/EI\lambda^2)_j & (b/EI\lambda)_j & (d\lambda)_j & a_j \end{bmatrix} \tag{3}$$

where the subscript v is omitted, $\lambda_j = \sqrt[4]{\mu_j\omega^2/EI_j}$; a_j, b_j, c_j, and d_j were given in Table 8-1e with the transverse dynamic parameter $\bar{\lambda}_j = \lambda_j l_j$. The evaluation of numerical values of the coefficients of (3) is exactly the same as shown in Problem 10.6.

Substituting (3) into (1) and then using the boundary conditions of (2) yield the frequency equation. As shown previously, the solution is $\omega = 213$ rad/s.

SUBSTITUTE FUNCTION METHODS

10.9. Derive the substitute function method of algebraic polynomials (10-3) and discuss its solution.

The differential equation of transverse vibration of a nonuniform bar (10-2) is

$$EI(x)X^{\text{iv}}(x) + 2EI'(x)X'''(x) + EI''(x)X''(x) - \mu(x)\omega^2X(x) = 0 \tag{1}$$

Let us consider a bar having the cross-sectional properties represented by

$$A(x) = A_L\left(a + \frac{bx}{s}\right)^{n-4} \qquad I(x) = I_L\left(a + \frac{bx}{s}\right)^n \qquad \mu(x) = \mu_L\left(a + \frac{bx}{s}\right)^{n-4} \tag{2}$$

where

$$a = \frac{1 - (c/s)\sqrt[n]{I_R/I_L}}{1 - c/s} \qquad b = \frac{\sqrt[n]{I_R/I_L} - 1}{1 - c/s} \tag{3}$$

where n is a real number and other notations refer to Fig. 10.2.

Substituting (2) into (1) yields

$$EI_L\left(a + \frac{bx}{s}\right)^n X^{\text{iv}}(x) + 2nEI_L\frac{b}{s}\left(a + \frac{bx}{s}\right)^{n-1} X'''(x)$$

$$+ En(n-1)I_L\left(\frac{b}{s}\right)^2\left(a + \frac{bx}{s}\right)^{n-2} X''(x) - \mu(x)\omega^2X(x) = 0 \tag{4}$$

Dividing (4) by $EI_L(a + bx/s)^{n-4}$,

$$\left(a + \frac{bx}{s}\right)^4 X^{\text{iv}}(x) + 2n\frac{b}{s}\left(a + \frac{bx}{s}\right)^3 X'''(x) + n(n-1)\left(\frac{b}{s}\right)^2\left(a + \frac{bx}{s}\right)^2 X''(x) - \left(\frac{q}{l}\right)^2 X(x) = 0 \tag{5}$$

in which $q = l\sqrt[4]{\mu_L\omega^2/EI_L}$.

To find the solution of (5), let $X(x) = e^z = a + bx/s$; then

$$\frac{dX(x)}{dx} = \frac{dX(x)}{dz}\frac{dz}{dx} = \frac{b/s}{a + bx/s}\frac{dX(x)}{dz} \tag{6, 7}$$

Let $\dfrac{dX(x)}{dz} = DX(x)$; then

$$\left(a+\frac{bx}{s}\right)\frac{dX(x)}{dx} = \frac{b}{s}DX(x), \quad \left(a+\frac{bx}{s}\right)^2\frac{d^2X(x)}{dx^2} = \left(\frac{b}{s}\right)^2 D(D-1)X(x), \quad \ldots,$$
$$\left(a+\frac{bx}{s}\right)^4\frac{d^4X(x)}{dx^4} = \left(\frac{b}{s}\right)^4(D-1)(D-2)(D-3)X(x) \tag{8}$$

Substituting (8) into (5) yields

$$\left(\frac{b}{s}\right)^4 D(D-1)(D-2)(D-3)X(x) + \left(\frac{b}{s}\right)^3 D(D-1)(D-2)X(x) + \left(\frac{b}{s}\right)^2 D(D-1)X(x)$$
$$+\frac{b}{s}DX(x) - \left(\frac{q}{l}\right)^4 X(x) = 0 \tag{9}$$

The solution of the fourth order differential equation is

$$X(x) = \sum_{m=1,4} C_m e^{\alpha_m z} = \sum_{m=1,4} C_m\left(a+\frac{bx}{s}\right)^{\alpha_m} \tag{10}$$

in which α_m should be determined from the coefficients of (9) and are in function of b, s, and q/l; C_m are integration constants and should be determined by using the boundary conditions of a given problem.

For a prismatic bar, $I_L = I_R$, $A_L = A_R$, and $\mu_L = \mu_R$; then from (3) $a = 1$, $b = 0$. Thus (5) becomes

$$X^{iv}(x) - \left(\frac{q}{l}\right)^4 X(x) = 0 \tag{11}$$

which is the differential equation of transverse vibration of a finite straight bar presented in Chapter 6.

10.10. Derive the substitute function method in terms of the exponential functions (10-5) and discuss its solution.

Let the variable cross section of the bar be represented by

$$A(x) = A_L\alpha e^{\beta x/s} \qquad I(x) = I_L\alpha e^{\beta x/s} \qquad \mu(x) = \mu_L\alpha e^{\beta x/s} \tag{1}$$

where
$$\alpha = \left(\frac{I_L}{I_R}\right)^{(c/s)(1-c/s)} \qquad \beta = \frac{\ln(I_R/I_L)}{1 - c/s} \tag{2}$$

Substituting (1) into the differential equation (1) of Problem 10.9 gives

$$X^{iv}(x) + \frac{2\beta}{s}X'''(x) + \left(\frac{\beta}{s}\right)^2 X''(x) - (q\tau)^4 X(x) = 0 \tag{3}$$

in which $q = \beta/s$, $\tau = (1/q)\sqrt[4]{\mu_L\omega^2/EI_L}$, and $q \neq 0$.

Let $X(x) = e^{ax}$; then (3) becomes

$$a^4 + 2qa^3 + q^2a^2 - (q\tau)^4 = 0 \tag{4}$$

Introducing $f = a/q$ into (4),

$$f^4 + 2f^3 + f^2 - \tau^4 = 0 \tag{5}$$

from which

$$f_1 = -\tfrac{1}{2}[1 + \sqrt{1+4\tau^2}] \qquad f_2 = -\tfrac{1}{2}[1 - \sqrt{1+4\tau^2}]$$
$$f_3 = -\tfrac{1}{2}[1 + \sqrt{1-4\tau^2}] \qquad f_4 = -\tfrac{1}{2}[1 - \sqrt{1-4\tau^2}] \tag{6}$$

Thus the desired result is

$$X(x) = c_1e^{qf_1x} + c_2e^{qf_2x} + c_3e^{qf_3x} + c_4e^{qf_4x} \tag{7}$$

which can exist only if $\tau < 1/2$.

For $\tau > 1/2$, (7) becomes

$$X(x) = c_1 e^{qf_1x} + c_2 e^{qf_2x} + c_3 e^{(-1/2)x}\,\mathrm{cs}(x) + c_4 e^{(-1/2)x}\,\mathrm{sn}(x) \tag{8}$$

where $\mathrm{cs}(x) = \cos[\sqrt{4\tau^2 - 1}(x/2)]$ and $\mathrm{sn}(x) = \sin[\sqrt{4\tau^2 - 1}(x/2)]$.

For $\tau = 1/2$, (6) becomes

$$f_1 = -\tfrac{1}{2}[1 + \sqrt{2}] \qquad f_2 = -\tfrac{1}{2}[1 - \sqrt{2}] \qquad f_3 = f_4 = -\tfrac{1}{2} \tag{9}$$

for which the solution is

$$X(x) = c_1 e^{qf_1x} + c_2 e^{qf_2x} + c_3 e^{qf_3x} + c_4 x e^{qf_3x} \tag{10}$$

For a prismatic bar $I_L = I_R$, $A_L = A_R$, and $\mu_L = \mu_R$; $\alpha = 1$, $\beta = 0$, and $q = 0$. Examining (4) gives

$$X^{\text{iv}}(x) - (0/0)\sqrt[4]{\mu_L\omega^2/EI_L}\, X(x) = 0 \tag{11}$$

where the second term is indeterminate. The exact limit of this term may be found using l'Hospital's rule: take the derivative of the numerator and denominator with respect to q and then let q approach zero as a limit. Finally (11) becomes

$$X^{\text{iv}}(x) - \frac{\mu_L\omega^2}{EI_L} X(x) = 0 \tag{12}$$

which is the differential equation of transverse vibration of a uniform bar.

10.11. Derive the Ritz method for the transverse vibration of a nonuniform bar.

Referring to Chapter 4, the maximum potential energy V and the maximum kinetic energy T of a nonuniform bar may be expressed as

$$V = \frac{1}{2}\int_0^l EI(x)\left[\frac{d^2X(x)}{dx^2}\right]^2 dx \qquad T = \frac{\omega^2}{2}\int_0^l \mu(x)X^2(x)\,dx \tag{1, 2}$$

Since $V = T$, equating (1) and (2) yields

$$\omega^2 = \frac{\displaystyle\int_0^l EI(x)\left[\frac{d^2X(x)}{dx^2}\right]^2 dx}{\displaystyle\int_0^l \mu(x)X^2(x)\,dx} \tag{3}$$

Let

$$X(x) = X_1g_1(x) + X_2g_2(x) + X_3g_3(x) + \cdots = \sum_m X_m g_m(x) \tag{4}$$

where $X_1, X_2, \ldots$ are unknowns; $g_1(x), g_2(x), \ldots$ must be chosen to satisfy boundary conditions. Since a finite number of terms can be used in (4), $X(x)$ cannot truly represent all possible shapes of deflection. Thus the frequency obtained in (3) will usually be higher than the real one. To overcome this difficulty one may choose (4) so as to make (3) a minimum. Mathematically the minimum of (3) requires

$$\frac{\partial(\omega^2)}{\partial X_m} = \frac{\displaystyle\int_0^l \mu(x)X^2(x)\,dx\,\frac{\partial}{\partial X_m}\int_0^l EI(x)\left[\frac{d^2X(x)}{dx^2}\right]^2 dx - \int_0^l EI(x)\frac{d^2X(x)}{dx^2}\,dx\,\frac{\partial}{\partial X_m}\int_0^l \mu(x)X^2(x)\,dx}{\displaystyle\left[\int_0^l \mu(x)X^2(x)\,dx\right]^2} = 0 \tag{5}$$

By (3), the numerator of (5) may be written as

$$\int_0^l \mu(x)X^2(x)\,dx\,\frac{\partial}{\partial X_m}\int_0^l EI(x)\left[\frac{d^2X(x)}{dx^2}\right]^2 dx - \omega^2\int_0^l \mu(x)X^2(x)\,dx\,\frac{\partial}{\partial X_m}\int_0^l \mu(x)X^2(x)\,dx = 0 \tag{6}$$

which becomes

$$\frac{\partial}{\partial X_m}\underbrace{\int_0^l \left\{EI(x)\left[\frac{d^2X(x)}{dx^2}\right]^2 - \omega^2\mu(x)X^2(x)\right\}dx}_{\tau} = 0 \tag{7}$$

Therefore the problem is reduced to finding τ and performing $\partial\tau/\partial X_m$.

10.12. Using the Ritz method derived in the preceding problem, find the fundamental frequency of the nonuniform beam of Fig. P-10.12. Given: constant thickness $b = 15$ in, $h = 15$ in, $l = 30$ ft, $\gamma = 490$ lbs/ft^3, and $E = 29 \times 10^6$ psi.

Let $A = bh$ and $I = bh^3/12$; then

$$A(x) = A(1 + x/l)$$
$$I(x) = I(1 + x/l)^3 \qquad (1)$$
$$\mu(x) = \mu(1 + x/l)$$

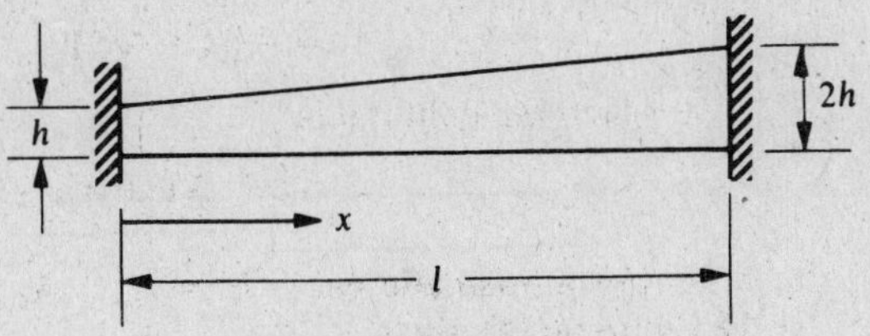

Fig. P-10.12

Select

$$X(x) = X_1 g_1(x) = X_0\left(1 - \cos\frac{2\pi x}{l}\right) \qquad (2)$$

where $X_1 = X_0$ represents the amplitude at $l/2$ (the true amplitude of the nonuniform beam is not at the center of the span); $g_1(x) = 1 - \cos 2\pi x/l$ which is chosen to satisfy the boundary conditions of $X(0) = X'(0) = X(l) = X'(l) = 0$.

According to (7) of Problem 10.12,

$$\int_0^l EI(x)\left[\frac{d^2X(x)}{dx^2}\right]^2 dx = \frac{2EI\pi^4 X_0^2}{l^3}\left[15 + \frac{9}{4\pi^2}\right] \qquad (3)$$

$$\int_0^l \omega^2\mu(x)X^2(x)\,dx = \frac{9\omega^2\mu l X_0^2}{4} \qquad (4)$$

and

$$\frac{\partial \tau}{\partial X_0} = \frac{\partial}{\partial X_0}\left[\frac{2EI\pi^4 X_0^2}{l^3}\left(15 + \frac{9}{4\pi^2}\right) - \frac{9}{4}\omega^2\mu l X_0^2\right] = 0 \qquad (5)$$

from which

$$\omega = \sqrt{\frac{\pi^2 EI}{\mu l^4}\left(2 + \frac{40\pi^2}{3}\right)} \qquad \text{or} \qquad \omega = 238 \text{ rad/s}$$

Supplementary Problems

LUMPED MASS METHODS

10.13 Using the stiffness matrix, find the fundamental frequency of the beam shown in Fig. P-10.13*a* by lumping the masses at quarter points (Fig. P-10.13*b*). Given: $l = 30$ ft, $h = 15$ in, $b = 15$ in, $E = 29 \times 10^6$ psi, and $\gamma = 490$ lb/ft^3. The thickness of the segments is a constant b.

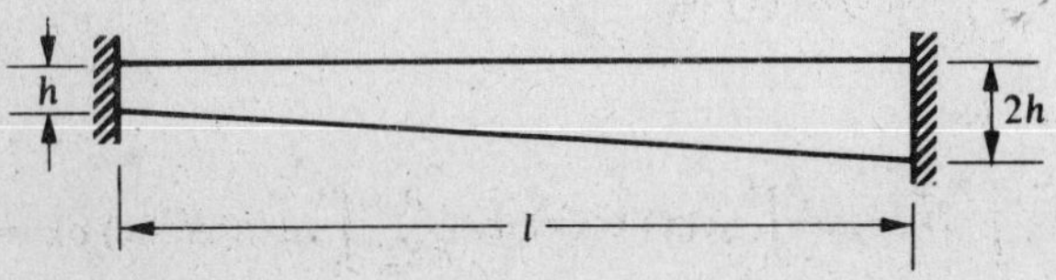

(*a*)

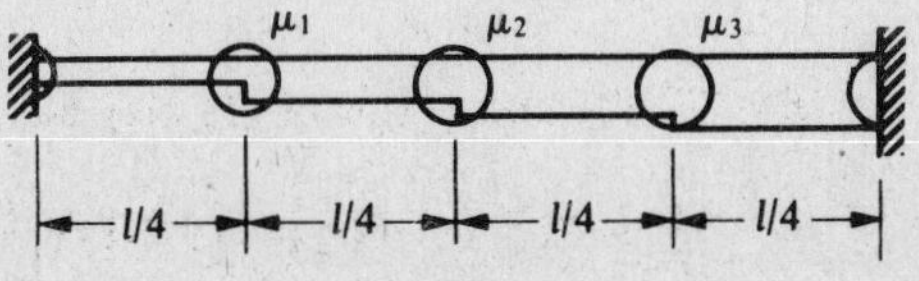

(*b*)

Fig. P-10.13

10.14. Rework Problem 10.13 using the flexibility matrix.

10.15. Rework Problem 10.13 using the transport matrix.

STEPPED MASS METHODS

10.16. Find the fundamental frequency of the nonuniform beam given in Fig. P-10.16. Let $l = 12$ ft, constant thickness $I_1 = 486\ \text{in}^4$, $I_2 = I_1/2$, $\mu_1 = 0.735\ \text{k-s}^2/\text{in}^2$, $\mu_2 = \mu_1/2$.

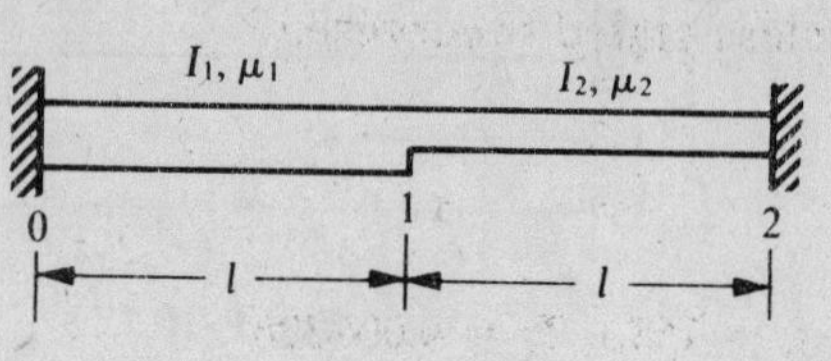

Fig. P-10.16

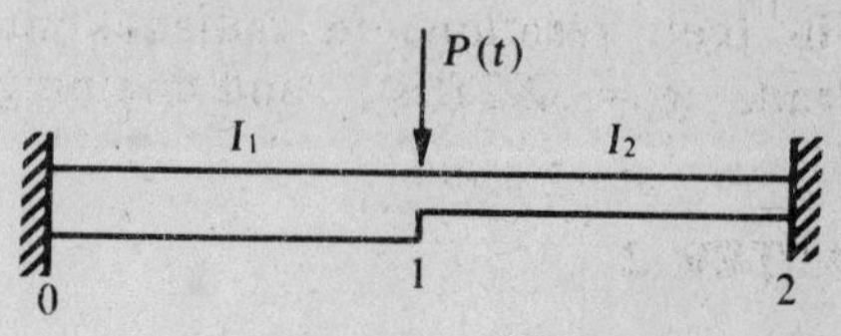

Fig. P-10.17

10.17. Using the stiffness method, find the moment at 0 of Fig. P-10.17 in which $P(t) = P\cos\Omega t$, $P = 15$ kip, and $\Omega = \pi$ rad/sec. The member properties were given in Problem 10.16.

10.18. Rework Problems 10.16 and 10.17 using the flexibility method.

10.19. Rework Problems 10.16 and 10.17 using the transport method.

SUBSTITUTE FUNCTION METHODS

10.20. Using the Ritz method, find the equation of fundamental frequency of Fig. P-10.20. Let $X_1g_1(x) = X_1(6l^2x^2 - 4lx^3 + x^4)$ and $X_2g_2(x) = X_2(20l^3x^2 - 10l^2x^3 + x^5)$.

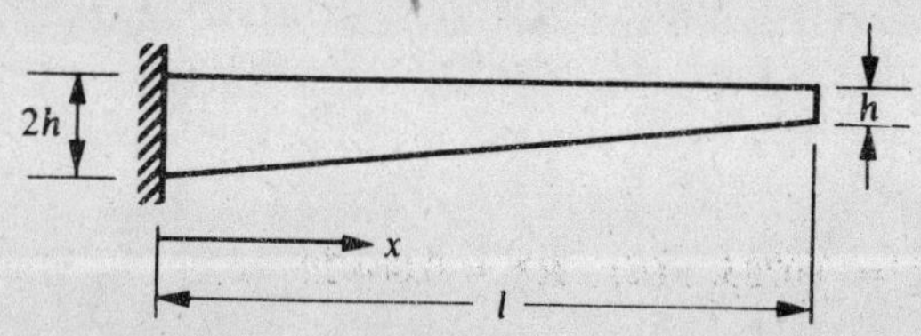

Fig. P-10.20

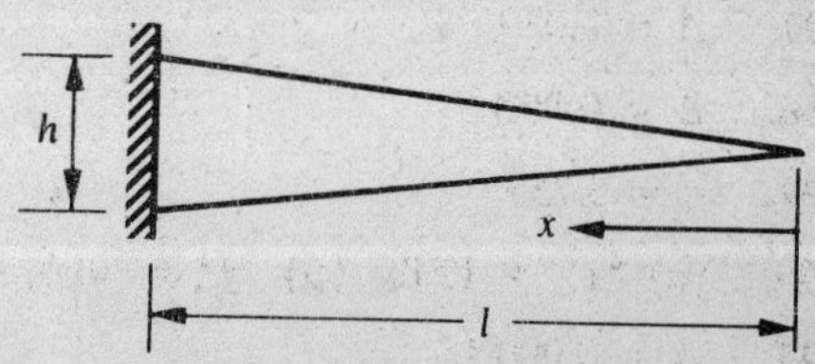

Fig. P-10.21

10.21. Find the equation of fundamental frequency of the wedge given in Fig. P-10.21 by using the Ritz method. Let the wedge thickness be a unit, $A(x) = hx/l$, $I(x) = (hx/l)^3/12$, $x_1g_1(x) = X_1(1 - x/l)^2$, and $X_2g_2(x) = X_2x/l(1 - x/l)^2$.

Answers to Supplementary Problems

Forces and shears, are in kips, moments in k-ft, masses in k-s^2/ft, stiffness in k/ft, circular frequencies in rad/s, natural frequencies in cycles/s, periods in seconds, velocities in ft/s, accelerations in ft/s^2, stresses in k/in^2, coordinates, segments, lengths, depths, widths, and deflections are in feet, rotations in radians, mass moment inertia in k-ft-s^2, angular spring constant in k-ft/rad, $g = 32.2$ ft/s^2, and damping in k-s/ft, unless stated otherwise.

CHAPTER 2

2.26. (a) $0.143\,42\,\ddot{v} + 141\,370.2\,v = 0,\quad \omega_v = 992.8,\quad f_v = 158,\quad T_v = 0.006\,328$

(b) $0.6454\,\ddot{\theta} + 1693.497\,\theta = 0,\quad \omega_\theta = 1.225,\quad f_\theta = 8.153,\quad T_\theta = 0.122\,65$

(c) $0.143\,42\,\ddot{u} + 184.076\,u = 0,\quad \omega_u = 35.826,\quad f_u = 5.702,\quad T_u = 0.175\,38$

2.27. $\mu L^2\ddot{\theta} + (kl^2 - \mu g L)\theta = 0,\quad f_u = \sqrt{(kl^2 - \mu g L)/\mu L^2}/2\pi$

2.28. (a) $\Delta = 0.0417\cos 4.7t + 0.0177\sin 4.7t$

(b) $M_B = M_c = 548.67$ (clockwise)

(c) $(12.422)(0.9112) + (274.08)(-0.041\,09) \approx 0$

2.29. $\sigma = 21.692$ (including the cage weight)

2.30. $\sigma = 35.19$

2.35. $X = 0.095\,05$

2.36. $P(t) = \dfrac{P}{2} - \displaystyle\sum_{m=1,3,5}^{\infty} \frac{4P}{m^2\pi^2}\cos\frac{m\pi t}{a}$

2.37. $\Delta = 0.059$

2.38. $\Delta = 0.059$

2.39. $c = 5.753$

2.40. (a) $c_{cr} = 12.0$, (b) $f^*_\Delta = 80.0$, (c) $T^*_\Delta = 0.0785$, (d) $R^*_\Delta = 0.004\,22$

2.41. $\Delta = 0.000\,15$

CHAPTER 3

3.11. (a) $\omega_1 = 19.53,\quad \omega_2 = 51.1$; (b) $T_1 = 0.322,\quad T_2 = 0.123$

(c) $\{X_1\} = \begin{bmatrix} X_{11} \\ 1.618X_{11}\end{bmatrix},\quad \{X_2\} = \begin{bmatrix} X_{12} \\ -0.618X_{12}\end{bmatrix}$

3.12. (a) $\omega_1 = 0,\quad \omega_2 = \sqrt{k(\mu_1 + \mu_2)/\mu_1\mu_2}$; (b) $T_1 = \infty,\quad T_2 = 2\pi\sqrt{\mu_1\mu_2/k(\mu_1 + \mu_2)}$

(c) $\begin{Bmatrix}\Delta_1 \\ \Delta_2\end{Bmatrix} = \begin{bmatrix} X_{11} & X_{12} \\ 1.618X_{11} & -0.618X_{12}\end{bmatrix}\begin{bmatrix} 1 + \dfrac{b_1}{X_{11}}t & 0 \\ 0 & \cos(\omega_2 t - \alpha)\end{bmatrix}$

3.13.
$$\begin{bmatrix} 0 & \mu_1 & 0 \\ 0 & 0 & \mu_2 \\ \rho_1 + \rho_2 & \mu_1 l & 0\end{bmatrix}\begin{Bmatrix}\ddot{\theta} \\ \ddot{u}_1 \\ \ddot{u}_2\end{Bmatrix} + \begin{bmatrix} -k_1 l & k_1 & -k_1 \\ k_1 l & -k_1 & k_1 + k_2 \\ k_\theta & -\mu_1 g & \mu_1 g\end{bmatrix}\begin{Bmatrix}\theta \\ u_1 \\ u_2\end{Bmatrix} = 0$$

3.14. $\omega_2 = 17.321,\quad \{X_2\} = \begin{bmatrix} X_{12} \\ -X_{12}\end{bmatrix}$

3.15. $\{\Delta\} = \{0.8823\quad 0.9301\quad 1.0608\}$

3.16. $[h] = \begin{bmatrix} 0.3784 & 0.6123 \\ 0.6123 & -0.3784 \end{bmatrix}$

$$\{\Delta\} = [h][\{h_{11}\Phi_{11}\} + \{h_{22}\phi_{11}\}]$$

$$= [h]\begin{bmatrix} -0.0188 \sin(37.6991\,t) + 0.0362 \sin(19.53\,t) \\ 0.0429 \sin(37.6991\,t) - 0.0317(\sin 51.1\,t) \end{bmatrix} = \begin{bmatrix} -0.00522 \\ -0.00798 \end{bmatrix}, \quad \text{at} \quad t = 1/4\,\text{s}$$

3.17. $\{\Delta\} = [h][\{h_{1k}\Phi_{1k}\} + \{h_{2k}\Phi_{2k}\}] \quad \Phi_{jk} = \dfrac{1}{\lambda_k}\displaystyle\int_0^t e^{-\gamma_k(t-\tau)} \sin \lambda_k(t-\tau) W_j(\tau)\, d\tau \quad [h] = \begin{bmatrix} 0.3784 & 0.6123 \\ 0.6123 & -0.3784 \end{bmatrix},$

$$\{\omega(\tau)\} = \begin{bmatrix} 100 \sin(12\,\pi\tau) \\ 80 \sin(12\,\pi\tau) \end{bmatrix}, \quad \{\gamma\} = \begin{bmatrix} 1.953 \\ 7.665 \end{bmatrix}, \quad \{\Delta\} = \begin{bmatrix} -0.0249 \\ -0.0561 \end{bmatrix}, \quad \{\lambda\} = \begin{bmatrix} 19.432 \\ 50.521 \end{bmatrix}$$

3.18. $\{\Delta\} = \begin{bmatrix} -0.00051 \\ -0.05336 \end{bmatrix}$

3.19. $\{\Delta\} = \begin{bmatrix} -0.0249 \\ -0.0561 \end{bmatrix}$

CHAPTER 4

4.20. (*a*) $m = 3j - r = 3(6) - 10 = 8$

(*b*) Let $\{\theta\} = [\theta_0 \;\; \theta_1 \;\; \theta_2 \;\; \theta_3 \;\; \theta_4 \;\; \theta_5]^T$, $\{q\} = [q_1 \;\; q_2 \;\; q_3 \;\; q_4 \;\; q_5 \;\; q_6]^T$,

$$[\theta \mid u_1 \;\; v_2]^T = [q \mid q_7 \;\; q_8]^T$$
$$[\theta \mid u_1 \;\; v_4]^T = [q \mid q_7 \;\; q_8]^T$$
$$[\theta \mid u_2 \;\; v_2]^T = [q \mid q_7 \;\; q_8]^T$$
$$[\theta \mid u_2 \;\; v_4]^T = [q \mid q_7 \;\; q_8]^T$$
$$[\theta \mid u_3 \;\; v_2]^T = [q \mid q_7 \;\; q_8]^T$$
$$[\theta \mid u_3 \;\; v_4]^T = [q \mid q_7 \;\; q_8]^T$$

4.21. $[\theta_1 \;\; \theta_2 \;\; \theta_3 \;\; u_1 \;\; v_3]^T = [q_1 \;\; q_2 \;\; q_3 \;\; q_4 \;\; q_5]^T$

4.22.

$$\left[\begin{array}{c} \theta_0 \\ \theta_1 \\ \theta_2 \\ \theta_3 \\ \theta_4 \\ \theta_5 \\ \hline u_3 \\ v_5 \\ \hline u_1 \\ v_1 \\ u_2 \\ v_2 \\ v_3 \\ u_4 \\ v_4 \\ u_5 \end{array}\right] = \left[\begin{array}{cccccccc} 1 & 0 & 0 & 0 & 0 & 0 & 0 & 0 \\ 0 & 1 & 0 & 0 & 0 & 0 & 0 & 0 \\ 0 & 0 & 1 & 0 & 0 & 0 & 0 & 0 \\ 0 & 0 & 0 & 1 & 0 & 0 & 0 & 0 \\ 0 & 0 & 0 & 0 & 1 & 0 & 0 & 0 \\ 0 & 0 & 0 & 0 & 0 & 1 & 0 & 0 \\ \hline 0 & 0 & 0 & 0 & 0 & 0 & 1 & 0 \\ 0 & 0 & 0 & 0 & 0 & 0 & 0 & 1 \\ \hline 0 & 0 & 0 & 0 & 0 & 0 & 1 & 0 \\ 0 & 0 & 0 & 0 & 0 & 0 & 0 & 0 \\ 0 & 0 & 0 & 0 & 0 & 0 & 1 & 0 \\ 0 & 0 & 0 & 0 & 0 & 0 & 0 & 1 \\ 0 & 0 & 0 & 0 & 0 & 0 & 0 & 0 \\ 0 & 0 & 0 & 0 & 0 & 0 & 0 & 0 \\ 0 & 0 & 0 & 0 & 0 & 0 & 0 & 0 \\ 0 & 0 & 0 & 0 & 0 & 0 & 0 & 0 \end{array}\right] \begin{bmatrix} q_1 \\ q_2 \\ q_3 \\ q_4 \\ q_5 \\ q_6 \\ q_7 \\ q_8 \end{bmatrix}$$

4.23.
$$\begin{bmatrix} \theta_1 \\ \theta_2 \\ \theta_3 \\ \hline u_1 \\ v_3 \\ \hline v_1 \\ u_2 \\ v_2 \\ u_3 \end{bmatrix} = \begin{bmatrix} 1 & 0 & 0 & 0 & 0 \\ 0 & 1 & 0 & 0 & 0 \\ 0 & 0 & 1 & 0 & 0 \\ \hline 0 & 0 & 0 & 1 & 0 \\ 0 & 0 & 0 & 0 & 1 \\ \hline 0 & 0 & 0 & -3/4 & 0 \\ 0 & 0 & 0 & 7/32 & 7/24 \\ 0 & 0 & 0 & 7/24 & -7/18 \\ 0 & 0 & 0 & 0 & 4/3 \end{bmatrix} \begin{bmatrix} q_1 \\ q_2 \\ q_3 \\ q_4 \\ q_5 \end{bmatrix}$$

4.24. $\{Q\} = [-66.67 \quad 66.67 \quad -96.0 \quad 96.0 \quad 0. \quad 0. \quad 20.0 \quad -44.0]^T$

4.25. $\{Q\} = [0. \quad 20.0 \quad 0. \quad -0.83 \quad 1.11]^T$

4.26. $\{Q\} = [-66.67 \quad 66.67 \quad -96.0 \quad 96.0 \quad 0. \quad 0. \quad 20.0 \quad -44.0]^T$

4.27. $4\mu l^2\ddot{\theta}_1 + 2\mu g l\theta_1 - l^2k(\theta_2 - \theta_1) = 0$
$4\mu l^2\ddot{\theta}_2 + 2\mu g l\theta_2 + l^2k(\theta_2 - \theta_1) = 0$

4.28. $\omega^4 - \left(1 + \frac{\mu_2}{\mu_1}\right)\frac{g}{l_1 l_2}[(l_1 + l_2)\omega^2 - g] = 0$

4.29. $\{\Delta\} = [0.2589 \quad -0.0776]^T$

4.30. $\{\Delta\} = [0.2589 \quad -0.0776]^T$

4.31. $\{\Delta\} = [0.3954 \quad 1.903]^T$

CHAPTER 5

5.19. (a) $\omega_u = \sqrt{2EA/\mu l}, \quad \omega_{v1} = \sqrt{12EI/\mu l^3}, \quad \omega_{v2} = \sqrt{24EI/\mu l^3}$
(b) $\omega_u = \sqrt{2EA/\mu l}, \quad \omega_{v1} = \sqrt{12EI/\mu l^3}, \quad \omega_{v2} = \sqrt{24EI/\mu l^3}$

5.20. $138.77\,u_3 = 129.88\,\omega^2 u_3$, that is, $[\Omega]\{\Delta\} = \omega^2[u]\{\Delta\}, \quad \omega = 1.0336$

5.21. $V_A = V_B = V_C = V_D = 0.636$

5.22. (a) $\omega_u = \sqrt{2EA/\mu l}, \quad \omega_v = \sqrt{24EI/\mu l^3}, \quad \omega_\theta = \sqrt{8EI/\rho l}$
(b) $\omega_u = \sqrt{2EA/\mu l}, \quad \omega_{v1} = \sqrt{3EI/\mu l^3}, \quad \omega_{v2} = \sqrt{24EI/\mu l^3}$

5.23. $\omega_1 = 2.709, \quad \omega_2 = 7.338$

5.24. $E = 30 \times 10^6$ psi, $\{\Delta \quad \delta_2 \quad \delta_5\} = [-0.004\,92 \quad 0.000\,421\,7 \quad 0.000\,421\,7]^T$
$\{M_1 \quad M_2 \quad M_3 \quad M_4 \quad M_5 \quad M_6\} = [-10.6748 \quad -6.4041 \quad 6.4041 \quad 6.4041 \quad -6.4041 \quad -10.6748]^T$

5.25. (a) $\omega_1 = 2.209, \quad \omega_2 = 4.45, \quad \omega_3 = 6.85$
(b) $[h] = \begin{bmatrix} 1.000 & 1.000 & 1.000 \\ 0.784 & -0.006 & -0.950 \\ 0.413 & -0.723 & 0.360 \end{bmatrix}$

5.26. $V_A = V_B = V_C = V_D = 0.636$

5.27. $\{\Delta\} = [1.632 \quad 1.057]^T$

CHAPTER 6

6.25. $\omega_m = \frac{m\pi}{l}\sqrt{\frac{EA}{\mu}}, \quad m = 1, 2, 3$ for both (a) and (b)

6.26. (a) $\frac{1}{k} = \frac{\cos \omega l\sqrt{\mu/EA} + k/(\omega\sqrt{EA\mu}) \sin \omega l\sqrt{\mu/EA}}{-\omega\sqrt{EA\mu}\,[\sin \omega l\sqrt{\mu/EA} + k \cos \omega l\sqrt{\mu/EA}]}$

(b) $\omega_m = \frac{m\pi}{l}\sqrt{EA/\mu}$, $m = 1, 2, 3$; (c) $\omega_m = \frac{(2m-1)\pi}{2l}\sqrt{EA/\mu}$, $m = 1, 2, 3$

6.27. Initial conditions: $u_{t=0} = \epsilon x$, $\dot{u}_{t=0} = 0$

$$u(x,t) = \frac{8\epsilon l}{\pi^2}\sum_{1,3,5}\frac{(-1)^{-1/2}}{m^2}\sin\frac{m\pi x}{2l}\cos\frac{m\pi ct}{2l}, \quad c = \sqrt{\frac{Eg}{\gamma}}$$

$u(x = l, t = 0) = \epsilon l$

6.28. $N_L = 56.40$, $u_R = 0.001\,88$

6.30. $u_{l/2} = 0.001\,739$, $N_{l/2} = 15.15$

6.31. $\cosh\lambda l\cos\lambda l = -1$, $\omega_n = (\lambda l)_n^2\sqrt{EI/\mu l^4}$,
$(\lambda l)_1 = 1.857$, $(\lambda l)_2 = 4.694$, $(\lambda l)_3 = 7.855$, $(\lambda l)_n \approx \frac{1}{2}(2n-1)\pi$, $n > 3$

6.32. (a) $\cosh\lambda l\cos\lambda l = 1$, $\omega_n = (\lambda l)_n^2\sqrt{EI/\mu l^4}$,
$(\lambda l)_1 = 4.730$, $(\lambda l)_2 = 7.853$, $(\lambda l)_3 = 10.996$, $(\lambda l)_n \approx \frac{1}{2}(2n+1)\pi$, $n > 3$

(b) $$\text{Det.}\begin{bmatrix} ak_R - b + dk_Lk_R - ak_L & bk_R - c \\ c + bk_L & d \end{bmatrix} = 0$$

(c) $\cosh\lambda l\cos\lambda l = 1$, λl same as given in (a)

6.33. $$v(x,t) = \frac{2Pl^3}{\pi^4 EI}\sum_{m=1,2,3}^{\infty}\frac{1}{m^4}\sin\frac{\pi l_1}{l}\sin\frac{m\pi x}{l}\cos\omega_m t$$

6.34. $V_L = 2.9397$, $M_L = 87.52$

6.35. $V_L = 2.9397$, $M_L = 87.52$

CHAPTER 7

7.18. $$\omega_m = \sqrt{\left(\frac{m\pi}{l}\right)^2\frac{AE}{\mu} + \frac{ku}{\mu}}, \quad \omega_1 = 2807.2, \quad \omega_2 = 5375.3, \quad \omega_3 = 7994.8$$

7.19. $$\begin{vmatrix} 1 & 0 & 1 & 0 \\ \cosh\lambda l & \sinh\lambda l & \cos\lambda l & \sin\lambda l \\ EI\lambda & k_\theta & -EI\lambda & k_\theta \\ EI\lambda\cosh\lambda l & EI\lambda\sinh\lambda l & -EI\lambda\cos\lambda l & k_\theta\cos\lambda l \\ -k_\theta\sinh\lambda l & -k_\theta\cosh\lambda l & +k_\theta\sin\lambda l & -EI\lambda\sin\lambda l \end{vmatrix} = 0, \quad \begin{matrix}(\lambda l)_1 = 1.4741, \ \omega_1 = 36.4; \\ (\lambda l)_2 = 3.1417, \ \omega_2 = 42.5.\end{matrix}$$

7.20. $\theta_L = -\theta_R = 2.254$, $V_L = -V_R = 846.47$

7.21. $\omega_1 = 33.43$, $\omega_2 = 111.34$, $\omega_3 = 235.06$ $\omega_4 = 403.87$
Note: The effect of N on ω may be approximately expressed as

$$\omega_m \approx \omega_{m,0}\sqrt{1 - N/N_{cr}}, \quad N_{cr} = (m\pi)^2 EI/l^2$$

$$\omega_{m,0} = \frac{(\lambda l)_{m,0}^2}{l^2}\sqrt{\frac{EI}{\mu}}, \quad \text{where } (\lambda)_{m,0} \text{ is given as follows:}$$

m	1	2	3	4	5	$m > 6$
$\lambda_{m,0}$	3.9266	7.0686	10.2102	13.3518	16.4934	$\frac{\pi}{4}(4m+1)$

7.22. $$\omega = \left\{\frac{EI}{\mu}\left(\frac{m\pi}{l}\right)^2\left[1 - \frac{N}{N_{cr}}\right] + \frac{k_v}{\mu}\right\}^{1/2}, \quad N_{cr} = (m\pi)^2EI/l^2, \quad \omega_1 = 34.1, \quad \omega_2 = 91.5, \quad \omega_3 = 201.6$$

7.23. $$y(x,t) = \sum_{2,4,6}^{\infty}(-1)^{m/2}\frac{Q}{l}\left(\frac{g}{EI\gamma A}\right)^{1/2}\sin\frac{m\pi x}{l}\left[\frac{cm\pi\cos\Omega t - cm\pi\cos cm\pi t}{(cm\pi)^2 - (\Omega)^2}\right] \quad \text{where } c = \sqrt{\frac{EIg}{A\gamma}}$$

7.24. Let $b_0 = (\gamma A/EIg)l^4\omega^2$ be the frequency parameter based on Bernoulli–Euler theory; then the frequency parameter, b, of Timoshenko theory can be expressed as $b_m = c_m b_{0m}$ for any mode m. Then b_0 and c_m are:

m	1	2	3
b_0	22.37328	61.67282	120.90339
c_m	0.9295	0.8508	0.7820

7.25. Let $b_0 = (\gamma A/EIg)l^4\omega^2$ and $b_m = c_m b_{0m}$ which are defined in Problem 7.24, then

m	1	2	3
b_0	22.37328	61.67282	120.90339
c_m	0.9605	0.9001	0.8320

CHAPTER 8

8.12. (*a*) $H_2^0 = J_2 U_{21}^0 J_1 U_{10}^0 H_0^0 \qquad H_2^0 = [0 \;\; 0 \;\; 0 \;\; U_2 \;\; 0 \;\; \theta_2]^T \qquad H_0^0 = [N_0 \;\; V_0 \;\; M_0 \;\; 0 \;\; 0 \;\; 0]^T$

$$J_1 = \begin{bmatrix} 1 & 0 & 0 & 0 & 0 & 0 \\ 0 & 1 & 0 & 0 & k_v & 0 \\ 0 & 0 & 1 & 0 & 0 & 0 \\ 0 & 0 & 0 & 1 & 0 & 0 \\ 0 & 0 & 0 & 0 & 1 & 0 \\ 0 & 0 & 0 & 0 & 0 & 1 \end{bmatrix} \qquad J_2 = \begin{bmatrix} 1 & 0 & 0 & k_u & 0 & 0 \\ 0 & 1 & 0 & 0 & 0 & 0 \\ 0 & 0 & 1 & 0 & 0 & 0 \\ 0 & 0 & 0 & 1 & 0 & 0 \\ 0 & 0 & 0 & 0 & 1 & 0 \\ 0 & 0 & 0 & 0 & 0 & 0 \end{bmatrix}$$

(*b*) $H_2^0 = U_{21}^0 J_1 U_{10}^0 H_0^0 \qquad H_2^0 = [0 \;\; 0 \;\; 0 \;\; u_2 \;\; v_2 \;\; \theta_2]^T \qquad H_0^0 = [N_0 \;\; 0 \;\; M_0 \;\; 0 \;\; v_0 \;\; 0]^T$

$$J_1 = \begin{bmatrix} 1 & 0 & 0 & 0 & 0 & 0 \\ 0 & 1 & 0 & 0 & 0 & 0 \\ 0 & 0 & 1 & 0 & 0 & k_\theta \\ 0 & 0 & 0 & 1 & 0 & 0 \\ 0 & 0 & 0 & 0 & 1 & 0 \\ 0 & 0 & 0 & 0 & 0 & 1 \end{bmatrix}$$

(*c*) $H_2^0 = \pi^{0b} U_{21}^b \pi^{ba} U_{10}^a \pi^{a0} H_0^0 \qquad H_2^0 = [N_2 \;\; V_2 \;\; M_2 \;\; 0 \;\; 0 \;\; 0]^T \qquad H_0^0 = [N_0 \;\; V_0 \;\; M_0 \;\; 0 \;\; 0 \;\; 0]^T$

$$\pi^{0b} = \begin{bmatrix} \cos\phi^{0b} & -\sin\phi^{0b} & 0 & 0 & 0 & 0 \\ \sin\phi^{0b} & \cos\phi^{0b} & 0 & 0 & 0 & 0 \\ 0 & 0 & 1 & 0 & 0 & 0 \\ 0 & 0 & 0 & \cos\phi^{0b} & -\sin\phi^{0b} & 0 \\ 0 & 0 & 0 & \sin\phi^{0b} & \cos\phi^{0b} & 0 \\ 0 & 0 & 0 & 0 & 0 & 1 \end{bmatrix}$$

(*d*) $H_2^0 = \pi^{0b} U_{21}^b \pi^{ba} U_{10}^a \pi^{a0} U_0^0 \qquad H_2^0 = [N_2 \;\; V_2 \;\; 0 \;\; 0 \;\; 0 \;\; \theta_2]^T \qquad H_0^0 = [N_0 \;\; V_0 \;\; 0 \;\; 0 \;\; 0 \;\; \theta_0]^T$

$$\pi^{a0} = \begin{bmatrix} \cos\phi^{a0} & \sin\phi^{a0} & 0 & 0 & 0 & 0 \\ -\sin\phi^{a0} & \cos\phi^{a0} & 0 & 0 & 0 & 0 \\ 0 & 0 & 1 & 0 & 0 & 0 \\ 0 & 0 & 0 & \cos\phi^{a0} & \sin\phi^{a0} & 0 \\ 0 & 0 & 0 & -\sin\phi^{a0} & \cos\phi^{a0} & \\ 0 & 0 & 0 & 0 & 0 & 1 \end{bmatrix}$$

The matrices $U_{RL,m}^0$ of the above (*a*), (*b*), (*c*), and (*d*) are given in (*4*) of Problem 8.1.

8.13. (*b*) Since $\begin{bmatrix} N_{10} \\ V_{10} \\ M_{10} \\ u_{10} \\ v_{10} \\ \theta_{10} \end{bmatrix} = \begin{bmatrix} \\ U_a \\ \\ \end{bmatrix} \begin{bmatrix} N_{01} \\ V_{01} \\ M_{01} \\ u_{01} = 0 \\ v_{01} = 0 \\ \theta_{01} = 0 \end{bmatrix}$, we have

$$\begin{bmatrix} N_{21} \\ V_{21} \\ M_{21} \\ u_{21} \\ v_{21} \\ \theta_{21} \end{bmatrix} = \begin{bmatrix} 0 \\ 0 \\ 0 \\ u_{21} \\ v_{21} \\ \theta_{21} \end{bmatrix} = \begin{bmatrix} \\ U'_b \\ \\ \end{bmatrix}_{[6\times 3]} \begin{bmatrix} N_{01} \\ M_{01} \\ V_{01} \end{bmatrix} \quad \text{from which} \quad \begin{vmatrix} u'_{b11} & u'_{b12} & u'_{b13} \\ u'_{b21} & u'_{b22} & u'_{b23} \\ u'_{b31} & u'_{b32} & u'_{b33} \end{vmatrix} = 0$$

(*c*) Since $H^a_{10} = \begin{bmatrix} N_{10} \\ V_{10} \\ M_{10} \\ u_{10} \\ v_{10} \\ \theta_{10} \end{bmatrix} = \begin{bmatrix} \\ U_a \\ \\ \end{bmatrix} \begin{bmatrix} \\ \pi^{a0} \\ \\ \end{bmatrix} \underbrace{\begin{bmatrix} N_{01} \\ V_{01} \\ M_{01} \\ u_{01} = 0 \\ v_{01} = 0 \\ \theta_{01} = 0 \end{bmatrix}}_{H^0_0}$, we have

$$H^0_2 = \begin{bmatrix} N_{21} \\ V_{21} \\ M_{21} \\ u_{21} \\ v_{21} \\ \theta_{21} \end{bmatrix} = \begin{bmatrix} N_{21} \\ V_{21} \\ M_{21} \\ 0 \\ 0 \\ 0 \end{bmatrix} = \begin{bmatrix} \\ U'_b \\ \\ \end{bmatrix}_{[6\times 3]} \begin{bmatrix} N_{01} \\ V_{01} \\ M_{01} \end{bmatrix} \quad \text{from which} \quad \begin{vmatrix} u'_{b41} & u'_{b42} & u'_{b43} \\ u'_{b51} & u'_{b52} & u'_{b53} \\ u'_{b61} & u'_{b62} & u'_{b63} \end{vmatrix} = 0$$

(*d*) Since $H^a_{10} = \begin{bmatrix} N_{10} \\ V_{10} \\ M_{10} \\ u_{10} \\ v_{10} \\ \theta_{10} \end{bmatrix} = \begin{bmatrix} \\ U_a \\ \\ \end{bmatrix} \begin{bmatrix} \\ \pi^{a0} \\ \\ \end{bmatrix} \underbrace{\begin{bmatrix} N_{01} \\ V_{01} \\ M_{01} = 0 \\ u_{01} = 0 \\ v_{01} = 0 \\ \theta_{01} \end{bmatrix}}_{H^0_0}$, we have

$$H^0_2 = \begin{bmatrix} N_{21} \\ V_{21} \\ M_{21} \\ u_{21} \\ v_{21} \\ \theta_{21} \end{bmatrix} = \begin{bmatrix} N_{21} \\ V_{21} \\ 0 \\ 0 \\ 0 \\ \theta_{21} \end{bmatrix} = \begin{bmatrix} \\ U'_b \\ \\ \end{bmatrix}_{[6\times 3]} \begin{bmatrix} N_{01} \\ V_{01} \\ M_{01} \end{bmatrix} \quad \text{from which} \quad \begin{vmatrix} u'_{b31} & u'_{b32} & u'_{b33} \\ u'_{b41} & u'_{b42} & u'_{b43} \\ u'_{b51} & u'_{b52} & u'_{b53} \end{vmatrix} = 0$$

8.14.

	Neglect axial deformation	Consider axial deformation
ω_1	314.497	312.109
ω_2	474.939	452.283
ω_3	873.185	657.993

8.15. $\omega_1 = 194.9986, \quad \omega_2 = 765.675, \quad \omega_3 = 1265.02$

8.16. See Figs. A-1 and A-2.

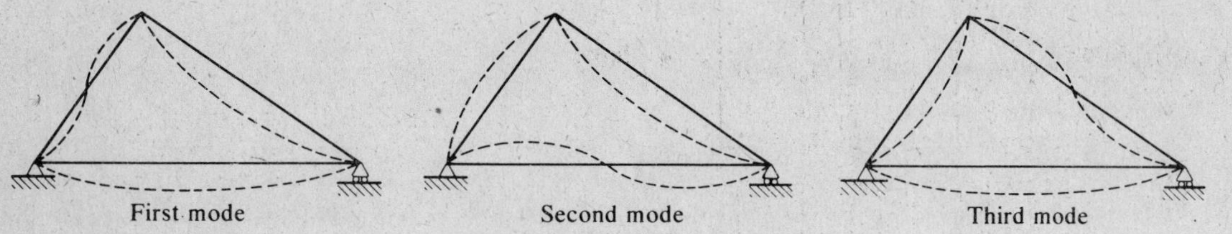

(a) Axial Deformation Neglected

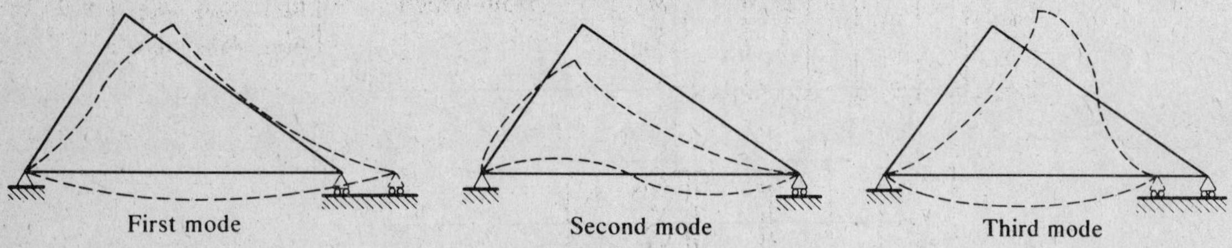

(b) Axial Deformation Considered

Fig. A-1. Mode Shapes of Problem 8.14

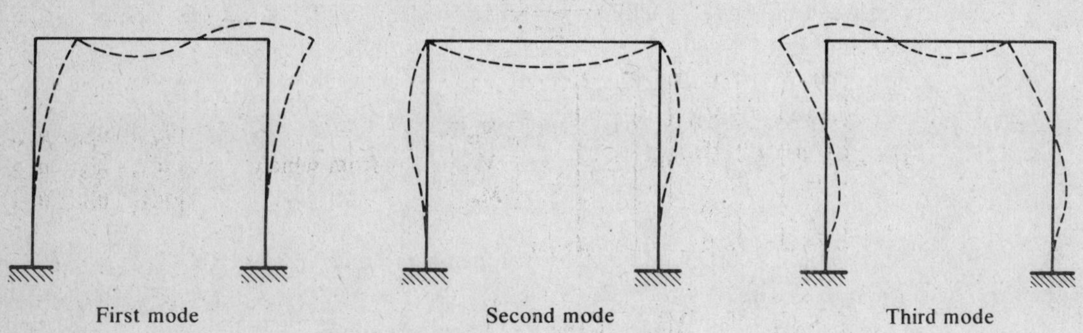

Fig. A-2. Mode Shapes of Problem 8.15

CHAPTER 9

9.22. (a) $\{\Delta_1\} = \{\Delta_{10} + \Delta_{12}\} = 0,\ \ \{\Delta_2\} = \{\Delta_{21} + \Delta_{23}\} = 0,\ \ \{\Delta_3\} = \{\Delta_{32} + \Delta_{34}\} = 0,\ \ \{\Delta_4\} = 0$

$$\underbrace{\begin{bmatrix} \Delta_1 \\ \Delta_2 \\ \hline \Delta_3 \\ \Delta_4 \end{bmatrix}}_{\{\Delta\}} = \underbrace{\left[\begin{array}{cc|cc} f_{11,1} + f_{11,2} & f_{12,1} & 0 & 0 \\ f_{21,1} & f_{22,2} + f_{22,3} & f_{23,3} & 0 \\ \hline 0 & f_{32,3} & f_{33,3} + f_{33,4} & f_{34,4} \\ 0 & 0 & f_{43,4} & f_{44,4} \end{array}\right]}_{[\phi]} \underbrace{\begin{bmatrix} \sigma_1 \\ \sigma_2 \\ \hline \sigma_3 \\ \sigma_4 \end{bmatrix}}_{\{\sigma\}}$$

where the subscripts ij of $f_{ij,k}$ refer to the nodal points 0, 1, 2, 3, and 4 and k refers to the member numbers between nodes, such as $k = 1$ for nodes $0 \sim 1$, $k = 2$ for nodes $1 \sim 2$, etc.; f_{ij} are given in (*7*), (*8*), and (*9*) of Problem 9.2.

(b) $\{\Delta_{01}\} = \{\Delta_0\} = 0,\ \ \{\Delta_{10} + \Delta_{12}\} = \{\Delta_1\} = 0$

$$\underbrace{\begin{bmatrix} f_{00,1} & f_{01,1} \\ f_{10,1} & f_{11,1} + f_{11,2} \end{bmatrix}}_{[\phi]} \underbrace{\begin{bmatrix} \sigma_0 \\ \sigma_1 \end{bmatrix}}_{\{\sigma\}} = -\underbrace{\begin{bmatrix} \Delta_{00,1} \\ \Delta_{10,1} + \Delta_{10,2} \end{bmatrix}}_{\{\Delta_0\}}$$

where $\begin{Bmatrix} \Delta_{00,1} \\ \Delta_{10,1} \end{Bmatrix}$ and $\{\Delta_{10,2}\}$ are load functions for members 1 and 2 respectively and correspond to Case (3) of Table P-9.20.

9.23. $(\lambda l)_1 = 3.1418, \quad \omega_1 = 2.0845$

9.24. Same as Prob. 8.14.

9.25. $\omega_1 = 12.636, \quad \omega_2 = 31.526, \quad \omega_3 = 47.704$

9.27. (*a*) $\{\sigma_0\} = \{\sigma_{01}\} = 0, \quad \{\sigma_1\} = \{\sigma_{10} + \sigma_{12}\} = 0, \quad \{\sigma_2\} = \{\sigma_{21} + \sigma_{23}\} = 0, \quad \{\sigma_3\} = \{\sigma_{32} + \sigma_{34}\} = 0$

$$\underbrace{\begin{bmatrix} k_{00,1} & k_{01,1} & 0 & 0 \\ k_{01,1} & k_{11,1} + k_{11,2} & k_{12,2} & 0 \\ 0 & k_{21,2} & k_{22,2} + k_{22,3} & k_{23,3} \\ 0 & 0 & k_{32,3} & k_{33,3} + k_{33,4} \end{bmatrix}}_{[K]} \begin{bmatrix} \Delta_0 \\ \Delta_1 \\ \Delta_2 \\ \Delta_3 \end{bmatrix} = 0$$

$$\text{Det.}\,(K) = 0$$

where the subscripts ij of $k_{ij,k}$ refer to the nodal points 0, 1, 2, 3, and 4 and k refers to the member numbers between the nodes, such as $k = 1$ for nodes $0 \sim 1$, $k = 2$ for nodes $1 \sim 2$, etc.; k_{ij} are given in (6) of Problem 9.10.

(*b*) $\{\sigma_1\} = \{\sigma_{10} + \sigma_{12}\} = 0, \quad \{\sigma_2\} = \{\sigma_{21}\} = 0$

$$\underbrace{\begin{bmatrix} k_{11,1} + k_{11,2} & k_{12,2} \\ k_{21,2} & k_{22,2} \end{bmatrix}}_{[K]} \underbrace{\begin{bmatrix} \Delta_1 \\ \Delta_2 \end{bmatrix}}_{\{\Delta\}} = -\underbrace{\begin{bmatrix} F_{10,1} + F_{10,2} \\ F_{20,2} \end{bmatrix}}_{\{F_0\}}$$

where $\{F_{10,1}\}$ and $\begin{Bmatrix} F_{10,2} \\ F_{20,2} \end{Bmatrix}$ are load functions for members 1 and 2 respectively and corresponding to Case (3) of Table P-9.26.

9.28. $(\lambda l)_1 = 3.1418, \quad \omega_1 = 2.0845$

9.29. $$\begin{bmatrix} (\bar\psi_2\bar\lambda_v)_1 EI_1/l_1 + (\bar\psi_2\bar\lambda_v)_2 EI_2/l_2 & (\bar\psi_3\bar\lambda_v^2)_2 EI_2/l_2^2 - (\bar\psi_3\bar\lambda_v^2)_1 EI_1/l_1^2 \\ \text{Sym.} & (\bar\psi_4\bar\lambda_v^3)_1 EI_1/l_1^3 + (\bar\psi_4\bar\lambda_v^3)_2 EI_2/l_2^2 \end{bmatrix} \begin{Bmatrix} \theta_p \\ v_p \end{Bmatrix} = \begin{Bmatrix} 0 \\ P \end{Bmatrix}$$

$M_L = \bar\psi_6\bar\lambda_v EI_1\theta_p/l_1 - \bar\psi_7\bar\lambda_v^2 EI_1 v_p/l_1^2$ where $l_1 = l_2 = l$

$\theta_p = -1.110 \times 10^{-3}, \quad v_p = -0.0477, \quad M_L = 235.26$

9.30. Same as Prob. 8.14.

9.31. $\omega_1 = 12.636, \quad \omega_2 = 31.526, \quad \omega_3 = 47.704$

CHAPTER 10

10.13. $\omega_1 = 218.654, \quad \omega_2 = 576.388, \quad \omega_3 = 985.249$

10.14. Same as Prob. 10.13.

10.15. Same as Prob. 10.13.

10.16. $$\begin{bmatrix} (\bar\psi_2\bar\lambda_v)_1 EI_1/l_1 + (\bar\psi_2\bar\lambda_v)_2 EI_2/l_2 & (\bar\psi_3\bar\lambda_v^2)_2 EI_2/l_2^2 - (\bar\psi_3\bar\lambda_v^2)_1 EI_1/l_1^2 \\ \text{Sym.} & (\bar\psi_4\bar\lambda_v^3)_1 EI_1/l_1^3 + (\bar\psi_4\bar\lambda_v^3)_2 EI_2/l_2^2 \end{bmatrix} = 0$$

$(\bar\lambda_v)_1 = 2.337\,25, \quad \omega_1 = 1.1536$

10.17. $\theta_p = 9.5047 \times 10^{-4}, \quad v_p = -0.0013, \quad M_L = 55.35$

10.18. Same as Problems 10.16 and 10.17.

10.19. Same as Problems 10.16 and 10.17.

10.20. $\omega = 3.378 \dfrac{h}{l^2} \sqrt{\dfrac{Eg}{\gamma}}$

10.21. $\omega = 1.5354 \dfrac{h}{l^2} \sqrt{\dfrac{Eg}{\gamma}}$

Selected References and Bibliography

Chapter 1 Concepts of Analysis

1.1 Clough, R. W. and J. Penzien, *Dynamics of Structures*, McGraw-Hill, New York, 1975.
1.2 Greenwood, D. T., *Principles of Dynamics*, Prentice-Hall, Englewood Cliffs, 1965.
1.3 Hohenemser, K. and W. Prager, *Dynamik der Stabwerke*, Springer, Berlin, 1973.
1.4 Koloušek, V., *Dynamics of Engineering Structures*, Butterworth, London, 1973.
1.5 Nowacki, W., *Dynamics of Elastic Systems*, Chapman, London, 1963.
1.6 Rubinstein, M. F., *Structural Systems—Statics, Dynamics and Stability*, Prentice-Hall, Englewood Cliffs, 1970.
1.7 Tuma, J. J., *Dynamics*, Quantum, New York, 1974.
I Tuma, J. J., *Schaum's Outline of Structural Analysis*, McGraw-Hill, New York, 1969.
II Tuma, J. J. and R. K. Munshi, *Schaum's Outline of Advanced Structural Analysis*, McGraw-Hill, New York, 1971.
III Tuma, J. J. and M. N. Reddy, *Schaum's Outline of Space Structural Analysis*, McGraw-Hill, New York, 1982.

Chapter 2 Systems with One Degree of Freedom

2.1 Churchill, R. V., *Operational Mathematics*, 2nd ed., McGraw-Hill, New York, 1958, pp. 35–45, 71–85.
2.2 Den Hartog, J. P., *Mechanical Vibration*, 4th ed., McGraw-Hill, New York, 1956, pp. 1–120.
2.3 Jacobsen, L. S. and R. S. Ayre, *Engineering Vibrations*, McGraw-Hill, New York, 1958, pp. 127–242.
2.4 Rogers, G. L., *Dynamics of Framed Structures*, Wiley, New York, 1959, pp. 59–64, 72–75.
2.5 Seto, W. W., *Schaum's Outline of Mechanical Vibrations*, McGraw-Hill, New York, 1964, pp. 1–27.
2.6 Spiegel, M. R., *Schaum's Outline of Laplace Transforms*, McGraw-Hill, New York, 1965, pp. 5–10, 88–90, 94–96.
2.7 Spiegel, M. R., *Schaum's Outline of Theoretical Mechanics*, McGraw-Hill, New York, 1967, pp. 86–102.
2.8 Timoshenko, S. P., D. H. Young, and W. Weaver, Jr., *Vibration Problems in Engineering*, 4th ed., Wiley, New York, 1974, pp. 1–142.
2.9 Tuma, J. J., *Engineering Mathematics Handbook*, 2nd ed., McGraw-Hill, New York, 1979, pp. 115, 214, 220–225.

Chapter 3 Systems with Several Degrees of Freedom

3.1 Biggs, J. M., *Introduction to Structural Dynamics*, McGraw-Hill, New York, 1964, pp. 85–147.
3.2 Cheng, F. Y., "Discussion of Dynamic Characteristics of Multistory Buildings," *Proc. ASCE, Journal of the Structural Division*, vol. 94, no. ST 8, August 1968.
3.3 Faddeev, D. K. and V. N. Faddeeva, *Computational Methods of Linear Algebra*, Freeman, San Francisco, 1963, pp. 226–289.
3.4 Hovanessian, S. A., and L. A. Pipes, *Digital Computer Methods in Engineering*, McGraw-Hill, New York, 1969, pp. 49–116.
3.5 Rogers, G. L., *Dynamics of Framed Structures*, Wiley, New York, 1959, pp. 113–148.
3.6 Zurmühl, R., *Matrizen*, 3rd ed., Springer, Berlin, 1961, pp. 150–225.

Chapter 4 Energy Methods

4.1 Cheng, F. Y., *Matrix Computer Methods in Structural Continuum Mechanics*, Extension Division Publication, University of Missouri, Rolla, 1970.
4.2 Greenwood, D. T., *Principles of Dynamics*, Prentice-Hall, Englewood Cliffs, 1965, pp. 229–280, 450–497.

4.3 Hurty, W. C. and M. F. Rubinstein, *Dynamics of Structures*, Prentice-Hall, Englewood Cliffs, 1964, pp. 7–17, 68–109, 260–271.
4.4 Meirovitch, L., *Methods of Analytical Dynamics*, McGraw-Hill, New York, 1970, pp. 45–100.
4.5 Pipes, L. A., *Matrix Methods for Engineers*, Prentice-Hall, Englewood Cliffs, 1965, pp. 210–287.
4.6 Spiegel, M. R., *Schaum's Outline of Theoretical Mechanics*, McGraw-Hill, New York, 1967, pp. 282–338.
4.7 Wells, D. A., *Schaum's Outline of Lagrangian Dynamics*, McGraw-Hill, New York, 1967, pp. 39–233.

Chapter 5 Matrix Methods

5.1 Cheng, F. Y., "Dynamics of Frames with Nonuniform Elastic Members," *Proc. ASCE, Journal of the Structural Division*, vol. 94, no. ST 10, October 1968.
5.2 Gillespie, J. W. and B. D. Liaw, *Dynamics of Structures, Part I, Frequency Analysis*, School of Civil Engineering Research Publication, no. 24, Oklahoma State University, 1964.
5.3 Pestel, E. C. and F. A. Leckie, *Matrix Methods of Elasto-Mechanics*, McGraw-Hill, New York, 1963, pp. 51–129.
5.4 Roger, G. L., *Dynamics of Framed Structures*, Wiley, New York, 1939, pp. 275–294.
5.5 Thompson, W. T., *Vibration Theory and Applications*, Prentice-Hall, Englewood Cliffs, 1965, pp. 228–251.

Chapter 6 Distributed Mass Systems, Basic Cases

6.1 Clark, S. K., *Dynamics of Continuous Elements*, Prentice-Hall, Englewood Cliffs, 1972, pp. 75–89, 140–165.
6.2 Felgar, R. P., "Formulas for Integrals Containing Characteristic Functions of a Vibrating Beam," Bureau of Engineering Research, Circular No. 14, University of Texas, 1950.
6.3 Pestel, E. C. and F. A. Leckie, *Matrix Methods of Elasto-Mechanics*, McGraw-Hill, New York, 1963, pp. 130–137.
6.4 Timoshenko, S. P., D. H. Young, and W. Weaver, Jr., *Vibration Problems in Engineering*, 4th ed., Wiley, New York, 1974, pp. 363–401, 415–447.
6.5 Tuma, J. J., *Engineering Mathematics Handbook*, 2nd ed., McGraw-Hill, New York, 1979, pp. 220–221, 224–225.
6.6 Volterra, E. and E. C. Zachmanoglou, *Dynamics of Vibration*, Merrill, Columbus, 1965, pp. 193–303.
6.7 Young, D. and R. P. Felgar, "Tables of Characteristic Functions Representing the Normal Modes of Vibration of a Beam," Engineering Research Series, no. 4, University of Texas, 1949.

Chapter 7 Distributed Mass Systems, Special Cases

7.1 Cheng, F. Y., "Vibrations of Timoshenko Beams and Frames," *Proc. ASCE, Journal of the Structural Division*, vol. 96, no. ST 3, March 1970.
7.2 Fuhrke, H., "Bestimmung von Balkenschwingungen mit Hilfe des Matrizenkalküls," Dissertation, Technical University Library, Darmstadt, 1953.
7.3 Huang, T. C., "Effect of Rotatory Inertia and of Shear Deformation on the Frequency and Mode Equations of Uniform Beams with Simple End Conditions," *Journal of Applied Mechanics*, vol. 28, no. 4, 1961.
7.4 Jacobson, L. S. and R. S. Ayre, *Engineering Vibrations*, McGraw-Hill, New York, 1958, pp. 496–505.
7.5 Rogers, G. L., *Dynamics of Framed Structures*, Wiley, New York, 1959, pp. 199–218.
7.6 Timoshenko, S. P., D. H. Young, and W. Weaver, Jr., *Vibration Problems in Engineering*, 4th ed., Wiley, New York, 1974, pp. 432–435.

7.7 Tuma, J. J., *Engineering Mathematics Handbook*, 2nd ed., McGraw-Hill, New York, 1979, pp. 224–225, 228–229.

Chapter 8 Dynamic Transport Method

8.1 Abramowitz, M. and I. A. Stegun, *Handbook of Mathematical Functions*, National Bureau of Standards, Washington, D. C., 1964, pp. 116–139, 142–185, 213–219.

8.2 Fuhrke, H., "Bestimmung von Balkenschwingungen mit Hilfe des Matrizenkalküls," *Ingenieur-Archiv*, vol. 23, 1955, pp. 329–348.

8.3 Fuhrke, H., "Bestimmung von Balkenschwingungen mit Hilfe des Matrizenkalküls," *Ingenieur-Archiv*, vol. 24, 1956, pp. 27–42.

8.4 Hessenberg, K., "Auflösung Linearer Eigenwertaufgaben mit Hilfe der Hamilton-Cayleyschen Gleichungen," Dissertation, Technical University Library, Darmstadt, 1941.

8.5 Hurty, W. C. and M. F. Rubinstein, *Dynamics of Structures*, Prentice-Hall, Englewood Cliffs, 1964, pp. 211–222.

8.6 Marquerre, K., "Vibration and Stability Problems of Beams Treated by Matrices," *Journal of Mathematics and Physics*, vol. 15, 1956, pp. 27–43.

8.7 Myklestad, N. O., "New Method of Calculating Natural Modes of Uncoupled Bending Vibration of Airplane Wings and Other Types Beams," *Journal of Aeronautical Sciences*, vol. 11, 1944, pp. 153–164.

8.8 Pestel, E. C. and F. A. Leckie, *Matrix Methods in Elasto-Mechanics*, McGraw-Hill, New York, 1963, pp. 130–152, 192–204.

8.9 Pilkey, W. D. and P. Y. Chang, *Modern Formulas for Statics and Dynamics*, McGraw-Hill, New York, 1978, pp. 8–99, 135–147.

8.10 Tuma, J. J., *Frequency Analysis of Three-Dimensional Bars*, Dissertation, University of Colorado Library, Boulder, 1967.

8.11 Tuma, J. J., *Engineering Mathematics Handbook*, 2nd ed., McGraw-Hill, New York, 1979, pp. 102, 220–221, 224–225.

Chapter 9 Dynamic Flexibility and Stiffness Methods

9.1 Blaszkowiak, S. and Z. Kaczkowski, *Iterative Methods in Structural Analysis*, Pergamon, Oxford, 1966, pp. 490–530, 579–581.

9.2 Cheng, F. Y. and W. H. Tseng, "Dynamic Matrix of Timoshenko Beam-Columns," *Proc. ASCE, Journal of the Structural Division*, vol. 99, ST 3, March 1973.

9.3 Koloušek, V., *Dynamik der Baukonstruktionen*, Bauwesen, Berlin, 1962, pp. 117–125.

9.4 Laursen, H. I., R. P. Shubinski, and R. W. Clough, *Dynamic Matrix Analysis of Framed Structures*, 4th U.S. Nat. Congress of Applied Mechanics, Berkeley, 1962, pp. 99–105.

9.5 Levien, K. W. and B. J. Hartz, "Dynamic Flexibility Analysis of Frames," *Proc. ASCE, Journal of the Structural Division*, vol. 89, no. ST 4, August 1963.

9.6 Paz, M., *Structural Dynamics*, Van Nostrand Reinhold, New York, 1980, pp. 388–401.

9.7 Tuma, J. J., *Frequency Analysis of Three-Dimensional Bars*, Dissertation, University of Colorado Library, Boulder, 1967.

9.8 Tuma, J. J., *Engineering Mathematics Handbook*, 2nd ed., McGraw-Hill, New York, 1979, pp. 105–106.

9.9 Velestos A. S. and N. M. Newmark, "Natural Frequencies of Continuous Flexural Members," *Trans. ASCE*, vol. 122, 1957, pp. 249–278.

Chapter 10 Variable Cross Section

10.1 Boeker, H. C., J. W. Exline, S. E. French, T. L. Lassley, J. T. Oden, H. S. Yu, and J. J. Tuma, *Flexibilities*, Vol. III, School of Civil Engineering Research Publication, no. 11, Oklahoma State University, Stillwater, 1963.

10.2 Cheng, F. Y. and M. E. Botkin, "Second Order Elasto-Plastic Analysis of Tall Buildings with Damped Dynamic Excitations," *Proceedings of the Specialty Conference on Finite Element Methods in Civil Engineering*, Canadian Society of Civil Engineers, 1972, pp. 549 ff.

10.3 Fiřt, J., "Vibration of Structures with Beams of Variable Cross Section," *Aplikace Mathematiky*, Czechoslovak Academy of Sciences, vol. 10, 1965, pp. 16–19.

10.4 Gillespie, J. W. and B. D. Liaw, "Frequency Analysis of Beams by Flexibility Method," *Proc. ASCE, Journal of the Engineering Mechanics Division*, vol. 90, EM 1, February, 1964.

10.5 Krinicki, E. and Z. Mazurkiewicz, "Free Vibration of Simply Supported Bar with a Linearly Variable Height of Cross Section," *Trans. ASME*, vol. 84, 1962, pp. 497–501.

10.6 Koloušek, V., *Dynamics of Engineering Structures*, Butterworth, London, 1973, pp. 143–185.

10.7 Thompson, W. T., "Matrix Solution for the Vibration of Nonuniform Beams," *Trans. ASME*, vol. 72, 1950, pp. 337–339.

Index